Warship 2025

The Future Fleet

Glasgow, United Kingdom
16-17 June 2025

ISBN: 979-8-3313-2657-9

Printed from e-media with permission by:

Curran Associates, Inc.
57 Morehouse Lane
Red Hook, NY 12571

Some format issues inherent in the e-media version may also appear in this print version.

The Royal Institution of Naval Architects
8-9 Northumberland Street
London, WC2N 5DA
United Kingdom

Phone: 020 7235 4622
Fax: 020 7259 5912

publications@rina.org.uk

Additional copies of this publication are available from:

Curran Associates, Inc.
57 Morehouse Lane
Red Hook, NY 12571 USA
Phone: 845-758-0400
Fax: 845-758-2633
Email: curran@proceedings.com
Web: www.proceedings.com

160 YEARS OF HISTORY,
KNOWLEDGE AND EXPERIENCE

THE ROYAL INSTITUTION OF NAVAL ARCHITECTS

Warship 2025: The Future Fleet

16-17 June 2025

Glasgow, United Kingdom

TOGETHER WE WILL SHAPE OUR FUTURE

Sponsored by:

BMT Ansys SSI

THE ROYAL INSTITUTION OF NAVAL ARCHITECTS

Warship 2025: The Future Fleet

16-17 June 2025

Glasgow, United Kingdom

THE ROYAL INSTITUTION OF NAVAL ARCHITECTS
8-9 Northumberland Street
London WC2N 5DA

Telephone: 020 7235 4622
Fax: 020 7259 5912

CONTENTS

[1] Work performed while at Delft University of Technology

THE UK NAVAL AUTHORITY GLIDEPATH TO CERTIFIED IMO DEGREE 4 AUTONOMY

A D Bain, Deputy Head – UK Naval Authority: Ships Division Team Leader, Bristol, UK and Chair international naval Safety Association – Surface Ship Committees

SUMMARY

The pathways and standards to certification are well travelled for crewed vessels, and are now modified for uncrewed. This paper touches on technology and design challenges faced in achieving IMO degree 4 autonomy. Readers should not fear regulatory barriers, as the UK Naval Authority welcomes further dialogue with any in the UK Defence Enterprise, and broader NATO allies. Working with the UK MCA, and the DSA-DMR, the regulatory spheres of authority are agreed and aligned to legal principles. The goal-based thinking enables staged certification of IMO degree 3 systems and onto degree 4. Reflections on early experiences and challenges to the paths lain, the choices of standards, risks and technology will be discussed. The Naval Authority's approach is adaptive to changing contexts of use, seeking to be responsive as technologies and new architectures emerge, each with the potential to alter what is materially 'safe to operate' and can be 'operated safely'. Despite the challenge, it is possible to certify machines that are remotely operated and even those that are expected to think for themselves.

NOMENCLATURE

AMofC	Accepted Means of Compliance
ANEP	Allied Naval Engineering Publication
COLREGs	Convention on International Regs for Preventing Collisions at Sea, 1972
CPA	Closest Point of Approach (CPA)
DMR	Defence Maritime Regulator within the Defence Safety Authority
IMO	International Maritime Organization
IOpC	Interoperability Operating Concept
KMHA	Key Material Hazard Area
MAP	Maritime Acquisition Publication
MASS	Maritime Autonomous Surface Ship
MoD	Ministry of Defence
NATG	Naval Authority and Technology Group
NSC	Naval Ship Code
NSubC	Naval Submarine Code
ROC	Remote Operations Centre
SDR25	Strategic Defence Review 2025
SIP	Software Integrity Policy
UUV	Unmanned Underwater Vessel

1. INTRODUCTION & CHALLENGES FACED

We engineers are naturally drawn to specific technologies or design challenges, informed by our backgrounds and disciplines. The challenge in achieving IMO Degree 4 autonomy[1] is often said to require new skills and new ways of thinking. This is the 4th RINA Autonomy Conference and RINA's technical committees for technology and for safety are already engaged in the topics herein. These committees keep close links to the Nautical Institute, to IMarEST's Special Interest Group (SIG) for autonomy and to regulators. Forums such as the UK Maritime Autonomous Systems Regulatory Working Group (MASRWG) have usefully contributed for some years [1]. Acknowledging the Defence sector has certain unique legal considerations, some insights and lessons are offered to the broader community.

The RN's Interoperability Operating Concept (IOpC) [2] talks of an inflection point between the Industrial Age and a new Information Age, where the Defence sector needs to retain a competitive edge. Keeping this edge requires competence to choose appropriate standards that enable procurement of the needed technologies. In understanding what is appropriate, it is important to reflect that autonomous shipping is not entirely new or

[1] **IMO Autonomy Degree four**: Fully autonomous ship: The operating system of the ship is able to make decisions and determine actions by itself.

novel. Experiences from hydrography, oceanography and seabed engineering provide helpful lessons and many vessels that already meet IMO autonomy degree 2[2] and 3[3] criteria [3], especially in use of Remote Operated Vessels (ROV). The broader transition from manned to unmanned machinery and use of autopilots provides lessons about IMO degree 1[4] & 2, if not degree 3. The Defence sector's own experience compliments and informs these other sectors. The policies leading up to UK's Strategic Defence Review (SDR25) challenge us to change our thinking, requiring a blend of the best old and new ideas, and the right mindset.

Autonomous systems need not remove the human completely. One vision envisages a blend of a manned system, supported by a swarm of loyal 'wingmen', (or perhaps ducklings or a school of fish?) to increase the mass of the naval task-force. Most nations find complex warships expensive and face pressures on crew numbers, with the range of design options and complexity expanding fast. As the economics of warfare change, the balance between large, manned ship concepts and small craft, as well as the balance of offensive and defensive countermeasures may change designs further. Such changes could make possible the sought crew reductions, or even fully un-crewed autonomous vessels a realistic prospect.

To date, technical challenges on conventional marinization topics have been as numerous as those on autonomy technologies. This lesson highlights how complex a warship capability is [27], and that they remain much more than a simple transport system or a simple missile barge, irrespective of whether they are reimagined as Maritime Autonomous Surface Ships (MASS) or Unmanned Underwater Vessels (UUV) in the future.

2. USING RISK-BASED DESIGN (RBD) STANDARDS

Standards[5] regularize good practices [4], by adopting proven or effective solutions and help reduce inconsistency. Adopting technologies from other sectors can reduce development costs allowing rapid introduction into service, but can also draw regulatory requirements for revalidation. Revalidation, testing or reassessment adds cost up-front, but can also help reduce overengineering (driving-in cost), or under-engineering (missing use-cases and lessons) that will cost a user (or a life) later. An important lesson from naval, cruise and offshore energy sectors is that different "use-cases"[6], infer different choices of standard and manifest in differing design or product solutions. By taking design concepts from other industries, it is highly likely they will have prioritized what is important to that sector. Modes of failure may differ, risks that are tolerated by the last customer, may not be for the new one. The significance of risk needs to be understood to focus on what is to be achieved.

For the Defence maritime operating environment, it has proven important to reflect on choices made, to set goals [5] and strategize what uses the MASS or UUV is to be applied to. This is called a Concept of Operations (CONOPS) in INSA Codes [6]. It is clear an autonomous vessel is likely to be used for the long-expressed "dull, dangerous and dirty", while other authors have extended this to also include "Deep, Dear and Duration", [7] but all such operations need to be outlined in the CONOPS.

---

[2] **IMO Autonomy Degree two:** Remotely controlled ship with seafarers on board. The ship is controlled and operated from another location. Seafarers are available on board to take control and to operate the shipboard systems and functions.

[3] **IMO Autonomy Degree three:** Remotely controlled ship without seafarers on board: The ship is controlled and operated from another location. There are no seafarers on board.

[4] **IMO Autonomy Degree one**: Ship with automated processes and decision support. Seafarers are on board to operate and control shipboard systems and functions. Some operations may be automated and at times be unsupervised but with seafarers on board ready to take control.

[5] Standards are "A document providing for common and repeated use, rules, guidelines or characteristics for activities or production of results", ISO [25]

[6] A methodology used in system analysis to identify, clarify and organize system requirements. Use of a specific situation in which a product or service could potentially be used, ISO [25].

3. ON REGULATION

3.1 DON'T FEAR THE REGULATOR

3.1 (a) Sandboxes and Play

For those that fear regulation prevents innovation, it must be remembered why regulations exist. Regulation[7] exists [8] to protect the public from unfettered risk, to regulate is thus to assure (or compel) control, so risk is managed. UK law focusses on risk being controlled by those who create it, rather than by the Regulator [25]. When the responsible (and the naïve but curious) manufacturers or their users proactively seek engagement, it is possible to discuss ideas, without ramifications. Such "sandboxing" enables innovation and experimentation through open discussion [9], complimented by some regulatory relaxation when exploring these ideas via controlled management of risk.

3.1 (b) Alignment

The UK's Maritime & Coastguard Agency (MCA) and the Defence Safety Authority's Defence Maritime Regulator (DSA-DMR) have an agreed demarcation, through the MOD-MCA Memoranda of Understanding [11]. Once DMR registers a MOD vessel [12] its mandate enables Naval Authority certification. The Naval Authority is chartered to certify compliance with standards and processes that reduce risk to life. Certification is defined through Naval Authority Rules [13] and by providing engineering advice and services to the maritime community in areas such as warship survivability and standards.

The demarcation between the MCA and DMR is informed by Defence and security interests, asset ownership and control of activity. Assessment is made of the intent behind established legal Dis-applications, Exemptions and Derogations (DED) [8], with attention to where the public is at risk of exposure to harm. In general, companies developing autonomous systems for their own use (incl. export), will have fewer if any DED, than assets under direct crown-control or activity for a crown customer. Demarcation typically assumes:

(i) there is a crown servant in control, called an "Accountable Person",
(ii) that the vessel is crown owned and/or operated, or
(iii) a service is provided for the benefit of UK Defence.

3.1 (c) The Difference in Maritime Regulation

Small companies working with a MOD customer, or directly with a regulator, can find the volume and complexity of paperwork challenging and may need Naval Authority support or to draw on skills from elsewhere in their supply network. Small companies have proven they can be more agile than the larger ones, with the best able to learn quickly. Sometimes the sandbox reveals a healthy safety culture other times the absence of one. The DSA-DMR continues to facilitate discussions and has published high-level guides on the more significant topics to be addressed for maritime systems [10].

Discussions have ranged from consideration of safety-critical software, how rugged or attrit-able autonomous maritime systems need to be, as well as the conduct of trials and basic occupational health & safety, especially for those new to a maritime environment. Thus the sandbox approach give opportunity for those experienced in the maritime enterprise to mentor those without experience in dealing with an unforgiving sea, while those offering new technologies teach traditionalists new approaches. Through systems thinking, it is possible to collectively succeed and support each other on the integration of these Capabilities.

3.1 (d) Application

Risk can be managed in a controlled sea-range area, or through developing evidence of the autonomous behaviour. Activity will be treated differently when it impacts the general public in open seas, or activities create risk to other UK seafarers or international shipping. If geographic ranges and MOD risk management controls are not used, it is best to engage with MCA. Any vessel in a navigable seaway (even while under test for UK or for Defence export), may not automatically have MOD legal protections. Where the conditions are

[7] A prescribed Rule or authoritative direction required of a statutorily empowered body. Civil Regulatory policies are highlighted in Defence Safety Regulations by the executive word 'must', to indicate a legal duty. [DSA01 Defence Regulation[8]]

met, DMR Registration and a Naval Authority Certificate may be achievable. The MOD-MCA MOU [11] and regular joint working helps coordinate regulatory oversight of all autonomous systems in UK waters, as well as offering a forum for sharing good-practice. Activities impacting other people or property, typically attract the major focus.

There is potential for standards and factors of safety to alter, where no risk to life exists [14] especially for an autonomous degree 4 vessel with no human onboard. Most certified vessels to date have applied small-craft vessel standards and have had associated risks managed in the standard fashion [15]. If any requirement is viewed as unnecessary, it can be tailored [21], [24] using the Standards Management process, but for now the Naval Authority approach is to assume:

- When a vessel presents a hazard to other mariners, it must meet COLREG, with a competent coxswain to provide overwatch, being sensible.
- Whenever a human is exposed to the risk of harm; such as when a maintainer, loader, replenisher comes aboard otherwise unmanned vessels the standards that protect life should be followed.

3.2 GOAL-BASED RULES AND WARSHIPS AT INSA

Rules and regulations are raised to protect the public, but occasionally the solutions may not work effectively for naval systems. For several decades, the Defence sector has developed goal-based regulations [5]. The evolution of the INSA Naval Ship Code [6], Submarine Code [16], [17] and Boat Code [18] facilitate international consensus on ways to achieve compliance with goals, functions and principles. Solutions are not mandated, rather they are selected according to the use-cases and standards plans. The UK recognizes the INSA Codes as Accepted Means of Compliance (AMofC) with the NA Rules [13], so each Code's goals, functions and assessment of risks become a UK Certification requirement. Requirements can be set, or tailored (removed with explanation). INSA develops Part 2 & 3 [19] to offer common solutions for compliance for each chapter, it records best-practices and shares lessons. Most nations develop their own solutions, or set levels for acceptance or adopt linked solutions, such as an endorsed set of classification society rules.

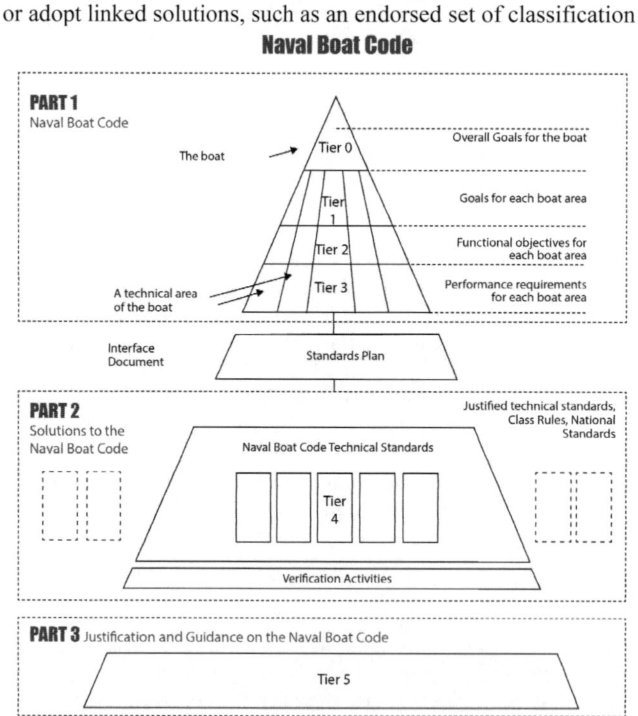

Figure 1: INSA Boat Code (Parts 1-3 Arrangements)

This approach offers a way to discuss equivalence and to use international procurement chains. The Naval Boat Code (NBC)[18] introduces the concept of the foundation safety standard, and the special function. Both foundation and the special functions can complement established technology standards, with the summative compliance helping build evidence and so achieve safety certification. The choice (fig.2) of solution, ability to alternate between standards or adapt requirements by "tailoring" requirements, so directly support innovation and consideration of more novel features. Decisions on standards selected and the extent of tailoring can be justified by the Safety Argument within a Safety Case. The NA Rules [13], Chapter 20 set out the process, supported by Maritime Acquisition Publication for Autonomous Systems [20].

It is for each Defence accountable person to strategize how they will fulfil their broad duty to demonstrate risk assessment, apply risk control and define thresholds of risk. To do this they need to be knowledgeable about new principles, system functions and solutions, and how risks are being controlled. Effort can be targeted by using known foundation standards (e.g. a known boat code), so that the 'special feature' of autonomy is the focus. Such tailoring and adaptation is already routine for manned vessels. The most widely used solutions are the classification societies rules, but international or national libraries of standards are also used.

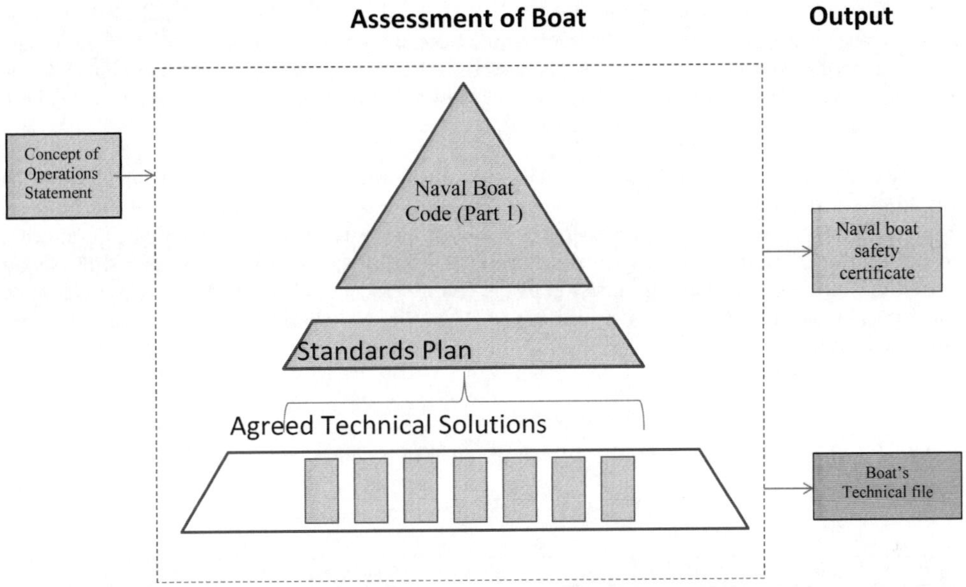

Figure 2: NBC Figure P1-0-1 main regulatory elements in the certification process of boats

Efficient certification requires dialogue, benchmarking and joint working. A broad range of solutions can be found when a vessel owner specifies special functions, on-top of a foundation safety standard or seek to replace elements of it (tailored to fit).

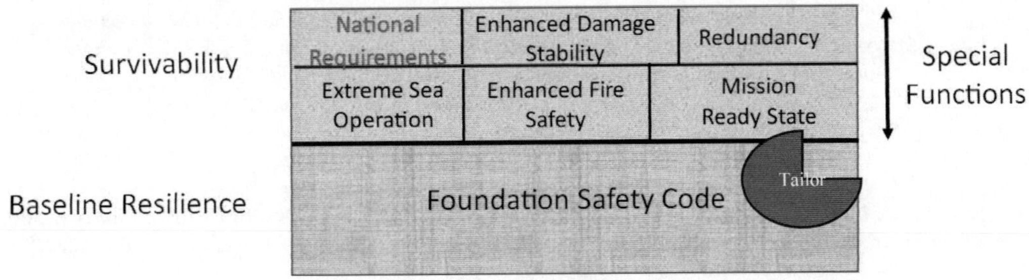

Figure 3: Relationship of Foundation Safety and Special Features Standards

The most widely chosen AMofC to date has been the Naval Boat Code (NBC) [18]. The approach has also allowed modified use of MCA workboat 3 [22] as a foundation. The ideas within the NBC have influenced the next version of the INSA NSC [19] which will use the same ideas of special features or functions (added or removed) for ships. Whilst we collectively improve our understanding of autonomy the importance of good relationships cannot be underestimated. The diversity of approaches and variety of autonomous machines has required the boundary of what is materially safe to operate and can be operated safely, to be kept in mind [5],[16].

4. FOLLOW THE STEPS TO CERTIFICATION

4.1 THE WORK OF THE NAVAL AUTHORITY

The Naval Authority requires a certification strategy to outline the operating concept, mission and roles (INSA CONOPS). This certification strategy should select the preferred foundation safety code or standard. Certified systems have successfully used MCA's Workboat 3 [22] and INSA Naval Boat Code [18], with larger ships able to use Special Purpose Ship (SPS) [23], SOLAS or Naval Ship Code (NSC) [19] as their foundation standard. The special features for specific safety or mission needs can be added, as required.

Early experiences show many common risks associated with bringing any new technology into service. Although flexible selection of standards has allowed new functions to be tested, it has occasionally resulted in unusual combinations. The more unusual the mix, or the greater the tailoring of the original baseline, the more essential risk assessment is to help assess the consistency of features back to the safety principles and goals. As lessons are identified, Naval Authority Notices (NAN) [21] are being published and with time NA guides will be formally updated and published in Maritime Acquisition Publications (MAP) or Defence standards to be make the lesson available across the UK Defence Enterprise.

The flexibility to add special functions has enabled progressive certification of over 50 IMO degree 3 boats so far. Novel ship-types require a longer conversation than compliance with established hull-forms and roles with familiar Accepted Means of Compliance (AMofC), but most concepts can be accommodated. Tailoring of requirements has also allowed a previously certified baseline vessels to be modified with autonomous features, or to be used in a new fashion. Through revision of the risk assessment, changes to any specific context of usage can be addressed agilely, while permitting new use-cases to be added and novel technologies or architectures to be trailed.

4.2 OPTIONS

The Naval Authority has tried to simplify the different choices of baseline safety standard as Ship Regimes

Table 1. The Ship Regime

Regime 0			**Regime 1**	**Regime 2**
Domestic	Non-IMO	IMO Convention	Auxiliary or Patrol	Warship
Type 0 Recreational Boats Type 1 Boats (Commercial baseline) Inland & Coastal Domestic Shipping River, pleasure boats and Workboats	Special Purpose Shipping (MCA Type VII / VIII) High-speed Craft (HSC) Coast-hopping International (MCA Coastal II to VI) Workboats, Superyachts	Short-Haul international trade (MCA Type II, VIII(T), IX Ocean Tugs etc) Long-Haul international Trade (MCA Type I, VII, XI, XII)	MCA-baseline + Military Deltas to deliver NSC-baseline + special features Requirement (*to blended Civil & Military International & National standards*)	MOD Requirement (*to blended Civil & Military International & National standards*)
Baseline Safety Standards, Tools & Analysis		>>>>>>> <<<<<<<	**Baseline Survivability Standards, Tools & Analysis**	

Having written the Certification Strategy and standards plan, the steps to achieve certification are as follows. First to follow the Regime [24] and registration path [12], having understood the broad mission or role of the vessel:

- A regime 0 or 1a boat, must comply with MCA workboat code 3, as the foundation standard. Ability to tailor is limited. A baseline compliant with this foundation standard allows up to IMO Degree 3 autonomy, (no more).
- A regime 0 or 1 ship must similarly comply with statute and alignment to a foundation baseline of a

UK domestic, Special Purpose [23], Short-hall SOLAS or long-haul SOLAS baseline Government Auxiliary. These red-ensign (chartered) or blue-ensign (auxiliary) vessels are likely to need a crew (no more than IMO degree 2 autonomy) [26].

- A regime 1b or 2 boat, would be expected to be compliant with NBC safety baseline, with one or more special functions or features (up to IMO Degree 4 autonomy, according to evidence). They can have a coxswain or be uncrewed.

A regime 2 or 3 ship would expect to have an NSC baseline, but could still use elements of SOLAS or other standards and multiple special functions. This would allow up to IMO Degree 4 autonomy, (according to evidence).

4.3 THE THREE CERTIFIED PATHS (SO-FAR) TO IMO DEGREE 4

Experience thus far, reveals three routes able to achieve certification. Each allows a safe environment, a safe system of supervision and a tested system:

4.3 (a) Route One: Certification based on use within a Cleared Range

The MASS or UUV may operate autonomously provided that no third-party is either within a range of 3 nautical miles (NM); or is shown to perform to a time & predicted range to Closest Point of Approach (CPA) with other traffic. If either caveat is breached, the MASS/ UUV must be capable of being taken back into direct operator control from a Remote Operations Centre (ROC) or by an embarked crew member.

4.3 (b) Route Two: Certification based of proven System Performance – Allowing Supervised Autonomy (D3)

MASS or UUV may operate autonomously (IMO Degree 4) or under operator supervision (IMO Degree 3) in an environment where third parties are also operating, as long as the system achieves, and can demonstrate either: it detects a non-charted object at adequate range and/or determines the CPA of the object.

Distinction is made between Open Waters and Constrained Waters. COLREG Rule 18 privileges determine the form of certification submission (e.g. COLREGs Rule 9 (Narrow channels), Rule 10 (Traffic Separation Schemes) or any local rules that apply). The autonomous system is then tested to show how the Commanding Officer and their Coxswain/ operator is informed. This type of MASS will detect objects at a range that is at least as good as a human operator in the primary conning position of the vessel (were it crewed) or equivalent. The vessel is typically equipped with sensors for situational awareness at least as good as equivalent crewed vessels. An acceptable CPA will be determined by factors incl.: Environmental conditions; Traffic density; Vessel characteristics; Vessel speeds; Aspect of vessels relative to one another; Proximity to navigational hazards.

4.4 (c) Route Three: Tailored Certification based on a Synthetic Environment Assurance Report

The COLREG application of the MASS or UUV is assessed by an independent third party (a Recognized Organization (RO) of the Naval Authority) using a synthetic environment and a report provided to the operator, and the NATG as the evidence toward Certification. This Synthetic Environment Assurance Service may have geographic or situational caveats, and allows the fullest possible range of autonomous operation. This method provides the operator with the maximum permissible freedom of operation. It may be applied for at any stage in a MASS's development. Contact the NATG for further details.

5. TECHNICAL DISCOMBOBULATION

5.1 REQUIREMENTS

Requirement sponsors seek to achieve IMO degree 4 autonomy quickly. However it may prove better to pick the degree of autonomy where it offers most value, rather than seek functions exclusively at the expense of other capabilities, that work satisfactorily. Even for degree 3 autonomy there are challenges, as remote control dislocates bridge crew from 1st party seamanship decisions aboard the vessel, and mission crew from the command centre. Engineers must engage and help with systems thinking, at component, system, warship and force design levels.

A crewed warship retains situational awareness using a command structure. The deconstruction of a warship to a system of systems discombobulates command-control across a task group or the world. The implications for each Defence line of Development (DLOD)[8] and component parts is still being considered by navies. The split between safe-to-operate and operate-safely split also becomes different.

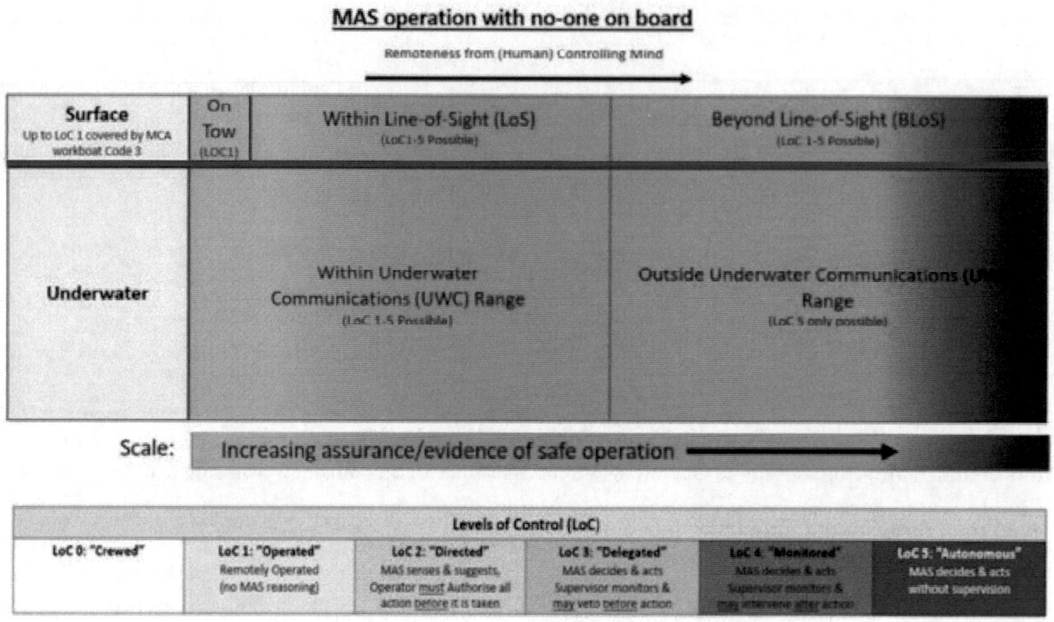

Figure 4: Levels of communication (DMR [10])

5.2 CONTROL AND ACOUNTABILITY

The DMR has recognised differing levels of control (DMR, Definitions [25]), and the need for more specificity (Fig-3), in the functions and technologies needed to step from IMO degree 3 to degree 4. As shown above (Fig-4), different autonomous functions could sit at different degrees. Compliance with:

- COLREG informs navigation and control of ship position & heading;
- the laws of armed combat informs control of the firing-chain, limiting levels of autonomy possible
- principles of machinery availability, internal damage control and supply of energy could become highly automated
- IMDG or military doctrine on use of a payload to deliver a mission function might define different degrees of autonomy.

Maritime-relevant solutions, such as in DNV's Rules for Autonomy & Remote Operation systems (AROS) can help illuminate how different levels of autonomy could work on the same vessel [28]. This standard has four defined function categories: navigation (NAV), engineering (ENG), operational (OPS), and vessel safety (SAFE). Other foundation safety standards for autonomy are emerging from Lloyds [29] and national centres [33], which can be applied directly to other special features.

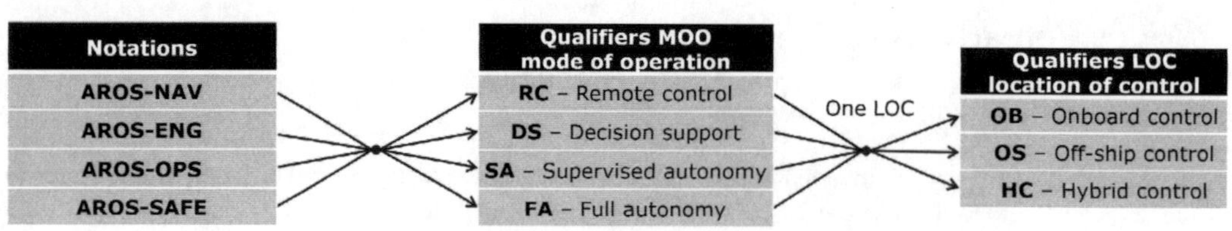

Figure 5: DNV Rules for Autonomy

[8] The DLODS are TEPIDOIL: Training, Equipment, Personnel, Information, Doctrine & Concepts, Organisation Infrastructure and Logistics

Current generations of Remote Control Offices (RCO) are based upon those used for aviation drones (first-generation), typically with up to 3 seats (a coxswain, cargo/mission and master or Commanding Officer (CO oversight). This is likely to evolve with experience.

6. THINKING ABOUT FUNCTION

6.1 NAVIGATIONAL FUNCTION

A Coxswain needs situational awareness to avoid navigation hazards, adapt to weather, steer a course past seafarers, set commands to engines, propeller and rudder, compliant with COLREG. Numerous additional sensors have needed to be added to gain the same situational awareness inside the RCO as can be achieved by a chair on a bridge. Demands on navigation aids make them safety-critical decision systems, priced to reflect this kind of type approval.

6.2 MISSION FUNCTION

For mission package personnel/ teams have needed to share mission tasking, to draw their own situational awareness data, to communicate with each other and task-group mission staff. The receipt and execution of mission orders has (until now) been delivered by housing multiple personnel in a large operations room. The demand on sensors reliance on safety or mission critical components has added cost and made systems more complex.

6.3 COMMAND FUNCTION

For the mission CO requires the ability to delegate command-control, seek instruction from the task-group and a means to apply orders.

Finally, depending on the mission's length, each vessel needs to sustain its course(s), engine speed, reliably consume fuel/energy and be available to receive instruction such as the actuation of fire-fighting or other self-repair instructions.

6.4 OPERATIONAL ANALYSIS & NEEDS: DETERMINING THE FUNCTION

Either autonomous systems will all be disposable with a level of attrition accepted, or designs will need to be rethought. It might prove easier to vary the form of automation or autonomy needed, since each function will need a level of safety (if people interact with it) or survivability (if mission-critical). These levels will be defined for each vessel and at the fleet level, noting at a system of systems, the force-designer will need to consider what form secure communications will take and consider how military threats like grey-warfare, cyber etc erode functions or alter war-fighting needs.

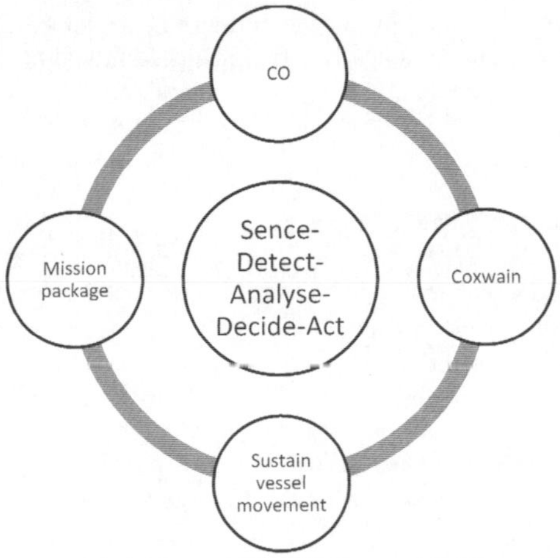

Figure 6: The Dislocated & Discombobulated Warship

Systems requirements engineers and industry partners will need to work with traditional maritime engineers to agree demarcation of effort and leads for each function type. For now, the maritime safety certified elements are limited to:

- the safe equipment and conduct of navigation compliant with COLREG,
- where a mission package is a risk to life
  - Temporary on-board (of resupply, maintainer, stevedore functions),
  - Cargo/ mission package incl. IMDG or
  - Weapon safety/clearance and firing-chain

Acknowledging this future complexity and the genuine fear of regulatory barriers, the NATG retains its passion for helping other's innovation and enable rapid insertion of new but safe technology. We seek to help assess and declare risk and are supported by LR, DNV, INSA, and Defence colleagues. There is much that is commendable in civil approaches, as well as cautionary lessons on over specification and complexity in the requirement.

The NA's Rules [17] list Acceptable Means of Compliance (AMofC), and these standards can be modified on application to us. Requirements of an autonomous design should align with the standards selected, and the solutions should conform to those standards. Assessment evidence and assumptions that work for manned vessels might need re-assessed when there is no longer a crew to deliver required functions. The application of any standard for a MAS/UUV should link back to the functional requirements, until such times as specific standards are developed by appropriate 'agencies' e.g. Class, IMO, NATO or MOD.

Tailoring should consider the overall function to be delivered and the goals of the rule or objective, including where a direct human input is removed, where that impacts protection systems or command functions, that may need to be developed to provide an equivalent capability. Standards that are to be tailored are to be recorded in:

- The project standard database or management plan;
- Any associated Class society Tailoring document (as appropriate);
- The Naval Authority Certification Strategy. Significant deviations may be noted in certification.

For new vessels, the place to begin is to follow the certification steps for autonomy as set-out in Naval Authority (NA) Rules, Chapter 20, the supporting MAP151 [20] with NA Notes (NAN) for testing unmanned system's [30],[31]. For the more sophisticated systems, we can test how the autonomous navigator behaves within a Synthetic Environment Assurance Service (SEAS). We have also published NAN that define the requirements for the intrinsic safety integrity of the system software [32], and we signpost good-practices on assessment of AI and sensors [33], with more to be published this year. The main baseline safety standards include current thinking on system recoverability [34] and special features, is central.

Where novel special features are added (or substituted in), or where there is no risk-to-life, hazards may be revisited. In this case, standards or solutions might be radically different e.g. lightweight structure, machinery availability, or choice of safety features e.g. a hypoxic atmosphere to reduce fire risk when no-one is aboard. The Naval Authority publishes many standards to address such novel risks as well as known risks and technologies.

Ideas on autonomous recoverability are also being worked up as follows:

- Features whose failure will not lead to danger
- Features whose failure could eventually lead to danger
- Immediate lead to danger no mission impact
- Features for Degraded mission to Lost mission
- Ideas on Essential safety function

6.5 DISENTANGLEMENT AND FUTURE CUSTOMER INFLUENCE

Selecting technology and designing a system to work, needs a designer (the naval architect & their multi-disciplinary team) to know what the vessel is for (Andrews, [27]) and to disentangle the requirement into workable technology solutions. The design challenges increase when the need changes, the functions are distributed or do not settle. Change impacts budget, partnering and the physical functions needed. It may be that the other navies and forces the RN works closest with may change, if so, then the budgets, the technologies used, and training will need to adapt. It is clear procurement is already changing in most countries, requiring navies and materiel support communities to innovate.

How any system, particularly a ship-platform is designed and bought impacts how they are certified. The three-part process (plan-approval; material survey and trial) is well established [6],[23] but by introducing new technologies, we cannot assume established processes will still work. The assurance community is uncovering new risks being introduced when procurement sequencing or emphasis changes. How autonomous systems will be used could be different than anticipated. This means more work on Remote Operating Centres (ROC) and task-group level requirements. The integration and trials of these systems are important to build evidence and trust. They need time to be conducted efficiently, integrated physically and in the virtual software world, and to be assured (software and AI itself needing its own kinds of checks). These can be addressed by professionalism and clear-eyed application of engineering principles.

7. CONCLUSIONS

We know that expensive, crewed platforms cannot easily (or quickly) be replaced in the short-term and expensive manned systems cannot be lost needlessly. Therefore, the importance of independent certification of critical and expensive manned or unmanned vessels will remain a necessary role for bodies like the Naval Authority. The UK Naval Authority has been working to facilitate transition from manned to unmanned systems and has laid a pathway to allow certification. It is adaptable and can be modified.

Clearly the ability to remotely control a vessel, or future sensor nets of multiple vessels will need to extra considerations, including network robustness, cyber and electronic warfare. It is possible when sufficient mass of vessels are bought that levels of attrition or survivability to conventional countermeasures may shift. While our fleets transition, we are seeking to gain mutual confidence in how these new technologies perform from the high-seas, to the geographic narrows and in busy sea-lanes.

We believe while designers face undeniable challenges that the path for certification can enable this significant shift in national (and international) ship design, in a scaled and proportionate manner, while still offering safety assurance and certification when requested.

It is therefore timely that you join us in our sandbox, engage in discussions between regulators, the regulated, the designers, builders and users. it is only by such engagement that we can build understanding, empathy and the competencies necessary to face the challenges we face.

8. ACKNOWLEDGEMENTS

I am grateful to all my colleagues in the Naval Authority who do the hard work in large and small craft and in writing this policy. I acknowledge the leadership of the DMR, who's foundations I had the honour to lay and to the MCA. Above all we acknowledge the international expertise of the classification societies, especially those we note as Recognised Organisations and to all other navies of INSA, who contribute diversity of view and expertise every time I meet them.

9. REFERENCES

1. MARSWG Guide retrieved 6/5/2025 from (https://www.maritimeindustries.org/resources/maritime-autonomous-systems-regulatory-working-group)

2. The MOD Integrated operating Concept accessed 20/12/2024 (https://www.gov.uk/government/publications/the-integrated-operating-concept-2025)

3. IMO Autonomous Shipping, retrieved 6/5/2025 from (https://www.imo.org/en/MediaCentre/HotTopics/Pages/Autonomous-shipping.aspx)

4. ISO Glossary (https://www.iso.org/glossary.html)

5. RUDGLEY et al., 'Development of a NATO Naval Ship Code' *RINA Safety Regulations and Naval Classification II*, 2005.

6. STANREC-4838 (ANEP-77) Naval Ship Code: Part-1: Goals, Functional Objectives & Performance Requirements, Edition H, version 4 (https://www.navalshipcode.org/publiccodes)

7. BMT (2023), An evolution of the 3 D of Robots: The 6D of Maritime Autonomy, accessed 1/5/2025 (https://www.bmt.org/insights/an-evolution-of-the-3-ds-of-robotics-the-6-ds-of-marine-autonomy)

8. UK MOD, Defence Safety Authority (DSA) 01.1 Regulations; retrieved 1/5/2025, from (https://www.gov.uk/government/publications/defence-safety-authority-011-regulations)

9. Gov.UK (new Approach to ensure regulators and regulation support growth (Centre for Connected and Autonomous Vehicles: Future of transport regulatory review: regulatory sandboxes, accessed 1/5/2025, from (https://www.gov.uk/government/consultations/future-of-transport-regulatory-review-regulatory-sandboxes/future-of-transport-regulatory-review-regulatory-sandboxes)

10. UK MOD, "DSA03 DMR Guide to Regulation of Maritime Autonomous Systems," version 3.1 (2024). retrieved 1/5/2025, from (https://assets.publishing.service.gov.uk/media/66cdacd1f04c14b05511b317/DSA03-DMR_Guide_to_Regulation_of_Maritime_Autonomous_Systems.pdf)

11. Memoranda of Understanding (MOU) between the Ministry of Defence (MOD) and the Maritime & Coastguard Agency (MCA) retrieved 20/4/202 from (https://www.gov.uk/government/publications/memorandum-of-understanding-between-the-ministry-of-defence-and-the-maritime-and-coastguard-agency)

12. DSA03-DMR Guide to Registration of MOD Shipping, Issued October 2023, retrieved 20/4/2024, from, (https://www.gov.uk/government/publications/defence-maritime-regulations)

13. UK MOD, "DSA03 DMR Naval Authority Rules for Certification of MOD Shipping," 2023 Edition. retrieved 1/4/2025, from (https://www.gov.uk/government/publications/defence-maritime-regulations)

14. UK MOD, "Def-Stan 00-055 Requirements for Safety of Programmable Elements (PE) in Defence Systems, Requirements and Guidance," Issue 5, 2021.

15. BS-ISO-31000-2009 Risk management — Principles and guidelines.

16. OLDERSHAUSEN, C., 'Using the Naval Submarine Code (NSubC – Allied Naval Engineering Publication 102) as means for Submarine Assurance during Design, Construction and Operation', *Warship 23*, 2023.

17. ANEP 102, Naval Submarine Code, Part-1: Goals, Functional Objectives & Performance Requirements, Ed. B, version 1, Oct 2022 (https://www.navalshipcode.org/publiccodes)

18. INSA Naval Boat Code, Ed. B, Vn1. (https://www.navalshipcode.org/publiccodes)

19. INSA Naval Ship Code parts 2 & 3 Edition I, vn1. (https://www.navalshipcode.org/publiccodes/)

20. UK MOD, Maritime Acquisition Publication 01-151 Certification of Autonomous Vessels (https://www.nsass.org.uk/Library)

21. UK MOD, Naval Authority Notice (NAN-08.1-2020): Naval Authority Requirements for the Certification of Maritime Autonomous Surface Ships

22. MCA, The Workboat Code, Ed. 3, (https://www.gov.uk/government/publications/the-workboat-code-edition-3)

23. MGN 674: Application of the Special Purpose Ships code, accessed 1/5/2025 (https://www.gov.uk/government/publications/mgn-674m-application-of-the-special-purpose-ships-code/mgn-674m-application-of-the-special-purpose-ships-code)

24. MAP 01-137 issue 1, Guide to Boat Safety Standards (https://www.nsass.org.uk/Library)

25. DSA03-DMR-SHIPPING, Regulatory Terms for DSA02-DMR Defence Maritime Regulations

26. DSA03-DMR-SHIPPING, A Guide to Red, White and Blue Ensigns, DMR, 1 October 2019

27. ANDREWS, D., 'The Expanding Scope of Ship Design Practice', 2024.

28. DNV, "Ship Rules, Part 4, Chapter 9," 2023.

29. Lloyd's Register, "Rules & Regulations for Classification of Naval Ships," July 2024.

30. NAN-07-2025 Tailored Approach for RPAS operating from Ships and boats (https://www.nsass.org.uk/Library)

31. NAN-1-2024: Naval Authority Requirements for Recognised Organisation Providing a Synthetic Environment Assurance service (https://www.nsass.org.uk/Library)

32. UK MOD, "NATG NAN09-2-2024: Software Integrity Policy (SIP)," (https://www.nsass.org.uk/Library)

33. York, 2025: The BIG Argument lunch, Centre for Assuring Autonomy, retrieved 9/5/2025, from (https://www.york.ac.uk/safe-autonomy/).

10. AUTHORS BIOGRAPHY

Allan Bain CEng, FRINA, FSaRS, RCNC holds the current position of Ships Division Team Leader within the UK Naval Authority, the body that holds a Charter from the Defence Maritime Regulator (DMR) for certification of all vessels be they ship, boat and autonomous. The groups sits within the UK MOD Submarine Delivery Agency. The Author is responsible for leadership in maritime safety certification and survivability. His previous experience includes founding the DMR in 2011, working in DSA HQ on strategic policy and audit, 2017-21, [and numerous projects across MOD ships and weapons].He is a former chairman of the RINA Safety Committee. He is the current chairman of the surface ship committees of the International Naval Safety Association (INSA), that are the sponsoring body for NATO STANREC-4838: The Naval Ship Code, Allied Naval Engineering Publication (ANEP-77), which together with its sister Naval Submarine Code (ANEP-102) for and Naval Boat Code are the foundation safety standards for allied warship safety (the SOLAS for warships). He is a Captain in the RCNC.

A FUNCTIONAL SAFETY APPROACH TO SUPPORT MARITIME AUTONOMY

A Barker MSc CEng FSP, I Groom Cdre MBE (Rtd.) CEng FIMarEST and **T Hussain**, Safeguard Engineering Ltd, UK.

SUMMARY

Maritime autonomy offers significant potential benefits for operational efficiency, safety and strategic advantage across both commercial and defence applications. However, realising maritime autonomy invokes substantial assurance challenges, particularly when integrating artificial intelligence and machine learning into safety-critical functions.

This paper explores a functional safety approach tailored to support maritime autonomy, building upon established safety standards while addressing the limitations of applying these traditional methods to artificial intelligence technology, given the dynamic and emergent behaviour characteristics of autonomous maritime operations. An illustrative example of a collision avoidance system is provided. Through this example, this paper outlines the combination of quantitative technical measures and qualitative organisational and governance considerations for the assurance of maritime autonomy through a lifecycle approach. This approach provides a structured, standards-aligned evidence stream specifically targeting artificial intelligence technology within maritime autonomous systems and complementing existing methods such as Systems Theoretic Process Analysis.

1. INTRODUCTION - An Overview of Maritime Autonomy

Levels of autonomy describe the extent to which an Artificial Intelligence (AI) system functions independently of human supervision and control. In the maritime sector, the International Maritime Organization (IMO) introduced the Degrees of Autonomy (DoA) framework as part of its scoping exercise for Maritime Autonomous Surface Ships (MASS) [1]. This framework categorises maritime levels of autonomy on a scale from fully crewed ships with some automated functions (Degree 1) to completely autonomous vessels operating without crew or remote human intervention (Degree 4).

Following this, the MASS industry has developed complementary Levels of Control (LoC) frameworks as an alternative to DoA [2]. The LoC framework ranges from fully crewed, e.g., 'traditional', vessels (LoC 0) to fully autonomous vessels (LoC 5) and has been adopted by the United Kingdom Defence Maritime Regulator (DMR) through UK defence guidance on regulating maritime autonomous systems [3].

In addition, classification societies have introduced further frameworks to guide certification levels for autonomous platforms [4]. For example, Lloyd's Register (LR) has created an Autonomy Level (AL) scale (AL0 to AL6) [5] while Det Norske Veritas (DNV) have established the Autonomous Readiness Operational Safety (AROS) framework to defines class notations with Modes of Operation (MOO) ranging from Remote Control (RC) to Full Autonomy (FA) [6].

The abundance of similar, but slightly different, frameworks for levels of maritime autonomy reflects the fundamental challenges in achieving widespread acceptance of a single autonomy framework; any chosen framework must be applicable to all stakeholders, and both current and future systems, resulting in constant evolution and adaptation [7]. We recognise that regardless of the framework used, whether defined as a degree, level, or scale, maritime autonomy is widely viewed as a critical enabler for future commercial and defence capabilities:

- The UK's Strategic Defence Review (SDR) highlights autonomy and AI as essential to achieving national defence objectives [8];
- The UK's maritime 2025 roadmap identifies the importance of smart shipping and autonomy in delivering a cleaner, safer, and more efficient maritime industry [9]; and,
- The Centre for Assuring Autonomy (CfAA) at the University of York shows that integrating digital technologies, AI, and autonomous navigation systems promises significant efficiency gains, cost reductions, and safety improvements [9].

Full autonomy then, under whichever framework is applied (e.g., 'Degree 4', 'LoC 5', 'AL6' or 'FA') is not merely a technological ambition. It represents a foundational capability required to support UK's defence and commercial maritime goals [11]. A natural question arises - if the benefits are so significant, why are Maritime Autonomous Systems (MAS) (Covering MASS and subsurface systems) not yet commonplace, replacing traditional crewed vessels at scale across the UK maritime sector?

Part of the answer lies in the complexity of the assurance challenges; assurance frameworks have not evolved rapidly enough to keep pace with technological advances, and they do not fully address the complex challenges posed by AI, Machine Learning (ML), and cybersecurity unique to autonomy [12, 13]. The existing 'traditional' assurance approaches assume deterministic and repeatable systems, or rely heavily on human operators for risk reduction, and thus are not suited to MAS [14].

While several autonomous vessels have been developed and trialled, most have relied on case-by-case, goal-based approaches within limited operational envelopes. Such bespoke approaches remain necessary in the absence of an overarching, standardised framework. The IMO's MASS Code is still under development, with adoption of a mandatory code expected by 2032 [15]. This paper does not propose abandoning goal-based approaches, which are fundamental for complex systems [16]. Instead, we outline an approach to adapt existing robust standards tailored specifically for autonomous systems. This approach aims to shift away from bespoke cases-by-case approaches towards greater harmony, auditability and scalability; supporting both industry and regulators and filling the current assurance gap.

We are also conscious that, in the near term, most MAS will likely operate in hybrid or transitional modes. These systems will blend supervised autonomy with remote or direct control, adjusting between levels of autonomy even during a single voyage. This, in turn, demands an incremental approach to assurance and a gradual expansion of the operational risk envelope. Accordingly, this paper proposes a pathway to navigate through the levels of autonomy (e.g., the various degrees, levels, and scales) to ultimately enable fully autonomous operations.

Finally, we wish to stress an important point. Our approach does not suggest an alternative to developing a safety case to provide overall assurance of 'safe' through a goal-based approach. Safety cases remain the foundational means of demonstrating system safety across many critical sectors, including nuclear, defence and aerospace [18]. There is growing interest in safety cases for AI systems, although much of the existing work focuses on preventing large-scale risks in general-purpose AI rather than safety-critical, integrated applications [19]. We recognise this and therefore present our approach as a method for developing the specific evidence required for the autonomous, AI, aspects of a maritime platform. This AI-related evidence will form an essential part of the overall safety case.

2. ASSURANCE APPROACHES FOR MARITIME AUTONOMY

2.1 DEFINITIONS AND EXISTING 'TRADITIONAL' APPROACHES

Autonomy is defined as a system capable of "modifying its intended domain of use or goal without external intervention, control, or oversight" [20]. Achieving autonomy requires AI technology, being the "technology that implements or enables an AI system" [20], consisting of:

- AI components, defined as "functional elements that construct an AI system" [20]; and
- AI systems, defined as "engineered systems that generate outputs" [20].

Assuring a MAS therefore requires assurance at multiple levels: the individual AI components, the integrated AI systems, and the resulting AI technology providing the autonomous behaviours.

In MAS, these AI components and AI systems will not function in isolation. By design, they will be required to operate in harmony with traditional hardware and software and interact with human operators throughout different operational stages (such as launch, recovery, and maintenance) and across the entire lifecycle. As a result, assurance depends not only on the technical properties of individual components (AI or otherwise), but also on the interactions, emergent behaviours, and broader social and ethical implications across the entire system lifecycle.

There is already a robust body of knowledge on how to avoid failures and demonstrating 'safe' in traditional hardware and software (e.g., non-AI) [21], and many established frameworks and standards are available for the assurance of complex systems. For example, the UK Ministry of Defence's Defence Standard (Def Stan) 00-056 [22] provides a comprehensive structure for integrating safety engineering into defence equipment and services [23]. This standard encourages the use of civil (non-defence), open, or other recognised good practice standards and promotes a goal-based approach, justified through a robust safety case.

A widely referenced open standard is International Electrotechnical Commission (IEC) 61508 [24], the international standard for the functional safety of electrical, electronic, and programmable electronic (E/E/PE) safety-related systems. IEC 61508 is broadly adopted as a sector-agnostic safety lifecycle framework. It introduces Safety Integrity Levels (SILs), which set derived safety targets depending on the level of risk. As SILs increase, so does the evidence and assurance activities that are required, using both qualitative and quantitative methods for software and hardware.

The framework is often illustrated using a traditional 'V' lifecycle model. On the left side, activities include hazard and risk analysis and allocation of functional safety requirements. On the right side, activities focus on realisation through design, installation, validation, operation, and maintenance. Figure 1 shows this V model hierarchy as applied to a maritime vessel, while Figure 2 provides an example flowchart demonstrating how IEC 61508 is implemented across its various parts.

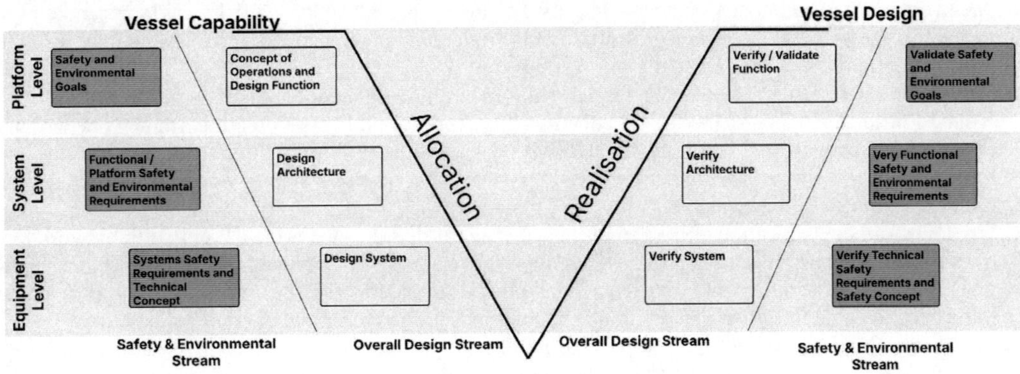

Figure 1: Systems V Model

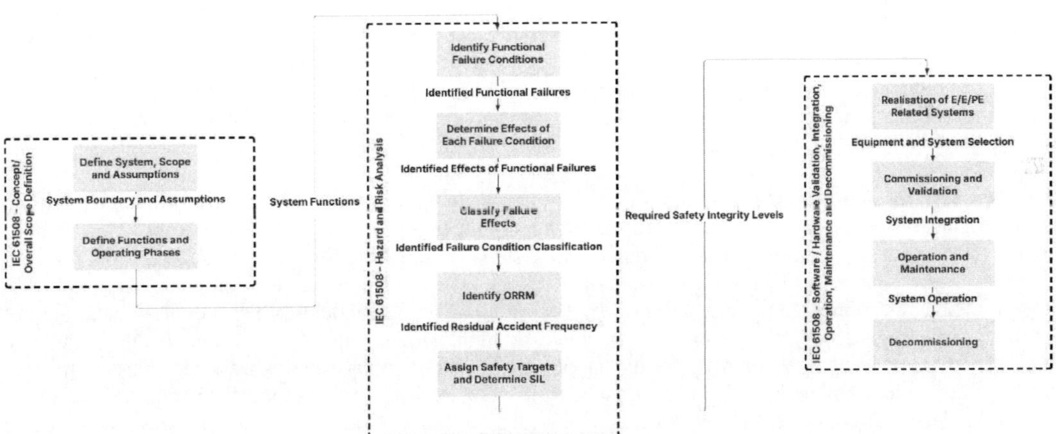

Figure 2: Example Flowchart Showing the Application of IEC 61508 (With Reference to Applicable IEC 61508 Parts)

However, traditional safety standards like IEC 61508 [24] were developed mainly for deterministic systems, where hazards are relatively stable and well understood. In contrast, MAS that incorporate AI and ML present new assurance challenges that these standards do not currently fully address [25, 26]. For example:

- Non-deterministic behaviours, continuous learning, and emergent properties make assurance significantly more complex;
- Traditional standards aren't suited to the agile development environment typically used for AI and ML systems; and,
- Unlike conventional systems, AI and ML based systems often involve complex hierarchies and iterative development across multiple organisations.

Despite these challenges, AI components and AI systems are increasingly used to perform safety-related functions in AI technology, replacing traditional software and hardware in maritime systems to enable true autonomy. For example, an autonomous vessel might rely on an AI system to integrate radar and camera data to make real-time collision avoidance decisions. The perception and decision-making algorithms in these systems can vary significantly depending on training data, algorithm weights, environmental conditions, unexpected objects, or emergent behaviours [26].

2.2 EMERGING ASSURANCE APPROACHES FOR AUTONOMY

In recognition of the challenges posed by AI components and AI systems, new approaches have been developed, and existing approaches adapted, to better address the complexity and associated assurance requirements:

- The latest version of Def Stan 00-056 [22] does not address AI but acknowledges the need for modern approaches to safety engineering to address increasingly complex systems. It highlights systems engineering-based techniques, such as Systems-Theoretic Process Analysis (STPA), which can be used to analyse complex systems where potential failures may not be immediately obvious and may result in emergent hazards, such as for autonomous systems;
- The IEC is currently drafting a new international standard titled Functional Safety – Framework for Safety-Critical E/E/PE Systems for Defence Industry Application [27]. Whilst this framework will not address AI directly, it will build on IEC 61508 [24], extending it to apply to defence applications and incorporating systems engineering principles to manage safety aspects in highly complex systems, such as autonomous platforms [27];
- UK Joint Service Publication (JSP) 936 [28] has been introduced as the principal policy framework for the safe and responsible adoption of AI within the UK Ministry of Defence. This policy sets out high-level overarching requirements for governance, ethical considerations, and safety of AI to provide an initial step towards integrating AI assurance into defence procurement and operational practices;
- The Safety-Critical Systems Club (SCSC) has published guidance on the assurance of autonomous systems incorporating AI and ML [26]. This guidance proposes tailoring assurance activities to system criticality and highlights the importance of evidence-based safety arguments.

While these initiatives provide a comprehensive overarching framework for the assurance of complex and autonomous systems, there remains a need to formalise specific assurance methods and techniques for the individual AI components of the AI system. Practical and detailed methodologies for conducting such hazard and risk analyses and deriving qualitative and quantitative safety requirements for AI components and AI systems remains underdeveloped.

This gap has been recognised across both industry and academia. In response, the Assurance of Machine Learning for Autonomous Systems (AMLAS) framework, developed by the CfAA at the University of York [29], provides a structured approach for integrating ML assurance into systems engineering lifecycles through hazard analysis, data management, model verification and validation, and evidence generation to support assurance cases. Although AMLAS is applicable to potentially safety-critical contexts, its methods are still evolving and are not yet tailored to the highest levels of assurance demanded by highly safety-critical environments and high SILs, such as MAS. The current focus of AMLAS has largely been on mitigating risks associated with ML-specific failures, for example, unintended behaviours arising from poor data quality or inadequate training and Large Language Models (LLMs), rather than on addressing broader systematic risks that may arise across the lifecycle of a system in safety-critical domains. There is a need for domain-specific and supplementary guidance for safe implementation in safety-critical systems [30].

2.3 ADAPTING A FUNCTIONAL SAFETY APPROACH FOR AUTONOMY

In recognition of the challenges posed by integrating AI components and AI systems in safety-critical systems, the IEC published Technical Report (TR) 5469 [31] in 2024. This report provides guidance on incorporating AI components and AI systems into functional safety frameworks. Rather than replacing established standards, such as IEC 61508 [24], IEC TR 5469 extends them to enable the inclusion of AI-specific characteristics within traditional safety lifecycles.

IEC TR 5469 [31] aligns with AI lifecycle concepts described in IEC 5338 [32] (AI System Life Cycle) and IEC 23053 [33] (Framework for AI Systems Using ML), integrating these as informative references to guide the extension of established standards such as IEC 61508, and the associated lifecycle standards of IEC 12207 for software development [34] and IEC 15288 for system lifecycle processes [35]. As a result, it provides additional assurance methods and techniques to address the unique properties of AI components and AI systems.

It is important to note that IEC TR 5469 [31] does not replace the requirement to address random hardware faults and systematic faults as outlined in IEC 61508 [24]. Instead, it supplements these requirements by introducing aspects necessary for assuring AI technology. From this perspective, the assurance approach follows a traditional functional safety lifecycle which is adapted to include AI components and systems. Through this approach:

- The allocation phase (e.g., hazard and risk analysis and safety target allocation - the left side of the V-model, as shown in Figure 1) remain guided by IEC 61508 [24]; however,
- The realisation phase (e.g., verification and validation through realisation, installation, validation, operation, and maintenance - the right side of the V-model, as shown in Figure 1) is modified to account for AI components and AI systems.

An updated example flowchart illustrating how IEC TR 5469 [31] interacts with the various parts of IEC 61508 [24] is presented in Figure 3.

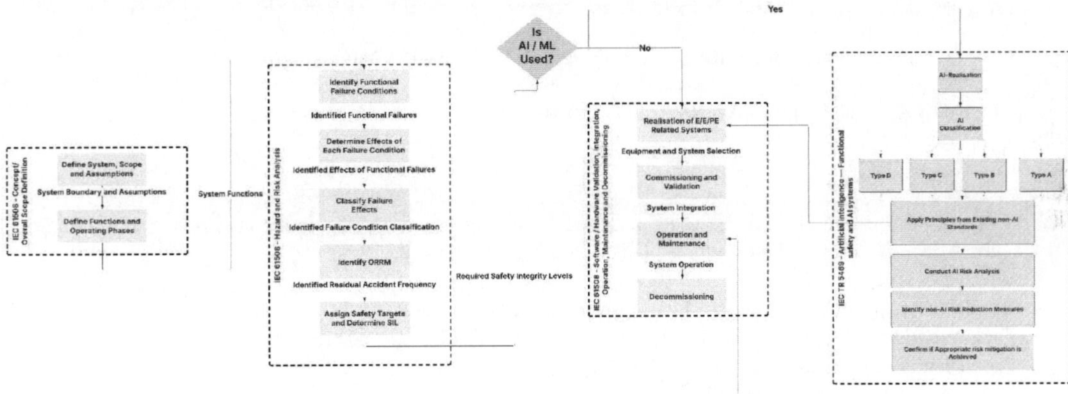

Figure 3: Example Flowchart Showing the Application of IEC 61508 / IEC TR 5469 (With Reference to Applicable IEC 61508 Parts and IEC TR 5469)

While IEC TR 5469 [31] is currently only a Technical Report and not a formal standard, the methods and techniques it outlines can be used to evaluate the suitability of AI components and systems for meeting safety targets derived from IEC 61508 [24]. This forms a necessary part of evidence to support the safety case for MAS that use AI. According to the Norwegian Research Centre for AI Innovation [36], IEC TR 5469 achieves this through:

- Describing the use of AI and ML in safety-related E/E/PE systems;
- Providing a classification scheme for the applicability of AI in safety-related E/E/PE systems;
- Explaining the properties of AI components and how they relate to safety within a functional safety context; and
- Detailing verification and validation methods and techniques tailored to AI and ML.

To support this, IEC TR 5469 aligns with the three-stage realisation principle of AI derived from IEC 22989 [20]:

1. Data acquisition - gathering representative, high-quality data;
2. Knowledge induction - learning from data, including model training;
3. Processing and output generation - producing decisions or actions from learned models.

This three-stage model provides a generalised process framework for AI. In some examples, this model has been further expanded into more detailed conceptual frameworks, such as the Sense-Understand-Decide-Act (SUDA) model proposed by the CfAA [18], but we choose not to do so in this work, to retain alignment with IEC TR 5469 [31]. The three-stage realisation principle from IEC 22989 [20] and the SUDA model are compared and shown to be compatible in Figure 4.

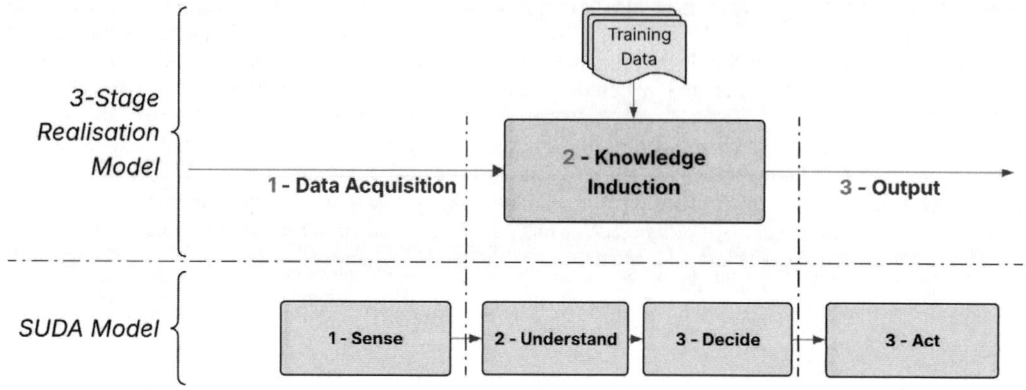

Figure 4:Three Stage Realisation Principle of AI / SUDA Model Comparison

3. APPLYING A FUNCTIONAL SAFETY APPROACH FOR AUTONOMY

3.1 INTRODUCTION

The three-stage realisation model can be used to derive acceptance criteria, assurance methods, and validation techniques for AI components and AI systems. This is achieved by considering:

- The desirable properties for each stage of the realisation model;
- The methods and techniques needed to achieve and verify those properties; and
- The acceptance criteria for assessing the adequacy of those methods and techniques.

An overarching 'AI safety' argument can then be constructed for the AI technology, covering each stage and demonstrating that all AI components and systems are safe for their intended use, through achieving the acceptance criteria for the desirable properties using the defined methods and techniques. This includes both quantitative (technical, numerical, performance-based) measures for AI elements and components but also qualitative measures at the organisational, governance, and system management levels in line with IEC 61508's [24] broader lifecycle framework fundamentals.

The rigour of the properties, methods, techniques and organisational arrangements increases in line with both:

- The safety targets (e.g., SIL's) derived from hazard and risk analysis on the left side of the V-model, guided by IEC 61508 [24]; and
- The AI technology 'Level' and 'Class', as defined in IEC TR 5469 [31].

3.2 AI TECHNOLOGY LEVEL

The AI technology 'Level' defines how AI is used and the extent of control within the system, as follows:

- Level A1 - The AI technology is used in a safety-relevant E/E/PE system, and automated decision-making of the system function using AI technology is possible;
- Level A2 - The AI technology is used in a safety-relevant E/E/PE system, but no automated decision-making of the system function using AI technology is possible; and,
- Level C - The AI technology is not part of the safety function but can have an indirect impact on system behaviour.

Additional levels (Level B1, B2) are defined where AI is used during the development of safety-relevant E/E/PE systems (e.g., as offline analysis or support tools) or where it is not part of the safety function and cannot have any impact on system behaviour (Level D). Given the focus on AI technology performing safety-related functions within MAS, these levels are not discussed further in this paper.

Because AI technology levels define the degree of control and influence AI technology exhibits, they can be directly mapped to frameworks such as the previously introduced DMR LoC, as illustrated in Figure 5.

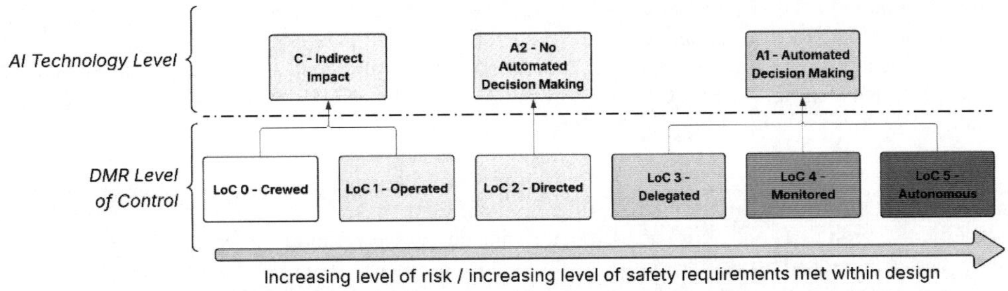

Figure 5: AI Technology Level / DMR Level of Control Mapping

3.3 AI TECHNOLOGY CLASS

The AI technology 'Class' categorises the feasibility of developing and assuring AI using existing functional safety standards, as follows:

- Class 1 - The AI technology is fully developed and reviewed using existing functional safety standards, such as IEC 61508 [24]. Established risk reduction concepts and techniques can be applied, leading to predictable and fully specified behaviour. This typically refers to 'traditional' AI technology, such as rule-based expert systems with no adaptive learning during operation;
- Class 2 - The AI technology cannot be fully developed and reviewed using existing functional safety standards alone. However, it is possible to identify additional complementary requirements, methods, and techniques (as outlined in IEC TR 5469 [31]). This typically applies to partially adaptive systems, such as ML-based perception modules, where risk can be managed through redundancy, operational restrictions, and fallback mechanisms;
- Class 3 - The AI technology cannot be adequately developed or reviewed using existing functional safety standards, and it is not possible to define sufficient complementary requirements, methods, or techniques to ensure acceptable risk reduction. This typically includes fully autonomous, adaptive, and opaque AI systems used in isolation, such as reinforcement learning (RL) controllers that learn online in open environments and continually adapt behaviour during operation.

Class 3 AI technologies are characterised by the absence of any currently known or accepted methods to achieve risk reduction to acceptable levels. Therefore, we would suggest they are not used in isolation in safety-related or safety-critical functions. This is not to say that Class 3 AI technology should be avoided altogether, rather efforts should be made to "reduce" the AI technology class through methods such as:

- Constraining the operational envelope to restrict online learning and adaptation in in-service environments;
- Introducing redundancy, where safety-critical decisions are overseen or validated by lower-classified (Class 1 or 2) systems or traditional E/E/PE; and
- Incorporating fallback mechanisms to ensure safe shutdown or transition to controlled degraded states in the event of unexpected behaviour (e.g., from LoC 5 to LoC 4 / 3).

3.4 DEMONSTRATING A FUNCTIONAL SAFETY APPROACH FOR AUTONOMY – QUANTITATIVE CONSIDERATIONS

To illustrate the practical application of a functional safety approach to maritime autonomy, we consider the example of an autonomous vessel's collision avoidance system, where tasks related to the three-stage realisation principle include perception (data acquisition), prediction (knowledge induction), and planning (processing and output generation) [37]. In this example, we focus on the prediction (knowledge induction) task, which is responsible for identifying objects to then generate a safe trajectory (through processing and output generation) that satisfies multiple objectives; advancing towards a destination, complying with navigational rules and avoiding collisions.

In this hypothetical example, we can assume that the system processes data, from the data acquisition stage, from onboard cameras using a perception algorithm based on a Deep Learning (DL) model, such as a Deep Neural Network (DNN) per the knowledge induction stage.

If we further assume that the system does not adapt online and that other inputs (e.g., radar) are utilised for redundancy and input fusion then, under IEC TR 5469 [31], this AI technology would be classified as Class 2. Based on this AI technology class and using the aforementioned definitions, we would apply established risk reduction concepts and techniques per IEC 61508 [24], supplemented by additional complementary requirements, methods, and techniques as outlined in IEC TR 5469 [31].

If the vessel operates in full autonomy (e.g., LoC 5), it would be categorised as AI Level A1, indicating automated decision-making without human intervention. Using the definitions of AI Level A1 and AI Technology Class 2, specific quantitative properties, methods and techniques, and acceptance criteria can be derived per IEC TR 5469 [31]. While full details are beyond the scope of this paper, a worked example focusing on the property of 'Specifiability' is presented in Table 1, following the format provided in IEC TR 5469.

Table 1. AI Technology Specifiability Properties, Methods and Acceptance Criteria

Stage	**Desirable Property**	**Requirements**	**Properties**	**Details (Acceptance Criteria)**	**Methods for Compliance**
2 – Knowledge Induction	Specifiability	Correctly identify diverse object types, unpredictable behaviours, and variable weather and sea states.	Specification of the Data Set	Amount of data (data set coverage)	- Manual creation - Active learning -Requirement traceability - Test plans - Compliance with IEC TR 24029 (Methods for Testing and Validating Robustness of AI/ML models)
			Type of Data	Object classes, object definitions, weather conditions, geographic domains, background environments	
			Specification of the labelling policy	Data annotation (labelling quality)	- Defined labelling guidelines - Random sampling and independent review of labels
			Number of Annotators	Use of independent annotators, random checks to confirm consistency and reduce bias	- Minimum number of annotators specified

3.5 DEMONSTRATING A FUNCTIONAL SAFETY APPROACH FOR AUTONOMY – QUALITATIVE CONSIDERATIONS

In addition to quantitative methods and techniques for individual AI elements and components, IEC TR 5469 [31] emphasises the importance of qualitative measures that address the organisational and managerial dimensions of AI safety, beyond just technical, quantitative validation of AI components. It formalises the requirement to establish a tailored AI management system, as specified in IEC 42001 [38], to ensure comprehensive oversight of governance, risk assessment, ethics and continuous learning throughout system operation aligned with the overall risk presented. To achieve this through the established management system, IEC TR 5469 [31] references, or signposts to, a set of qualitative standards and guidance documents to be followed for the example collision avoidance system at the identified Class and Level including:

- IEC 38507 (Guidelines for governance, oversight, accountability and transparency in AI systems) [39] to demonstrate that adequate organisational structures and decision-making frameworks are in place to oversee AI technology development;
- IEC 42005 (AI System Impact Assessment) [40] to address the non-functional risks associated with AI, including:
    - Model bias;
    - Inconsistent behaviour; and,
    - Fail-unsafe scenarios and potential unintended system-level effects.
- IEC TR 24028 (Trustworthiness in AI) [41] to demonstrate that AI systems are trustworthy, explainable, and reliable and that identified biases have been systematically mitigated.

IEC TR 5469 [31] highlights the importance of including structured feedback mechanisms such as post-deployment monitoring, incident reporting and stakeholder feedback loops within the operational framework. These mechanisms support the ongoing validation of safety assumptions and enable the system to adapt to emerging behaviours or changes in the environment. Their inclusion ensures that safety assurance evolves in response to operational experience, reinforcing the integrity of the safety case over time.

3.6 DEMONSTRATING A FUNCTIONAL SAFETY APPROACH FOR AUTONOMY – A TRANSITIONAL APPROACH

As highlighted earlier in this paper, most MAS are expected to operate in hybrid or transitional modes, blending full autonomy (LoC 5) with remote (LoC 2) or direct human control (LoC 1) at different stages of operation.

As illustrated in Figure 5, increasing the LoC towards full autonomy expands the operational risk envelope and reduces external, outside of design, risk controls, such as those provided by human intervention and external Other Risk Reduction Measures (ORRM). From an AI technology perspective, the required safety target will correspondingly increase when transitioning from AI Level A2 (human-authorised decision-making) to AI Level A1 (fully automated decision-making), which necessitates higher levels of technical assurance and greater rigour in both qualitative and quantitative safety measures.

Rather than attempting to adopt AI Level A1 or LoC 5 from the outset, the mapping of LoC to AI Levels (as shown in Figure 5) demonstrates how the level of assurance rigour of the AI technology can and therefore should incrementally increase in parallel with the level of autonomy. This staged approach enables the demonstration of an acceptable level of safety at each step, tailored to the level of risk presented at that stage of operational maturity.

By applying a functional safety approach, incorporating standards such as IEC 61508 [24] and IEC TR 5469 [31], both qualitative (e.g., governance, organisational readiness, management systems) and quantitative (e.g., verification, validation, robustness testing) methods can be progressively identified, demonstrated and strengthened. This supports not only immediate operational safety but also the development of a documented strategy and roadmap to achieve future acceptance criteria required for increased SIL, AI technology level and LoC as the MAS advances toward fully sustained autonomous operations.

In simple terms, this approach can be summarised using a "crawl, walk, run" strategy; vessels initially operate at low LoC and in well-defined operational envelopes, supported by robust external risk reduction measures for AI technology and conservative assumptions. As assurance evidence is generated and confidence in the AI technology increases, these constraints can be systematically relaxed, to expand the AI technology level and LoC, thereby enabling the gradual expansion of the operational scope of a vessel to become a MAS.

4. CONCLUSION

This paper has examined the integration of a functional safety approach to support the assurance of MAS, with a particular focus on adapting established standards such as IEC 61508 [24] supported by IEC TR 5469 [31]. Through the proposed mapping of LoC to AI technology levels, it is possible to incrementally demonstrate acceptable safety at each developmental and operational stage for the associated AI technology, including AI components and AI systems.

This staged strategy supports a "crawl, walk, run" transition model, enabling conservative initial operations within bounded risk envelopes and allowing for gradual expansion as confidence in AI technologies grows and reliance on ORRM decreases. Crucially, this paper highlights how assurance for maritime autonomy must encompass not only quantitative engineering measures but also qualitative organisational and governance dimensions of a lifecycle approach. In doing so, this approach does not propose abandoning existing goal-based safety case methodologies. Instead, it aims to provide a structured, standards-aligned evidence stream that addresses the unique challenges presented by AI and autonomy, thereby complementing and reinforcing the evidence within the overall safety cases for MAS.

4.1 RECOMMENDATIONS FOR FUTURE WORK

While this paper argues that the proposed functional safety approach can provide a robust foundation for AI assurance in maritime autonomy, further work is needed to fully integrate these methods within comprehensive safety case frameworks. What is clear from our work, is that there is much work involved in the assurance of MAS; the structured evidence stream presented here offers specific support for AI aspects of MAS, but it is not a light touch. Given this, we recommend future work must focus on providing tailoring and practical guidance usable by both industry and regulators. This can be achieved by exploring how this AI technology focused evidence stream can be integrated with system-level analyses derived from STPA. In our view, while STPA can be used to support identifying hazards from complex interactions and emergent behaviours, it often lacks mechanisms to convert these insights into verifiable requirements for AI systems and components. The methods outlined in this paper are proposed to fill this gap by defining specific properties, methods, and acceptance criteria for AI technologies following the application of STPA, thereby complementing STPA outputs and strengthening the overall argument within safety cases.

Additionally, there is significant potential to further align an approach with the UK DMR's guide to the regulation of MAS. The DMR advocates a tailored, goal-based approach that supports gradual progression from lower to higher LoC levels. By combining a functional safety strategy for the supporting AI technology that scales the assurance rigour with increasing autonomy, and integrating it with STPA and systems-engineering approaches, industry can provide traceable, auditable evidence to meet DMR and emerging international regulatory expectations.

Finally, we identify that this further work should also focus on developing detailed guidance to unify these approaches into a single safety case framework tailored to MAS aligned with DMR policy, Defence Standard 00-056, and evolving international standards such as the forthcoming IMO MASS Code whilst building on academic frameworks. Extending these methods to develop bespoke assurance frameworks for Class 3 AI technologies, which currently lack robust verification pathways, is critical to support the maritime industries autonomous aims. This includes investigating adaptive operational constraints, real-time monitoring, fail-safe mechanisms, and hybrid assurance architectures that combine deterministic supervisory controls with adaptive, learning-based components for those online dynamic learning AI systems and components.

5. ACKNOWLEDGEMENTS

The authors wish to thank Dr Paul Hogan, Safeguard Engineering Ltd, for their contribution and support in the development of this work.

6. AUTHOR BIOGRAPHIES

Alexander Barker holds the current position of Principal Consultant at Safeguard Engineering Ltd. The author has a Masters in Safety, Risk and Reliability Engineering, is a Chartered Engineer and is an accredited Functional Safety Professional. He is currently responsible for leading safety, risk and reliability projects, with a particular focus on digital technology. The Author's previous experience includes safety and assurance for UK Ministry of Defence uncrewed surface and subsurface vessels, and commercial remotely operated and autonomous surface vessels.

Ian Groom MBE holds the current position of Principal Consultant at Safeguard Engineering Ltd. The author is a Chartered Marine Engineer responsible for technical support and advice to defence projects and programmes, with a particular focus in the maritime domain. The Author's previous experience includes an extensive career as a decision maker in the UK Royal Navy, latterly as the Defence Maritime Regulator (DMR) for the UK MOD, where he supported the registration of defence autonomous vessels.

Thayyab Hussain holds the current position of Graduate Consultant at Safeguard Engineering Ltd. The author has a Masters in Physics and is an accredited AI Risk Management Practitioner. He is currently responsible for supporting safety and environmental projects, with a particular focus on software safety, cybersecurity, AI risk management and legislative compliance.

7. REFERENCES

1. IMO, 'MSC.1/Circ.1638 – Regulatory Scoping Exercise for the Use of Maritime Autonomous Surface Ships (MASS)', *International Maritime Organization*, 2021.

2. SMI, 'Maritime Autonomous Ship Systems (MASS)', *UK Industry Conduct Principles and Code of Practice, Version 8*, November 2024.

3. Defence Safety Authority, DSA03-DMR, Guide to Regulation of Maritime Autonomous Systems, Version 3.1, August 2024.

4. RODSETH, O et al., 'A Criticism of Proposed Levels of Autonomy for MASS', *Proceedings of the 33rd European Safety and Reliability Conference* (ESREL 2023), 2023.

5. Lloyd's Register, 'Out of the box – Implementing Autonomy and Assuring AI', 2023.

6. Det Norske Veritas, 'Ensuring the Safety of Autonomous Shipping', 2025.

7. HOWARD, G., CHILCOTT, J., and PARKIN, P., 'Unmanned and autonomous maritime – the challenges of assurance', Conference Proceedings of INEC, 2020.

8. UK Government, 'The Strategic Defence Review 2025 – Making Britain Safer: Secure at home, strong abroad', 2025.

9. Department for Transport, Maritime 2050: Navigating the Future, UK Government, 2019.

10. Centre for Assuring Autonomy, University of York, Future of Autonomous Maritime Operations Report, 2023.

11. DE VOS, J., HEKKENBERG, R G., and VALDEZ BANDA, O A., 'The Impact of Autonomous Ships on Safety at Sea – A Statistical Analysis', *Reliability Engineering and System Safety*, 2021.

12. PAYNE, A., and STEHR, A., 'Is Regulation really the barrier? Exploring the Opportunities and Challenges in Certifying Maritime Systems with Increased Automation and Autonomy', *Conference Proceedings of INEC*, 2024.

13. IMO, 'MSC-FAL.1/Circ.3/Rev.1 - Guidelines on Maritime Cyber Risk Management', 2021.

14. DIERMET, S., MILLET., GROVES, J., and JOYCE, J., 'Safety Integrity Levels for Artificial Intelligence', *Critical Systems Labs Inc*., 2023.

15. IMO, Development of a Non-Mandatory MASS Code and Roadmap, IMO Documents MSC 107/5/1 and MSC 107/5/2, 2023.

16. LEVESON, N G., 'Engineering a Safer World: Systems Thinking Applied to Safety', *MIT Press*, 2012.

17. Ministry of Defence, Defence Standard 00-056: Safety Management Requirements for Defence Systems, UK MOD, 2022.

18. Centre for Assuring Autonomy, University of York, 'The BIG Argument: Assuring Safety of AI Systems', 2022.

19. Safety-Critical Systems Club, 'Guidance on the Assurance of Machine Learning in Safety-Related Applications', Version 2, 2021.

20. IEC 22989, Information Technology — Artificial Intelligence — Concepts and Terminology, 2022.

21. HOBBS, C., 'Embedded Software Development for Safety-Critical Systems', 2nd ed.

22. Ministry of Defence, Defence Standard 00-056: Safety Management Requirements for Defence Systems, UK MOD, 2022.

23. HAMILTON, V., 'A New Concept in Defence Safety Standards: The Revised UK Defence Standard 00-56' *ACS Workshop on Tools and Standards*, 2005.

24. IEC, IEC 61508: Functional Safety of Electrical/Electronic/Programmable Electronic Safety-Related

25. Systems, International Electrotechnical Commission, 2010.

26. ISO, ISO 21448: Safety of the Intended Functionality (SOTIF), International Organization for Standardization, 2019.

27. Safety-Critical Systems Club, 'SCSC Guidance on the Assurance of Autonomous Systems', Version 1, 2022.

28. INGE, J., et al., 'Engineering Safety into Complex Defence Systems', *Proceedings of the 2023 International Conference on System Safety*, 2023.

29. UK Ministry of Defence, JSP936 – Dependable Artificial Intelligence (AI) in Defence, Part 1 – Direction, 2024.

30. HABLI., et al., "The BIG Argument for AI Safety Cases", *Centre for Assuring Autonomy'*, 2025.

31. LAHER, et a., 'Review of the AMLAS Methodology for Application in Healthcare', 2022.

32. IEC, IEC TR 5469: Functional Safety – Guidance for AI and ML Elements, International Electrotechnical Commission, 2024.

33. IEC, IEC 5338, Information Technology – Artificial Intelligence – AI System Life Cycle Processes, 2023.

34. IEC, IEC 23053, Framework for Artificial Intelligence (AI) Systems Using Machine Learning.

35. IEC, IEC 12207, Systems and Software Engineering – Software Life Cycle Processes, 2017.

36. IEC, IEC 15288, Systems and Software Engineering – Systems Life Cycle Processes, 2023.

37. Norwegian Research Centre for AI Innovation, Review of AI Regulations and Governance, 2022.

38. Stanford University, "AI-Based Vehicle Motion Planning: Foundations and Challenges", Stanford AI Lab, 2022.

39. IEC, IEC 42001, Information Technology – Artificial Intelligence – Management System, 2023.

40. IEC, IEC 38507, Information Technology – Governance of IT – Governance Implication of the Use of AI by Organizations, 2022.

41. IEC, IEC 42005, Information Technology – AI System Impact Assessment, 2025

42. IEC, IEC TR 24028, Information Technology - Artificial Intelligence – Overview of Trustworthiness I Artificial Intelligence, 2020.

Warship 2025: The Future Fleet: Smart Technology, Sustainability and Autonomy
16th – 17th June 2025
Glasgow, UK

A RISK AGGREGATION APPROACH TO SUPPORT RISK ASSESSMENT FOR FUTURE CREWED AND UNCREWED VESSELS

A Barker, I Groom and **P Hogan**, Safeguard Engineering Ltd, UK

SUMMARY

This paper addresses the need for risk aggregation within Risk-Informed Decision Making (RIDM) for crewed and uncrewed vessels. It defines, explores, and applies risk aggregation methods, evaluating their applicability based on risk characteristics. We identify and formalise three aggregation approaches: Type 1 (same risk type, independent), Type 2 (different risk types, independent) and Type 3 (dependent risks). Using practical examples for each type, we demonstrate the technical feasibility of risk aggregation to present the overall risk to support RIDM, while highlighting significant challenges and limitations. Consequently, we argue that aggregated risk approaches should support, rather than replace, detailed risk assessments and contextual information in RIDM. We recommend that future risk aggregation strategies be guided by top-down policy frameworks to ensure consistency, transparency, and robustness in RIDM for future crewed and uncrewed vessels.

1. INTRODUCTION

1.1 DEFINING RISK

There are various definitions of risk [1], with ongoing debate over the merits and rationale of each [2, 3, 4]. For the purposes of this paper, we define risk as a 'combination of consequence and likelihood'. This definition is consistent with United Kingdom (UK) Ministry of Defence (MOD) submarine [5] and surface ship [6] safety management procedures.

When considered in isolation, risks are termed single risks [7, 8]. Single risks are categorised by type, e.g., safety, environmental and operational, each with different consequence measures [9]. Combining two or more single risks is termed overall risk [2]. The process of combining single risks to present the overall risk is termed risk aggregation [8].

1.2 DEFINING RISK ASSESSMENT

Risk assessments determine single risk consequence and likelihood using qualitative, semi-quantitative, or quantitative methods [10]:

- Qualitative methods use descriptive values (e.g., 'minor').
- Quantitative methods use numeric values (e.g., '3 deaths).
- Semi-quantitative methods uses both descriptive and numerical values in ranges (e.g., 'Critical / 1 to 10 deaths').

For maritime vessels, single risks are typically assessed semi-quantitatively per UK MOD surface ship [6] and International Maritime Organization (IMO) [11] guidance. The assessed risks are compared to risk tolerability frameworks [8], which establish maximum tolerable levels of risk. For safety risks, these levels are established by regulatory authorities such as the UK Health and Safety Executive (HSE). However, for other risk types, no such regulatory frameworks currently exist. In these cases, organisations often develop their own or utilise industry guidance.

In the UK, the tolerability framework for safety risk is established by the HSE through Reducing Risks, Protecting People (R2P2) [12]. This framework establishes the:

- Basic Safety Level (BSL) - the highest level of safety risk that is tolerable if shown to be ALARP (As Low As Reasonably Practicable); and,
- Basic Safety Objective (BSO) - the level of risk considered broadly acceptable if shown to be ALARP.

These levels apply to both individual and societal risk. Although these terms may appear analogous to single and aggregated risk, they have different meanings:

- Individual risk refers to the probability that a specific person will experience harm within a defined time period [13].
- Societal risk refers to the probability of events causing harm to multiple individuals within a population or group [13].

1.3 DEFINING RISK INFORMED DECISION MAKING

Risk assessment is typically performed by a risk assessor, separate from the decision maker who acts on the results [14]. This process of making decisions on the presented risk information is termed Risk-Informed Decision Making (RIDM) [15,16]. In a UK defence context, the decision maker is referred to as the Accountable Person (AP) [5, 6]. Throughout this paper, we use the term 'AP' to refer the RIDM decision maker in both defence and commercial maritime vessel contexts.

Maritime vessels inherently carry significant risks due to their function and operation. APs must make strategic, financial, and operational decisions based on the presented safety, environmental, and operational risk types. For crewed vessels, first-party safety risk is typically dominant, driven by the presence of onboard crew and regulatory frameworks (e.g., HSE R2P2 [12]). However, as vessels adopt uncrewed and autonomous systems [17], it is generally perceived that first-party safety risks decrease with the removal of onboard crew [18], while third-party safety, environmental, and operational risks may increase due to reduced human intervention in hazardous scenarios. Therefore, RIDM requires APs to consider single risk and the overall risk, and with the increase in uncrewed vessels increasingly across all risk types, which results in a requirement for aggregated risk. In addition to this reason, David [8] identifies several further requirements to aggregate risk:

- To account for different risk types arising from the same event.
- To assess tolerability against regulatory frameworks such as BSL and BSO for individual and societal risk.
- To compare total exposure with apportionment targets.
- To understand the platform's overall risk.
- To evaluate system-of-systems risk across multiple platforms.
- To consider dependent events across systems.

In addition to David's [8] points, we emphasise that, in the context of crewed and uncrewed vessels, understanding aggregated risk enables a more meaningful comparison of the relative risks between these platform types.

Beyond the requirements, recent research underscores the value of risk aggregation in supporting RIDM [19, 20, 21, 22]. Reflecting this, UK MOD APs increasingly advocate the need for risk aggregation in RIDM, as evidenced by the UK MOD Royal Navy (RN) [23] and Military Aviation Authority (MAA) [24] direction.

This emphasis is not new. Prior policy, review and guidance recognises the need for aggregated risk in RIDM, including:

- Defence Equipment and Support (DE&S) Safety and Environmental Protection (S&EP) Leaflet 02/2011, requires understanding of "aggregated risk posed by the whole system" to avoid exceeding tolerability criteria or allocating the entire risk budget to a single project [25].
- The Nimrod Review recommends appropriate resource allocation to reduce overall risk [26].
- Tolerability of Risk from Submarines (TOR(SM)), outlines aggregation methods for submarines, both new-build and in-service [5].

Yet despite this policy foundation and supporting research, practical implementation remains limited. We argue this is largely due to the complexity of aggregation methods, challenges in application and the lack of mature practical guidance on suitable approaches. A clear gap exists between risk aggregation research, policy, and practice.

This paper seeks to help bridge that gap by developing a strategy for risk aggregation and highlighting the complexities and challenges of adopting such approaches in support of RIDM for future crewed and uncrewed vessels.

2. THE CONCEPT OF RISK AGGREGATION

2.1 A CONCEPTUAL FRAMEWORK FOR RISK AGGREGATION

Like "risk," the term risk aggregation is defined differently across sectors, including safety [8] and finance [27]. Additionally, terms such as cumulative risk, risk profile, hazard footprint, and integrated risk picture are sometimes used interchangeably [8]. This has created much confusion, aptly captured in the title of Bjørnsen et al.'s paper, *"Risk aggregation: What does it really mean?"* [7]. Despite ongoing debate [20, 21, 22], no universal definition has emerged.

While not seeking to expand the debate, we note that this confusion hampers the closing of the aforementioned risk aggregation research-policy-practice gap. Therefore, a conceptual framework is necessary to underpin this papers discussion. Drawing on the literature [7,8,20] and ISO guidance [10, 28], we define risk aggregation as 'The process of combining different risk information from two or more single risks through a risk aggregation process, to gain an understanding of the bigger risk picture'. This paper explores risk aggregation processes that may be used to achieve this understanding of the bigger risk picture and discusses application for crewed and uncrewed vessels.

Risk aggregation methods fall into two distinct forms, although the two forms are often conflated in the available literature [20]:

- Visual aggregation – visually displaying multiple single risks together, such as in matrices, tables or dashboards.
- Mathematical aggregation – combining risks through mathematical methods, potentially across assessors, types, scenarios, systems, or assessment methodologies.

This paper focuses on mathematical aggregation in the context of maritime vessels. Visual methods, such as risk matrices, are widely used and extensively critiqued [29, 30]. We do not wish to add to this, but it is important to emphasise that visual methods merely display multiple single risks without quantifying an aggregated overall risk or capturing interactions and dependence between the risks [31]. Aggregation can be attempted qualitatively (e.g., 'by eye' [8]); however, such approaches are neither accurate nor robust. Crucially, declaring a single risk as tolerable within a matrix does not guarantee that overall individual or societal risk is tolerable. While visual tools can aid RIDM, they do not constitute risk aggregation.

Mathematical aggregation has gained interest [2, 22, 32] and encompasses various methods [8, 20, 22]. However, there remains practitioner confusion when and how these methods can be applied. This confusion arises from ambiguous guidance, for example, the ISO standard on risk assessment techniques [10] states that "qualitative or semi-quantitative measures of risk cannot be directly aggregated" but states that they can be "combined" and that "relevant data about different risks can be brought together to assist decision makers." This ambiguity leaves practitioners without clear guidance, reinforcing the research-policy-practice gap.

The absence of a universally accepted definition of risk aggregation, coupled with uncertainty around the availability and suitability of methods, has led some authors to suggest: "As a general rule, one should avoid the need for aggregation" [33, p. 26]. While this is a strong position, it is largely informed by the inherent challenges in risk aggregation (discussed in the following section). Acknowledging this caution and the associated challenges, we argue that the specific drivers for risk aggregation in the context of crewed and uncrewed vessels (discussed in the previous section) justify exploring available approaches to help bridge the research–policy–practice gap.

2.2 THE CHALLENGES OF RISK AGGREGATION

The challenges of risk aggregation are well illustrated by its use, and misuse, in the financial sector. In the financial sector, standardised aggregation methods are well established [27], shaped by lessons learnt from the Gaussian copula (Equation 1).

$$\Pr[T_A < 1, T_B < 1] = \Phi_2(\Phi^{-1}(F_A(1)), \Phi^{-1}(F_B(1)), \gamma)$$

Equation 1

The Gaussian copula was used to aggregate collateralised debt obligations (CDOs) into a single metric by combining individual default probabilities (A and B) through a dependency parameter (γ). Its apparent simplicity appealed to decision-makers and led to widespread adoption, supporting the trading of millions of CDOs, often presented through colour-coded dashboards. However, the copula masked underlying uncertainty and failed to capture real-world interactions and dependencies between seemingly independent CDOs [8]. This misapplication contributed to systemic failure and played a key role in the 2007–2008 financial crisis, earning the Gaussian copula the name "the formula that killed Wall Street" [34].

This example rightly raises caution about aggregating risk, and more broadly reducing complex systems to single metrics. However, as we have highlighted, to support RIDM risk aggregation should not necessarily be avoided. Based on the example of the Gaussian copula, and in the context of crewed and uncrewed vessels, we identify five key challenges in the application of risk aggregation methods which must be addressed to bridge the research-policy-practice gap:

- Loss of contextual information – There is a desire to view risk assessment as objective, but there are always underlying assumptions, dependencies and biases [35]. Aggregating across multiple risk assessments compounds distortions. Therefore, aggregation methods must transparently document assumptions, dependencies, and assessor information for the underlying risks.
- Risk interactions and dependencies – Different risks may occur simultaneously, because of another risk, or be inadvertently captured more than once, especially where different processes or assessors are used. This is further compounded in 'system of systems', such as the use of 'swarm 'uncrewed vessel technology [36]. Therefore, aggregation methods must explicitly account for dependencies.
- The illusion of safety - complex formulas and metrics can falsely appear to 'prove' safety. While useful for decision support, aggregated risk values should never substitute engineering judgement or detailed risk understanding. Therefore, aggregation methods must be framed as a support tool for RIDM, not a standalone decision-making mechanism.
- Overconfidence in visualisations – colour-coded dashboards (e.g., Red-Amber-Green) may falsely imply precision or compliance if unsupported by robust analysis [37]. Therefore, aggregation methods resulting in risk visualisations must be clear, understandable and evidence based.
- Different risk types - Different types of risk may be measured on incompatible consequence scales, especially for crewed and uncrewed vessels, making their consequences difficult to compare directly. Therefore, aggregation methods must use a clearly defined and consistent consequence scale, or a justified conversion method, to ensure comparability across risk types.

In light of these challenges, this paper seeks a careful balance: avoiding falling into the trap of the aforementioned challenges whilst identifying risk aggregation strategies for crewed and uncrewed vessels. We recognise both the inherent drivers and benefits of risk aggregation for RIDM, alongside the inherent challenges of implementation and interpretation. Bridging the research-policy-practice gap requires addressing these tensions. The following section develops a guiding strategy, grounded in these challenges, to support risk aggregation for crewed and uncrewed vessels.

3. RISK AGGREGATION STRATEGY FOR CREWED AND UNCREWED VESSELS

3.1 INTRODUCTION

The choice of an appropriate mathematical risk aggregation method depends on key characteristics of the single risks:

- Whether risks are assessed qualitatively, semi-quantitatively, or quantitatively.
- Whether risks are of the same risk type.
- Whether risks are dependent or independent.

Given the scope and focus of this paper, clear boundaries must be drawn. Risk assessments for vessels are typically conducted using semi-quantitative methods, e.g., following either UK MOD [6] or IMO [11] surface ship risk assessment guidance. As such, this paper focuses on mathematical aggregation of semi-quantitative risks and does not attempt to develop a strategy for aggregating purely qualitative or purely quantitative risks. This focus reflects:

- Aggregating qualitative risks may be possible, via expert judgement, interviews, or consensus frameworks [10]. However, it is generally not recommended for robust decision-making due to subjectivity and lack of standardisation [38]. We therefore do not wish to explore these methods further.

- Aggregating quantitative risks is possible, using established methods, such as FN curves, Expected Value (EV), and Risk Integral (RI) for independent risks, and copula-based or multicriteria frameworks for dependent risks [20, 22]. As such methods are well established [8, 16, 32], we do not wish to explore these methods further.

We therefore focus on developing a strategy to aggregate semi-quantitative single risks for crewed and uncrewed vessels. We recognise there are various methods to do so, including adapting the aforementioned methods for quantitative risks, and the suitability of each method depends on the single risk characteristics. To determine the suitability of these methods, we provide the following definitions:

- We define aggregation of risks that are of the same risk type and independent as 'Type 1' risk aggregation.
- We define aggregation of risks that are of different risk types and independent as 'Type 2' risk aggregation.
- We define aggregation of risks that are dependent as 'Type 3' risk aggregation.

We highlight that it is critical that the dependence of single risks be examined carefully before any aggregation method is selected or applied. If risks are incorrectly assumed to be independent, then applying a Type 1 or Type 2 approach will not only be mathematically invalid, it may also produce dangerously misleading results. In practice, risk practitioners often assume independence for simplicity, but do not always verify this assumption [30]. This is a dangerous assumption to make, see previous discussion on the Gaussian Copula. Vessels are inherently complex and tightly coupled systems and therefore some risks are likely to be dependent. The integration of multiple subsystems and increasingly autonomous technologies to support uncrewed vessels only amplifies this complexity and resulting dependency [35].

3.2 RISK AGGREGATION STRATEGY – TYPE 1 RISK AGGREGATION

Type 1 risk aggregation aggregates single risks of the same risk type assuming independence to present the overall risk. To illustrate Type 1 risk aggregation, four example semi-quantitative first-party safety risks were assessed for both a crewed and an uncrewed vessel, using the UK MOD surface ship risk assessment guidance and Risk Classification Matrix (RCM) [6], as shown in Table 1.

Table 1. Example Safety Single Risks for Crewed and Uncrewed Vessels

Vessel	Risk	Consequence	Likelihood	Risk Type	Risk Class
Crewed	Risk 1 (R_1)	Disastrous	Incredible	Safety	D
	Risk 2 (R_2)	Critical	Incredible	Safety	D
	Risk 3 (R_3)	Marginal	Remote	Safety	C
	Risk 4 (R_4)	Critical	Improbable	Safety	C
Uncrewed	Risk 1 (R_1)	Negligible	Incredible	Safety	D
	Risk 2 (R_2)	Negligible	Incredible	Safety	D
	Risk 3 (R_3)	Negligible	Remote	Safety	D
	Risk 4 (R_4)	Negligible	Improbable	Safety	D

In this RIDM scenario, it is visually apparent to the AP that the uncrewed vessel exhibits lower overall safety risk, given the lower classification of single risks, e.g., quantity of C vs D class risks between the respective vessels. However, such qualitative aggregation (e.g., 'by eye') is inadequate for larger risk sets or formal decision-making as it lacks traceability and quantification. A formal mathematical aggregation method is therefore required for Type 1 risk aggregation.

We have selected the RI method (Equation 2) as an appropriate approach for this demonstration, though other methods are available [20]. The RI method is derived from the EV method [39], which calculates aggregated risk as the area under a frequency-fatality (f–N) curve. The RI method extends EV by incorporating a scale aversion index (a) that assigns greater weight to higher consequence events, thereby reflecting societal aversion to catastrophic accidents [12]. The HSE recommends a value of 1.4 for this scale aversion index [40], and this value is applied in the present example for both crewed and uncrewed vessels. However, we highlight that further research is advised to determine an appropriate scale aversion index specific to both crewed and uncrewed vessel contexts.

$$RI = \sum fN^{a}$$

Equation 2

As previously stated, the RI method is intended for aggregating quantitative risks to provide a measure of societal risk [40]. However, we apply it to aggregate the semi-quantitative safety risks in Table 1, with adaptions to suit the risk characteristics for the crewed and uncrewed vessels as outlined in the following paragraphs.

Because the RI method requires numeric values for both consequence and likelihood, it is necessary to derive equivalent fatality values where the assessed consequence of the single risks represents an injury rather than fatality (e.g., crewed vessel R_3 and uncrewed vessel R_1, R_2, R_3 and R_4). The UK MOD surface ship RCM provides numeric bounds for fatal consequences (e.g., 'critical' is '1 to 10 deaths'), but no equivalents for non-fatal consequences (e.g., major, marginal, negligible). To address this, we adopt equivalent fatality factors from IMO guidance [11]:

- Major: 0.1 to <1 fatality.
- Marginal: 0.01 to <0.1 fatality.
- Negligible: 0.001 to <0.01 fatality.

This mapping allows non-fatal consequences to be expressed on a fatality-equivalent scale, enabling subsequent aggregation. It is important to identify that such equivalence is subjective, and, given the scope of this paper, the conversion factors used to equate different consequences of different risk types have not been considered in detail at this point in time. Different stakeholders might disagree, and even within the referenced IMO guidance the chosen equivalence is a policy decision and should not be made by the individual assessor. Although we have applied the IMO guidance in this example, we understand there is no one correct fatality-equivalent scale suit all scenarios. One should treat these factors as assumptions and therefore document them in RIDM and conduct sensitivity analysis, noting that altering the fatality–injury equivalence or introducing risk aversion for large accidents would affect the scale [40]. We therefore declare that any aggregated risk is fragile to the weighting scheme applied in a fatality-equivalent scale and highlight that further research is recommended to determine an appropriate fatality-equivalent scale tailored to both crewed and uncrewed vessel contexts.
Figure 1 presents an updated RCM from the UK MOD surface ship risk assessment guidance integrating these equivalent fatality intervals.

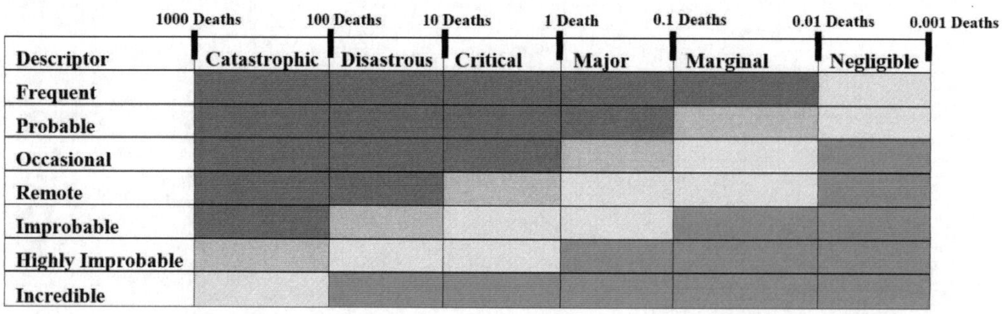

Figure 1: Updated MOD Surface Ship Risk Assessment Guidance RCM Showing Consequence (Equivalent Fatality) Intervals

Because the UK MOD surface ship RCM [6] is semi-quantitative, both consequence and likelihood are expressed as upper and lower bounded intervals rather than exact numeric values. Therefore, the use of semi-quantitative assessment methods, such as risk matrices, introduces uncertainty or 'fuzziness' for the assessed single risks [33]. This uncertainty is exacerbated further by inherent flaws in the construction of risk matrices, such as rank reversals and poor resolution. Given the scope of this paper, we do not expand further on these flaws but highlight that this uncertainty in the single risks is compounded when the single risks are aggregated [30].

Therefore, we apply the RI method to produce an interval-based output, bounded by the upper and lower limits of the aggregated risk distributions, rather than a single deterministic value. This interval representation is illustrated on an fN basis in Figure 2 for the crewed vessel's identified risks from Table 1, showing the maximum, midpoint, and minimum values for each risk.

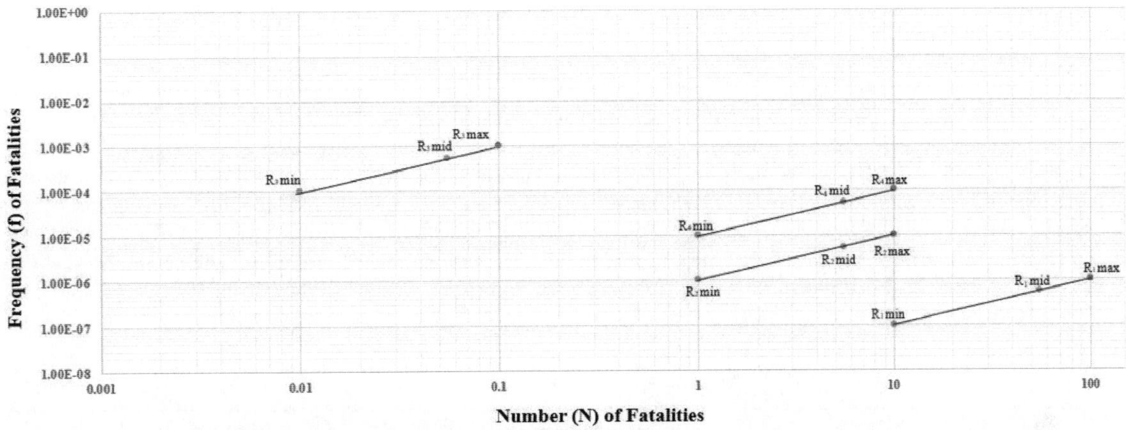

Figure 2: fN Representation for Crewed Vessel Risks

It is important that this uncertainty is clearly communicated to support RIDM, ideally in a form more accessible to APs than the technical representation in Figure 2. Reporting a single aggregated risk value without acknowledging these bounds could misrepresent the underlying uncertainty. While academic techniques have been proposed to provide a single value when aggregating semi-quantitative risks, such as fuzzy set theory, interval number theory, or probability density functions [2], these methods remain theoretical with practical challenges around error propagation and traceability [41]. Therefore, considering the challenge of 'loss of contextual information' identified in the previous section, we recommend retaining and reporting interval bounds to ensure transparency and preserve uncertainty within the aggregated risk values.

The resulting aggregated risk values, using the RI method, for the crewed and uncrewed vessels are shown in Table 2. As expected, 'by eye', the uncrewed vessel exhibits a lower overall risk. While this approach offers interim value for aggregating risks of the same risk type, and can be expanded further for larger risk sets, we recognise it carries significant limitations; there is a risk of hiding the underlying assumptions of the scale aversion index and the fatality-equivalent scale, and both can materially influence the resulting aggregated risk value. We therefore highlight that further research is recommended to determine an appropriate approach to present the underlying assumptions to the AP in RIDM along with the overall risk values.

Table 2. RI Aggregated Risk for Crewed and Uncrewed Vessels

Vessel Type	**RI Value**	
	Upper	**Lower**
Crewed	3.434×10^{-3}	1.367×10^{-5}
Uncrewed	1.588×10^{-6}	6.322×10^{-9}

3.3 RISK AGGREGATION STRATEGY – TYPE 2 RISK AGGREGATION

Type 2 risk aggregation aggregates single risks of different risk types assuming independence to present the overall risk. For reasons highlighted earlier in this paper, Type 2 aggregation is particularly relevant for uncrewed vessels, especially in RIDM scenarios that compare crewed and uncrewed vessels. In these scenarios, APs are likely required to consider single and overall risk covering safety, environmental and operational risk types in the decision-making process.

To illustrate Type 2 risk aggregation, ten example semi-quantitative risks were assessed for both a crewed and an uncrewed vessel covering safety, environmental and operational risk types. The safety risks were assessed using the UK MOD surface ship risk assessment guidance and RCM [6]. As the RCM assesses safety risks only, an equivalence framework was developed to align consequence categories across safety, environmental, and operational risk types.

This equivalence framework draws on approaches from available literature [5, 9] and uses past precedent for equivalence established by the IMO [11] and Bureau Veritas (BV) [42] to map the consequences from different risk types onto a common scale. Table 3 presents the equivalent safety, environmental and operational consequences.

We acknowledge that this equivalence framework is inherently subjective and has limitations; in current form, it treats all risk types as equally weighted and assumes equivalence between consequence categories. Alternative mapping, such as conversion to equivalent monetary value (e.g., as proposed by Bao et al. [43]), may be more appropriate and, critically, while we have made equivalences to facilitate risk aggregation this does not resolve differences in risk tolerability thresholds across risk types (although we have classified all risks using UK MOD surface ship risk assessment guidance [6] for the purposes of this example). Given the scope of this paper, we have developed this equivalence framework to support the development of a strategy for Type 2 risk aggregation, but highlight that further research is recommended to determine an appropriate equivalence framework for crewed and uncrewed vessels.

Table 3. Equivalence Framework for Consequence Types

Consequence	**Safety**	**Environmental**	**Operational**
Catastrophic	More than 100 Deaths	Prohibition of Operation (International)	Loss of Asset
Disastrous	10 to 10 Deaths	Prohibition of Operation (Local)	Withdrawal from Operations for Repair in Dockyard
Critical	1 to 10 Deaths	Restriction of Operation	Withdrawal from Operations for Repair in Naval Base
Major	Permanent Injury / Illness	Nuisance Duration >1 Month	Withdrawal from Operations for Repair at Sea
Marginal	Recoverable Injury / Illness	Nuisance Duration 1 Month to 1 Week	Temporary Reduction in Capability
Negligible	Major Injury / Illness	Nuisance Duration Up To 1 Week	No Reduction in Capability, but Some Reconfiguration Required

Table 4 presents the ten example risks assessed for the crewed and uncrewed vessels, covering safety, environmental, and operational risk types.

Table 4. Example Safety, Environmental and Operational Single Risks for Crewed and Uncrewed Vessels

Vessel	Risk	Consequence	Probability	Type	Risk Class
Crewed	Risk 1 (R_1)	Major	Remote	Environmental	C
	Risk 2 (R_2)	Major	Improbable	Environmental	C
	Risk 3 (R_3)	Critical	Highly Improbable	Environmental	C
	Risk 4 (R_4)	Disastrous	Incredible	Safety	D
	Risk 5 (R_5)	Critical	Highly Improbable	Safety	C
	Risk 6 (R_6)	Marginal	Remote	Safety	C
	Risk 7 (R_7)	Critical	Improbable	Safety	C
	Risk 8 (R_8)	Major	Incredible	Operational	D
	Risk 9 (R_9)	Marginal	Improbable	Operational	D
	Risk 10 (R_{10})	Negligible	Improbable	Operational	D
Uncrewed	Risk 1 (R_1)	Major	Remote	Environmental	C
	Risk 2 (R_2)	Critical	Improbable	Environmental	C
	Risk 3 (R_3)	Critical	Highly Improbable	Environmental	C
	Risk 4 (R_4)	Negligible	Incredible	Safety	D
	Risk 5 (R_5)	Negligible	Incredible	Safety	D
	Risk 6 (R_6)	Negligible	Remote	Safety	D
	Risk 7 (R_7)	Negligible	Improbable	Safety	D
	Risk 8 (R_8)	Major	Incredible	Operational	D
	Risk 9 (R_9)	Major	Improbable	Operational	C
	Risk 10 (R_{10})	Critical	Improbable	Operational	C

In this RIDM scenario, it is not immediately clear which vessel presents a lower overall risk. A risk matrix could be used to visualise the ten single risks (Tables 5 and 6). A comparison of the matrices could be used to visually infer lower overall risk for the uncrewed vessel based on fewer Class C risks. However, noting the issues of visual risk aggregation highlighted earlier in this paper, such a 'by eye' interpretation lacks rigour and traceability. As with Type 1 aggregation, a mathematical aggregation method is required to support RIDM.

Table 5. Risk Matrix for Crewed Vessel Showing Single Risks

Descriptor	Catastrophic	Disastrous	Critical	Major	Marginal	Negligible
Frequent						
Probable						
Occasional						
Remote				R1	R6	
Improbable			R7	R2	R9	R10
Highly Improbable			R3, R5			
Incredible		R4		R8		

Table 6. Risk Matrix for Uncrewed Vessel Showing Single Risks

Descriptor	Catastrophic	Disastrous	Critical	Major	Marginal	Negligible
Frequent						
Probable						
Occasional						
Remote				R1		R6
Improbable			R2, R10	R9		R7
Highly Improbable			R3			
Incredible				R8		R4, R5

The RI method used for Type 1 aggregation can also be used for Type 2 aggregation. As before, because the input data of the single risks is semi-quantitative, the resulting aggregated risk values are interval-based. The resulting aggregated risk values, using the RI method, for the crewed and uncrewed vessels are shown in Table 7.

In this case, despite a visual aggregation suggesting lower overall risk for the uncrewed vessel, mathematical aggregation identifies that the uncrewed vessel exhibits a higher overall risk. This outcome reflects the influence of magnitude of consequences of the single risks, amplified by the applied scale aversion index (a) of 1.4. This example demonstrates how Type 2 risk aggregation provides a more accurate picture of the overall risk than visually aggregating the single risks or presenting the single risks in isolation.

While this approach offers interim practical value for aggregating different risk types, we recognise it carries significant limitations. In addition to the limitations of consequence equivalence framework, 'boiling down' the multi-dimensional risk types may hide the type and respective domain which drives the overall risk. We therefore highlight that further research is recommended to determine an appropriate weighting framework for different risk types.

Table 7. RI Aggregated Risk for Crewed and Uncrewed Vessels

Vessel Type	RI Value	
	Upper	Lower
Crewed	4.790×10^{-3}	1.907×10^{-5}
Uncrewed	6.378×10^{-3}	2.539×10^{-9}

3.4 RISK AGGREGATION STRATEGY – TYPE 3 RISK AGGREGATION

Type 3 risk aggregation aggregates single risks that are dependant to present the overall risk. Noting that if the single risks are of different risk types then an equivalence framework, as identified for Type 2 risk aggregation, can first be used to map the consequences onto a common scale. Dependency of single risks can be categorised as either:

- Probabilistic interaction, where the likelihood of one affects another [14]; or,
- Consequential interaction, where the consequence of one affects another [44].

Several mathematical methods are proposed in the literature to account for dependence of single risks, including:

- Copula-based models, which capture statistical dependence between the single risks; and,
- Multi-criteria evaluation framework, which utilise fuzzy logic, Choquet integrals, or multiple linear regression models to model dependence between the single risks.

These methods, as discussed by Bao et al., [20] and Hassel et al., [22], are inherently quantitative in nature and require significant computational resources. Advanced techniques, such as Monte Carlo simulation, are used to model scenarios and quantify dependencies and interactions over time. As such, these methods rely on single risks being quantitatively assessed.

Applying these methods to semi-quantitative risks, as would be required for typical vessel RIDM, presents challenges. Similarly to Type 1 and Type 2 risk aggregation, the semi-quantitative nature of the single risks results in an interval-based output, rather than a deterministic value. While technically feasible, in practice the literature highlights limited appetite for deploying such complex probabilistic models in real-world settings due to concerns around traceability, interpretability, and escalating complexity [32]. We highlight that further work is required to develop practical quantitative methods for semi-quantitative dependent risk aggregation; however, this lies beyond the scope of this paper.

An alternative approach involves subjectively identifying dependencies through structured review of the single risks [20]. This method is inherently qualitative in nature and requires expert judgement and Suitably Qualified and Experienced Personnel (SQEP) input to quantify dependencies and interactions over time. This judgement can be supported by tools such as Fault Tree Analysis (FTA), Event Tree Analysis (ETA), or double-layer network analysis [43]. Bao et al. [43] propose a guiding question set for this method, which can be used to determine dependency of single risks, as follows:

- Does the occurrence of one risk affect the likelihood of another?
- Does the occurrence of one risk affect the consequence of another?
- Does the simultaneous occurrence of multiple risks trigger a greater consequence than their individual effects?

In this example, we have applied these questions to the risks identified in Table 4 for crewed and uncrewed vessels to highlight the following, illustrative, dependencies:

- For the crewed vessel the occurrence of Risk 5 increases the likelihood of Risk 4. Therefore, a likelihood factor is applied to quantify the likelihood increase. In this example, by two likelihood intervals from 'Incredible' to 'Improbable'.
- For the uncrewed vessel the occurrence of Risk 3 increases the consequence of Risk 4. Therefore, a consequence factor is applied to quantify the consequences increase. In this example, by one consequence interval from 'Critical' to 'Disastrous'.

The remaining risks are therefore deemed to be independent and aggregated using the Type 2 risk aggregation approach. The resulting aggregated risk values, applying the RI method and adjusted for dependence, for the crewed and uncrewed vessels are shown in Table 8. When dependencies are accounted for, the crewed vessel exhibits a higher overall risk compared to the assumption of independence of the crewed vessel single risks, and as such a higher overall risk than the uncrewed vessel.

This example illustrates how modelling dependencies can materially influence the aggregated risk values and therefore the overall risk. However, we highlight that qualitative adjustments to reflect dependence, such as applying the aforementioned semi-quantitative interval shifts in likelihood and consequence, introduce inherent subjectivity and uncertainty. Beyond obvious direct dependencies easily identifiable by SQEP (e.g., failure of a power system leading to system-wide shutdown), there are often latent or systemic interactions such as the effects of reduced staffing, deferred maintenance, or organisational constraints, which can indirectly influence multiple risk pathways. These 'hidden dependencies' may not be identified by qualitative judgement alone.

While this approach offers interim practical value for aggregating dependent risks, we recognise it carries significant limitations. We therefore advise caution in relying on such qualitative adjustments for RIDM at this point in time; we highlight that further research is required to develop robust, transparent, and validated methods for qualitatively aggregating dependent semi-quantitative risks however this currently lies beyond the scope of this paper.

Table 8. RI Aggregated Risk for Crewed and Uncrewed Vessels Considering Dependency

Vessel Type	RI Value (Assuming Independence)		RI Value (Modelling Dependence)	
	Upper	**Lower**	**Upper**	**Lower**
Crewed	4.790×10^{-3}	1.907×10^{-5}	2.819×10^{-2}	1.779×10^{-4}
Uncrewed	6.378×10^{-3}	2.539×10^{-9}	6.787×10^{-3}	4.282×10^{-5}

4. DISCUSSION

The preceding section demonstrates that selecting an appropriate risk aggregation approach depends on the characteristics of the single risks, specifically dependence, type, and assessment method. We defined three approaches: Type 1, Type 2 and Type 3 risk aggregation, and practically applied the RI method as an illustrative strategy to aggregate risks for crewed and uncrewed vessels. This work helps bridge the research-policy-practice gap, showing how aggregated risk can support RIDM.

This work provides a guiding strategy for 'how' risk aggregation can be conducted, complementing the earlier rationale for 'why' it should be conducted for crewed and uncrewed vessels. Whilst we demonstrate that risk aggregation is technically feasible, each approach introduces distinct challenges affecting the usability and interpretation. These distinct challenges align with the broader challenges of risk aggregation identified earlier e.g., loss of contextual information, modelling dependencies, accounting for different risk types and the illusion of safety. Throughout this paper, we have identified these distinct challenges and provided recommendations for further research. A summary of the distinct challenges is as follows.

Type 1 risk aggregation appears mathematically simple but poses key challenges:

- Aggregating the interval-based inputs of semi-quantitative single risks yields interval-based outputs, requiring transparent communication to support RIDM.
- Selecting equivalence factors introduces assumptions that can materially influence aggregated risk.
- The RI method depends on the chosen scale aversion index, which directly affects the aggregated risk.

Type 2 risk aggregation invokes the same challenges as Type 1, and additionally:

- Requires a consequence equivalence framework to map different risk types onto a common scale, embedding assumptions and ethical, tolerability, and value judgments.

Type 3 risk aggregation invokes the same challenges as Type 1 and Type 2, and additionally:

- Quantitative aggregation methods to model dependency require quantitative inputs, limiting applicability to semi-quantitative risks.
- Qualitative adjustments to model dependency introduce subjectivity and additional uncertainty.

5. CONCLUSION

Given the collective magnitude of these distinct and broader challenges to risk aggregation for crewed and uncrewed vessels, we argue that aggregated risk values should currently be framed as a decision-support tool to present overall risk, not as stand-alone for RIDM; aggregated risk can be used to inform and contextualise an overall risk, not replace detailed risk assessments. In addition, the challenges, limitations and uncertainties must be clearly documented to mitigate the risks of RIDM, and the highlighted recommendations for further research should be advanced to support effective risk aggregation application for crewed and uncrewed vessels.

Holistically, we identify that risk aggregation inherently operates as a bottom-up process, combining single risks to construct an overall 'bigger' risk picture. However, we argue that an effective risk aggregation strategy must be guided top-down by policy to ensure consistency, validity, and relevance for RIDM and to ensure the aforementioned challenges are adequately addressed. Bottom-up harmonisation is impractical; practitioners and assessors at lower system or platform levels lack authority to set cross-project methods or scales for vessels. As such, risk aggregation must be governed by a top-down goal-based framework driven by the AP. Such a framework should be developed to establish reporting requirements, equivalence assumptions, and aggregation parameters to support RIDM.

Beyond aggregation for RIDM, such a framework could also enable risk apportionment across system functions. Figure 3 illustrates how this process integrates into a combined physical and digital Systems 'V' lifecycle, with risk apportionment during left-side system definition and risk aggregation during right-side verification. In the context of uncrewed and autonomous vessels, this lifecycle could extend to digital models and virtual testing environments (as shown by the upper inverse 'V').

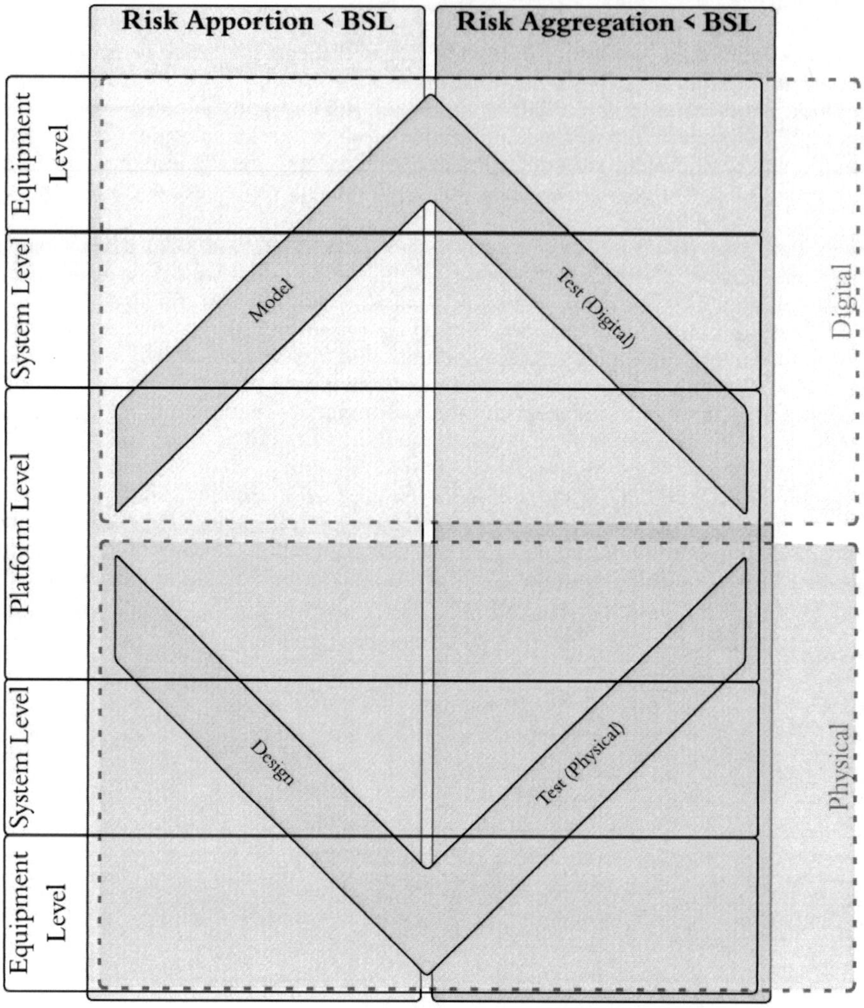

Figure 3: Digital and Physical Systems V Lifecycle Diagram Including Risk Apportionment and Risk Aggregation.

6. ACKNOWLEDGEMENTS

The authors thank Adrian Payne, Chief Technical Officer of Safeguard Engineering Ltd, for their valuable contributions and support.

7. REFERENCES

1. AVEN, T., 'The Risk Concept – Historical and Recent Development Trends', *Reliability Engineering and System Safety*, 2012.

2. BAO, C., JIANPING, L., and DENGSHENG, W., 'Risk Matrix: Rating Scheme Design and Risk Aggregation', *Springer*, 2022.

3. KAPLAN, S., and GARRICK, B., 'On the Quantitative Definition of Risk', *Risk Analysis*, 1981.

4. LOWRANCE, W W., 'The Nature of Risk', in *Societal Risk Assessment*, General Motors Research Laboratories, 1980.

5. MINISTRY OF DEFENCE, 'MAP01-126 - Tolerability of Risk from Submarines (TOR(SM)) – Principles and Guidelines for Setting Targets and Assessing Safety Risk in Submarines', 2013.

6. DE&S SHIPS OPERATING CENTRES, 'Leaflet 5 - End to End Safety Risk Management Process', 2020.

7. BJØRNSEN, K., and AVEN, T., 'Risk Aggregation: What Does It Really Mean?', *Reliability Engineering and System Safety*, 2019.

8. DAVID, S R., 'Safety Risk Aggregation: The Bigger Picture', *Safety and Reliability*, 2009.

9. BOLBOT, V., THEOTOKATOS, G., MCCLOSKEY, J., VASSALOS, D., BOULOUGOURIS, E., and TWOMEY, B., 'A Methodology to Define Risk Matrices – Application to Inland Waterways Autonomous Ships', *International Journal of Naval Architecture and Ocean Engineering*, 2022.

10. ISO, 'Risk Management – Risk Assessment Techniques (IEC 31010:2019)', *British Standards Institution*, 2019.

11. IMO, 'Revised Guidelines for Formal Safety Assessment (FSA) for Use in the IMO Rule-Making Process', *MSC-MEPC.2/Circ.12/Rev.2*, 2018.

12. HSE, 'Reducing Risks, Protecting People – HSE's Decision-Making Process', *Health and Safety Executive*, 2001.

13. HSE, 'Guidance on ALARP Decisions in COMAH', *SPC/Permissioning/37*, 2013.

14. ACKERMANN, F., HOWICK, S., QUIGLEY, J., WALLS, L., and HOUGHTON, T., 'Systemic Risk Elicitation: Using Causal Maps to Engage Stakeholders and Build a Comprehensive View of Risks', *European Journal of Operational Research*, 2014.

15. ZHAO, D., et al., 'Stakeholder Perceptions of Risk in Construction', *Safety Science*, 2016.

16. BANI-MUSTAFA, T., ZENG, Z., ZIO, E., and VASSEUR, D., 'A New Framework for Multi-Hazards Risk Aggregation', *Safety Science*, 2020.

17. BIGGS, W., ASTLE, J., and MCINTYRE, J., 'Integrating Maritime Autonomous Systems (MAS) into the Future Force: How the UK MAPLE Architecture is Being Developed to Enable Generation After Next Capabilities', *Conference Proceedings of INEC*, 2022.

18. DE VOS, J., HEKKENBERG, R G., and VALDEZ BANDA, O A., 'The Impact of Autonomous Ships on Safety at Sea – A Statistical Analysis', *Reliability Engineering and System Safety*, 2021.

19. CHAPMAN, M., BONALI, R., SMITH, K., YANG, I., PAVONE, M., TOMLIN, C., 'Risk-Sensitive Safety Analysis Using Conditional Value-at-Risk', *IEEE Transactions on Automatic Control*, 2022.

20. HASSEL, H., CEDERGREN, A., and TEHLER, H., 'Strategies for Aggregating Risk Information in a Societal Safety Context', *Safety Science*, 2022.

21. BROWNE, T., TAYLOR, R., VEITCH, B., HELLE, I., PARVIAINEN, T., KHAN, F., and SMITH, D., 'A General Method to Combine Environmental and Life-Safety Consequences of Arctic Ship Accidents', *Safety Science*, 2022.

22. BAO, C., CAI, M., LI, J., ZHENG, Q., WU, D., and MENG, Q., 'Risk Aggregation Considering Probabilistic and Consequential Interactions: A General Formulation with Computational Cost Handling', *Risk Analysis*, 2024.

23. COMSURFLOT, 'Active Risk Management', 2025.

24. MINISTRY OF DEFENCE, 'MAA RA 1210 – Ownership and Management of Operating Risk (Risk to Life)', Issue 8, 2024.

25. DE&S, 'Safety and Environmental Protection Leaflet 02/11 – ALARP in a Military Equipment Capability Context', Issue 2.0, 2013.

26. HADDON-CAVE, C., 'The Nimrod Review – An Independent Review into the Broader Issues Surrounding the Loss of the RAF Nimrod MR2 Aircraft XV230 in Afghanistan in 2006', *The Stationery Office*, 2009.

27. BASEL COMMITTEE ON BANKING SUPERVISION, 'Principles for Effective Risk Data Aggregation and Risk Reporting', 2013.

28. ISO, 'Risk Management – Vocabulary (ISO 31073:2022)', *British Standards Institution*, 2022.

29. COX, L A., 'What's Wrong with Risk Matrices?', *Risk Analysis*, 2008.

30. THOMAS, P., BICKEL, E J., BRATVOLD, R B., 'The Risk of Using Matrices', *SPE Economics and Management*, 2014.

31. MANSON, P., 'Uncommon Sense: A Review of Challenges and Opportunities for Aggregating Disaster Risk Information', *International Journal of Disaster Risk Reduction*, 2019.

32. MENG, C., BAO, C., and MENG, Q., 'Overview of Risk Aggregation Approach in Different Risk Scenarios', *Procedia Computer Science*, 2022.

33. DUIJM, N I., 'Recommendations on the Use and Design of Risk Matrices', *Safety Science*, 2015.

34. DEKKER, S W A., 'Drifting into Failure: Complexity Theory and the Management of Risk', in *Chaos and Complexity Theory for Management: Nonlinear Dynamics*, IGI Global, 2013.

35. LEVESON, N G., 'Engineering a Safer World: Systems Thinking Applied to Safety', *MIT Press*, 2012.

36. ALQUDSI, Y., and MAKARACI, M., 'UAV Swarms: Research, Challenges and Future Directions', *Journal of Engineering and Applied Science*, 2020.

37. KELLY, T., 'Managing Complex Safety Cases', *Department of Computer Science, University of York*, 2003.

38. COX, L., BABAYEV, D., and HUBER, W., 'Some Limitations of Qualitative Risk Rating Systems', *Risk Analysis*, 2005.

39. ALE, B., BURNAP, P., and SLATER, D., 'On the Origin of PCDS – Probability Consequence Diagrams', *Safety Science*, 2015.

40. HIRST, I., and CARTER, D., 'A Worst Case Methodology for Obtaining a Rough but Rapid Indication of the Societal Risk from a Major Accident Hazard Installation', *Journal of Hazardous Materials*, 2002.

41. BARKER, A., 'Navigating Safety Risks: The Role of Risk Assessor Experience on the Research–Practice Gap in Risk Aggregation', 2023.

42. BUREAU VERITAS, 'NI641 – Guidelines for Autonomous Shipping', *Bureau Veritas*, 2019.

43. BAO, C., LI, J., and WU, D., 'A Fuzzy Mapping Framework for Risk Aggregation Based on Risk Matrices', *Journal of Risk Research*, 2018.

44. LIU, J Y., and WANG, Z F., 'Non-Additivity Analysis on Scheduling Risks Based on Bayesian Networks', *Systems Engineering – Theory & Practice*, 2011.

8. AUTHOR BIOGRAPHIES

Alexander Barker holds the current position of Principal Consultant at Safeguard Engineering Ltd. The author is responsible for leading safety, risk and reliability projects, with a particular focus on digital technology. The Author's previous experience includes safety and engineering management roles across the UK Ministry of Defence and large engineering organisations.

Ian Groom MBE holds the current position of Principal Consultant at Safeguard Engineering Ltd. The author is responsible for technical support and advice to defence projects and programmes, with a particular focus in the maritime domain. The Author's previous experience includes an extensive career as a decision maker in the UK Royal Navy, latterly as the Defence Maritime Regulator (DMR) for the UK MOD.

Dr. Paul Hogan holds the current position of Safety Consultant and Non-Executive Director at Safeguard Engineering Ltd. The author is responsible for technical support and advice to projects and bidding activities. The Author's previous experience includes over 35 years of practical application of safety engineering across a range of significant programmes.

ADVANCED NUMERICAL SIMULATION OF HYDRO-ELASTIC VESSEL STRUCTURAL RESPONSE TO ASSIST SUSTAINABILITY ASSESSMENTS

B Cartwright, University of Newcastle / Pacific ESI Pty Ltd, Australia, **M Renilson**, Australian Maritime College, University of Tasmania, Australia, and **T Magoga**, Defence Science Technology Group, Melbourne, Australia

SUMMARY

Naval and other vessels critical for defence activities may from time to time be subject to unexpected, possibly extreme, sea-state or operational conditions that potentially have adverse effects on vessel safety and continued operational effectiveness. In order to ensure effective sustainment of the fleet, the degree to which this may occur requires accurate tools for the assessment of structural effects, not easily obtained using either conventional physical modelling such as that done in wave tanks or conventional numerical simulation done with Computer Aided Engineering software tools. Herein a novel approach is described for numerically simulating sea-state conditions using Smoothed Particle Hydrodynamics (SPH), interacting directly with conventional but detailed elastic Finite Element analysis (FEA) for simulating the often non-linear vessel structural response. Various calibration studies showed good correlations against results from physical models in towing tanks and against conventional numerical models. On that basis, SPH-FEA simulations for frigate-type vessel responses under head-on sea-states with wave heights to 6m and 10% steepness were performed. Some typical results are presented. Although quite computationally intensive in its current development form, on-going studies indicate that significant savings in computational time can be achieved with appropriate refinement of the SPH modelling parameters.

1. INTRODUCTION

The procurement of new naval vessels is a lengthy and protracted process, which can end up taking longer than anticipated. When delays occur it may become necessary to extend the life of the existing vessels in the fleet that they are planned to replace, as there are usually no alternative vessels available to the navy.

All vessels, particularly relatively lightweight military ones, suffer degradation with time such as: fatigue; corrosion; and local damage during their operation. This can make it challenging to extend their lives beyond that originally planned. In addition, if major maintenance is required to repair such degradation, then not only is that expensive, but it takes the ship out of operation for an extended period.

Hence, a reliable assessment of the level of degradation that is acceptable for a given ship is vital to ensure that unnecessary excessive maintenance is not carried out. This will ensure that the ship maintenance costs, and the downtime when it is not available, are both kept to a minimum.

In order to achieve this, it is necessary to be able to predict the maximum stress that will occur at the various critical locations, taking into account the actual level of degradation, on the ship when it is operating in a seaway. If it can be shown that these levels are below acceptable values then maintenance can be deferred. By monitoring the progress of such degradation with time, it is also possible to predict in advance when a ship will need to undergo major maintenance.

In addition, if it is necessary to continue to operate the ship, possibly due to a replacement not being available, for example, then limitations can be placed on its operational conditions, based on acceptable sea state, vessel speed, and/or headings to the waves.

In order to do this it is necessary to have accurate tools for assessment of the structural effects, not easily obtained using conventional physical modelling such as in wave tanks or using conventional numerical tools. Herein a novel approach is introduced and discussed for numerically simulating sea-state conditions using Smoothed Particle Hydrodynamics (SPH), interacting directly with conventional but detailed elastic Finite Element analysis (FEA) for simulating the often non-linear vessel structural response.

Various calibration studies have shown good correlation against results from physical models in towing tanks and against conventional numerical models. On that basis SPH-FEA simulations for a generic frigate-type vessel responses under head-on sea-states with wave heights to 6m and 10% steepness were performed.

2. SPH-FEA AS A STRUCUTRAL RESPONSE PREDICTION TOOL

2.1 THE ANALYSIS CONCEPT – "CAR-CRASH OF THE OCEAN"

Every new car on the road is required to demonstrate a certain level of occupant survivability through physical testing in controlled crash scenarios. Physical testing is expensive, and so the reliance on passing such physical testing at the first test is paramount. For this reason, the automotive industry is the leading industrial sector for the evolution of design through numerical simulation (Ref 3.). Not only does the automotive industry use numerical simulation for design and efficiency modelling, but before the physical crash test is attempted, each new car design undergoes extensive "virtual" testing and refinement through numerical models such that there is a high degree of confidence of passing the crash test. This numerical testing reduces the risk of failing a physical crash test.

The key aspect of these numerical crash tests is that the response of the vehicle is dictated by the physics of the event and the way the vehicle structure (and its occupants) react to those physical forces generated by the crash event. The results can be reviewed and analysed in detail and in slow motion to understand how the structure is deforming or failing. This detailed analysis can provide insight to design improvements that can be trialled in a subsequent numerical crash test for comparison.

This physics-based numerical simulation technique has been applied to study the response of a frigate in waves. In much the same way as the car-crash is conducted, a finite element (FE) model of a ship can be set on a course to "crash" into waves of predefined amplitude and wavelength, and the response of the ship can be measured in terms of the forces that act on it due to interacting with the waves. In the simulations performed to date, only uniform height waves of a single frequency of have been considered, and so the crash-event is the repeated encounter of the ship with uniform height waves, in other words, a steady response is achieved.

For the numerical simulations performed here, the ship is a conventional structural FE model of a ship, and the waves are modelled using the mesh-free Smoothed Particle Hydrodynamics (SPH) technique that allows for ease of modelling splash, wave-breaking, complex flows and fluid-structure interaction (Liu and Zhang, 2019).

A numerical simulation employing the as-built ship would provide a prediction of the response of the as-built ship to the prescribed wave scenario. This could be used to validate a design. Alternatively, the wave state could be quite extreme in the SPH environment, and the ship speed also extreme, allowing a truly insightful ship response to be predicted.

The analysis could also be performed on a modified FE model of the ship, where the modification represents a change in structural detail that accounts for:

1) Plate thinning due to corrosion
2) Localised cracks due to fatigue
3) Localised dents/damage due to a collision/grounding/mooring accident
4) Modified structure or mass distribution to reflect a capability upgrade

It is acknowledged that any or all the above structural detail changes may be necessary for a life of type extension.

The outcome of the numerical analysis is the heave, pitch and bending moments at specific locations in the ship, as well as the stress/strains throughout the ship.

A comparison of the allowable stress at a specific location in the ship compared to the predicted stress from the simulation technique described here will be useful in the decision-making processes to support sustainability.

2.2 IMPLEMENTATION OF THE SPH-FEA TECHNIQUE

The generic commercial explicit software suite Virtual Performance Solution, VPS, from ESI Group France was used for this analysis. The software is heavily parallelised to enable use of high-performance computing systems.

The waves developed in SPH using a shallow-tank moving-floor technique described in Vivanco et al (2021). The tank had a test length of 1000m, a width of 100m and a depth of 20m. Due to the moving-floor wave generation technique, waves develop simultaneously over the entire length of the tank. The waves were generally less-well formed in the upstream 150-200m of the tank. This resulted in about 800 m of consistent wave profiles for ship-wave interactions.

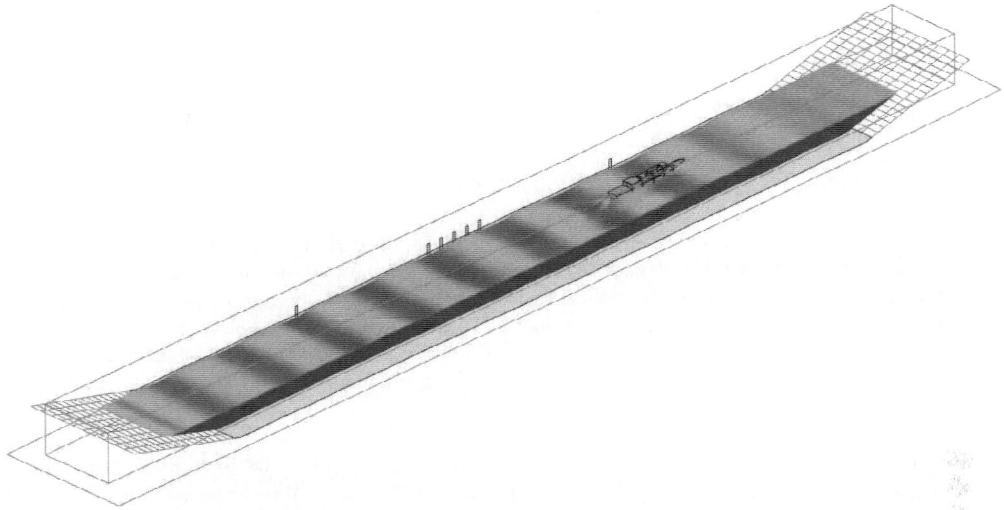

Figure 1: SPH-FEA simulation of a ship near the end of its run in the wave tank.

The ship model was a structural model of the ship previously used for static and fatigue analysis (Magoga and Dwyer, 2018). This model had also been tested as a segmented model in waves in a physical towing tank, and hence bending moment, as well as heave and pitch, data was available (Morrison et al, 2010).

The interaction between the ship and the waves is handled by a standard industrial contact-interface. The interface works by applying a force between each SPH particle that comes in contact with the ship. The sum of these individual contact forces is the net force on the ship. Accuracy of the contact forces acting between the ship and the water were verified by calculating the hydrostatics of the ship in still water, including verifications of moment to change trim.

The solution employs an explicit time-marching scheme where a complete structural analysis is performed approximately one thousand times a second, based on the material properties and minimum element size of the FE ship model. The interaction of the SPH and the FEA can occur at every time step, or at a reduced rate determined by the timestep of the SPH.

A simulation duration of between 80 and 100 seconds was typically used to generate the results presented here. This enabled approximately 10-15 wave-encounters which ensured a consistent steady state oscillatory ship response was achieved before the ship response was measured. The ship commenced from the downstream end of the tank and progressed into the oncoming waves. Measurements of ship response were made when the ship response was steady, and this typically corresponded to the ship being at speed in the central region of the tank.

2.2 COMPUTATIONAL EFFORT REQUIRED

The computational effort of the simulation was heavily dependent on the volume of the tank and the diameter of the SPH used to fill the tank. The computation effort is a compromise between SPH size and economic solution times.

Much of the research work presented here was conducted on a 54-core workstation at the offices of Pacific ESI, Sydney Australia. For these simulations an SPH size of 1.2 m resulted in about 1.8 million SPH particles and enabled a single test case to be solved in approximately 24 hours.

As will be shown in the next section, a smaller SPH diameter allowed greater correlation to experimental results. However, the smaller diameter SPH required considerably greater computational effort.

Although the number of cores used was not revealed, it was demonstrated by our research partners that solution times of 24 hours per run were achievable with up to 15 million SPH particles on their high performance computing facility, and a single test case employing 51 million SPH particles was also run in a practical timeframe.

3. REPRESENTATIVE RESULTS

The reader is referred to Cartwright (2025) for a more comprehensive set of simulation results. For the purpose of this paper a few examples will be shown.

3.1 HEAVE AND PITCH AS A FUNCTION OF SPH PARTICLE SIZE

Figures 2, 3 and 4 show the experimental and simulated response of the ship at 18 knots in nominal 3m waves, in non-dimensional heave, pitch and midship bending moment for a range of wavelengths. In Figures 2a, 3a and 4a the simulated response is shown for SPH sizes at 1.2m, 0.6m and a single result using 0.4m SPH. Figures 2b, 3b and 4b show the non-dimensional heave, pitch and bending at 18 kts and λ/L=1, for 3m wave height as a function of particle diameter in simulation. It is noted that the heave and pitch result obtained using 0.4m SPH particles is almost indistinguishable from the result using 0.6m SPH, yet significantly greater CPU effort is required for the 0.4m SPH simulations.

For the conditions used here, it is considered that accuracy achieved with the 0.6m SPH is a good compromise between SPH size, computational effort and correlation to experimental results.

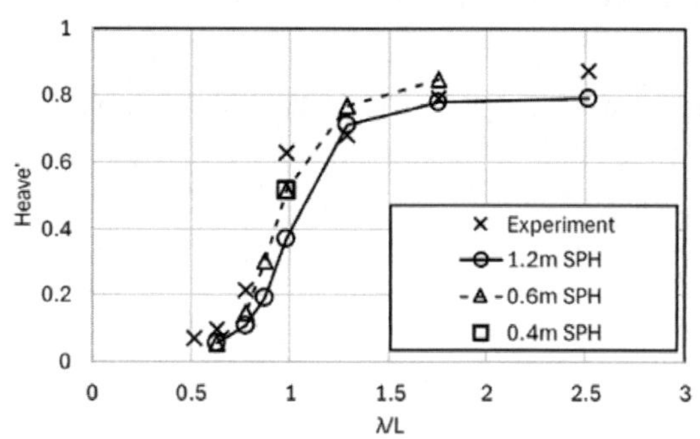

Figure 2a: Experimental and simulation non-dimensional heave at 18 kts, for a 3m wave height at various wave lengths, for various particle sizes.

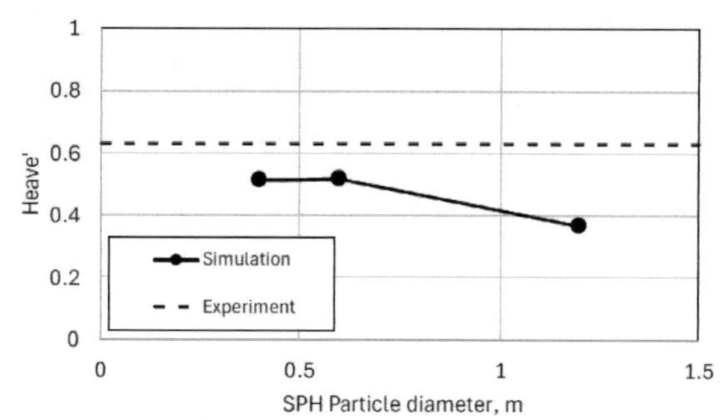

Figure 2b: Experimental and simulation non-dimensional heave at 18 kts and λ/L=1, for 3m wave height as a function of particle diameter in simulation.

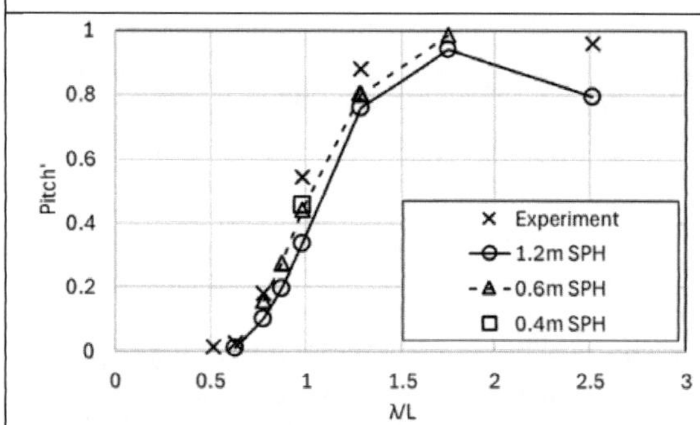

Figure 3a: Experimental and simulation non-dimensional pitch at 18 kts, for a 3m wave height at various wave lengths, for various particle sizes.

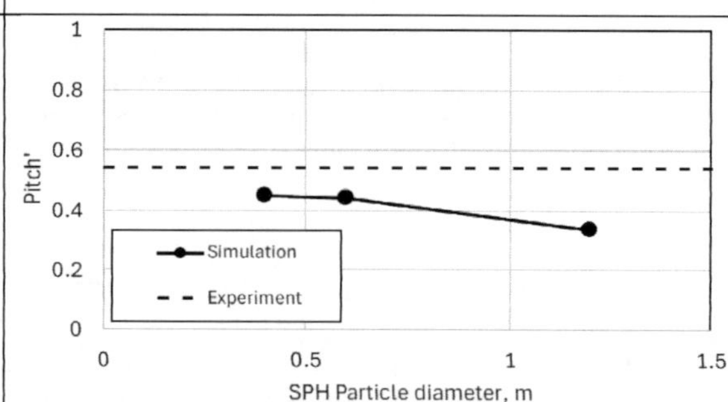

Figure 3b: Experimental and simulation non-dimensional pitch at 18 kts and λ/L=1, for 3m wave height as a function of particle diameter in simulation.

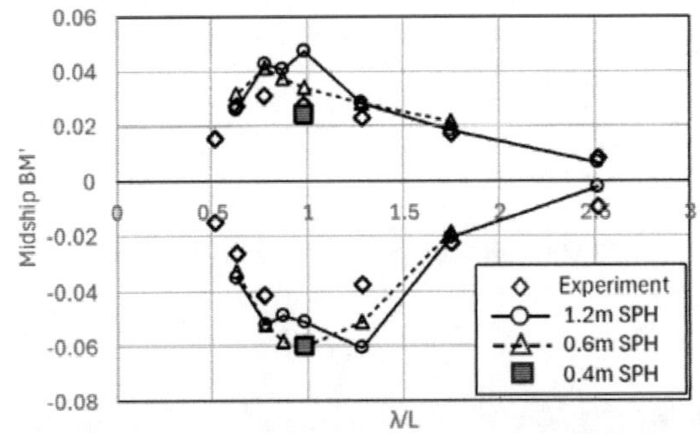

Figure 4a: Experimental and simulation non-dimensional bending moments at 18 kts, for a 3m wave height at various wave lengths, for various particle sizes.

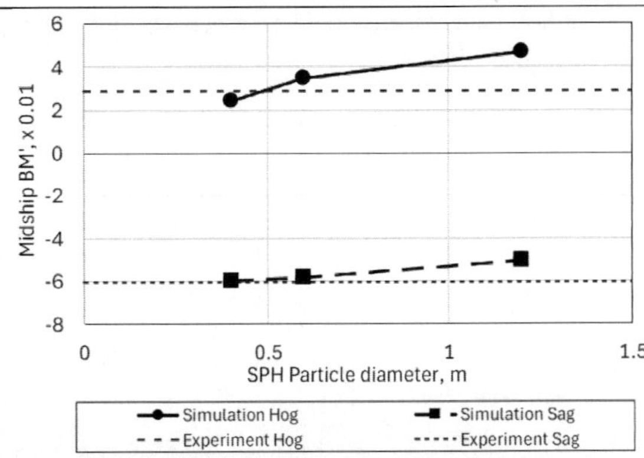

Figure 4b: Experimental and simulation non-dimensional bending moments at 18 kts and λ/L=1, for 3m wave height as a function of particle diameter.

3.2 HEAVE AND PITCH AS FUNCTIONS OF WAVE STEEPNESS

Figures 5 and 6 show that the response of the frigate in heave and pitch as predicted by the SPH-FEA method shows a similar relationship to experimental results as wave steepness increases, for an SPH diameter of 1.2m. We could perhaps expect this correlation between simulation and experiment to improve with smaller diameter SPH particle, as shown in Figures 2b, 3b and 4b, however the bending moment is more heavily influence by hull shapes such as bow flare and transom flat in larger waves,

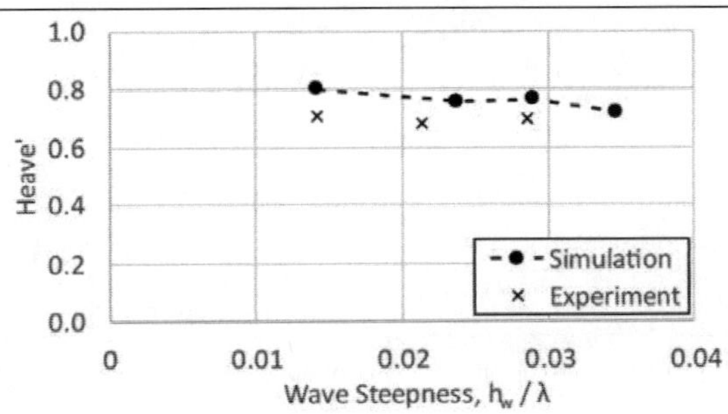

Figure 5: Non-dimensional heave as a function of wave steepness, for Fr=0.28 and λ/L =1.25.

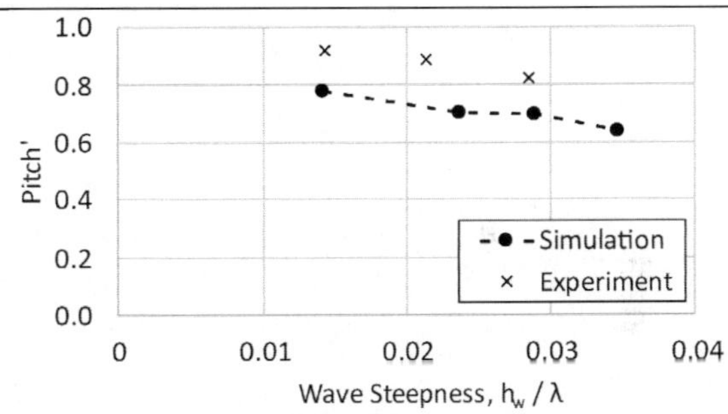

Figure 6: Non-dimensional pitch as a function of wave steepness, for Fr=0.28 and λ/L =1.25.

4. SUITABILITY FOR PREDICTION OF RESPONSE IN LARGE WAVES

The results presented in Cartwright 2025, and reproduced here, illustrate that the SPH-FEA analysis technique has good predictive capability for the response of the frigate at speed in large waves. Specifically, the numerical technique has shown quite acceptable correlation to the physical segmented model testing conducted at high speed and large wave heights – conditions that are known to impact heavily on fatigue life and stress condition of the ship.

Having demonstrated that the SPH-FEA technique is able to make reliable predictions of the motions and bending moments in a ship under severe operating conditions, it is proposed that stress-state within the ship would also be reliable. It is then further proposed that a SPH-FEA analysis of a ship model that has been modified to represent some real degraded physical condition of a ship structure would result in a trustworthy prediction of the stress state of the degraded ship in that specific operating condition.

4.1 METHODOLOGY FOR PREDICTING STRESS DUE TO PLATE THICKNESS DIMINUTION

This proposal was demonstrated in Cartwright et al 2022, where the stress response in a hull bottom plate section was studied for a specific speed, wavelength and wave height, for various levels of plate diminution, as a representation of material lost due to corrosion. In that specific case, the stress in a thin athwartship strip of hull plate adjacent the keel in a midship location was studied for identical wave conditions, for thicknesses of 100%-20%, as shown in Figure 7.

Figure 8 shows the actual stress divided by the allowable stress at that specific location, as a function of time, for different remaining plate thicknesses. The nominal original plate thickness in this case was 15mm. It is interesting to note the asymmetric stress response with time, showing a sharp peak in the sagging stress, associated with bow-flare immersion in large waves, and the more gradual hogging stress as the wave crest passes midship. It is also interesting to note that the sagging stress becomes more peaked as the plate thickness is reduced.

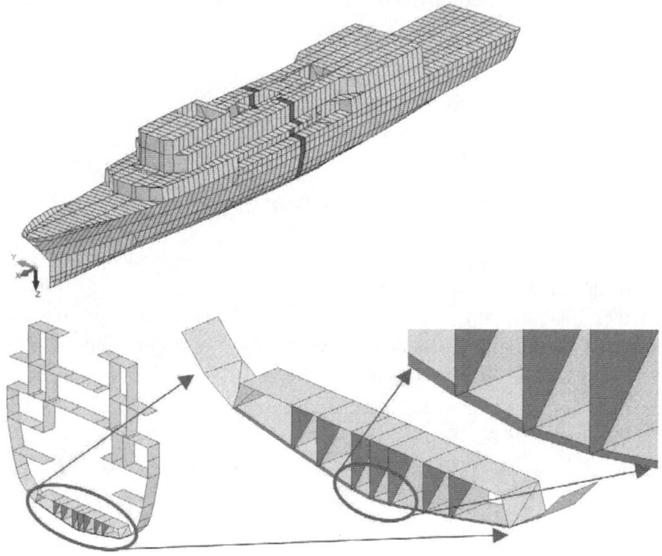

Figure 7: Detail of the hull plate that was reduced in thickness to represent uniform corrosion.

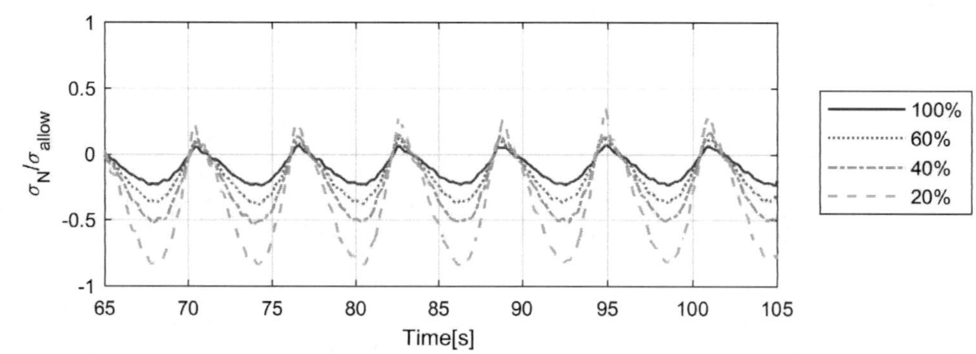

Figure 8: Normal stress, in longitudinal direction, at reference structural location, for 100%, 60%, 40% and 20% remaining thickness of plate (for FN=0.18, L/λ=1.0, and H/L=0.028)

4.2 METHODOLOGY FOR RELATING PLATE DIMINUTION TO CALENDAR TIME

The previous section showed how the stress can be predicted within the ship structure for a given operating condition, as a function of the remaining plate thickness. From a point of view of estimating remaining ship life, or planning maintenance, it would be useful to predict the calendar time at which a stress level within the ship might exceed a specific value. This next section proposes a methodology to do this.

Figure 9 is a representation of the stress levels of Figure 8 expressed as a function of the diminution of the plate thickness, expressed as percent remaining plate thickness. It is interesting to note for this hypothetical case that a considerable stress margin is still present at the point of minimum plate thickness set by class.

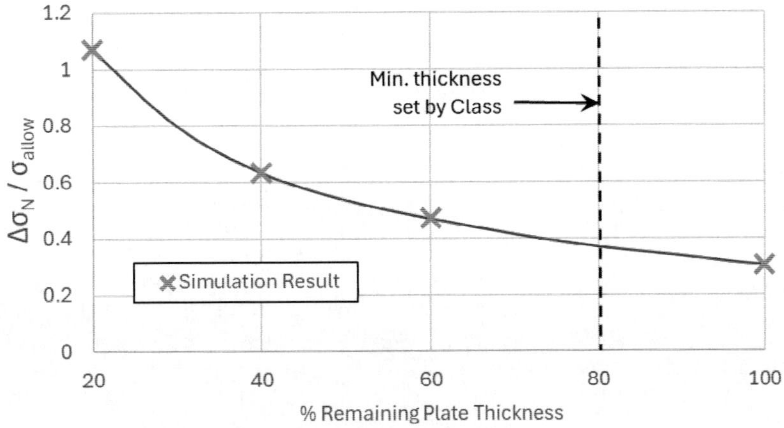

Figure 8: Normal stress, in longitudinal direction, at reference structural location, as a function of remaining plate thickness (for FN=0.18, L/λ=1.0, and H/L=0.028).

The relevance of the thinned hull bottom plate is that a correlation of plate thickness lost with time due to corrosion is possible, based on actual hull surveys and other corrosion observations in the marine environment (Melchers et al 2024). After an initial period of about 4 years of rapid corrosion, the corrosion rate steadies to a linear rate. This characteristic is typical of many steels in various environments, the specific rate of material lost being dependent upon factors such as the type of steel, the salinity, the flow conditions and the temperature. For representative values of steel in moderate temperatures, a typical corrosion rate for a steel might be that shown in Figure 9, as developed for typical parameters presented in Melchers et al (2024), commencing with a plate of 15mm thickness to correlate with "% Remaining Plate Thickness" as presented above.

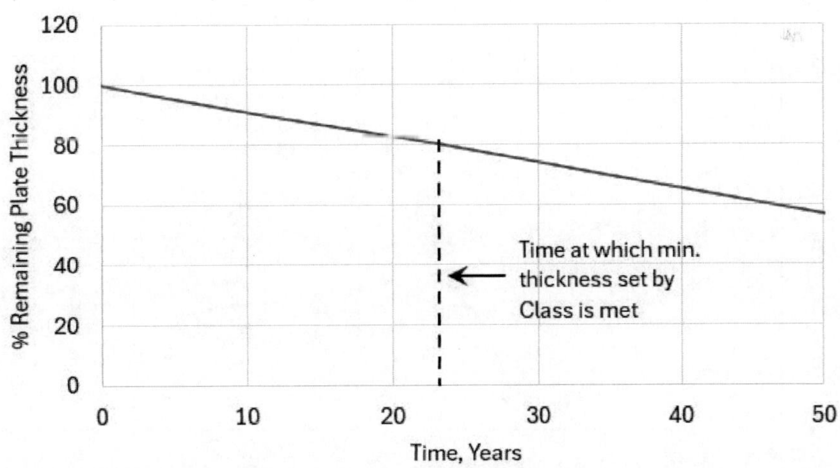

Figure 9: Representative curve showing the percent of remaining plate thickness as a function of time, for a mild steel plate in salt water intertidal conditions, which could be related to localised corrosion in a ship.

Knowing the influence of the remaining plate thickness on the stress in the ship when operating at a given speed in a particular sea state, as shown in Figure 8, and knowing the typical reduction in plate thickness due to corrosion as a function of calendar time, as shown in Figure 9, it is possible to develop the relationship between the stress in the plate and the calendar time, as shown in Figure 10.

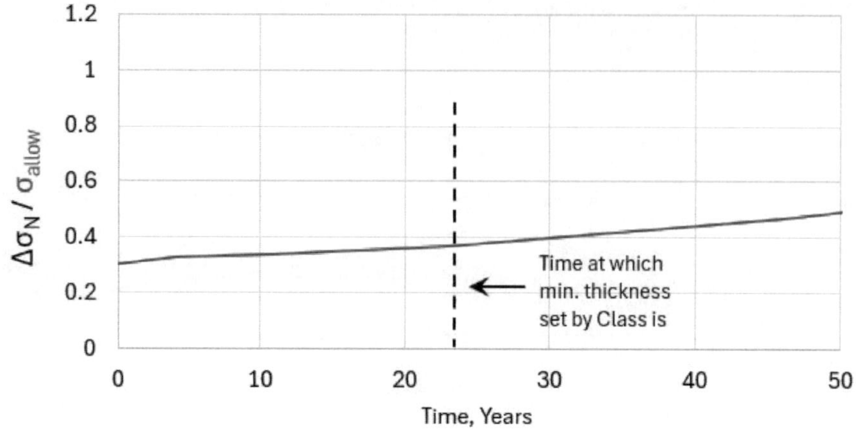

Figure 10: Actual over Allowable stress as a function of time, for the location and operating conditions of interest.

From Figure 10 it is possible to estimate the rate at which the maximum stress at a specific location in the ship will increase with calendar time. This can inform planning or decision making about the need or otherwise to conduct major maintenance to prevent the maximum stress from reaching the allowable value.

4.3 INFLUENCE OF OPERATING SPEED AND WAVE HEIGHT ON STRESS LEVEL

The effect of operating speed on the stress at a specific location in the ship can be assumed to be represented by the curves in Figure 11. Here it proposed that for a given hull plate diminution, an increase in ship speed, or Froude speed number Fn, will result in greater stresses in the location interest, or similarly that reduction in Fn should decrease the stress. Thus, such curves may dictate slower speeds for an aged vessel that is showing significant hull plate diminution.

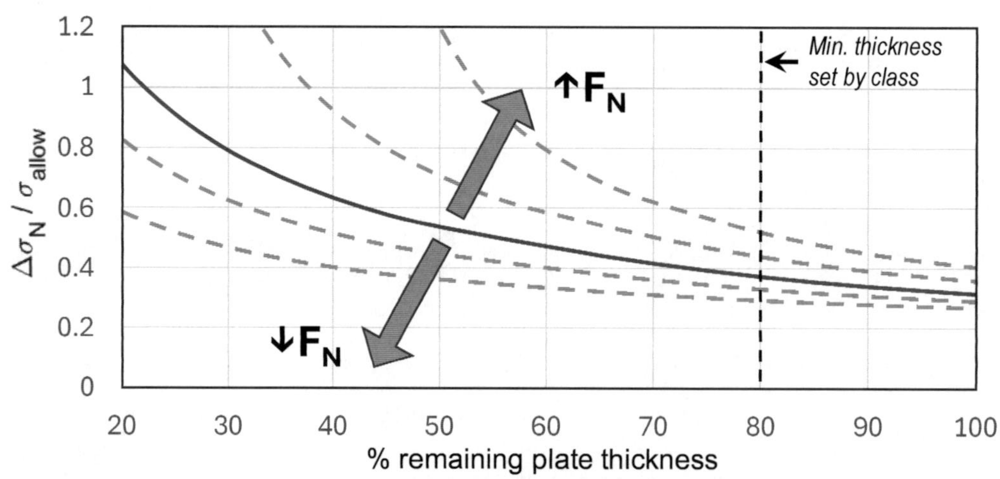

Figure 11: Proposed effect of ship speed, Fn, on the stress as a function of remaining plate thickness, for the location and operating conditions of interest.

Further it is proposed that a similar effect may result from the effect of wave height, as shown in Figure 12. Here it is proposed that for a given wavelength and ship speed, that operating in higher wave heights would result in an increase in stress at the location of interest, and operating at lower wave heights would result in lower stress.

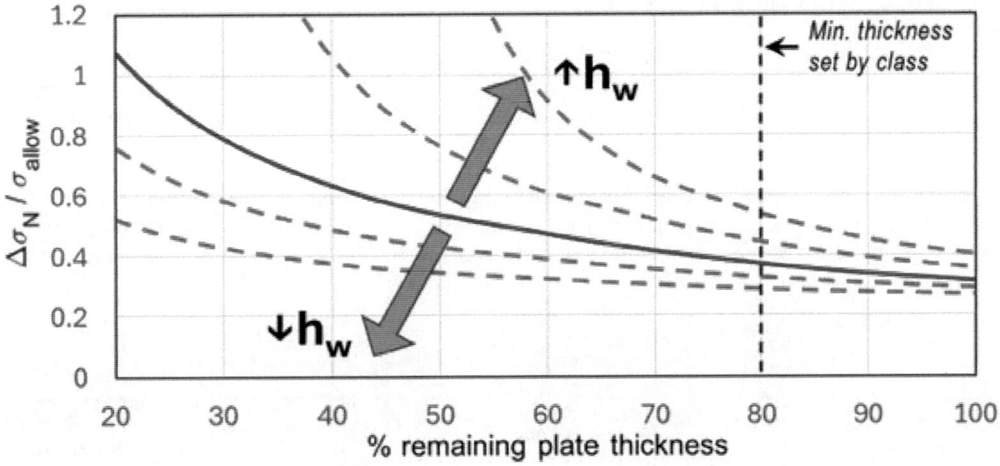

Figure 12: Proposed effect of wave height, hw, on the stress as a function of remaining plate thickness, for the location and operating conditions of interest.

5. DISCUSSION

5.1 PREDICTION OF MAINTENANCE SCHEDULE

Using the approach discussed above, it is possible to estimate the stress throughout the ship as a function of time due to the response from waves, considering diminution of plate and stiffener thickness due to corrosion. This can be compared with the permitted allowable stress. The procedure to conduct this analysis is given in the flowchart shown in Figure 13.

The novelty to Figure 13 is to repeat the process with the uppermost loop that accommodates degradation in the ship with time. This upper loop should result in a stress state that exceeds the allowable stress, at some time in the future. This can then be used to determine the best time to conduct remedial maintenance. It will prevent costly maintenance being undertaken well in advance of when it is necessary, thereby reducing maintenance costs, and increasing vessel availability.

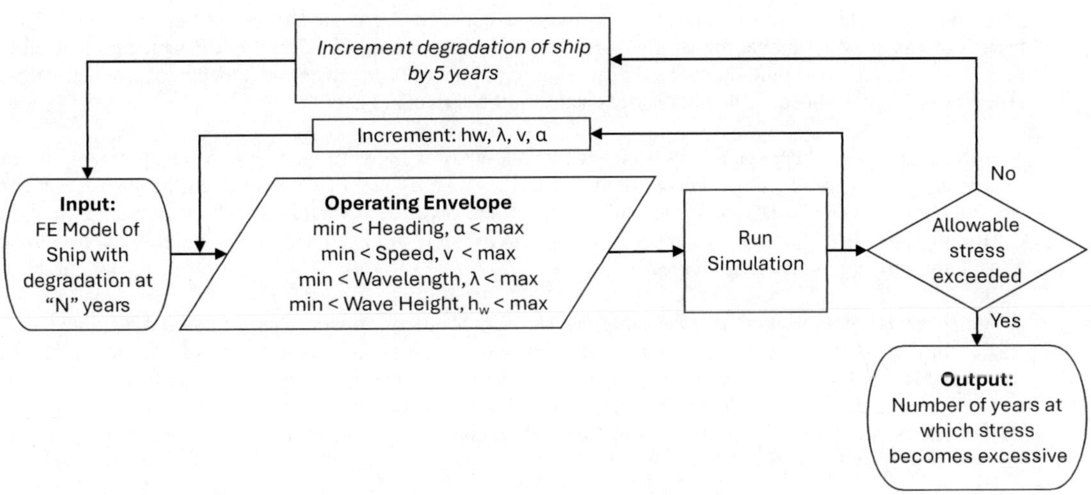

Figure 13: Flowchart of a process to assess the condition of the ship with increasing degradation across the range of range of operating conditions.

5.2 PREDICTION OF SAFE OPERATIONAL LIMITS

The maximum stress will be a function of Fn, and wave height, possibly following the dependency shown in Figures 11 and 12 respectively. Thus, if it is not possible to dock a vessel for the maintenance recommended by the procedure above, an alternative may be to impose operational limits on a vessel to prevent the stresses exceeding the allowable maximum limit.

The process to assist in defining acceptable operational limits is described by the flowchart shown in Figure 14. The process is identical to the previous process, except here the condition of the ship is the current known condition of the ship, and the output is the range of speeds as a function of heading, wavelength and wave height for which the defined stress limits have not been exceeded.

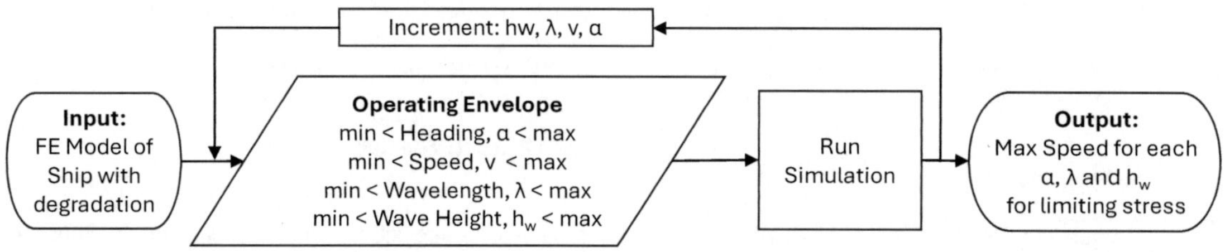

Figure 14: Flowchart of a process to assess the safe operating conditions of a ship with degradation.

5.3 COMPUTATIONAL EFFORT REQUIRED

It is acknowledged that the computational effort to conduct the SPH-FEA analyses of the entire operating envelope of a ship with sufficient accuracy is not trivial. With the facilities currently available to the authors, the task would in fact not be feasible in a meaningful timeframe.

However, the purpose of this paper is to propose a concept that might one day be feasible, due to the ever-increasing computational capacity that is affordable, or deemed justifiable for a specific costly asset. Further these concepts may also benefit from the advancement of optimisation and machine-learning techniques, or similar, that can prioritise computational effort on the degradation and operating conditions that are important to decision making based on the resulting stress state of the ship (Majidayn et al, 2025).

6. CONCLUDING REMARKS

Previous studies have shown quite good correlation between segmented physical model test results and prediction made by numerical modelling using the SPH-FEA technique. The SPH-FEA process is useful for providing an alternative prediction capability to more acceptable or conventional techniques, specifically for large waves or high speeds, when high stress states in the ship could result.

A process is proposed here that relates accepted rate of corrosion models to a time-dependent increase in stress in critical locations of the ship. The SPH-FEA provides a high-level of confidence of the stress-state in the ship at a specific operating condition with a predefined level of degradation. This enables a predictive capability as to when corrosion at a specific location in the ship will become critical, and this information can assist long-term maintenance strategies.

A complimentary interpretation of the same results set is to impose operational limits on a degraded ship to ensure high stress states are not realised. In this case the acceptable operating conditions are those by which the SPH-FEA analysis predicts stress states below some threshold, and defines not-allowable operating conditions as those when the threshold is exceeded. Using this basis to limit the operation conditions of the ship might extend the life of the ship, or at least reduce the risk of damage until such time that maintenance can be carried out.

It is acknowledged that the computational effort to conduct the study as proposed might not be practical today. The authors feel the embedding the SPH-FEA in an analysis tool to support maintenance decisions is worthwhile, and that such a tool may become more feasible as computer capacity increases, and machine-learning or similar critical decision-making ship degradation condition choices can be automated to reduce the analysis load.

8. ACKNOWLEDGEMENTS

The authors acknowledge the support provided to this work through the Australian Research Council through Linkage Projects LP160100391 and LP190101283.

9. REFERENCES

1. CARTWRIGHT, B., 'The Analysis of Ships in Waves using Mesh-Free and Finite Element Methods', *PhD Thesis, University of Newcastle, Australia*, 2025.

2. CARTWRIGHT, B., RENILSON, M., MCGUCKIN, MAGOGA T., MULCAHY, L., and MELCHERS, A C., 'Towards a Cost-Effective Approach to Extending the Life of Naval Surface Ships', *International Maritime Conference*, 2022.

3. https://www.automotive-technology.com/articles/role-of-simulation-technologies-in-automotive

4. LIU, M., and ZHANG, Z., 'Smoothed particle hydrodynamics (SPH) for modeling fluid-structure interactions', *Sci. China-Phys. Mech. Astron*. 62, 984701, (2019), https://doi.org/10.1007/s11433-018-9357-0

5. MAGOGA, T., and DWYER, D M., 'Fatigue Life as a Variable in Assessing Naval Ship Flexibility', *Naval Engineers Journal*, No. 130-3, 2018.

6. MAJIDIYAN, H., ENSHAEI, H., HOWE, D., and GUBESCH, E., 'Part A: Innovative Data Augmentation Approach to Enhance Machine Learning Efficiency - Case Study for Hydrodynamic Purposes', *Appl. Sci. 2025*, 15, 158. https://doi.org/10.3390/app15010158

7. MELCHERS, R E., JEFFREY, R., CHAVES, I A., and PETERSEN, R B., 'Predicting corrosion for the life estimation of ocean and coastal steel infrastructure', *Materials and Corrosion*, 1-14., 2024, https://doi.org/10.1002/maco.202314201

8. MORRIS, B., HUTCHISON, C., THOMAS, G., PHELPS, B., and LILIENTHAL, T., 'Hydroelastic Response of a Segmented Frigate Model', *PACIFIC 2010, 27-29, Sydney, Australia*, 2010.

9. VIVANCO, I., CARTWRIGHT, B., ARAUJO, A L., GORDILLO, L., and MARIN, J F., 'Generation of Gravity Waves by Pedal Wavemakers', *Fluids 2021*, 6(6), 222; https://doi.org/10.3390/fluids6060222. 2021.

10. AUTHOR BIOGRAPHIES

Martin Renilson moved from Scotland to Tasmania in 1983 to work at the Australian Maritime College, where he established the Ship Hydrodynamics Centre in 1985, and the Department of Naval Architecture and Ocean Engineering in 1996, with the first naval architecture and then ocean engineering degrees at AMC.

In 2001 Martin moved to the UK to work at QinetiQ as Technical Manager, Maritime Platforms and Equipment. In 2007 he returned to Tasmania and conducted work for a range of clients in the maritime field. In 2011 Martin was appointed the inaugural Dean of Maritime Programs at the HCT in the UAE. Martin returned to Tasmania in 2015, and was appointed Adjunct Professor at AMC/UTas.

Bruce Cartwright has over 25 years' experience in numerical simulation. He studied mechanical engineering in Adelaide, composite materials in London, and ship theory in Launceston. He held a position of Guest Scientist with the German Aerospace Research Agency (DLR) in Germany for one year through the Australian Cooperative Research Centre for Advanced Composite Structures.

Bruce has diverse work experiences, ranging from a part-time professional trumpet player to pay his way through university, a vibration analyst at a power station, a boatbuilder for America's Cup yachts, a consultant engineer with Pacific Engineering Systems International (Sydney) and as a Research Associate with University of Newcastle, through whom this work was done. Bruce is interested in pursuing further applications of the mesh-free techniques for maritime applications.

Teresa Magoga is currently a Discipline Lead Global Naval structures at Defence Science and Technology Group in Melbourne, Australia.

MARITIME AUTONOMY – A VISION OR REALITY

B Twomey, BEng, CEng, FIET, FIMarEST, MIMechE, Kongsberg Maritime, UK

SUMMARY

There are several claims relating to the benefits of using autonomous functions within the maritime sector. These range from a reduction in the human contribution to system hazards, enhancing safety, improving efficiency through optimisation of the ship design, and reduction in operating costs associated with a reduction in on board personnel. Unfortunately, there is very little evidence to support the above claims. What we can say with certainty, is an autonomous ship with no personnel on board will remove people from harm. What we cannot claim is autonomy will 'enhance safety', as there are currently no defined 'levels of safety' within the International Maritime Organisation (IMO) instruments, so a claim of 'enhanced safety' would be difficult to verify.

This paper will explore the principles of 'Inherent Safety, Fault Tolerance and Layers of Protection' in the context of the Maritime Autonomous Infrastructure (MAI), and how these principles influence the design within a dynamic safety case, challenging legal framework and changing stakeholder community depending on the operational requirements of the MAI.

1. INTRODUCTION

While conventional ships benefit from a long-established framework of prescriptive regulations, classification rules, and national legislation that support consistent evaluation and certification of design, construction, and operation, the MAI introduces fundamentally different challenges. In contrast, the MAI presents a complex socio-technical system and unprecedented regulatory and operational challenges due to its distributed, cross jurisdictional nature and reliance on emerging technologies. Addressing these challenges necessitates a shift from traditional rule-based compliance towards a comprehensive assurance/safety case approach.

Such an approach enables the systematic justification that the MAI, when operating within defined boundaries, is 'Acceptably Safe' to operate in a defined context of use, despite the absence of prescriptive regulations. By framing the MAI as a system-of-systems (SoS), this paper proposes a structured methodology that provides a defensible foundation for certification and stakeholder assurance, within a cross-jurisdictional and technologically dynamic environment.

2. WHAT IS A MARITIME AUTONOMOUS INFRASTRUCTURE (MAI)

The long development phase in the production of the IMO mandatory code for Maritime Autonomous Surface Ships (MASS), and the absence of a globally accepted set of requirements, has necessitated extensive engagement with individual national regulators for vessel developers and operators to ensure compliance with local requirements.

This reliance on national frameworks can lead to regulatory divergence among authorities, complicating certification processes, potentially slowing down industry adoption, and introducing uncertainties when arguing safety of the Maritime Autonomous Infrastructure (MAI), across different jurisdictions. To address these challenges, this section introduces a new conceptual framework defined as follows:

The 'Maritime Autonomous Infrastructure (MAI)' is a comprehensive and multifaceted socio-technical set of elements consisting of physical, digital and tangible assets, that support the safe operation of autonomous or remotely controlled functions of a vessel.

The constituent element(s) of the MAI may include, but are not limited to:

- Vessel/ship systems, including essential, and non-essential services.
- Land-based remote operations centre (ROC).
- Connectivity solutions (e.g. satellite and terrestrial communications infrastructure)
- Human – includes all relevant stakeholders.

- Port and harbour services, including Pilotage and Vessel Traffic Services (VTS).
- Emergency response services
- Supporting land-based infrastructure and services.

Furthermore, as the MAI will necessarily include elements that extend beyond the boundaries of the vessel to encompass a wider socio-technical system, the precise configuration and identification of the elements must be derived through systematic engagement with relevant stakeholders, recognising that its composition may evolve throughout a single voyage due to technical, operational or regulatory requirements.

2.1 MAI UNIQUE RISKS

This paper identified a range of unique risks associated with the MAI by conducting a detailed analysis of individual elements and their potential impact on safety, and the environment in the event of system failure or degradation of the function, for example:

- **The loss of situational awareness**: This could present a potential risk if the system cannot perceive or interpret its environment which could result in collision or grounding.
- **The loss of connectivity**: This risk shall be analysed with respect to the context of use.
- **Inability to regain control**: This is use-case dependent and could result in an intolerable risk where remote control of the function is required to ensure safety.
- **Inability to isolate autonomy**: This may lead to significant safety implications when humans are interacting with the technologies on board the vessel.
- **Inability to verify autonomy isolation**: This is a potential risk during maintenance operations or attempting rescue/recovery of a vessel
- **Inability to safely reinstate autonomy**: this may put undue pressure on remote operators and complicate risk mitigations during fallback operations.
- **Inability to identify a decrease in medical fitness of a remote operator**: Unlike a conventional ship, where a decrease in medical fitness and the ability of the individual to discharge the responsibility of their position is normally assessed by the colleagues on board.
- We can reasonably expect remote operators to leave the ROC at the end of their shift. This may or may not make it more difficult to detect fitness impairment that could potentially lead to an intolerable risk. It should; be noted that the European Aviation Safety Agency (EASA) have addressed this risk and provides help to air traffic controllers in deciding whether expert aero-medical advice is needed.[2]
- **Not taking a SoS perspective on the three primary elements**: A failure to consider the three primary elements of the MAI which includes the vessel, ROC and connectivity solutions as a SoS could make the claim of ''Acceptably Safe' to operate in a defined context of use' difficult to achieve.
- **Failure to detect degradation of the MAI functions** – Detecting functional degradation within the MAI may necessitate the correlation of data across multiple elements of the MAI. If such data correlation is not adequately addressed within the LPM, the claim of 'Acceptably Safe' may not be justifiable.
- **Cyber security**: A malicious cyber-attack may have an impact on safety. Although this is not a unique risk for the MAI, the attack surfaces can increase when operating a vessel remotely from a ROC.

Given that the list of unique risks associated with the MAI is not exhaustive, the process of hazard identification under 'normal, reasonably foreseeable abnormal and emergency conditions', must remain a continuous and iterative activity throughout the entire lifecycle process. This ensures that emerging risks are promptly identified, assessed and managed as the operational context evolves for the MAI.

3. THE MAI LIFECYCLE PROCESS MODEL (LPM)

The proposed LPM as illustrated in Figure 1, builds upon established standards, and good practices such as those defined in IEC 61508, ISO 3100, ISO 15288, ISO 17894, ARP 4754/4761, ISO 2626 and ISO 21448. The model provided a systematic framework for identifying the elements of the MAI that are important to safety and serves as a framework and methodology that provides a structured, coherent and traceable decision-making process that promotes safety assurance and facilitates the development of the compelling body of evidence and demonstration of ''Acceptably Safe'' in the context of the MAI.

Although the LPM is presented as a structured linear framework, it is designed to be inherently adaptive and iterative, allowing for continuous assessment and refinement throughout its implementation. This adaptability is particularly critical in responding to operational changes such as alterations in the vessels port of arrival or an unexpected loss of connectivity on the planned route that may have an impact on control of the vessel's functions. This iterative approach takes into consideration the dynamic and evolving safety assurance process, changing regulatory requirements, jurisdictional differences and the complexities when transitioning between voyage or operational conditions.

Furthermore, the application of the LPM necessitated a context specific approach, as each project is anticipated to generate a distinct set of requirements, shaped by the risk tolerability thresholds agreed by the relevant stakeholders, the stakeholder requirements and regulatory obligations. For instance, port authorities may impose specific operational restrictions on the operation of MAI within their jurisdiction, or may specify a requirement for additional external support, such as tug assistance, during berthing and unberthing or may indeed introduce an outright prohibition of autonomous operations. Similarly, pilotage requirements may introduce additional operational requirements, particularly if conducted remotely, for example, port authorities may require a clearly defined communication framework and declaration of predefined fallback states before vessels of this type are permitted entry.

The application of the LPM, will support the development of:

'*a compelling body of evidence that without bias, makes a claim for both the 'safe and unsafe' characteristics of a system when used for a defined use case in a given environment'*. SCSC-159 [2].

A compelling body of evidence refers to a structured, traceable and comprehensive set of data, analysis, and justifications that collectively substantiate the claims regarding the safety and operation of the MAI within a defined context of use. Such evidence should include but not limited to, the outcome of hazard identification, structured risk assessments, verification and validation results derived from simulations and testing, documented evidence demonstrating compliance with identified standards and regulatory requirements, and a clearly articulated design rational linking system requirements to the implemented safety functions.

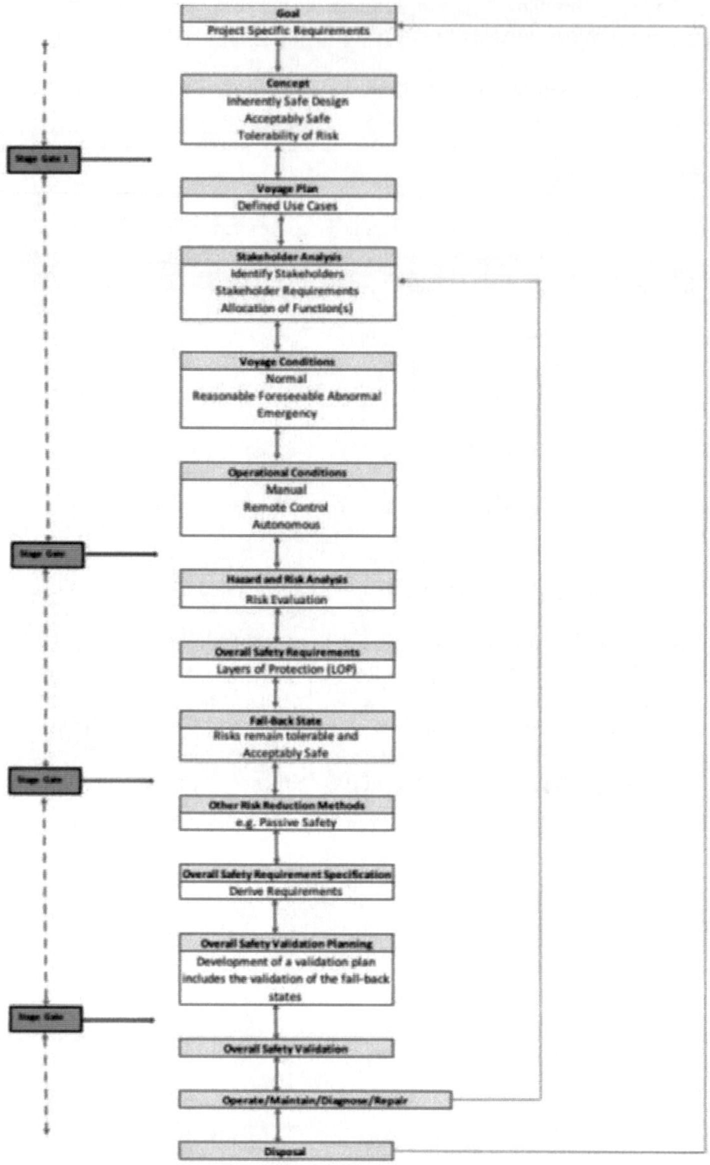

Figure 1: Lifecycle Process Model (LPM) for the MAI

4. OVERALL SAFETY REQUIREMENT

The development of the 'Safety Process Model' (SPM) for the MAI, is illustrated in Figure 2 The primary function of the SPM within the context of the MAI is to provide a structured methodology for deriving the safety requirements, determining whether a claim of ''Acceptably Safe'' has been substantiated, verifying the stakeholder requirements have been fulfilled, and supporting the creation and evaluation of the compelling body of evidence for regulatory submission and review.

The 'SPM' comprises of several core components, each addressing a distinct aspect of the safety/assurance/safety requirements for the MAI. Collectively, these components form a systematic framework for identifying, evaluating and verifying the outcome of each stage of the LPM with emphasis on independent verification, prior to regulatory submission.

To enhance credibility and confidence in the documented safety report, the SPM requires an Independent Safety Assessment (ISA) to be carried out, which aligns nuclear sectors [3], the UK Defence Security Authority [4] and BS EN 61508 [5]. This assessment ensures that the safety aspects of the MAI have been subjected to rigorous and continual oversight by a third party with no financial interest or connection to the project. The ISA serves to identify potential biases or oversight in the safety assessment process whilst ensuring the overall Goal, as defined in the LPM has been effectively achieved.

This approach represents a fundamental change from conventional crewed vessel acceptance processes, which typically rely on acceptance from a recognised Classification Society based on compliance with predefined and established rules. Given that the MAI consists of elements that extend beyond the scope of the traditional maritime domain, the requirement for independent verification is a critical enhancement in the assurance process for the MAI.

An additional requirement in the SPM is the emphasis on an iterative approach in the development of the compelling body of evidence, requiring greater collaboration among stakeholders. This approach could require revisions to existing contractual frameworks, particularly given the complexity and cross-disciplinary nature of the MAI. Twomey [6] has observed that; '*The marine sector, with its fragmented approach to system integration, will need to adopt a process that is not solely controlled by commercial and contractual agreements'*. While this issue lies outside the direct scope of this paper, it remains a relevant consideration within the context of the MAI.

The 'Safety Assessment 'element embedded within the SPM links stakeholder requirements to the 'Safety Approach', informs the selection of the 'Safety Analysis' methodology, and influences the 'Operational Considerations' for the MAI. While the SPM does not prescribe specify methodologies, tools or techniques for the safety analysis, it requires that these methods be explicitly defined, justified, and agreed upon by the relevant stakeholders. The methodologies must consider the complexity of the MAI and whether established techniques provide sufficient evidence that the MAI is ''Acceptably Safe'' to operate in a defined context of use. If traditional methods are deemed inadequate, as stated by Leveson, advanced techniques such as Safety Assurance of autonomous systems in Complex Environments (SACE) [7] and System Theoretic Process Analysis (STPA) [8] may need to be considered.

The operational considerations will be defined by the company and the relevant stakeholders. Safety functions will need to be agreed to ensure the risks do not become intolerable throughout the MAI lifecycle. The output from the 'Safety Assessment 'influences the way in which the MAI can be used, whether the voyage conditions can be achieved and how the International Safety Management Code (ISM) needs to consider the operation of the MAI at a SoS Level.

The 'Safety Report' generated from the 'Safety Assessment' should provide sufficient documentary evidence in sufficient scope and detail to support the conclusions reached and provide an adequate foundation for independent verification and regulatory reviews.

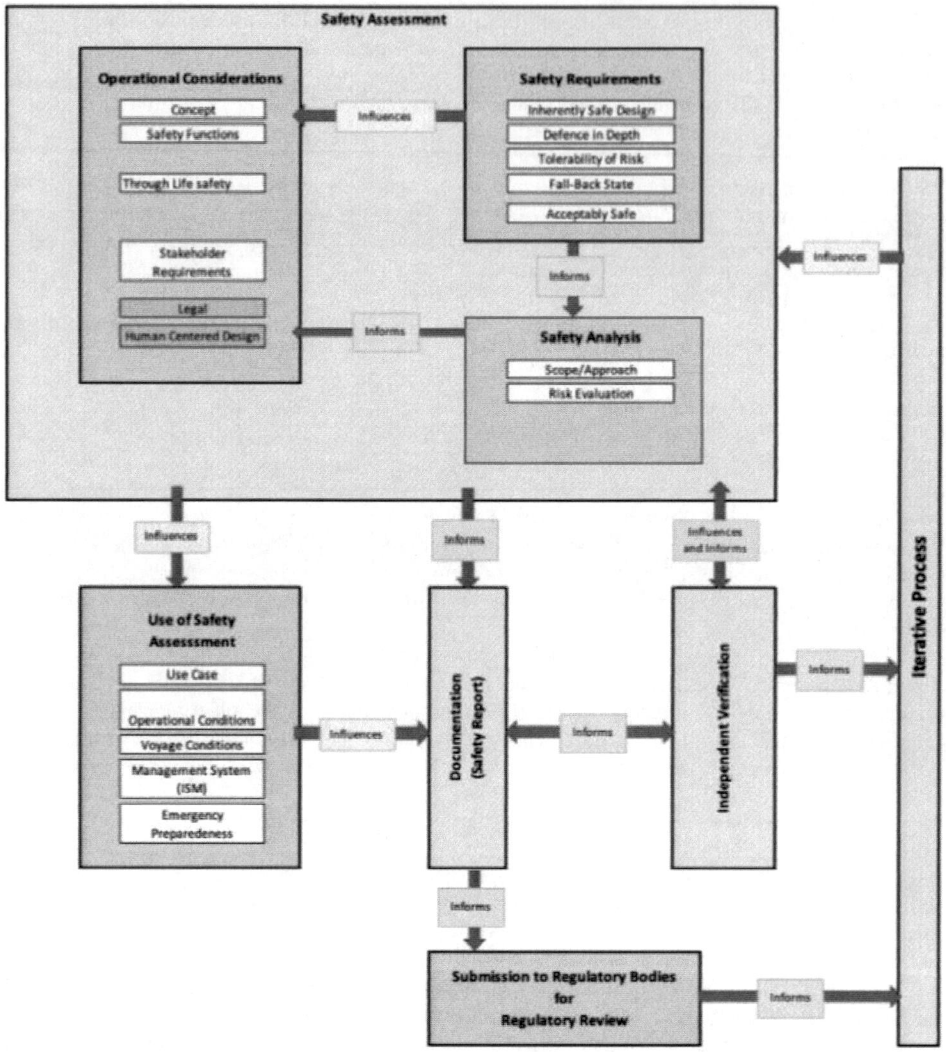

Figure 2: Safety Process Model (SPM) for the MAI.

5. THE CONCEPT OF 'ACCEPTABLY SAFE'

In the absence of prescriptive requirements specific to the MAI, the concept of what constitutes ***'Acceptably Safe'*** must be collaboratively derived by the relevant stakeholders and agreed with the relevant regulatory authorities. The development of this definition is an important step in constructing a credible and defensible assurance/safety case for the MAI.

A review of existing literature across various domains, highlights how the notion of 'Acceptably Safe' is context dependent, and often determined by domain specific criteria. For example, the International Civil Aviation Organisation (ICAO) defines an Acceptable Level of Safety Performance as:

'*The level of safety performance agreed by the State authorities to be achieved for the civil aviation system in a State*'.[9]

This emphasises the role of the regulatory authorities in setting performance requirements.

Similarly, ISO 21448, asserts that "*An acceptable level of safety for road vehicles requires the avoidance of unreasonable risk caused by every hazard associated with the intended functionality and its implementation, especially those not due to failures"* [10]. Furthermore ISO 17894 states that "*The required level of safety shall be realized by appropriate activities throughout the lifecycle"* which reinforces the necessity for lifecycle safety practices to be applied [11].

Leveson recommends a proactive approach that embeds safety considerations from the outset of system development, stating that the need to "*create techniques and approaches. that emphasise building safety into a system from the very beginning*"[8].

From a regulatory perspective the Institute of Mechanical Engineers (IMechE) provides clear guidance on the principles of As Low As Reasonably Practicable (ALARP). which offers a structured framework to ensure the risks are reduced to a level that satisfied legal requirements [12]. Additionally, Habli explains that *safety cases created must take into account the broad operational context to ensure that different kinds of hazards, both immediate and long-term, are addressed in an integrated and consistent manner. In this way, the safety cases are grounded in the established principles of system safety that are expected for all safety-related* applications'[13].

Drawing upon these diverse yet complimentary sources, it is evident that, safety cannot be defined in absolute terms for the MAI. Instead, safety must be articulated through stakeholder derived criteria based on their tolerability of risk. These criteria must be justified and integrated into the assurance/safety process to ensure that the MAI can be demonstrated as '***'Acceptably Safe''*** within a defined context of use.

In conclusion and as illustrated in Figure 3, while the overarching safety objective (Goal G0) of ensuring that (X) is ''Acceptably Safe' to operate for a defined use case in a given environment' applies equally to both conventional ships and the MAI, however, the assurance methodologies required to substantiate this claim differ substantially. For conventional vessels, assurance is achieved through compliance with existing prescriptive regulatory frameworks such as those found in SOLAS and related codes and conventions. Safety claims are substantiated through objective evidence demonstrating adherence to these regulatory requirements(D1), enabling the regulatory authorities to issue the necessary certificated for operation. This process is relatively straightforward as the deterministic nature or prescriptive regulations lends itself to clear verification procedures, but it does not involve the development of a safety argumentation process. While compliance with prescriptive requirements is necessary, it does not guarantee a claim of having a safe system.

In contrast, the assurance process for the MAI necessitates a more adaptive, stakeholder driven and argument-based approach as the regulatory landscape is still evolving and lacks substantive guidance. Consequently, the assurance evidence must be constructed through a systematic and iterative approach that involves the derivation of requirements (G4), the comprehensive identification and mitigation of hazards (G5), and demonstrable compliance with defined requirements, international standards, and applicable best practices (G6). This process is reinforced through an ISA which adds a layer of credibility and helps to address stakeholder concerns.

The divergence between these two assurance pathways illustrates the challenges, complexity and novelty of validating the assurance/safety argument for a socio-technical system such as the MAI. It also identifies the necessity for a multi-disciplinary and cross sectoral engagement strategy to effectively manage the risks, especially where legacy regulatory frameworks are not yet equipped to address the complexities of the MAI. This divergence requires an evolving assurance/safety model to accommodate the use of remotely controlled or autonomous functions within the maritime domain.

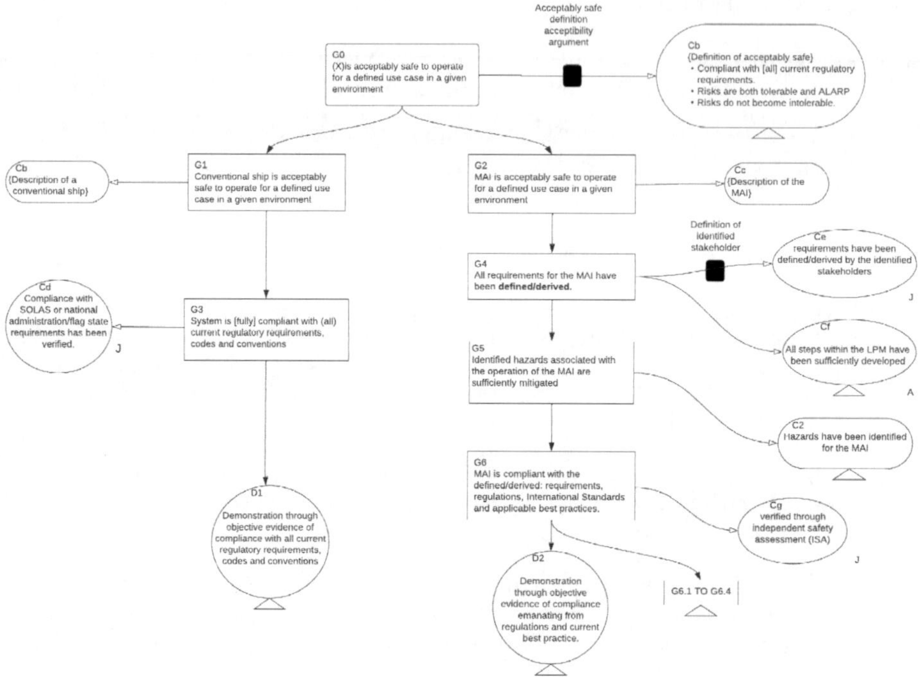

Figure 3: Conceptual framework for defining 'Acceptably Safe' in the context of a conventional ship and the MAI

6. LAYERS OF PROTECTION

The purpose of introducing the principles of 'Layers of Protection (LoP)' into the MAI is:

> '*to provide a series of independent barriers aimed at preventing faults in the first instance and ensuring appropriate protection and mitigation of accidents in the event that protection fails.*' [3]

This paper proposes the development of the Layers of Protection framework as illustrated in Table 1, is specifically designed for application within the MAI. The table consists of five distinct layers of protection, intentionally designed to be as independent as possible to maximise their effectiveness and provides a structured, scalable and stakeholder adaptable framework for implementing LoP within the MAI, acknowledging that the number of layers will be refined by the relevant stakeholders based on specific operational and regulatory requirements.

While the principles of layers of protection is a well-established principle its explicit definition and structured application within the MAI is a unique contribution to this thesis. The removal of trained crew eliminates a key human layer of defence, placing increasing reliance on system design to prevent, contain, and respond to faults. Within this context, the development of the LoP framework is not merely a theoretical exercise, it is a practical necessity for maintaining control and avoiding hazards escalating to an emergency condition.

This structured integration of layers of protection into the MAI lifecycle not only addresses existing regulatory gaps but also establishes a foundation upon which future autonomous or remotely controlled operations can be assessed to ensure appropriate risk mitigation strategies are acceptable to the relevant stakeholders.

Table 1. Layer of Protection Model for the MAI (LoP)

Layers of Protection				LPM Step	
Level	Objective	Defence/Barrier	Guidance	Concept	Voyage Condition
Level 1	Prevention of abnormal operation and failures by design	Conservative design, high quality in construction, maintenance and operation in accordance with appropriate safety criteria, engineering practices and defined quality standards	Compliance with specified standards	Inherently Safe Design	Normal
Level 2	Prevention and control of abnormal operation and detection of failures	Control, limiting and protection systems, other surveillance features and operating procedures to prevent and minimise damage from failure	Compliance with specific standards and outcome from the risk analysis of the system under consideration	Fault Tolerant	RFAE
Level 3	Control of faults within the design basis to protect against escalation to an accident	Engineering safety features, multiple barriers and accident or fault control procedures	Compliance with specific standards and outcome from the risk analysis and requirements from the shipowner/company	Fault Tolerant	RFAE
Level 4	Control of severe ship or infrastructure conditions, in which the design basis may be exceeded, including protecting against further fault escalation and mitigation of the consequences of severe accidents	Additional measures and procedures to protect against or mitigate fault progression and for accident management	Severe ship or infrastructure conditions could include fire, flooding or evacuation of a ROC due to environmental conditions, tsunami, earthquake, terrorist attack etc	Fault Tolerant/Emergency	RFAE/Emergency (Tolerable)
Level 5	Mitigation of accident consequences through emergency responses	Emergency control and on-off site emergency response (e.g. salvage vessels, fire fighting tugs, land based emergency services, etc)	An emergency condition will be declared by the owner/operator. Example Vessel out of operation (not under command) and drifting towards an object, evacuation of a ROC with no possibility of transfer of control to another ROC/remote operator, etc	Emergency Condition	Emergency (Intolerable)

The following section builds upon this foundation by introducing the MAI Design Envelope (MAI-DE), which formalises the integration of Concept, Voyage Conditions, and Layers of Protection into a unified framework that supports design, assurance, and regulatory engagement throughout the MAI-LPM.

7. INHERENTLY SAFE DESIGN

This section examines the principle of inherently safe design and its application within the MAI. The principle, well established in the process engineering and nuclear sectors, prioritises the elimination of hazards at the design stage rather than relying on procedural controls or reactive mitigation. Its relevance to crewless maritime operations lies in the absence of onboard trained personnel who have historically acted as a layer of protection able to identify, interpret, and respond to emerging failures before they escalate into intolerable risks.

In the context of the MAI, and without human intervention, requires a fundamental shift from a compliance driven approach to a design process that require hazard elimination over hazard control and design philosophies that prioritise inherent safety. This represents a fundamental transformation in maritime safety thinking, one that is not yet widely recognised in existing regulatory requirements but is essential for the successful introduction of maritime autonomy and remotely operated vessel functions.

This paper proposes the use of inherently safe design principles within the MAI, thereby making a unique contribution to the field of maritime autonomy. It advances the argument that autonomous or remotely controlled vessel safety cannot be achieved by simply substituting human oversight with automation or remote control; instead, safety requirements must be embedded systematically from the outset. This shift in design philosophy is not only necessary but foundational to supporting the 'Acceptably Safe' claim under evolving regulatory and operational conditions.

The objective of an inherently safe design is to reduce the number of events that result in a loss or degradation of a function below acceptable levels. As stated by Kletz T in his 1978 article[14]:

'*What you don't have can't leak*', and '*before we estimate the probability and consequence of a hazard and compare them with a target, we should ask if the hazard can be eliminated*'.

A similar principle is used by the Office for Nuclear Regulations in their document Safety Assessment Principles for Nuclear Facilities: [3] states:

'The underpinning safety aim for any nuclear facility should be an inherently safe design, consistent with the operational purposes of the facility. An 'inherently safe design' is one that avoids radiological hazards rather than controlling them. It prevents a specific harm occurring by using an approach, design or arrangement which ensures that the harm cannot happen.'

Likewise, the UK Health and Safety Executive document '*Reducing Risks, Protecting People'* [15] advocates the principle of an *Inherently safer design:*

'Adoption of the principles of inherently safer design is particularly important where the consequences of plant or system failure are high. HSE will press for the incorporation of inherently safer design features, where these are possible, to reduce the reliance on engineered safety systems or operational procedures, to control risk'

Where it is not reasonably practicable to achieve an inherently safe design, the MAI must demonstrate *fault tolerance* which is defined by the IEC as:

'the ability of a function to continue to perform in the presence of a fault' [5].

The application of the inherently safe design principle is particularly relevant for the MAI, as the layer of protection afforded by the onboard crew will no longer be available on a crewless vessel.

Notes:

- Inherent safety is not the same as 'passive safety'. Where an inherently safe design is not achievable, the design should be fault tolerant'
- Passive safety - Providing and maintaining a safety function without the need for an external input such as actuation, mechanical movement, supply of power or operator intervention.

8. THE MAI DESIGN ENVELOPE (MAI-DE)

This section described the development of the Maritime Autonomous Infrastructure Design Envelope (MAI-DE) which integrates the concept phase, voyage conditions, and Layers of Protection principles of the LPM providing the relevant stakeholders with a structured method to determine whether the layers of protection and the design basis of the safety systems are sufficient to support a claim of ''Acceptably Safe'' under defined operational conditions.

As illustrated in Figure 4, the MAI-DE should preferably be operating as:

- Inherently Safe Design – Normal – Level 1, shown as Green
- If this cannot be achieved, levels 2 and 3, shown as Yellow and Brown, will detect and control deviations or activate safety features/
- If the fault escalates to level 4, shown as Red is about limiting consequences and preventing escalation, whilst ensuring the risks do not become intolerable.
- A level 5 event would necessitate emergency response and external intervention

The relevant stakeholder's tolerability of risk will dictate how many layers and what protection is required, which should be agreed and submitted to the regulatory authorities for regulatory approval. *Question: is a ship sinking a tolerable risk – one owner believes it is which informs the design and the assurance/safety case presented to the regulatory authorities for acceptance?*

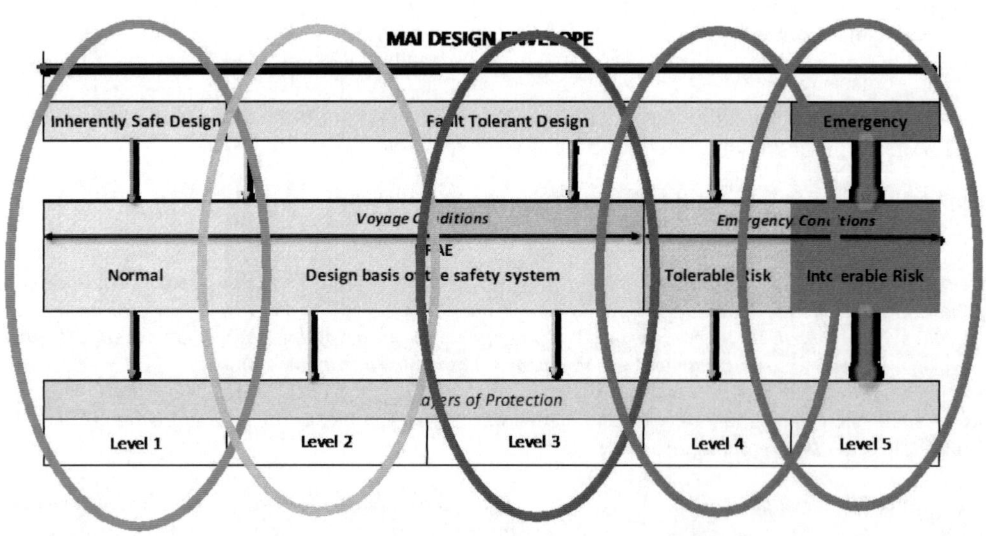

Figure 4: MAI – Design Envelope (MAI-DE)

9. CONCLUSIONS

The journey towards the Maritime Autonomous Infrastructure (MAI) represents a profound shift in how safety, assurance, and regulatory oversight are conceptualised and operationalised in the maritime domain.

While conventional crewed vessels benefit from a long-standing framework of prescriptive regulation and classification rules, the MAI introduces a distributed, socio-technical system characterised by emergent behaviours, jurisdictional fragmentation, and the absence of onboard human intervention. This renders traditional compliance approaches insufficient for substantiating safety claims in autonomous or remotely controlled operations.

This paper has addressed this gap by developing a comprehensive and systematic assurance methodology for the MAI, encompassing a **Lifecycle Process Model (LPM)**, **Safety Process Model (SPM)**, **Layers of Protection (LoP)**, and a **Maritime Autonomous Infrastructure Design Envelope (MAI-DE)**. Collectively, these models offer a structured framework for identifying and managing risk, substantiating the safety argument, and generating a compelling body of evidence to support the claim that the MAI is 'Acceptably Safe' within a defined operational context.

A unique contribution of this work lies in the explicit incorporation of safety engineering principles—inherent safety, fault tolerance, and independent layers of protection into the design and operational validation of the MAI. The structured integration of these principles is essential, not only due to the absence of a trained crew, but also to address the limitations of current regulatory instruments which do not yet define 'acceptable safety levels' for autonomous systems. The framework proposed here recognises the evolving nature of both operational requirements and regulatory expectations and supports continuous and adaptive assurance throughout the system's lifecycle. Furthermore, by situating the MAI as a '**system-of-systems**', this paper challenges conventional notions of vessel-centric regulation and underscores the need for **cross-disciplinary engagement**, **context-specific risk tolerability definitions**, and **independent safety assessments (ISA)**. It also demonstrates that safety cannot be claimed through compliance alone; rather, it must be argued, evidenced, and independently verified.

As the maritime sector advances toward greater autonomy, the models and methodologies proposed herein provide a defensible and adaptable foundation for stakeholder collaboration, regulatory engagement, and long-term operational assurance. In doing so, this paper contributes not only to the academic discourse on maritime autonomy but also to the practical development of safety cases that are robust, transparent, and responsive to the unique challenges posed by the MAI.

As a closing remark, we acknowledge that the development of the IMO MASS Code is under development, and each autonomous or remotely operated vessel project requires a case-by-case negotiations with the national regulatory authorities. This approach can result in regulatory divergence and uncertainty for developers and operators. A historical parallel may be drawn from a comparison between five European Ports' Cordes [16] analysed the maritime law codes from 1255 to 1276, wherein he concluded that:

'the rules to be analysed are evidently the result of negotiations conducted by the parties involved: that is to say regulations on which the professional merchants and seafarers had agreed upon amongst themselves'

10. REFERENCES

1. EASA., 'Acceptable Means of Compliance (AMC) and Guidance Material (GM) to Part ATCO.MED Medical requirements for air traffic controllers', *Issue 1.* 13 March 2015.

2. Safety Critical Systems Club., 'Assurance Case Guidance Version 1: The Assurance Case Working Group (ACWG)'

3. Office for Nuclear Regulations Safety Assessment Principles for Nuclear Facilities 2014 Edition, Revision 1 (January 2020)

4. DSA02-DMR: Defence Maritime Regulations For Health, Safety and Environmental Protection 2025 Edition

5. BS EN 61508: Functional safety of electrical/electronic/ programmable electronic safety-related systems - General requirements

6. TWOMEY, B., 'System Design and Integration: Encyclopedia of Maritime and Offshore Engineering', *Volume 3 Section* 119 - ISBN: 978-1-118-47635-2, July 2018.

7. HAWKINS et al., 'Guidance on the Safety Assurance of Autonomous Systems in Complex Environments (SACE)', *Assuring Autonomy International Programme (AAIP) University of York*

8. Leveson: Engineering a safer world: systems thinking applied to safety. ISBN 978-0-262-01662-9: 2011

9. International Civil Aviation Organization 9589: Safety management manual ISBN 978-92-9258-552-5

10. ISO 21448 :2022 - Road vehicles — Safety of the intended functionality

11. ISO 17894:2005: - Ships and marine technology — Computer applications — General principles for the development and use of programmable electronic systems in marine applications

12. IMechE - ALARP for engineers a technical safety guide. - Safety and Reliability Group – May 2024

13. HABLI., 'The BIG Argument for AI Safety Cases: Centre for Assuring Autonomy', *University of York*

14. Kletz Trevor Kletz Compendium - His Process Safety Wisdom Updated for a New Generation 2021, Pages 73-114: https://doi.org/10.1016/B978-0-12-819447-8.00003-1

15. UK Health and Safety Executive - Reducing risks, HSE's decision-making process. First published 2001 - ISBN 0 7176 2151 0

16. CORDES Conflicts in 13th Century Maritime Law: A Comparison between five European Ports (2020) Oxford U Comparative L Forum 2 at ouclf.law.ox.ac.uk

11. AUTHORS BIOGRAPHY

Bernard Twomey Beng, CEng, FIET, FIMarEST, MIMechE leads maritime regulatory development within Kongsberg Maritime. He brings extensive experience from across the maritime sector, including roles as the Global Head of Electrotechnical Systems at Lloyd's Register and as maritime assurance specialist at Rolls-Royce Maritime. Bernard serves as an advisor to the Danish Maritime Association at the IMO and participates in the technical committees of two leading classification societies. In addition to his industry roles, he is a visiting professor and visiting fellow at several UK Universities. He holds a BEng in Electro-Mechanical Power Engineering from Loughborough University of Technology and is currently completing his PhD at the University of York, focusing on the assurance of complex systems.

A GLOBAL VIEW OF THE IMPLICATIONS OF WAVE CLIMATE EXTREMES ON SUSTAINABLE SHIP DESIGN AND OPERABILITY: A REVIEW

C S Earlie, E Gebauer, C Ingram, D Boyce, T Matthewson and **A Kimber,** BMT Ltd, UK

SUMMARY

Under a changing climate, with the potential for increased storminess and the likelihood of more extreme atmospheric temperatures, it will become increasingly prudent to consider the implications of extreme wave and weather conditions on maintaining safe and effective operations for the required lifespan of a vessel. In this paper, we explore the current status of operational design versus oceanographic and meteorological variables within ship design and operability, with a focus on wave characteristics and cyclones as these are of concern for a ship's stability, structural integrity, operations and manoeuvrability. We then illustrate the latest IPCC (AR6) projections and provide evidence from scientific literature for oceanic extremes (significant wave height, extreme wave height, wave power, storm intensity, storm frequency, storm tracks and wind speeds) across the globe for the typical lifespan of a vessel. By comparing existing approaches to the use of oceanographic variables in ship design criteria, with how these variables may change under future climates, we will provide recommendations for continued development and further research towards maintaining safe (structural and stability) and effective (energy efficient/economical) operations within the context of the physical effects of climate change.

NOMENCLATURE

ρ	Density of water (kg m^{-3})
H_s	Significant Wave Height (m)
T_z	Mean wave period (s)
T_{sd}	Standard deviation of wave period (s)
H_x	Extreme design wave height (m)
E	Wave energy (Nm^{-2})
g	Acceleration due to gravity (9.81ms^{-2})
h	Water depth (m)
k	Wave number ($\frac{2\pi}{L}$) where L = wave length (m)
RCP	Representative Concentration Pathway
SSP	Shared Socioeconomic Pathway
IPCC	Intergovernmental Panel on Climate Change

1. INTRODUCTION

In the face of a changing climate, the maritime industry must adapt to the weather conditions of the future to ensure the safety and efficiency of ship operations. This paper explores the current status of ship design and operability in relation to oceanographic and meteorological variables, highlighting the importance of integrating climate resilience into naval ship design. By examining the latest projections from the Intergovernmental Panel on Climate Change Sixth Assessment Report (IPCC AR6) and reviewing scientific literature on oceanic extremes, we aim to contribute to the wider understanding of how future climate scenarios may impact ship design criteria and operational limits.

The rationale for this paper stems from the need to address the implications of climate extremes on sustainable ship design and operability. With the potential for increased storminess and more variable atmospheric temperatures, how these conditions will affect the safe and effective operations of vessels throughout their lifespan warrants consideration. As western economies are actively seeking to modernise their naval fleets to address evolving security challenges and maintain maritime superiority, there are opportunities to enhance the durability and performance of naval assets in the face of potentially changing environmental conditions.

1.1 CURRENT STATUS OF SHIP DESIGN IN THE CONTEXT OF WAVE CLIMATE PROJECTIONS

The United Kingdom Royal Navy has announced a three-phase approach to climate change and sustainability, with plans since 2022, to build 150 vessels over the next 30 years (Ashbridge and Beard, 2022). The Ministry of Defence (MoD) Defence Equipment and Support (DE&S) organisation has three key environmental commitments. The first being 'Mitigating and Adapting to Climate Change' which includes 'Implementing interventions to reduce greenhouse gas emissions and enable operation in a climate changed environment.' (Defence Equipment and Support, 2024). The three commitments are to be delivered and assessed against multiple outcomes, the most relevant being Outcome 4: ensure changing environmental conditions resulting from, for example, climate change and environmental degradation are recognised, and due diligence is applied when procuring, designing, and supporting Products Systems and Services (PSS) to ensure that it is resilient to changing environmental conditions and can therefore maintain operational capability (Defence Equipment and Support, 2024). Other Navies are also seeking to modernise; under the United States Navy's 2025 shipbuilding plan, total fleet size will increase from 295 today to 390 in 2054 (Congressional Budget Office, 2025), and in 2024 the Australian Government announced plans for 55 new vessels (Australian Government, 2024). Importantly, these modernisation processes offer many opportunities to incorporate climate resilience, ensuring that ships are suitable for their entire lifespan.

Operational ship design describes the process through which ships can be designed for efficient and safe operation for their lifespan, currently a period of typically 25 to 50 years. Severe weather events can contribute to loss of life, injury and damage to ships and in extreme cases, loss of the vessel. During a study of naval accidents during the 1945 to 1988 period, it was determined that there had been 65 publicly available documented accidents involving adverse weather conditions, impacting 107 different ships (Arkin and Handler, 1989). To ensure ships are sufficiently resilient, naval architects and marine engineers design for specific operational requirements and to a specific range of environmental conditions. This process has become more regulated as design standards have changed with time, evolving through operational experience, lessons learned, and technological advancements. For example, prior to the 1980s, the consideration of environmental conditions in United States Naval ship design was quite limited (Bales, 1983). Technological advancements have understandably had a significant impact on the ship design process. Since the 1990s, the testing of ship resilience has been increasingly aided by computer analysis and can be carried out throughout the design process (Ashe *et al.,* 2006). However, as the maritime domain is particularly susceptible to the effects of climate change, there is a level of uncertainty regarding the suitability of the existing environmental conditions used for design criteria. There is concern regarding rising sea levels, storm frequency, storm severity and altered wave patterns, which may pose threats to naval infrastructure and operations (Mihailov and Ichim, 2024). The effects of climate change may be compensated for by adjusting routes based on improved weather forecasts (Bitner-Gregersen *et al.,* 2018). For naval ships it is not always possible to adjust routes or operations based on forecasts, which still hold levels of uncertainty (Petacco and Gualeni, 2023). In fact, naval ships are often required to operate in locations outside of shipping routes and in areas where limited ship operations are conducted.

The standards for operational ship design are set by a variety of organisations. Environmental loads do not substantially differ between the naval and commercial sectors, but the operational profiles do (Ashe *et al.,* 2006), with the requirement for military vessels to be capable of withstanding higher sea states without compromising their military or support capabilities. Globally, the International Maritime Organisation (IMO), a specialised agency of the United Nations, are the standard-setting authority for the safety, security and environmental performance of international shipping. Every commercial ship should have a registered flag state which enacts the IMO regulations, which can authorise Classification Societies to act on their behalf, carrying out statutory survey and certification duties. Additionally, individual Classification Societies develop and apply technical standards for the design, construction and survey of ships. There are currently more than 50 of these societies worldwide (IACS, 2025). Lloyd's Register is one example of these societies, which published the first set of naval rules for the design, construction and classification of ships in 1998 (Ashe *et al.,* 2006). Reference to climate change is not systematically included in Classification Societies rules, most likely due to uncertainties associated with climate change projections (Bitner-Gregersen *et al.,* 2018), yet to enable improved design and operational criteria, these uncertainties need to be quantified, and updates made to current guidance (Bitner-Gregersen *et al.,* 2018).

Currently, many navies rely on external organisations for ship design and construction, making the regulation of design through Classification Societies even more important (Ashe *et al.,* 2006). The North Atlantic Treaty Organisation (NATO) is a key organisation, who act as an intergovernmental transnational military alliance and publish the Naval Ship Code. The Naval Ship Code, developed by the International Naval Safety Association (INSA), is applicable to all surface naval craft, coastguard, and other vessels operated by a government which are not for commercial purposes. It provides a consistent framework to parallel the IMO and process of flag states for commercial ships. Each commercial ship should have a registered flag state which can authorise Classification Societies to act on their behalf, carrying out statutory survey and certification duties. The goal-based framework within the Naval Ship Code allows for ships to be certified as safe to operate (International Naval Safety Association, 2019). Individual naval administrations also set their own standards, often building upon NATO frameworks. These present an opportunity to implement specific protocols and personalised approaches. In the United Kingdon, the Defence Standards, a set of technical specifications, are published by the MoD (UK Defence Standardization, 2025).

Key guidelines for ship design in the context of wave climate are the Allied Naval Engineering Publication 77 Part 1 Edition G Version 3 (ANEP 77), the NATO Standardisation Agreement No. 4194 (STANAG 4194) Edition 1 and the Lloyd's Register Rules and Regulations for the Classification of Naval Ships (NSR 2024). ANEP 77 is part of the Naval Ship Code, outlining goals, functional objectives and performance requirements, dated July 2019. The Naval Ship Code has been developed in consideration of risks and issues that may impact on safety (International Naval Safety Association, 2019). This includes the development of high-level goals that need to be addressed through the application of technical standards (International Naval Safety Association, 2019). For example, Regulation 3 in Part 1 Chapter II 'Structural Design' outlines the following goals:

> *3.13 As a minimum it is mandatory to assess the structural capacity of all ships when subjected to the following demands:*
> *3.13.1 Above water: Wind, air temperatures (high and low), ice accretion, solar radiation.*
> *3.13.2 Sea surface: Waves, green seas, ice navigation, ship motions (including slamming).*

STANAG 4194, a longstanding document, is aimed to establish a common source of wave and wind data for NATO operational areas. It also provides NATO sea state summaries (Table 1) to categorise sea conditions for the North Atlantic Ocean (North Atlantic Treaty Organization and Military Agency for Standardization, 1983). These values were produced using a 20 year hindcast climatology database that was in development by the United States Navy from 1976 (North Atlantic Treaty Organization and Military Agency for Standardization, 1983).

Table 1. NATO sea state numeral table for the open ocean North Atlantic (North Atlantic Treaty Organization and Military Agency for Standardization, 1983)

Sea State Number	**Significant Wave height (m)**		**Sustained Wind Speed (Knots)**		**Percentage probability of sea state**	**Modal Wave Period (sec)**	
	Range	Mean	Range	Mean		Range	Most probable
0 -1	0 – 0.1	0.05	0 - 6	3	0.7	-	-
2	0.1 – 0.5	0.3	7 - 10	8.5	6.8	3.3 – 12.8	7.5
3	0.5 – 1.25	0.88	11 - 16	13.5	23.7	5.0 – 14.8	7.5
4	1.25 – 2.5	1.88	17 - 21	19	27.8	6.1 – 15.2	8.8
5	2.5 - 4	3.25	22 - 27	2.5	20.64	8.3 – 15.5	9.7
6	4 - 6	5	28 - 47	37.5	13.15	9.8 – 16.2	12.4
7	6 - 9	7.5	48 - 55	51.5	6.05	11.8 – 18.5	15.0
8	9 - 14	11.5	56 - 63	59.5	1.11	14.2 – 18.6	16.4
>8	>14	>14	> 63	> 63	0.05	18.0 – 23.7	20.0

NSR 2024, dated January 2024, Volume 1 Part 5 Chapter 2 outlines the environmental conditions including wind, wave, currents, ice and thermal conditions that should be applied in the derivation of loads (Lloyd's Register, 2024). Oceanographic conditions are particularly important for use in calculations of ship structural strength and stability. Waves have the largest impact on load and response calculations in ship design (Bitner-Gregersen *et al.*, 2013). The environmental conditions used are derived from hindcast data, long term measurements and theoretical studies complemented with service experience (Lloyd's Register, 2024). The sea state definitions match that of STANAG 4194 (Table 1). Wave environmental data depends on the service area (SA) that the ship will operate in (Table 2). For example, SA1 refers to ships that have unrestricted global operations and SA2 is primarily for ships that will operate in tropical and temperate regions, these extents are shown in Figure **1** (Lloyd's Register, 2024).

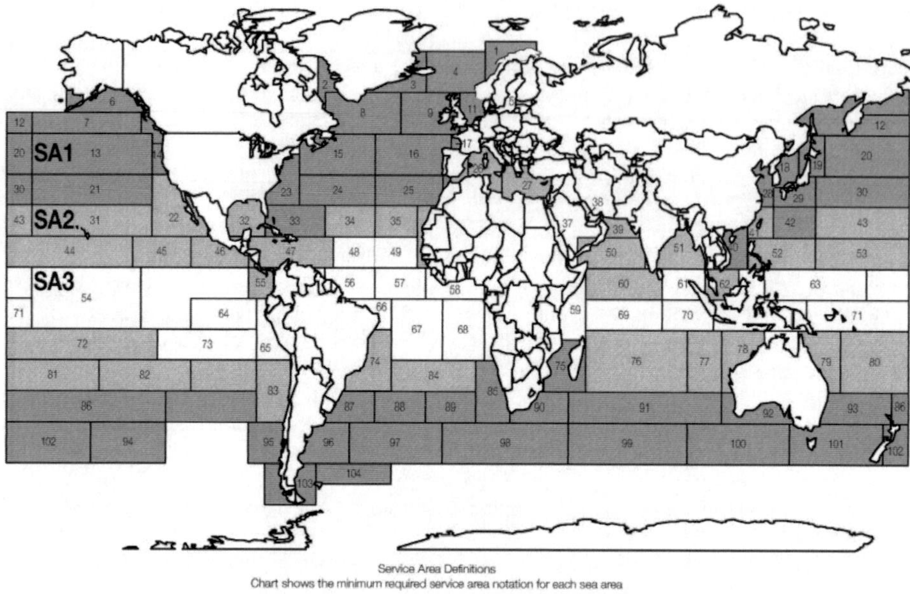

Figure 1: Map showing sea areas (numbers) and corresponding Service Areas (SA) (Lloyd's Register, 2024)

Table 2. Environmental wave data for each service area (Lloyd's Register, 2024)

Service Area Notation	**Wave height for the service area, m**	**Mean wave period, seconds**	**Standard deviation of wave period, seconds**	**Extreme design wave height, m**
	H_s	T_z	T_{sd}	H_x
SA1	5.5	8.0	1.7	18.5
SA2	4.0	7.0	1.7	13.5
SA3	3.6	6.8	1.7	9.5
SA4	2.5	6.0	1.5	6.0
SAR	*To be specially considered*			

1.2 CURRENT STATUS OF OPERATIONAL LIMITS AND THRESHOLDS IN THE CONTEXT OF WAVE CLIMATE PROJECTIONS

In the context of naval ships, operational limits are the predefined thresholds that define the boundaries within which a ship can safely operate. These limits can include environmental conditions, vessel motion and structural stresses. When these thresholds are exceeded, the ship may be unable to maintain its stability, performance, and its overall integrity may be at risk. These limits are usually set during the design phase and then used onboard to guide operational processes, linking the two phases. Ship design criteria assume that the crew possesses a predetermined level of expertise (Bačkalov *et al.,* 2016). Additionally, during the design phase, assessments on ship stability are carried out to fulfil requirements set by design rules (Petacco and Gualeni, 2023). Traditionally, these assessments are often based on static equilibrium conditions, and investigations on ship motion in waves focus primarily on the vertical motions (Petacco and Gualeni, 2023).

One example of guidelines on the operational limits and thresholds for vessels are the Second-Generation Intact Stability Criteria (SGISC), which were approved by the IMO in 2020. This second generation of guidelines were required due to advancements in computer technology which had occurred following the 2008 Intact Stability Code (International Maritime Organization, 2020). The SGISC includes information on vulnerability criteria, direct stability failure assessment and operational measures (limitations and guidance) (International Maritime Organization, 2020). Operational guidance (sitting within operational measures) is particularly relevant to the current status of limits and thresholds. To summarise, it identifies a set of assumed dangerous situations for each sea state (Petacco and Gualeni, 2023). These assumed situations are defined by the combination of environmental parameters such as wave height and the vessel condition (Petacco and Gualeni, 2023). It is important to note that within the SGISC, this concept is covered in more detail and has various approaches.

The implications of climate change go beyond the physical design of the ship, as vessel safety also depends on the operational processes in place. These operational processes guide crew in their decisions and actions when faced with extreme conditions. Modern technology enhances these processes. Decision support systems can be integrated with extreme weather warning criteria to offer timely, actionable advice, enabling crews to react quickly and appropriately during challenging scenarios (Bitner-Gregersen *et al.*, 2013; Bačkalov *et al.*, 2016). Real-time sensor inputs further enhance onboard safety and operability by continuously measuring key environmental factors such as wave height, wind speed, and the ship's motion. These sensors provide immediate feedback that informs crew decisions but can be used to adjust operational strategies if required (Ferrier *et al.*, 2022). Consequently, operational processes can be adjusted dynamically based on real-time environmental conditions and predictive analytics. This capability is critical in the face of extreme weather or changing climate conditions.

1.3 GLOBAL WAVE CLIMATE PROJECTIONS

When considering sustainable ship design and operability in the context of how oceanic states are and will become altered under a changing climate, we can consider a number of different oceanic variables and parameters that enable us to quantitatively and qualitatively describe changes to wave characteristics. Due to the influence wave climate has on the fate of the coastal zone, infrastructure, offshore energy, global transport and operations, numerous global and regional studies have been conducted in the last few decades since AR5 (IPCC, 2014), and the recent AR6 (IPCC, 2023), attempting to forecast how wave characteristics are likely to alter with climate change. Here, we assess how the literature (mostly since IPCC AR5 and the recent AR6) has influenced our understanding of global wave climate.

Some global wave climate studies have focussed on changes to significant wave height as a measure of changing wave climate (the average height from trough to crest, of the highest-one third of the waves (H_s)), for extreme wave climate and average wave climate, including return values. Others have characterised how wave energy flux or wave power (the rate at which wave energy is carried by moving waves) and the intensity, frequency and tracks of tropical cyclones have varied over the last few decades and how they are projected to vary in the future. The IPCC reports provide recommendations in terms of large-scale global adaptation and resilience for a wide range of climate related risks across a variety of different sectors. The IPCC AR6 also provides recommendations for the transport and shipping sector (Jaramillo *et al.*, 2022), which we refer to here, as no recommendations are made for the Defence sector.

The latest Intergovernmental Panel on Climate Change, Physical Science Basis Sixth Assessment Report highlighted how anthropogenic climate change has altered a range of observed oceanic variables over the last 30-40 years (IPCC, 2023). In the context of intensity of extreme wave climate changes, extreme wave heights have been found to have increased in the Southern and North Atlantic Oceans by around 1.0 $cmyr^{-1}$ and 0.8 $cmyr^{-1}$ (Young and Ribal 2019) since the mid- 1980's (Collins *et al.*, 2019). It was found with high confidence that observed extreme sea level events associated with tropical cyclones has increased the intensity of multiple extreme events and associated cascading impacts. The report found with low confidence that a poleward migration of maximum tropical cyclone intensity in the North Pacific has been related directly to anthropogenically-forced tropical expansion. Under RCP 8.5 (Representative Concentration Pathway high emissions scenario) the report projects, with high confidence, an increase in significant wave height (H_s) across the Southern Ocean, tropical eastern Pacific and Baltic Sea (with medium confidence), yet a decrease in the North Atlantic and Mediterranean Sea. A global and regional scale wind-wave review study (Morim *et al.*, 2018) found there was a general consensus across 91 different sources, that H_s would increase over the Southern Ocean, tropical eastern Pacific and Baltic Sea and decrease in the North Atlantic and Mediterranean. Sea ice loss was also found to lead to a positive trend in significant wave height in the Arctic between 1994-2014 (Stopa *et al.*, 2016; Thompson *et al.*, 2016). Overall, there is general agreement in the literature on how significant wave height is projected to change with climate change.

1.3 (a) Mean and significant wave height

There are clear regional dependencies of annual averaged and extreme wave hight changes from present to future (Mori *et al.,* 2010). It is anticipated that generally, the Southern Ocean, Eastern Tropical Pacific and Arctic Oceans will face the greatest changes in wave height. Mean wave height is projected to increase by 1-3 cmyr^{-1} in the Southern and Arctic Oceans with extremes increasing by >10 cmyr^{-1} in the Arctic (Casas-Prat *et al,* 2024; Jaramillo *et al.,* 2022; Morim *et al.,* 2018). By 2100, the mean wave height is projected to rise by 5-10% in the Southern Ocean and Eastern Tropical Pacific and by >100% in the Arctic Ocean. Wave heights are projected to increase at mid-latitudes and in the Antarctic Ocean, decreasing at the equator (Mori *et al.,* 2010). There is also overall agreement in the literature that reductions in mean wave height are projected, by up to 10% in North Atlantic and North Pacific with regional variability and uncertainty (Casas-Prat *et al.,* 2024). Analysing changes in H_s return values compared with annual mean H_s values, enables a view on how extreme waves undergo changes that may be linked to climate change (Lobeto *et al.,* 2021). Return values for H_s reflect the same pattern globally. Under the RCP8.5 scenario, the highest increase in extreme H_s is projected in the Southern Ocean with local increases in over 2 m regardless of return period. In the tropical North Pacific, a decrease in extreme H_s with local decreases of 1.5 m. In the SE Indian, NW Atlantic and NE Pacific Oceans and increase in extreme H_s return values, however a decrease in annual mean H_s was also projected (Lobeto *et al.,* 2021). A detailed study (Mori *et al.,* 2010) predicting future wave climate based on the atmospheric general circulation model and global wave model focuses on present and future climate for both average and extreme wave conditions. Exploring smaller scale regional patterns, they found that wave heights are likely to increase at mid-latitudes and in the Antarctic Ocean, decreasing at the equator. They also project that using the normalised difference between the future and present climate (future minus present divided by present), the future mean *Hs* for the Pacific Ocean will decrease along the equator by 7%, corresponding to a 0.1 m decrease in average *Hs.* In the latitude range of 30˚N – 40˚N in the North Pacific and North Atlantic will decrease by 7%, corresponding to a 0.15 m decrease in average *Hs*. Conversely, in the Antarctic, there is likely to be an increase in average *Hs* by 6-9%, corresponding to a 0.2 m increase (Figure 2). The results showed a latitudinal dependence in the future daily wave climate, where mid-latitudes will experience lower mean wave heights and higher wave heights at higher latitudes.

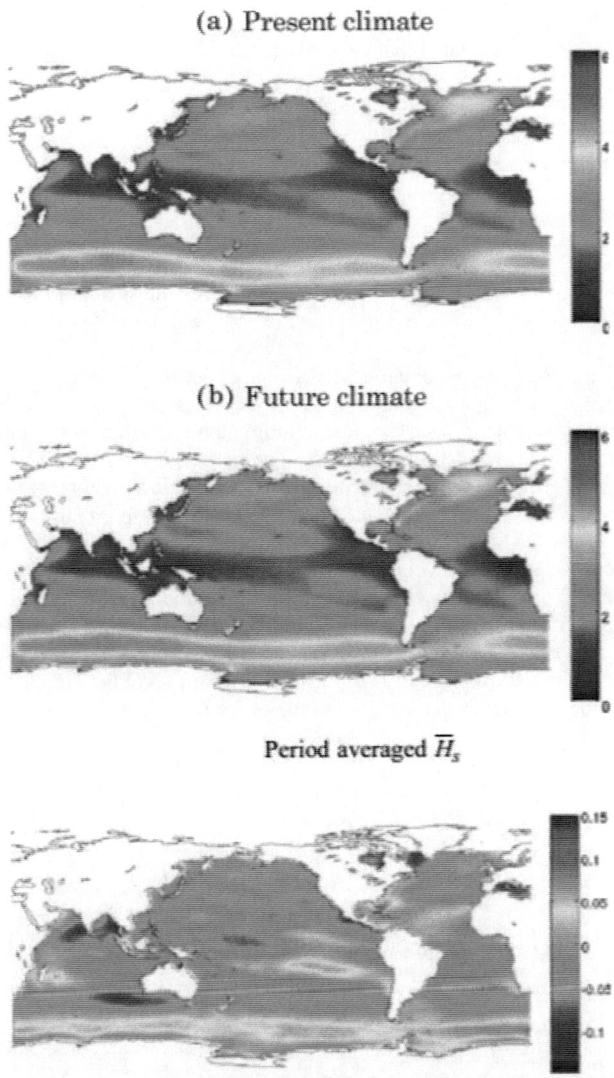

Figure 2: Present climate H_s *(m)*, Future Climate H_s *(m)*, and the difference of the mean H_s *(m)* between future minus present climate normalised by present climate (Mori *et al.,* (2010)).

1.3 (b) Wave energy Flux/ Wave Power

Using wave power or wave energy flux, as opposed to significant wave height alone, has been suggested as being a more appropriate means of characterising long term wave behaviour (Reguero *et al.,* 2019) as it measures the transport of energy by air-sea exchanges and therefore takes into account factors that employ wave motion, for example sea surface temperature. Wave power is a function of the square of the wave height (eqn 1), therefore a 1% annual increase in significant wave height will translate to a 2% increase in wave power per year. For example, multi-model ensemble of wave climate projections indicates changes in mean H_s of 7.1% globally, 10% in the Southern Ocean, which equates to an increase in wave power of 20%. Another example, using W10 (wind speed measured at 10 m height above surface) an observed 0.54% annual increase in upper ocean warming over the past 70 years has led to a 0.5 m increase in H_s causing a boost in wave power by 32.8 kWm^{-1} (Chen, 2024). This implies if current warming trends continue, this will lead to greater energy levels in the world's oceans.

$$P = EC_n \qquad \textit{eqn 1}$$

where wave energy E is given by

$$E = \frac{1}{8}\rho g H^2 \qquad \textit{eqn 2}$$

And wave group velocity (C_g) is related to the speed of individual waves C_n by

$$C_g = C_n \qquad \textit{eqn 3}$$

Where the parameter n increases from 0.5 to 1 in deep to shallow water and is given by

$$n = \frac{1}{2}\left[1 + \frac{2kh}{\sinh{(2kh)}}\right] \qquad \textit{eqn 4}$$

The parameter n describes how deep-water waves travel at twice the speed of the wave group ($n=0.5$) and shallow water waves propagate at the same speed as the wave group ($n=1$) (Masselink and Hughes, 2003).

As wave power measures the transport of energy (EC_n), upper ocean warming: a consequence of anthropogenic global warming is creating waves with greater power. It is therefore a valuable climate change indicator, and the long-term signal of systematic sea surface temperature increase behaves similarly to the increase in global wave power, yet a global time series of such a valuable variable is undetermined (Reguero *et al.,* 2019). Wave climate projections for the end of century indicate that sea surface temperature warming in the tropical Pacific will lead to increases in wave heights and wave energy levels at mid-high latitudes over Southern Ocean and Central North Pacific as SAM ENSO and NAO intensify. Measuring wave energy flux also requires information on wave period and as wave periods have also been increasing over the period of 1985-2008, this will influence the measure of global wave power, which was found to increase globally by 0.4% per year over this period (Figure 3). Similar results were found by Chen (2024) where an increase of 0.54% annually in global wave power was found between a 70-year period between 1957-2020. Looking forward, this study projects a 30% increase in wave energy flux at a 100-year return level for the majority of coastal areas in the southern temperate zone and higher in the open ocean, by the end of the century.

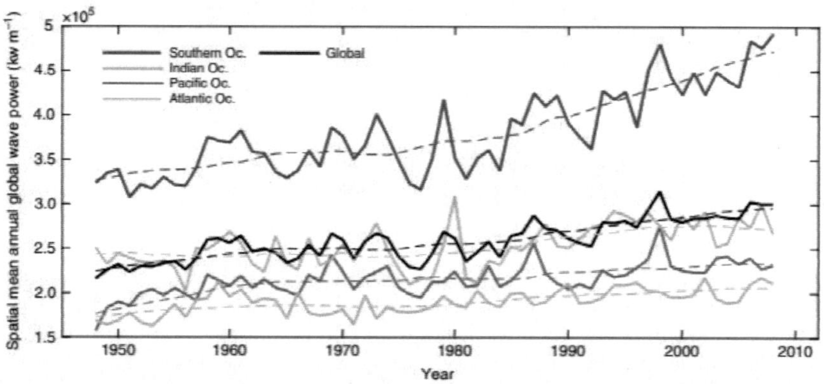

Figure 3: Spatial mean annual wave power (kwm^{-1}) calculated globally and by ocean basin. Dashed lines show 10-yr moving average, solid lines indicate each time series (Reguero *et al.,* 2019).

Using the longest reanalysis available 1940-2022 (ERA5) taken from multiplatform satellite altimeters, Lui *et al.,* (2024) evaluate the long-term variability of global wave power and the underlying mechanisms. They highlight, as others, a consistent increase in global wave power over this period, stemmed from the influence of global climate change and the intensification of Antarctic Oscillation events. This study highlights the increases projected in the southern hemisphere, where low pressure systems moving southwards are likely to generate an increase in global wave power 2.6 times larger than the global average increase in wave power. Interestingly, examining the wind-wave and swell contributions to wave power, they find the wind-waves are more important in the Southern Ocean than in low and mid-latitude regions, whereas the swell component of the wave power is less important.

1.3 (c) Tropical Cyclone Intensity and Frequency

As well as general global wave climate, variability in tropical storm tracks and intensity play a crucial role in an understanding of sustainable ship design and operability. Warming of the ocean surface by human causes is likely to lead to more powerful tropical cyclones (Knutsen *et al.,* 2021). Along with precipitation rates increasing with more moisture content in the atmosphere due to global warming, rapid intensification, the poleward migration of the latitude of maximum intensity and the slowing of forward motions of tropical cyclones have been observed. Although most climate models however project a decrease in the total number of tropical cyclones per year, the proportion of category 4 and 5 tropical cyclones is too likely to increase (Knutsen *et al.,* 2021). Under RCP 8.5, global atmospheric trends, which drive such changes were found to show that the northern hemisphere mid latitude jet has migrated polewards since the 1980's leading to poleward migration of TC's as well as poleward migration of storm tracks seen in the southern hemisphere over the last few decades. There is high likelihood that the mid latitude jet will continue to shift polewards and strengthen, leading to subsequent strengthening of storm tracks in the southern hemisphere by 2100. There has been a marked increase in the proportion of intense tropical cyclones (TC's) over the last 40 years and there is high confidence that this proportion is likely to continue to increase, yet the global number of TC's is likely to decrease or remain the same. The proportion of major TC's (Cat 3-5) and the frequency of rapid intensification has increased over the last 40 years. The average location of the peak TC wind intensity, which has consequences for wind-wave interaction, has likely migrated polewards in the western North Pacific since the 1940's. There is high confidence that as average TC peak wind speeds will increase with global warming, the peak wind speeds of the most intense TC's will increase, leading to growth of larger waves. (IPPC, 2023). In a first attempt to quantify the magnitude of expected changes in extreme wave heights generated by tropical cyclones, Grossman-Matheson *et al.,* (2024) explore how historical and future tropical cyclone tracks generate wind fields and use this to force a model to estimate global H_s as a result of tropical cyclones across all tropical cyclone basins. They estimate 1–100-year return periods for *Hs* for historical and future changes under SSP 585 high emissions scenario. They project that in the next 25 years (by 2050) there will be an increase in the 100 yr H_s by up to 1.5 m (10%) in the Eastern and Western Pacific, Southern Indian and South Pacific Ocean basins, yet no consistent trend of increase in the North Atlantic and a decrease in the North Indian Ocean, largely the Bay of Bengal (Figure 4).

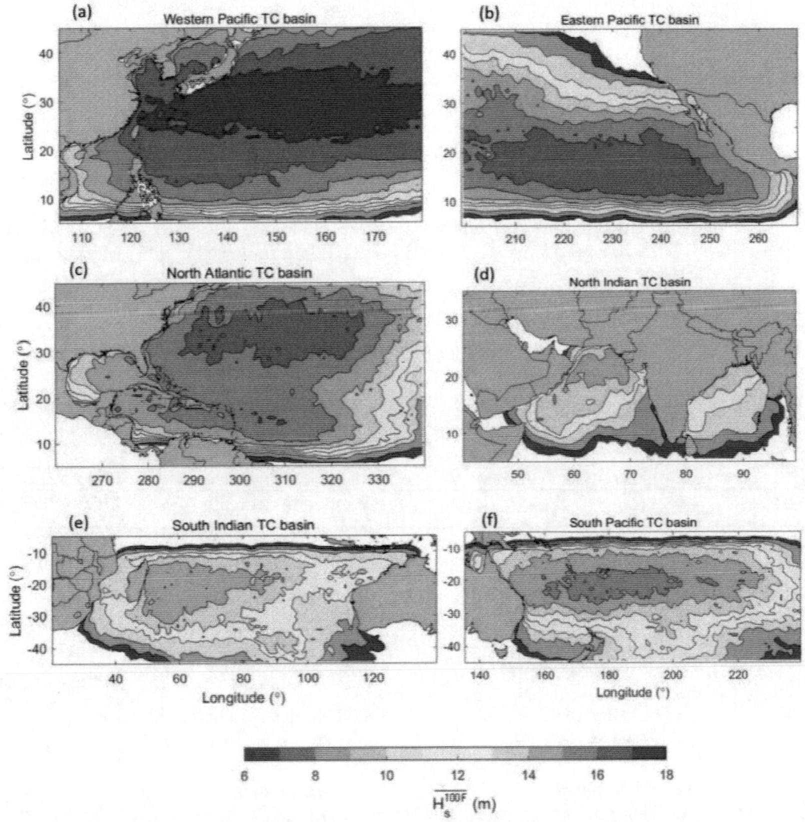

Figure 4: Contour values of projected future tropical cyclone ensemble median, 100-yr significant wave height (by the end of 2050 for SSP585) for each tropical cyclone basin (Grossmann-Matheson *et al.,* 2024).

2. DISCUSSION

Overall, there tends to be agreement within the literature that wave climate will increase and/or decrease in the same regions/oceanic basins (Table 3). Taking a global view on how wave climate is likely to alter under climate change naturally comes with a degree of uncertainty and generalisation. That said, it is understood, broadly speaking that average significant wave height and as a result, wave power are on the increase and will continue to increase in certain regions, i.e. Southern Ocean, Tropical and North Eastern Pacific and the Arctic Oceans. The degree of which varies regionally, but the greatest increase is likely to occur in the Southern Ocean (1-3 cmyr^{-1}). Conversely, it is agreed that a decrease in annual/average significant wave height is likely in the North Pacific, North Atlantic, Mediterranean Sea and Indian Ocean, however extreme significant wave heights in these regions will increase. Increases in extreme significant wave heights will be seen in all basins, apart from the North Pacific, where a decrease in 1.5 m is projected. The region facing the greatest extreme significant wave heights is again the Southern Ocean, where an increase in over 2 m is projected. Global wave power is likely to increase across all ocean basins with climate change, with the greatest increase 2.6 times more than any other basin, seen in the Southern Ocean.

Table 3. Summary of how wave climate is likely to change under climate change across different literature sources.

Oceanic Basin	Sea State Variables	Key findings	Reference
North Pacific	Future Mean H_s	Decrease closer to equator of 7% (0.1 m) 30°N-40°N Projected to decrease by up to 10 % / 1.5 m	Mori *et al.*, 2010; Casas-Prat *et al.*, 2024
	Extreme H_s	Projected to decrease by up to 1.5 m	Lobeto *et al.*, 2021
	North Eastern Pacific Extreme H_s	Projected to increase	Lobeto *et al.*, 2021
	Annual H_s	Projected to decrease	Lobeto *et al.*, 2021
South Pacific	Tropical Eastern Pacific H_s	Projected to increase By 2100 mean wave height projected to rise by 5-10%	IPCC; Morim *et al*, 2018; Mori *et al.*, 2010
North Atlantic	Extreme wave heights (over the last 30-40 years)	Increased by 0.8 cmyr^{-1}	Young and Ribal 2019; Collins *et al.*, 2019
	H_s	Projected to decrease by up to 10 %	IPCC; Morim *et al*, 2018; Casas-Prat *et al.*, 2024
	Extreme and Annual H_s	Projected to increase	Lobeto *et al.*, 2021
Mediterranean Sea	H_s	Projected to decrease	IPCC; Morim *et al*, 2018
Southern Ocean	Extreme wave heights over the last 30-40 yrs	Increased by 1 cmyr^{-1}	Young and Ribal 2019; Collins *et al.*, 2019
	H_s	Projected to increase	IPCC; Morim *et al*, 2018
	Mean wave height	Projected to increase by 1-3 cmyr^{-1} By 2100 mean wave height projected to rise by 5-10%	Casas-Prat *et al.*, 2024; Jaramillo *et al.*, 2022; Morim *et al.*, 2018; Mori *et al.*, 2010
	Extreme H_s	Projected to increase by over 2 m	Lobeto *et al.*, 2021
	Wave power	Increase in wave power since 1948 Projected to increase 2.6 times larger than average global wave power increase	Reguero *et al.*, 2019

Arctic	H_s	Projected to increase	IPCC; Morim *et al*, 2018
	Mean wave height	Projected to increase by 1-3 cmyr^{-1}	Casas-Prat *et al.*, 2024; Jaramillo *et al.*, 2022; Morim *et al.*, 2018
	Extreme mean wave height	Projected to increase by >10 cmyr^{-1} By 2100 mean wave height projected to rise by >100%	Mori *et al.*, 2010
Indian	South East Indian Ocean Extreme H_s	Projected to increase	Lobeto *et al.*, 2010
	Annual mean H_s	Projected to decrease	Lobeto *et al.*, 2010

Both wave height and wave power are of concern for a ship's structural stability and integrity, operations and manoeuvrability. As noted previously, NATO Standardisation Agreement No. 4194 (STANAG 4194) among others, set out guidelines for design and operation standards for vessels. These are based on sea states as a function of wave height, wind speed and wave period. As wave power incorporates parameters such as wave speed and wavelength, not only significant wave height, it accounts for wave period, which has been increasing in line with wave height and is projected to continue to do so. Wave power is therefore a key indicator of climate change and is likely to be the predominant influence on propulsion and manoeuvrability and therefore efficiency and resilience and has a greater influence on the long-term maintenance and lifespan of a vessel. In light of this, and the projected global increase in wave power, these authorities may need to, in the future, address the influence of not only wave height and period but wave power, to determine guidelines for ship design (hull design and integrity, propulsion and stabilisation systems) and operations. The bending of a ship's hull on a wave has over time become updated to our increased understanding of waves. With the potential for more powerful and more frequent extreme waves to occur in the future, it will therefore be necessary for the characteristics used in global strength to be revisited. This will fall most likely to the classification societies that publish such rules. It may also be necessary to consider the characteristics used in the fatigue life assessments, where changes in the wave period and potentially wave power may impact. As well as considering future designs by adapting the rules sets, where current vessels are expected to remain in service for significant periods with operations that expose them to these more extreme conditions, it may be necessary to re-evaluate their strength and fatigue lives to derive any further restrictions that may be necessary, if these changes are realised.

Tropical cyclones and extreme sea states play a crucial role in ship operations and planning, as well as feeding into the limitations of ship design and materials. From a stability assessment point of view, the assessments and particularly those conducted on naval vessels (i.e. deterministic rather than probabilistic) do not directly utilise wave data. However, the use of wave and wind heeling moments to assess some criteria indirectly includes characteristics which have a historical basis. Whilst the conditions considered are not the extreme, the potential for vessels to operate within higher sea state conditions on a more frequent basis would be a matter to consider. Globally, it is agreed that warming of the ocean surface as a result of anthropogenic causes is likely to cause more destructive and powerful storms. Global models however, demonstrate that the overall number of tropical cyclones is projected to reduce, yet the frequency and intensity of more powerful category 4 and 5 tropical cyclones are likely to increase. A general poleward migration of storm tracks, particularly in the Southern Hemisphere has been observed over the last 40 years and is likely to continue. In addition to this, the peak wind speeds associated with tropical cyclones in this region are projected to increase, leading to growth of larger waves. In light of this, as such extreme events will tend to evolve in the future, it will become vital to review structural design codes, for example NATO Standardisation Agreement No. 4194 (STANAG 4194), equipment, materials and redundancy. In terms of operational shifts, factors which will need reviewing in the near future include operational routing and planning, training protocols, intelligent systems and models for forecasting conditions to be able to deal with the potential for more severe events globally.

Sea state requirements for launch and recovery from Naval assets has led to the advancement of techniques such as the MoD/University of Exeter's Quiescent Period Prediction (QPP) program, where technologies are in development to use ship deployed wave sensors to measure the sea surface several hundred metres in advance of the vessel. These measurements can feed into short-term deterministic wave models which pre-propagate the wave field to the ship's location (Ferrier *et al.*, 2022). The development of such technology will become more and more vital to a ship's operations as wave climate potentially alters. In addition to *in situ,* short-term forecasting techniques, longer term predictions of extreme events and more robust quantification of wave climate projections will become crucial. That said, such predictions at regional scales are limited in the response of global climate models to greenhouse gas emissions and other forcing. Improving resolution of climate models may reduce some uncertainties as well as ensembles or much longer climate model runs to incorporate extreme events, and nonlinear changes (IPCC, 2023).

A study in 2021 by Losada *et al.,* exploring climate change related risks to ports worldwide identified that under a high-end warming scenario, Ports in the Pacific Islands, Caribbean Sea and Indian Ocean are likely to be at extremely high risk by 2100, whereas those in African and Arabian Peninsula are expected to experience very high risk. This however takes into account coastal risks associated with climate change in terms of sea-level rise and coastal flooding, not only wave overtopping and potential erosion. An investigation into the Marine Casualties and Incidents Reports issued by the IMO concluded that over 30% of the commercial shipping accidents from 2001 to 2010 were caused by poor weather, with an additional 25% being unexplained (Zhang and Li, 2017). As satellite earth observation technological advancements are made, along with global models and machine learning and AI approaches (Juan and Valdecanos, 2024), vast improvements in more accurate predictions of wave and weather patterns are in development. A recent study by Casas-Prat *et al.,* (2024) makes recommendations for resolving global scale climate change impacts on wave climate including improving modelling capabilities, enhancing spatio-temporal resolution and coverage of observations, for data products to be more homogenous and a better sampling of uncertainty with larger ensembles. It is also recommended that integration of real-time monitoring of weather conditions, wave heights, and ship performance with decision support systems are developed further, to provide actionable recommendations to the crew during extreme weather events. The importance of these processes becomes even more pronounced when considering that performance requirements vary by ship type. For example, combat vessels must be able to withstand environmental conditions without any decrease in the operational ability (Ashe *et al.,* 2006). Maintaining operational readiness in the face of climate change is a relatively new challenge but there has already been a shift in naval operations from the use of static design limits towards adaptive thresholds which may better capture changing environmental hazards.

3. CONCLUSIONS

In this paper we explore the current status of ship design and operability in relation to oceanographic and meteorological variables, highlighting the importance of integrating climate resilience into naval ship design. Following the latest IPCC AR6 (IPCC, 2023) report and subsequent scientific literature exploring the premise of an increasingly changeable global wave climate, we re-visit the current international military vessel design standards and specifications and operational criteria and provide a review of these in light of potentially increasing global wave heights, wave power and tropical cyclone intensity as a result of anthropogenically influenced climate change. Current design standards tend to account for extreme significant wave heights and periods according to the wave climate over the last 30-40 years. Looking towards the future however, it is understood that average significant wave height and wave power have been increasing and will continue to increase in certain regions, i.e. Southern Ocean, Tropical and North Eastern Pacific and the Arctic Oceans. The greatest increase is likely to occur in the Southern Ocean (1-3 $cmyr^{-1}$) where the greatest extreme significant wave heights is likely to increase by over 2 m and wave power will increase by 2.6 times more than the global wave power increase in any other ocean basin. Warming of the ocean surface as a result of anthropogenic induced climate change is likely to cause more destructive and powerful storms and although the overall number of tropical cyclones is projected to reduce, the frequency and intensity of more powerful category 4 and 5 tropical cyclones is likely to increase. Such storms are likely to experience a general poleward migration in terms of storm tracks, particularly in the Southern Hemisphere. In addition to this the peak wind speeds associated with tropical cyclones in this region are projected to increase, leading to growth of larger waves. The potential for such changes to global wave conditions in the future calls for further research and review of current naval ship design standards, codes and operational requirements. It is recommended that as climate change continues to affect maritime conditions into the future, the use of adaptive processes, enhanced by real-time data and innovative technology will be essential in ensuring that naval ships remain resilient.

4. REFERENCES

1. ARKIN, W M., and HANDLER, J., 'Neptune Paper No. 3: Naval Accidents 1945 - 1988.', *Washington, United States of America: Greenpeace/Institute for Policy Studies.* https://uploads.fas.org/2014/05/NavalAccidents1945-1988.pdf, 1989.

2. ASHBRIDGE, S., and BEARD, A., '*Greening Defence: Transforming the Royal Navy for a Climate-Affected world*', https://www.rusi.org/explore-our-research/publications/commentary/greening-defence-transforming-royal-navy-climate-affected-world, 2022.

3. ASHE, G., *et al.* 'Committee V.5 Naval Ship Design: Volume 2', *Southampton, United Kingdom of Great Britain and Northern Ireland: International Ship and Offshore Structures Congress*, 2006.

4. Australian Government (2024) Naval Shipbuilding and Sustainment Plan: Evolving the Enterprise. Accessed via 2024 Naval Shipbuilding and Sustainment Plan | About | Defence [ONLINE 2025].

5. BAČKALOV, I. *et al.,* 'Improvement of ship stability and safety in intact condition through operational measures: challenges and opportunities,' *Ocean Engineering*, 120, pp. 353–361. https://doi.org/10.1016/j.oceaneng.2016.02.011, 2016.

6. BALES, S L., 'Designing ships to the natural environment,' *Naval Engineers Journal*, 95(2), pp. 31–40. https://doi.org/10.1111/j.1559-3584.1983.tb00574.x, 1983.

7. BITNER-GREGERSEN, E M. *et al.,* 'Ship and offshore structure design in climate change perspective', *SpringerBriefs in climate studies*, https://doi.org/10.1007/978-3-642-34138-0, 2013.

8. BITNER-GREGERSEN, E M., *et al.,* 'Climate change and safe design of ship structures,' *Ocean Engineering*, 149, pp. 226–237. https://doi.org/10.1016/j.oceaneng.2017.12.023, 2018.

9. CASAS-PRAT, M., HEMER, M A., DODET, G., MORIM, J., WANG, X L., MORI, N., YOUNG, I., ERIKSON, L., KAMRANZAD, B., KUMAR, P., MENÉNDEZ, M., and Feng, Y., 'Wind-wave climate changes and their impacts', *Nature Reviews Earth & Environment*, vol. 5, no. 1, pp. 23-42. https://doi.org/10.1038/s43017-023-00502-0, 2024.

10. CHEN, W B., 'Analysing seven decades of global wave power trends: The impact of prolonged ocean warming', *Applied Energy*, 356, 122440, 2024.

11. COLLINS, M., SUTHERLAND, M., BOUWER, L., CHEONG, S M., FRÖLICHER, T., JACOT DES COMBES, H., KOLL ROXY, M., LOSADA, I., MCINNES, K., RATTER, B., RIVERA-ARRIAGA, E., SUSANTO, R D., SWINGEDOUW, D., and TIBIG, L., 'Extremes, Abrupt Changes and Managing Risk. In: IPCC Special Report on the Ocean and Cryosphere in a Changing Climate', [PÖRTNER, H O., ROBERTS, D C., MASSON-DELMOTTE, V., ZHAI, P., TIGNOR, M., POLOCZANSKA, E., MINTENBECK, K., ALEGRÍA, A., NICOLAI, M., OKEM, A., PETZOLD, J., RAMA, B., WEYER, N M., (eds.)]. *Cambridge University Press, Cambridge, UK and New York, NY, USA*, pp. 589–655. https://doi.org/10.1017/9781009157964.008, 2019.

12. Congressional Budget Office (2025) An analysis of the Navy's 2025 shipbuilding plan. Accessed via An Analysis of the Navy's 2025 Shipbuilding Plan | Congressional Budget Office [ONLINE 2025].

13. Defence Equipment and Support (2024) DE&S Safety and Environmental Protection Leaflet 18/2023 Accessed via 20241023 - S&EP Leaflet 18 2023 V2.1 - O.pdf [ONLINE 2025].

14. FERRIER, B., TAYLOR, R J., BELMONT, M R., CHRISTMAS, J., and FEDRICK, L., 'Resolving the Impact of Ship's design on deck and well-deck launch and recovery operations by applying of Quiescent Period Prediction (QPP)', *Warship 2022: Designing Warships for the Future,* 2022.

15. GROSSMANN-MATHESON, G., YOUNG, I R., and MEUCCI, A., *et al,* . 'Global changes in extreme tropical cyclone wave heights under projected future climate conditions.' *Sci Rep* ***14***, 31797 . https://doi.org/10.1038/s41598-024-82892-9, 2024.

16. IACS (2025) International Association of Classification Societies; IACS Members, Accessed via IACS Members - Shaping Maritime Safety & Sustainability - IACS [ONLINE 2025]

17. International Maritime Organization, *Sub-Committee on Ship Design and Construction (SDC 7), 3-7 February 2020*. https://www.imo.org/en/MediaCentre/MeetingSummaries/Pages/SDC-7.aspx, 2020.

18. International Naval Safety Association (2019) *ANEP-77 Part 1: Naval Ship Code: Goals, Functional Objectives and Performance Requirements, Edition G Version 3*. North Atlantic Treaty Organization.

19. IPCC, (2014): *Climate Change 2014: Synthesis Report. Contribution of Working Groups I, II and III to the Fifth Assessment Report of the Intergovernmental Panel on Climate Change* [Core Writing Team, R.K. Pachauri and L.A. Meyer (eds.)]. IPCC, Geneva, Switzerland, 151 pp.

20. IPCC (2023): Climate Change 2023: Synthesis Report. Contribution of Working Groups I, II and III to the Sixth Assessment Report of the Intergovernmental Panel on Climate Change [Core Writing Team, H. Lee and J. Romero (eds.)]. IPCC, Geneva, Switzerland, pp. 35-115, doi: 10.59327/IPCC/AR6-9789291691647.

21. JARAMILLO, P., KAHN RIBEIRO, S., NEWMAN, P., Dhar, S., DIEMUODEKE, O E., KAJINO, T., LEE, D S., NUGROHO, S B., OU, X., HAMMER STRØMMAN, A., WHITEHEAD, J., 'Transport. In IPCC, 2022: Climate Change 2022: Mitigation of Climate Change. Contribution of Working Group III to the Sixth Assessment Report of the Intergovernmental Panel on Climate Change', [SHUKLA, P R., SKEA, J., SLADE, R., AL KHOURDAJIE, A., VAN DIEMEN, R., MCCOLLUM, D., PATHAK, M., SOME, S., VYAS, P., FRADERA, R., BELKACEMI, M., HASIJA, A., LISBOA, G., LUZ, S., MALLEY, J., (eds.)]. *Cambridge University Press, Cambridge, UK and New York, NY, USA*. doi: 10.1017/9781009157926.012, 2022.

22. JUAN, N P., and VALDECANTOS, V N., 'A novel model for the study of future maritime climate using artificial neural networks and Monte Carlo simulations under the context of climate change', *Ocean Modelling,* 190, https://doi.org/10.1016/j.ocemod.2024.102384, 2024.

23. KNUTSON, T R., CHUNG, M V., VECCHI, G., SUN, J., HSIEH, T L., and SMITH, A J P., 'Science Brief Review: Climate change is probably increasing the intensity of tropical cyclones', *In: Critical Issues in Climate Change Science, edited by: Corinne Le Quéré, Peter Liss & Piers Forster.* https://doi.org/10.5281/zenodo.4570334, 2021.

24. Lloyd's Register (2024) *Rules and Regulations for the Classification of Naval Ships: LR-RU-004.*

25. LIU, J., LI, R., MEUCCI, A., and YOUNG, I., 'Increasing wave power due to global climate change and intensification of Antarctic Oscillation', *Applied Energy,* 358, 122572, 2024.

26. LOBETO, H., MENENDEZ, M., and LOSADA, I J., 'Future behaviour of wind-wave extremes due to climate change', *Nature: Scientific Reports*, 11:7869. https://doi.org/10.1038/s41598-021-86524-4, 2021.

27. MASSELINK, G., and HUGHES, M G., 'Introduction to coastal processes and geomorphology', *Hodder Education, London UK*, 2003.

28. MIHAILOV, M E., and ICHIM, M., 'Enhancing Romanian naval resilience to climate change: Advanced functional materials for green defence strategies,' *Journal of Environmental Science Studies*, 7(2), p. 17. https://doi.org/10.20849/jess.v7i2.1466, 2024.

29. Ministry of Defence (2021), Climate Change and Sustainability Strategic Approach, Accessed from Ministry of Defence Climate Change and Sustainability Strategic Approach - GOV.UK [ONLINE 2025].

30. MORI, N., YASUDA, T., MASE, H., TOM, T., and OKU, Y., 'Projection of extreme wave climate change under global warming', *Hydrological Research Letters* **4**, 15-19, 2010.

31. MORIM, J., HEMER, M., CARTWRIGHT, N., STRAUSS, D., and ANDUTTA, F., 'On the concordance of 21st century wind-wave climate projections', *Glob. Planet. Change,* 167, 160–171, doi:10.1016/j.gloplacha.2018.05.005, 2018.

32. North Atlantic Treaty Organization and Military Agency for Standardization (1983) *STANAG No. 4194: Standardized wave and wind environments and shipboards reporting of sea conditions*.

33. PETACCO, N., and GUALENI, P., 'The influence of ship stability in waves on naval vessel operational profiles,' *Journal of Ocean Engineering and Marine Energy*, 9(4), pp. 681–695. https://doi.org/10.1007/s40722-023-00291-0, 2023.

34. REGUERO, B G., LOSADA, I J., and MENDEZ, F J., 'A recent increase in global wave power as a consequence of oceanic warming', *Nature Communications*, 10:205, https://doi.org/10.1038/s41467-018-08066-0, 2019.

35. STOPA, J E., ARDHUIN, F., and GIRARD-ARDHUIN, F., 'Wave climate in the Arctic 1992–2014: seasonality and trends', *The Cryosphere*, 10(4), 1605–1629, doi:10.5194/tc-10-1605-2016, 2016.

36. THOMPSON *et al.*, 'Emerging trends in the sea state of the Beaufort and Chukchi Seas', *Ocean Modelling*, 105, 1-12, 2016.

37. UK Defence Standardization (2025), Guidance UK Defence Standardization, Accessed from UK Defence Standardization - GOV.UK [ONLINE 2025].

38. YOUNG, I R., and RIBAL, A., 'Multiplatform evaluation of global trends in wind speed and wave height', *Science*, 364(6440), 548, doi:10.1126/science.aav9527, 2019.

39. ZHANG, Z., and LI, X M., 'Global ship accidents and ocean swell-related sea states,' *Natural Hazards and Earth System Sciences*, 17(11), pp. 2041–2051. https://doi.org/10.5194/nhess-17-2041-2017, 2017.

6. AUTHOR BIOGRAPHIES

Dr Claire Earlie (lead author): Principal Consultant (Coastal) BMT Ltd

Claire is a coastal geomorphologist with a 14 years' experience in nearshore hydrodynamic and morphological measurement and modelling. As a field and remote sensing scientist, her experience lies in measuring and predicting how coastlines respond to oceanic forcing, particularly under extreme storm conditions, under rising sea levels. Prior to joining industry as a consultant, Claire was a Lecturer in Coastal Processes at Cardiff University and then went on to work at the NASA/Caltech Jet Propulsion Laboratory in Pasadena, CA, as a Sea Level Scientist, using in situ measurements and satellite Earth observations to provide technical assessments and advice on the coastal impacts of climate change to a global range of clients, authorities and agencies. She is now a Principal Consultant at BMT where her work focusses on the prediction and measurement of the influence of climate change and extreme oceanic conditions to the coastal, offshore energy and maritime industries.

Eliane Gebauer: Consultant (Flood Risk) BMT Ltd

Eliane is a consultant in the Flood Risk and Resilience Team at BMT, specialising in climate change science and policy. She has experience working with Lead Local Flood Authorities across various London boroughs. Eliane's expertise spans the development of flood alleviation schemes, TUFLOW modelling, strategic policy development, and stakeholder engagement.

Catherine Ingram: Principal Naval Architect BMT Ltd

Cathy is a Principal Naval Architect with particular expertise in stability analysis for naval vessels, and hydrodynamic assessment of platforms. In the 20 years she has spent in the industry, she has worked on multiple naval and auxiliary vessel designs, as well as provided technical support to a broad range of in-service naval platforms.

Dr Daniel Boyce: Principal Consultant (Flood Risk) BMT Ltd

Daniel is an environmental scientist specialising in the science and information behind flood risk, water resource management and climate change. He has over 20 years' experience in academia, Environment Agency and consultancy. He leads the Flood Risk and Resilience team to deliver flood resilient community projects.

Tom Matthewson: Head of Environmental Services Capability BMT Ltd

Tom is an environmental scientist with over 30 years' experience and is the BMT technical authority for environmental and sustainability studies for marine and coastal projects. He leads a team of consultants across EIA and permitting, environmental management and sustainability, and flood and coastal risk serving the ports and maritime defence, oil and gas/ CCUS, and coastal and flood management sectors.

Andy Kimber: Chief Ship Design Naval Architect BMT Ltd

Andy is the Chief Naval Architect for Surface Ship Design at BMT. He has 35 years of experience as a Naval Architect in the defence industry, covering concept studies, design delivery and in-service support. He has held senior technical positions in the QEC design team, Tide Class (leading the design) and the Type 31 bid team in the UK and has been involved in projects for the US, New Zealand, Australia, Canada, Norway and Finland amongst others. He has also been involved in projects considering fleet mix balance, options for capability delivery and procurement approaches as well as in-service support and ship certification activities for the Royal Navy.

HOW IMPORTANT IS IT THAT STAKEHOLDERS SHOULD BETTER UNDERSTAND NAVAL SHIP DESIGN?

Professor D Andrews, University College London, UK

SUMMARY

It would seem that many of the important decision makers and influencers involved in the design and acquisition of complex naval vessels have a poor appreciation of what is involved in designing naval ships and submarines. As a frequent author over several decades of papers to RINA Warship conferences and indeed of some extensive RINA journal/Transaction papers and beyond, I remain frustrated at the apparent lack of understanding as to what is the nature of the task and required skills that naval ship designers need to exhibit, right from commencing a complex warship project. This "ship design blindness" seems to apply even to other engineering disciplines involved but clearly is the case with naval operators and senior managers of the wider aspects of through life naval ship acquisition. This lack of understanding of the complex nature of such 'ship design' practice seems to be made worse by what appears, to this very experienced naval ship designer and more recently as an extensive writer on the topic, to be further compounded by such individuals' lack of appreciation. This applies to the extent such that this is not seen by most stakeholders as an impediment to achieving 'better ship design practice' – however one might seek to define this term.

A classic instance of this misunderstanding of the product we are designing is in the misuse of the term 'platform' and its juxtaposition with that of 'payload' in regard to naval ship design, whether such terminology has meaning for commercial transportation vessels or not. Furthermore, the term continues to be misapplied despite the author strenuously disabusing users of the term at warship conferences over more than the last decade. A second example was a description, by the Nobel Prize winner Herbert Simon, an early writer on artificial intelligence, who in writing on engineering design at the most sophisticated level used a naval ship design example. I consider his case was totally false: not just in the description he used but also how the actual 'design authority' for such complex vessels was and should continue to be exercised by ship designers.

The historic example just referred to, led the author to explore whether this misleading understanding of naval ship design imbued by "naval persons" (to use Churchill's description of himself to another "former naval person", namely President Roosevelt) can mean 'bad design choices' are made by influencers, which the practicing ship design fraternity (ourselves) seem to find difficult to counter. So why might this be? Is it because the 'influencers' are not prepared to defer or even listen to and then take on design informed opinions or is it (and even also) due to the ship designers not being confident or prepared or able to adequately explain, to those other NSD stakeholders, the technical implication of such key choices. I believe the issue is a combination of both parties being at fault and neither willing to change the joint mindset. Thus, my message below is intended for both sets of those collaborators involved in future NSD with sustainability and AI to the forefront.

1. INTRODUCTION

"He who imagined a ship, towers considerably above him who imagined innate ideas."

Voltaire

(quoted by Peter Noble and Mark Tipping in SNAME Marine Technology Jan 2022 "Boundary Pusher: Brunel's Great Ships")

"Strategy should govern the types of ships to be designed. Ship design, as dictated by strategy, should govern tactics. Tactics should govern the details of armament."

Admiral Jackie Fisher, 1904

Quoted by D K Brown (1991) (But see Section 6)

1.1 SHIPS MEAN SEA POWER

It is considered appropriate to address the (lack of) understanding of the nature of naval ship design (NSD) by considering why ship design is so important to naval warfare. This ought to be obvious but with the change from naval warfare being fleets of ships with guns to the complexity of modern naval warfare ranging across the air above, the sea (on and below) and amphibiosity from littoral to 1000s of nm inland with the full gamut of ISTAR (Gates, 2005), means sea power is so diverse, that one might (just) be forgiven in neglecting that naval ship design (NSD) is key to it. Rather than having a ship designer (or even a thinker as eminent as Voltaire) say so, I refer to a book by Admiral of the Fleet the Lord Hill-Norton, who was both First Sea Lord and then CDS in the height of the Cold War (Hill-Norton and Dekker, 1982), whose book is entitled "Sea Power". I have selected that publication because this topic is then considered in ten chapters, eight of which are the single word of a naval ship type, from Battleship to Submarine, with the exception of "Commando", which denotes a chapter on amphibious warfare – thus the key to Sea Power is essentially naval vessels. Thus, it would seem that exercising Sea Power is dependent on properly doing NSD – so why is there a problem in recognising its importance? The rest of this introductory section considers a series of issues that highlight why there does appear to be a problem in the stakeholder community giving sufficient recognition to understanding NSD.

1.2 THE VERDICT ON NSD FROM THE 1983 FALKLANDS CAMPAIGN

The definitive publication for popular consumption of this strange, late in the Cold War, short and limited exercise of sea power by a now second-rate naval force, but still considered a major power, fighting against a regional military junta armed with some modern weapons, was jointly authored by a well-respected popular military historian and a political journalist (Hastings and Jenkins, 1983). On page 229 having said "The British won the battle for San Carlos which was the decisive struggle of the war..." they concluded that Admiral Woodward (in command of the Task Force) had successfully asserted: "…command of the sea… (and so deserves) … credit (as do) his men." The authors then say:

"It was only the ship designers of the Royal Navy, who for all the excuses made for them by the heads of their service, can have derived little satisfaction from the Battle of San Carlos"

This unsubstantiated conclusion was quite uncalled for. It is clearly directed at the RCNC, who as civil servants could not directly respond and was made before the official enquiry (see Section 4.3) and ignores the then First Sea Lord (1SL), Admiral Leach's belief that four times the DD/FF losses subsequently incurred were possible in any opposed amphibious recovery of the Falklands Islands. On page 319 Hastings and Jenkins say: "Ship designers have been forcibly reminded (that) it is essential to distribute vital installations…" but these authors do not make the obvious inference: to distribute the many "vital installations" ships need to be bigger, which the other stakeholders - including shipbuilders – would immediately assume will be more expensive - at least – to build. This issue is discussed further in Section 5 regarding the Type 23 Frigates' design history. This Falklands' War assumption of the apparent inadequacies of "the ship designers of the Royal Navy" is a prime example of wider stakeholders showing woeful ignorance of nature and practice of NSD, which this current paper sets out to address.

1.3 THE "SHORT-FAT" SHIP SAGA

The second "case study" is that of the "short fat ship" UK controversy of the early 1980s, which I suspect many British naval officers, without the benefit of in depth naval architectural theory or subsequent practice in ship design, seemed to feel showed a resistance to innovation by the naval constructor fraternity. Leaving aside the clear evidence that historically the British naval record is one of innovative technical adoption rather than resistance to it (Brown, 2000), any novel proposal has to be assessed through a clear demonstration that such a proposal is technically justified and good value for money. So, any new ship design proposal needs to follow the scientific practice of being presented, preferably in peacetime in an open forum (such as the RINA Transactions), to a suitably broad range of informed professionals able to objectively assess its merits from the evidence presented.

The proponents of the "short fat frigate" proposal never presented their design configuration to the naval architecture discipline, as can be seen from the written discussion to the then Controller of the Navy, Admiral Sir Lindsay Bryson's paper on "The Procurement of a Warship" in the RINA Transactions (Bryson, 1984). That "short fat" proposal was presented in the Royal Navy's Naval Review, the internal journal for RN officers to write "provocative views", but that article was not in a form that the proposal could be technically assessed.

In fact, the hull form drew directly on existing small craft practice, namely the published UK National Physical Laboratory's (NPL) guidance on such (Marwood Bailey, 1969) planing forms. This hull-form is said in the NPL report to be valid only up to 200 tons displacement, but even then, not at typical frigate speeds applicable to such displacement vessels, which are typically greater than ten times heavier than the NPL limit. Elementary knowledge of the effect of Froude Number on a ship's wave making resistance would have ruled out such a form for frigate size vessels at their maximum speed. Furthermore, the dynamic stiffness of a frigate sized vessel, given such a wide beam and any practical weight distribution, would have been such that it is highly likely that this would produce a very stiff roll motion such that personnel, weapons and helicopters could not operate effectively from such a "fat" vessel.

The whole saga was more an example of the low status given to professional engineering in UK, certainly when compared to our continental *Ingenieur* equivalents, plus the manner in which "miracle cures" are grasped by peacetime democratic governments, when under fiscal pressure and when are they not? In this case it was doubly ironic that the first scientifically (if not technically) educated British Prime Minister fostered a government attitude that positively distrusted professional advice, especially if the experts were in public service. Furthermore, neither the Royal Navy nor the wider UK MoD system seemed to strongly support its own ship design professionals, when (as I observed from comments by my NATO colleagues at the time) no professional ship designer could understand that lack of official support for the professional engineering advice on this technically dubious proposal. This said more than a little about the nature of the UK's decline from its former position of being the leading industrial power throughout the Nineteenth Century. My comments on the short-fat saga were asked for and published in an invited chapter to the official centenary book on the R.N. Naval Review publication (Andrews, 2013), yet never challenged. A situation in marked contrast, I would hazard to assert, to that which would likely to have been the case for such an issue, and would have been strongly debated in the US Naval community, not least through the Proceedings of the Naval Institute and probably in the technical organs of SNAME.

1.4 POOR STAFF REQUIREMENTS AND THE WRONG INFERENCE

In his very insightful introductory chapter to his book "The World's Worst Warships" (Preston, 2002) the popular writer on naval ships, Antony Preston, pointed out: "There are more bad staff requirements than there are bad ship designs". This reveals how important it is to get the requirements right, otherwise ship designers can be fully technical competent, but will design the "wrong ship". In pointing this out to ship design students, I got the response: "so it's the naval staff's fault more than ours" (i.e., the designers), to which I responded: you don't understand my term Requirements Elucidation. (See detailed justification in Andrews (2011) and expounded with examples in Andrews (2018).) A poor set of requirements means the ship designer, despite having the ship design knowledge, has not fully explored the initial 'outline requirements' to see if they are coherent and where they can be improved to get a technical and operational balance, that is also coherent and, importantly, affordable. This should then be part of an ongoing dialogue with the naval staff. Such a sustained dialogue is difficult to conduct in a competitive design exercise should the eventual concept design, to match the emergent set of requirements, be produced by a competitive process, which has so often been the case since the turn of this new millennium, rather than by "the ship designers of the R.N." (to use Hastings and Jenkins term). This dialogue is one of the main reasons why the Concept Phase should be conducted or, at least, directly managed "in-house", so the government achieves the best possible concept/requirements mix, uninfluenced by commercial/contractual conflicts. Such conflicts rarely produce the right mix in elucidating what is essentially the "wicked problem" (Andrews, 2018), namely working out what is wanted (and shown to be realistic). Moves in recent decades to acquiring concepts based on just requirements driven specifications have led to another version of the wrong ship design. (See US Navy's assertion of Navy (i.e. NAVSEA) led ship acquisition (Winter, 2007).) So, my response to the students was poor staff requirements are as much the fault of the ship designers as of the naval staff.

1.5 SIMON's VIEW OF NSD "BY COMMITTEE"

Herbert Simon was a Nobel Prize winner no less and an early writer on artificial intelligence (and author of the design classic "The Science of the Artificial," (Simon, 1969)) who, in writing on engineering design at the most sophisticated level, decided to use a naval ship design example. He found a description as to how (in a post-WWI example) British naval ship design seemed to be practiced and used this as his key instance of what Simon called "design by committee" (Simon, 1973). I will show his case was totally false: not just in the description he used, but also how the actual 'design authority' for such complex vessels was and, largely, continues to be exercised.

Simon (1973) took as his "description of complex design processes that take place in organisations" that written by the British Admiralty's senior administrator (Murray, 1923) "of designing and producing a new battleship". He quotes Murray extensively (pp. 193 to 194) saying the design process: "…start(s) with the 1SL and his ACNS…" then the "naval constructors …formulate provisional schemes outlining the kind of ship desired together with forecasts of the size and cost involved by different arrangements", which of course is actual ship design. But Simon then goes on to list the various departments consulted (actually by the Director of Naval Construction (DNC)'s design team) only to conclude: "Some fourteen departments are concerned in the settlement of the final detailed arrangements." Simon's analysis (of a falsely implied 'collective endeavour') is that "Each small phase of the activity appears to be quite well structured, but the overall process meets none of the criteria we set down for WSPs." (WSPs is Simon's term for well-structured problems, which he invoked in contrast to "ill-structured" ones, as the basis of the main argument of this AI focused 1973 paper.)

However, as both an example of "design by committee' (i.e., Murray's 14 Admiralty directorates) and of the (1923 battleship) NSD process, this is totally misleading. Firstly, neither 1SL/the Board of Admiralty started the design process without the direct involvement of the Director of Naval Construction (DNC), given he sat with the Board (until the1950s) whenever technical issues were discussed (Brown, 1983). So, the other 13 named directorates (from Navigation to Torpedoes and Naval Equipment) do no more than provide (important) specialist advice – there is no corporate design team that Simon's term "Design by Committee" implies. Rather the actual design of the overall <u>ship</u> could only be <u>created</u> by DNC's 'Battleship Section' (i.e. the various design studies' "size and cost" to ensure a naval architecturally 'balanced' configuration). Furthermore, as Simon describes it: "An initial stage of laying down general (and tentative) specifications…" is not how the naval architects (and they alone) incorporate and coordinate specialist inputs to <u>their</u> ship design. (See Brown (1983) for detailed description of early-stage ship design by the RCNC over the century from their inception to 1983.) Finally, when DNC sought the Board's approval to proceed with acquisition, it was the DNC himself who presented <u>the</u> design with which <u>he</u> was satisfied to the Board and only after he had personally (for every RN ship design on his watch) physically signed off that design's Book of Calculations. This clear personal responsibility for producing each design is no longer so for R.N. ship (and submarine) designs and that loss of clarity is due to the rampant reorganisations in the centralised ("purple") Ministry of Defence, which has been the situation since it allowed the erosion of the DNC's dominant role from the late 1960s. In contrast, regardless of who actually produces each ship design, it is the Technical Director (and Chief Naval Architect) in NAVSEA (the responsible US Navy Command under the Chief of Naval Operations (CNO), who still signs off every US Navy ship design (Keane, 2018).

1.6 THE FALSE USAGE OF THE TERM PLATFORM

I have in several publications, and quite vociferously at most recent RINA Warship conferences, berated fellow authors when they have used the term "platform" to denote a naval vessel (surface ship or submarine). As I said in a major paper to a largely American naval design forum (Andrews, 2022), it seems that naval combat engineers (and their close engineering colleagues, systems engineers) have gone seriously wrong in applying that term to warships (putting them alongside aircraft, tanks, trucks and even motorbikes) and in so doing demonstrated poor systems practice. In contrast there are software designers who call themselves "systems architects", who have declared that they have followed naval architects with a systems architectural focus to designing (Maier, 1996)). While a naval combatant can provide a physical platform for complex vehicles/platforms, such as helicopters (see Farrell et al, 1972)) or UXVs, it is wrong to use the term "Platform" to describe a naval vessel as such. Furthermore, it is a whole system of systems and the further split of the 'platform' into 'payload' plus 'platform' (i.e. combat system plus the 'rest of the ship') is a very bad design and acquisition mind-set. It implies the former is good and must be maximised at the expense of the latter. (This might be justifiable in certain merchant ship design, such as containerships and bulkers, where the earning 'payload' is distinct from the core ship (i.e., hull, machinery and outfit), the same cannot be said for a highly integrated naval vessel, where much of the 'non-payload', including most of the support to the crew, is directly part of the fighting capability or contribute military capability.)

Firstly, the so-called "platform half" of the (overall) 'platform' (itself a tautological nonsense – and terms like IPMS perpetuate this falsehood) provides flotation/survivability and mobility, both of essential military worth. That 'ship half' may be some 50% in cost of acquisition terms, but is far more demanding of the total internal volume. The 'ship half" also provides the infrastructure for the personnel (who are largely 'fighting' the ship) together with the supporting ship services (such as electrical power and chilled water) without which (along with the bulk of the crew) the combat elements could not function. Thus, there is no unnecessary 'overhead' to be minimised as is implied by the false term 'platform'. This typical combat system and procurement policy mind-set has led to false justifications to reduce ship capabilities in vital features, such as those contributing to sustainability, survivability, supportability, adaptability, robustness and mission flexibility with through life cost consequences beyond the individual design. The example of the "margin-less and short life" Type 23 Frigate programme is discussed in Section 4 of Andrews (2018) and summarised here in Section 5, as are the consequence of the Type 42 'platform' inadequacies in the Falklands Campaign. In the latter design, as a result of minimising the so-called 'platform half', when the combat system elements failed to function and SHEFFIELD was hit, the inadequacies of this "cut price" design meant the ship could not be saved (Manley, 2015; Andrews, 2022). All of which shows the utter nonsense of this arbitrary sub-division, which amounts to not seeing the warship as a discrete military capability, uniquely bounded by its mobile envelope providing sustainability and survivability.

1.7 OUTLINE OF HOW THE ISSUE WILL BE ADDRESSED

Having, in the five examples above, given instances of why there appears to be such a poor appreciation of the nature of NSD, this paper now explores further this overall belief that NSD is hide bound and as C.S. Forester's (1942) quote (see Andrews, 2022) implies, it is the men onboard (as has been so historically, if no longer solely the case in many navies) that make up for poor ship design? This means we should enquire as to how this mindset has come about. Maybe it is we, as the designers, with whom the fault lies, in that everyone with an interest in NSD seems not to have been challenged or has acquired this general perception of the responsibility for 'bad ship designs' lies with 'the ship designers' (as Hastings and Jenkins brutally state above). However, I would argue the reality is more that: -

(a) (N)SD is a collaborative effort, both in crucial ESSD (primarily, but not solely the naval ship designers interacting with the naval staff) and then working up a design (ending with the final detail being rightly developed in the production focused shipyard). There may be 100 or so technical authorities and the other stakeholders who have a big say in specific ship designs, but still the execution lies with the naval architectural team for conceiving the overall ship concept and then maintaining each design's cohesion and technical balance right to the design's completion (see Andrews, 2022). Sometimes the operator's input could be fully supportive to the ship designers, as was Admiral Fisher with the 1905 Dreadnought, (Brown, 1983), sometimes operators and naval staff have significantly ignored ship design advice (see Andrews (2021) and discussion below in Section 5).

(b) The naval ship designers (as Systems Architects (Maier (1998); Andrews (2022)) have better appreciated than have most naval staff officers and administrators (like Murray)/ scientists/politicians and, certainly, than have historians and the media (such as Hastings), the challenge of bring together all the stakeholders and better explaining their impossible expectations. (See Rydill's (1984) comment on the RCNC's failing the RN for not producing the impossible: "better ships for less money".)

(c) There is a need to better understand what is unique about NSD. This has been seen by this author as his task in recent years, once immediately outside the RCNC (i.e., since 2000 at UCL), through writing on the nature of ship design, especially early-stage ship design (ESSD) (see Andrews, 2018).

This issue is expanded on in the rest of the paper by first considering how the naval culture's view of ship design, particularly considering the British naval experience of WWII, can be seen to have led to the Falklands verdict above and the other falsehoods on NSD, already flagged up in this introductory section.

2. NAVAL HISTORIANS' TYPICAL VIEW OF NAVAL SHIP DESIGN

2.1 INTRODUCTION TO APPENDIX A

In order to better understand the question posed in this paper's title, I have reviewed at Appendix A, the comments made by a range of eminent professional historians of naval history to see how they have addressed the performance of the Royal Navy and of its ships. I have particularly focused on the last world war, given it lasted almost six years and was fought over all the major seas and oceans and their crucial coastlines. Since WWII there have been some considerable exercises of sea power, to use Admiral Hill-Norton's term, but 1939-45 was the last time ships designed for the full intensity of naval warfare (with perhaps the exception of the strange and quite short campaign to recover the Falklands Islands in 1983) were extensively 'tested' to their full design intent. Given the question posed, it is believed the fraught relationship between naval ship designers and the principal stakeholders, strongly involved in (but not professionally practicing) naval ship design, to a significant degree has been influenced by the somewhat limited consideration of NSD by almost all the cohort of historians, who have addressed naval warfare. Some nine books (beyond the somewhat wider issues drawn on in the introductory section) have been reviewed to see, in general, how superficially NSD has been investigated, and thus whether the main stakeholders, thus 'educated', have been led to disregard the importance of NSD. Or is it that the wider stakeholders with often very significant involvement in NSD decision making, (like most naval historians) have all but disregarded on naval ship design issues that they should defer to the only profession educated and practicing the designing and through life technical responsibility for such complex vessels?

2.2 CONCLUSIONS ON HISTORIANS' VIEW OF NSD FROM CONSIDERING THE NAVAL ASPECT OF WWII

So, what can be concluded from the historians? Apart from the summary in Section A.10 (Rodger, 2024), both popular and academic naval historians have largely ignored or, worse, seem ignorant of the role played by the RCNC, in particular in WWII. Was it part of the nation's disappointment at the RN's collective "failure" to produce a decisive second Trafalgar in WW1, even if it did defeat the U-boats and enabled the Empire (at the price of ending the Empire's economic world dominance) to be the source of the Kaiser's defeat? It would seem that in WWII the RCNC performed very effectively (Brown, 1983), being led by the DNC of the time ((Goodall) with some inspired assistants (such as Baker's contribution to combined operations and what he, and other constructors, provided pre-D-Day to the amphibious assault and the Mulberry Harbours). The RCNC provided and then technically supported "the best fleet" they could without the industrial might then shown by America, once they joined forces post-Pearl Harbour (even if Anglo-American naval cooperation was at times blunted by USN Admiral King's animosity to the RN in his key position as CNO).

3. A NAVAL CONSTRUCTOR'S VIEW OF THE PERFORMANCE OF RN SHIPS IN WWII

Only Rodger (2024) of the academic and popular historians' books reviewed in Appendix A, largely addressing the naval campaigns of WWII, considered naval ship design alongside the specific performance of the British fleet. This had been essentially designed, procured and up-dated by Goodall and his RCNC colleagues. Rodger both acknowledges and copiously references the key two books covering WWII by my hydrodynamics lecturer, one time boss (1978-80) and late friend, David Brown (DKB), who was a former Vice President of the Institution and frequent contributor to RINA Warships (and RINA Transactions from 1980 to 2010).

3.1 BROWN'S HISTORY OF THE RCNC

The first of DKB's two detailed technical histories that covered WWII from the perspective of the RCNC, was written (while still in service so a few 'punches were pulled' with regard to getting government (security) clearance and to individuals still alive at that time) to mark the 100 years since the Order in Council setting up the RCNC as a Royal Corps to be: "…the Civilian Officers charged with the Shipbuilding duties in the Department of the Controller of the Navy, and in Your Majesty's Dockyards; …" (Brown, 1983).

Most of the chapters cover each successive Head of RCNC and Director of Naval Construction (DNC), effectively from Sir William White to the then Head in 1983, with four further chapters covering the RCNC activities beyond the main design role, including a final chapter on "Versatility". In that chapter Brown covers three main areas (I draw attention to this aspect since I believe it shows the real "systems integrator" nature of a designer of complex vessels.):

- Constructors on loan to other governments.
- Some unusual work (including airships and tanks in WWI and the acquisition of the Mulberry Harbour and PLUTO in WWII).
- Some Constructors' subsequent careers beyond UK MoD, including in academia and the maritime industry, home and abroad.

However, given this section focuses on responding to Section 2 (and Appendix A) on 'the judgement of naval historians to the performance of R.N. ships in WWII', this is covered in Chapter 6 of Brown's history (1983) which is a counter to Section 2 and Appendix A as an overview, with his later more detailed technical coverage of the run up to and then British ships' performance in WWII, and is addressed in his Section 3.2 (Brown,

2000). The centenary history's Chapter 6 covers Sir Stanley Goodall's tenure as DNC from 1936 to 1944 (and also his vital wartime role as Assistant Controller (Warship Production) 1942-45). DKB's view is that Goodall was "probably the most outstanding warship designer of all time" and he clearly dominates the 2000 book, which includes significant extracts from Goodall's official diaries (see Appendix 1 of Brown (2000) for British Library listing). The most significant of these extracts, some of which were also used by Rodger (2024), include a concluding paragraph on each year from 1937 to 44, plus nine points in his 1944 "farewell manifesto" which are still relevant, not least comments on how poorly Goodall felt the Corps efforts were recognised by government in general. Goodall took over an impossible situation following post WWI cuts, combined with an aging fleet in need of modernisation, "a depressed shipbuilding industry" and a "reduced ", under paid, under promoted RCNC, with cancelled designs they knew were badly needed. This immense re-armament effort was behind the saga of the design error in the Hunt Class small AA destroyers (see summary of DKB's detailed analysis in Andrews (2021)). Brown (2000) states 971 major warships were built during the War with 1700 requisitioned merchant ships and trawlers converted (and 300 US built ships also joined the R.N.) to show the measure of Goodall's and the RCNC's achievement.

3.2 "NELSON to VANGUARD: Warship Design and Development 1923-1945"

Brown (2000) covers the long run up from 1920s in a comprehensive technically detailed history of British WWII naval designs and their (and the RCNC) performance in WWII, with Goodall's clear dominance as DNC under the 3rd SL (firstly, Admiral Sir Reginald Henderson and then Admiral Sir Bruce Fraser) but also the contributions of some of the younger constructors, notably (the future Sir Rowland) Baker and his innovative landing ships and craft, including the LSD (a much better designation than the NATO one of LPD) (Brown, 1983; 2000).

After the post-WWI introductory chapter, DKB has eight chapters, one on each major ship type, from battleships to submarines (Goodall had persuaded his predecessor (Johns) to put him in charge of submarines as he felt his submarine knowledge was negligible) and "Miscellaneous Vessels", followed by three distinct chapters on: Modernisations, Updates and Scrapping; Wartime Damage; and Production and Repair. There are also 20 short appendices including: "Instructions for the DNC"; "Seakeeping" and "Damage to RN Armoured Hangar Carriers" (both latter aspects, DKB argues, were where RN ship designs were better than even their only competitive naval designs – those of the WWII US Navy). All these appendices are still well worth reading today.

3.3 BROWN's OTHER RELEVANT SHIP DESIGN PUBLICATIONS

DKB was not just the author of a set of historical books on naval ship designs from the pre-ironclads to beyond "NELSON to VANGUARD" (see Section 3.5), he also authored (sometimes with co-authors) a large number of technical papers on naval ship design and detailed technical issues, published in technical journals, including in RINA Transactions (all these are mentioned in Andrews (2010) 150 year review), where he was also an incisive contributor to published discussions; RINA Warship conferences and to US naval journals (many are referenced in his "future" book addressed in Section 3.5).

Three publications were particularly personal to DKB:

- DKB's comprehensive biography of William Froude (Brown, 2006), *pace* DKB's quote to us as his students: "Froude discovered 90% of naval architecture, and we have subsequently spent our efforts dealing with the remaining 10%";

- DKB's detailed article on "Sir Rowland Baker" (Brown, 1995) drawing on Baker's experience captured by DKB and recorded in both books covered in Sections 3.1 and 3.2 with regard to Baker's early career (see also Andrews (2022) covering Baker's view of (naval) ship design);

- DKB's late paper to RINA (Brown, 1993) on "History as a Design Tool" with an extensive discussion (and author's response) from DKB's many admirers, who lauded his unique combination of naval architectural professionalism and unmatched knowledge of ship design history.

So, the three opening remarks in the third of the above (Brown, 1993) are both good in highlighting the need to be cautious and in noting justifications. Thus: "the eternal verities" (e.g., the sea; ships' responses; explosive impacts; design team behaviours); "changes" (e.g., construction and new materials; much better habitability at sea and increased equipment sophistication); and "the whole truth" (e.g., identifying relevant facts (see

WWII Flower Class reputation); data needs to relate to the rapid rate of change; yet management of change is a lesson from history (particularly many WWII lessons for modern naval ship designers). DKB's many pithy insights and those of many (not all?) discussers to that paper are too extensive to summarise here – 'though his conclusion: "but often the verdict of history is correct". On that, I take DKB to mean the verdict from a professionally informed design and technical analysis, rather than most "historians'" limited conclusions on NSD outlined in Appendix A. One of DKB's conclusions is regarding "the paper mountain" and its "mindless destruction", which now has been replaced by reliance on biased electronic records and likely to be further distorted by AI foraging.

3.4 BROWN ON WHAT IS A GOOD DESIGN? AND WWII LESSONS

One of DKB's lessons from history (his Section 2.2 of Brown (1993)) is "What is a Good Design?" and is a justification for that lesson from history TRINA paper, but given the focus of this overall section is on British naval design in WWII, its challenging of the views of naval historians on that sustained conflict that is pertinent. So, I mainly want to consider the last chapter of Brown's (2000) WWII analysis, which in part I drew upon in my general presentation to a US NSD audience (Andrews, 2022), where I argued for a current and more nuanced view of good (general) ship design. That paper went beyond just NSD but was still limited to complex (service) vessels (see Andrews (2018) for this design distinction vis a vis ships which are part of a wider commodity transportation system (see Erichsen, 1998)).

Chapter 12 of Brown, (2000) is entitled "What is a Good Design?", although this could be seen as the issue behind the detailed technical discussions throughout that book and in particular with regard to itemising survivability under the extremes for those RCNC designs detailed in DKB's Chapter 10 (i.e., "Wartime Damage"). That catalogue can be seen as a clear test of a major aspect of "good NSD design" under the circumstances of six years of all out, oceans' wide warfare and as the ultimate test of a major naval power. In fact, this turned out to be its victorious, if extremely hard-fought, swan song for the R.N. as the world's most powerful and successful navy for the last two centuries preceding WWII.

Chapter 12 of Brown (2000) is divided up into five main sections: Foretelling the future; Quality versus quantity; Summary of specific problems; Good designs; and Fleet balance. All are pertinent to 'good NSD design' and the chapter should be read by all practicing ship designers (and all naval officers – if only). The specific problems DKB identified included: machinery (where USN was better served), guns (too many calibres), top speed (over emphasised at the expense of cruising speed), poor habitability, and seakeeping. On the latter DKB is critical but "most other navies had far worse problems" and he says this is the one aspect, when compared to USN ships, RN designs were better. On British ship design, DKB also mentions welded structures, where the shipbuilders were conservative and resistant and then he provides comments on specific designs: the KGV Battleships (he considered "under-rated") and submarines (both classes were "goodish"); armoured hangars in the Fleet Carriers ("magnificent technical feat but bigger hangars would have been better"– especially if the RAF pre-War hadn't prevented carrier aircraft development (nothing changes); all types of cruisers were largely well designed as were the several Emergency classes of destroyers; the Black Swan sloops he considers excellent and Loch Class frigates are given DKB's ship design "pride of place", including "clever and simple" structural design.

Andrews (2022) discusses DKB's conclusion on "good design", namely, this should be seen in overall terms:

"British Navy's ships' performance in World War II (where in comparison he considers only the US Navy performed comparatively well (sometimes better)) and concludes the (British) fleet balance (that word again) was "probably the best that was achievable within the UK's marine industries capability to deliver". Overall Brown remarks that the best *(ship)* design according to "some naval architects *(is the design)* that meets the *(formal)* requirement at a reasonable price", but Brown disputes that as being far too narrow a definition. He proposed that a better definition lay in that by the 1842 UK Parliamentary Committee on Naval Ship Design which ends: "...to endeavour to produce the best effects with given means". This to me, rightly, broadens the basis of what is "good" but sensibly leaves the issue qualitative, reflecting Rydill's (1966) NSD rebuttal of Systems Engineering and also this author's belief that ship design remains a subtle and sophisticated multi-facetted human process."

This summary of "good design" is another justification for both the stakeholders in NSD to better listen to naval ship designers and for the latter to see their role to be beyond providing just a simple ship synthesis and then correctly working up its naval architecture. Rather, designers must also be more strategic and take the lead, rather than being pushed into acquiescence on major design decisions, well beyond a narrow focus on ship safety - see Baker, whom I quoted extensively in Andrews (2022). Furthermore, on the design issues I would include those of the doyen of USN designers (Bob Keane (2018 and 2022)) who rightly identified the need in executing "good ship design" to also include consideration of the "Fuzzy half of (*naval*) Ship Design".

3.5 BROWN'S DESIGN LESSONS IN "THE FUTURE BRITISH SURFACE FLEET"

For those who might feel that a focus on WWII (with battleships, armoured fleet carriers and cruiser design issues to be of questionable current relevance) has little appropriate to 21st Century NSD, I would respond that my historical review in Appendix A and Section 3 remains worthwhile and pertinent to the paper's title question. However, I also want to show how DKB learnt well the lessons from WWII in his design proposals for a Future Fleet (Brown, 1991). That book gives DKB's perspective as he retired from the RCNC and very largely drew on his time as Head of Concept Design, when this paper's author was his lead Naval Architect (1978-80), and then in DKB's last appointment as Deputy Chief Naval Architect in the UK MoD. It can also be seen as a "future fleet" follow on to Brown and Moore ((2003), which provides design details of many post-WWII axed designs and the evolution of other designs, eventually developed and which formed the Cold War navy. (A less technically detailed but a more "blow by blow" series of recollections by members of the RCNC, each of whom led on specific built designs covering "30 Years of Naval Construction" (Betts, 2018), is given in Betts' up-dating of DKB's first 100 years history (1983). Betts edit of those recollections provide a more nuanced record alongside Brown and Moore (2003) and both publications can be seen as providing the basis for DKB's "Future Fleet" book. They also show how the "ship designers of the Royal Navy" (to use Hastings' 1983 Falklands designation) actually fought hard to achieve Brown's "best effects with given means" against what sometimes seemed to be a Kafkaesque undermining of their intent to provide the 'best for the R.N.' by the 'wider MoD system.

Brown's (1991) future British surface fleet (sub-titled "Options for medium-sized navies") has 11 chapters and 4 specific appendices, with many references as footnotes but lacks the excellent annotated bibliography of Brown (2000). There are some excellent artists' impressions and line drawings of "future ships" from our 1978-80 concepts. The early chapters on strategic overview and ship roles plus "resources and constraints" are followed by more technical chapters on: design philosophy; technical features; and cost; before tackling: the introduction of advanced technology; fleet mix; future (novel) warships; design and build; and ending with conclusions. The technical features comments emphasise survivability and seakeeping, where the latter (with less militarily sensitive analyses and trials data more readily publishable) shows how modern engineering analysis can provide a numerically justifiable demonstration that the hull designs of WWII escorts produced some pretty effective 'seakeepers' on relatively small hulls. On introducing advanced technology (where DKB was a long-term proponent of high speed and novel hull forms) at page 116, he lists nine issues agreed with the USN concerning the difficulties in introducing new technology into warships (see Appendix B), which are damming and add to the concern caught in this paper's title. Such that 'control' in introducing new technologies into something as sophisticated as a modern warship requires understanding the overall sense of "balance" (see Andrews, 2022) and requires exercising design control (Winter, 2007) that only ship designers can technically accomplish.

Summarising much that is highly relevant today from the 1991 book's conclusions, they range from there being "no perfect solution" due to the ever-rising expectations and commensurate cost impact (which ought to be properly based on cost of ownership not just questionably accounted initial cost, i.e., UPC) to the fact that, despite UK MoD trying act otherwise: 'Ships really are different'. The latter is something the navy seems poor at insisting is a fact, yet one that ship designers can readily demonstrate (Andrews, 2018) to the wider stakeholders (who too often are direct competitors, thus WWII animosity between RN and RAF remains as pernicious to this day in the 'purple centre'). "Design skills and experience are scarce and diminishing" (page 174 of Brown, 1991) when coupled with the woeful impact of MoD 'reforms' tracked by Brown (1983) and Betts (2018), yet again have been ignored by both wider stakeholders and, sadly, the R.N. itself (see Andrews, 2013). All this would ring true to Goodall (Brown, 1983, 2000).

What does this mean for understanding how little the other direct disciplines involved in ship design (RN officer corps and other design engineers) and the wider NSD stakeholders still seem to reveal that they have such a poor appreciation of the importance of NSD to Sea Power? This can then be coupled with their inability to recognise just how little they understand of the practice of whole vessel NSD. If this was acknowledged it ought to result in better credence being given to the views of the only true NSD practitioners – the ship designers?

4. COMMENTS ON NSD FOLLOWING THE 1983 FALKLANDS CAMPAIGN

Despite the fact that the previous two sections have emphasised NSD perceptions from the extensive WWII experience of the Royal Navy, it is also worth considering (given the direct quote at Section 1.2) the limited, but completely successful, naval campaign to liberate the Falkland Islands from Argentinian occupation in Spring of 1983. There are views some 40 years on from WWII of how a relatively more recent British fleet performed from a NSD perspective. After recapitulating Andrews (2022) analysis of the loss of the two destroyers and two frigates in the campaign, the other two British ship losses are discussed before subsequent sub-sections consider the various publications on that naval campaign, other than that by Hastings and Jenkins (1983) at Section 1.2, for insights into how their perception of NSD compares to those of the set of WWII, histories, when addressing the question that the paper's title poses.

4.1 SUMMARY OF THE RN FALKLANDS LOSSES

Table 1 reproduces Table 6 from Andrews (2022) which summarises the principal warships lost in the 1983 Campaign: the two T42 and two T21 escorts as the primary cases that Hastings and Jenkins (1983) invoked. Also lost to Argentinian air attack were the LSL Sir Galahad and the container ship Atlantic Conveyor. The LSLs were RFAs and as such did not have the resilience of fully capable warships, being designed in industry as back up logistics support to the fully NSD LPDs. The latter Ship Taken Up From Trade (STUFT) vessel was a vulnerable merchant ship (with unfortunate consequences for the troops once ashore given its Chinooks were lost with it). Brown's (1983) Postscript: "The Corps in the Falklands" spells out how quite a few RCNC members were directly involved in rapidly modifying the 60 STUFT vessels necessary to get the Marine Commandos and 5th Brigade ashore to recapture Stanley and end the occupation. Brown's various books show this was a classic RCNC task to design some military capability into STUFT ships very quickly, *pace* 1700 merchant ships in WWII. Yet again the message is if such "commercially designed hulls" are attacked then, as noted in Section A.3 for the WWII Battle of the Atlantic, merchant ships sink rapidly under lethal attack. So, for a NSD investigation Table 1 just covers the true warship losses.

Summarising the fuller consideration of these losses in Andrews (2022), firstly, the Type 42 design was conceived as "the smallest ship to take the Sea Dart AAD missile system to sea" (Purvis, 1978). That concept of AAD against Soviet air attack was inadequate against the Argentinian sea skimming Skyhawk launched Exocets and 1000 lb. bombs and so it could be argued that the MoD "ship designers" should not have produced what the Naval Staff asked for? The comparison with the design it replaced, the Type 82, of which only one was built: HMS BRISTOL, is stark. The Type 82 had decent margins for upgrade, whereas the T 42s proved very difficult and expensive to upgrade, subsequent (post 1983) air defence additions degrading the ships' overall performance. Also, the T 82 had 7 fire pumps, Type 42s just 4 and only one survived HMS SHEFFIELD being hit, so the ship burnt out and the consequences of fire-fighting induced flooding later sank the ship in the South Atlantic (Manley 2015).

Table 1. Summary of the four RN Combatants sunk in the Falklands Campaign

Date	Ship	How Sunk	Operational Deficiency	Design Deficiency
4th May 1982	SHEFFIELD Type 42 AAD Close escort to Carriers	Exocet ASM exploded amidships, 2 days later capsized from flooding	Designed against attack from high flying Soviet aircraft. Hit when under "radar silence/poor EW"	Lack of CIWS; Cheap ship design solution with inadequate survivability/insufficient redundant auxiliary power/fire pumps
21st May 1982	ARDENT Type 21 Light Frigate	Close naval screen for gun fire Two bombs aft Fire destroyed ship	Highly exposed to protect landing forces.	Lack of CIWS. Aluminium super structure burnt out. Limited survivability in ship design
24th May 1982	ANTELOPE Type 21 Light Frigate	Close naval screen for gun fire Two bombs in hull which exploded on diffusing	Highly exposed to protect landing forces.	Lack of CIWS. Diffusing inert bomb in hull led to catastrophic explosion
25th May 1982	COVENTRY Type 42 AAD Close escort to Carriers	Three air launched bombs hit along ship's length	Designed against high Soviet aircraft. Acting as gate keeper, turned to avoid attack but made worse	Lack of CIWS; Cheap design solution with inadequate survivability/insufficient redundancy but bigger ships might still be no better against this level of damage

The Type 21 was not a normal (for that time) in-house (RCNC) design but was acquired quickly to fill the gap between the very successful Leander Class frigates and the fully capable, PDMS equipped Type 22s. The Type 21s were modified from a commercially designed export family of corvette designs (Usher and Dorey, 1982) and lacked the margins to be re-armed mid-life (unlike the previous Leander Class). Despite the MoD stretching this Vosper's' initial design (Daniel, 2004), it still retained "unacceptable" features, such as the whole superstructure being aluminium yet fully structurally contributing (Brown & Moore, 2003). So, the two ships (out of 7 T21s in the campaign) were badly hit when stationary while supporting the troop landings in San Carlos Water: HMS ARDENT caught catastrophically on fire from several 1000–pound bombs and HMS ANTELOPE, having been hit with two unexploded bombs, one subsequently exploded on defusing causing a fire that detonated the main magazine. Probably any frigate size vessel would have succumbed to the damage sustained by ANTELOPE and COVENTRY, but for other two ships it could be argued that better designs might have survived.

Several of the older and larger County Class Destroyers and the Type12/Leander Class frigates were severely damaged but survived, possibly proving that "First Rate" designs (Brown & Andrews, 1980) more than justify that relatively small extra bit of capability in their ship designs. Too often such 'margins' have been denigrated as "Gold Plating" (a term of abuse used by some admirals and the finance authorities in MoD and Treasury that pushed the T 42 design and T 21 purchases), but no ship is ultimately survivable if unlucky, as was the last ship in Table 1 (see also Section 4.4).

4.2 FROM RAWSON'S MEMOIR

Rawson's (2006) autobiography as a senior RCNC member, covers the Falklands campaign briefly, when he was the Chief Naval Architect (CNA) at two-star level, taking much of the professional burden of an emasculated DNC (see pages 123-125 for his view of the RCNC contribution to that campaign). Rawson was already fighting the absurd issue of the "short-fat frigate" (see Section 1.3). On that, his memoir makes sober reading of the malfunctioning of most of the British government (MoD and beyond) and the media, which subsequently again abused Ken Rawson over his committee's criticism predicting the loss of two Ro-ro Ferries when he was chair of the RINA Safety Committee. In both the "short-fat" and Ro-ro ferry cases, the "establishment" showed its ignorance and contempt for engineering professionalism (and as such is highly relevant to the title of the current paper).

4.3 FROM MEEK'S MEMOIR

Marshall Meek's (2003) autobiography is by a former RINA President and "the father of the modern containership" (Andrews, 2010), and relevant to any NSD discussion of the Falklands campaign, in that as a senior (but non-warship designer) naval architect on the MoD's Defence Scientific Advisory Committee (DSAC): the independent body of experts to the UK MoD. Thus, Meek was asked to chair a post-Falklands DSAC working group on what he, in the midst of the "short-fat" DSAC involvement (see Part V of Meek (2003)), provocatively called: "What was wrong with our ships"). With all due respect to Marshall, whom I personally liked in our RINA dealings, this showed yet again from the phrasing of the question to Meek's team, the basic animosity of MoD to its own ship designers. This seems unchanging from the burden born by Reed (see HMS CAPTAIN in Andrews, 2021), Goodall and now Rawson, and has been perpetuated by many "wider stakeholders", such as Murray (1923) in Section 1.5, simply not understanding the proper nature of NSD. That the Falklands C-in-C, Admiral Sir John Fieldhouse, had already answered "We didn't find out anything that we didn't already know (about our ships performance) in the Campaign" was then, slighted by Meek's phrase reflecting almost exactly Forester's (1942) insight in "The Ship" that:

"...all combine to curse the designer... men on the winning side will take all the credit to themselves and the losers – such of them that survive – will blame the designer all over again."

Thus, Meek's comment: "That the naval personnel came out somewhat better than the ships." However, he did recognise: "...the RCNC would have wanted many of these improvements, *[which Meek flagged up, see his Page 215]* but were continually thwarted by cost restriction in their implementation."

4.4 CAPTAIN HART DYKE R.N. – THE REALITY OF NAVAL WARFARE

As Commanding Officer of HMS COVENTRY on picket duty outside San Carlos Water during the British Forces landing on the main Falklands Island, Hart Dyke (2007) vividly describes his ship being hit with three 1000 lbs bombs, two of which exploded and (as the 10 colour photos in his book vividly show) the ship rapidly sank - in 20 minutes. While the book is largely about Hart Dyke's clear case of PST, due to his deeply felt burden at the loss of 19 of his crew, from a NSD this showed that two such explosions and (see his figure on page 255) the flooding of seven central WTCs resulting in much of the necessary Reserve of Buoyancy (RoB) being lost, meant capsize was inevitable. It serves as an important reminder of "the Price of Admiralty" (Section A.1) but also the nonsense of Hastings and Jenkins (1983) criticism and that while increasing a ship's size (such as T82 vis a vis T42) can mitigate some weapon effects, any ship can be overwhelmed beyond its designed maximum flooded incident (see UK NES 109 (1990) and the usual warship design limit of any three WTCs being breeched). Whether, as with the quite different damage case of SHEFFIELD, the larger and more capable T82 might have survived such a major fire and flooding (as I suspect it could from SHEFFIELD's single Exocet hit) was a design example from which the Falklands Task Force was spared. To a naval ship designer it is sobering that a well-respected military historian as Hastings laid the blame for ships sunk in battle solely with the "ship designers", rather than the constraining remit to design "cheap and nasty" and Hastings typically and like Meek, echoed CS Forester's (1942) quote (in the last paragraph and spelt out in full at the head of Andrews, 2022).

4.5 O'BRIEN BOOK – A HISTORIAN's VIEW – LESSONS FOR NSD?

I want to conclude this section with reference to a post Falklands review of naval technology in the Twentieth Century and beyond (O'Brien, 2001), not to quote from my own contribution (as the author of the book's last chapter: Andrews, 2001), which has anyway been succeeded by my key 2018, 2022 and 2024 papers, in particular, but that by Till (2001) listing his "implications for the R.N.":

- Gaining and exploiting sea control;
- Maritime power projection;
- Expeditionary operations;
- Maritime diplomacy;
- Good order at sea: defined as: "securing maritime transport; securing maritime resources; Maritime sovereignty; Protecting the marine environment; and maintaining maritime requirements in good order.

While the first four of Till's implications are good strategic roles for a maritime power (and one the UK financial stagnation is currently putting in peril), the last is the basis behind any approach to NSD and should be behind answering the question posed by this paper's title. This then needs to be coupled with securing and maintaining the intellectual capability to exercise the practice of NSD. Before I state how that needs to be done (before it is lost) I want to pose, in the next section, a critique of the main R.N. Cold War naval ship design (at least in numeric terms), which was conceived, put into service and now reaching the end of its R.N. use - the Type 23 Class Frigates. This I believe shows how the attitude shown by the stakeholders in British NSD has ignored the wishes of the (1883) Victorian Order in Council for a "Civilian Corps" to be "charged with Shipbuilding duties" (see Section 3.1), which I take to be responsible for British NSD, as it was in its first century of existence. That is to be in a manner akin to that fairly recently spelt out by the then US Secretary of the Navy, Doug Winter, following several "hands-off" USN ship acquisitions, that the US Navy must reassert its 'control of NSD', expressed in Winter's "Getting Shipbuilding Right" (2007). Andrews (2022) highlighted Winter's three NSD recommendations:

Item 1. "... re-assert its control of the entire shipbuilding acquisition process"; (Winter goes on to say this covers the "selection of key trade-offs – performance, crew size, logistics support, cost and schedule".);

Item 2. "... the lead systems integrator should be the Navy (which means the USN appropriate command (NAVSEA) whose equivalent is the MoD ship project/IPT) – not the (individual project) contractor";

and (more of a pertinent detail) at Item 4. "... the Navy needs independent cost estimates for trade-offs/decisions that we make".

Such "taking back control" leaves the main detailed engineering task, rightly, with industry's resources (funded by the IPT) but then the 'Navy/MoD' needs the wider stakeholders to support provision and maintenance of the skills and experience of NSD by government. That direction and overall control and integration is currently withering away in the UK, when clearly our maritime capability most needs it.

5. POLITICAL PRAGMATISM or POOR DESIGN DECISIONS? (TYPE 23 CASE STUDY)

5.1 THE TYPE 23 FRIGATE

In several papers (Andrews, 2018, 2022, 2024) I have referred specifically to the Type 23 Frigate, for which I was the lead concept designer (under DKB as Head of the Concept Design Section in Forward Design Directorate led by Ken Rawson (see Section 4.2)). In my Table 1 of Andrews (2022) summarising the 18 submarine and ship design programmes in which I had (to varying degrees) a significant involvement in my 27 years as a constructor before becoming a "real" academic in 2000, I gave the key lesson from the (1978-80) Type 23 concept phase as: "Do not design to precise OA scenarios." Now the ISD of the First of Class was in 1989, being just as the Cold War ended on which the T23 Operational Analysis (OA) was totally predicated. As what became the main frigate class in the RN, the ships were used as General-Purpose combatants with a medium gun and PDMS, and only exercising its Towed Array (TA) and Merlin in its specific ASW role, when needed. Clearly the design lacked margins for necessary world-wide adaptation. Table 1 (of Andrews, 2022) is a good summary that I wish to expand upon here as a case study to answer my title's question and whether the design finally built was a "bad design" (in DKB's terms) or just the best possible outcome resulting from a series of pragmatic decisions/changes in a challenging time for a major navy in economical driven decline? Kana et al., (2016) argue that the issue with the USN Littoral Combatant Ship (LCS) stems from its "complex decision space landscape" but I wonder whether it is another excessive and thus inappropriate use of OA. In that case the obsession with high speed forced the adoption of extreme ship technologies that the later "LCS restructuring" (Warship Technology, 2016) has then tried to restore to a more appropriate naval combatant capability, while 'been saddled' with that obsession with top speed?

As my major publication on ESSD (Andrews, 2018) states: in the concept phase the major design decisions are made. In this case the need for a new T23 class resulted not from a new weapon system (as usually the case) but from a new sensor – the Towed Array (passive hydrophones) to detect Soviet submarine transiting the GRIUK gap, so there was clear role, the only problem was this was met (in concept) too precisely. It was easy for me and my small concept team to use the emerging OA on how the TA would be used, to develop the major ship design driver, that of the ship being ultra quiet, by accommodating a new type of ship propulsion system, CODLAG, for the ship to "sprint and (when deploying the TA) drift". This gave the two operational speeds and the requisite endurance on a small frigate hull form able to keep on station in those inhospitable seas throughout the year. A helicopter platform (sensible use of the term) was required to prosecute detected submarines, and the question of a garage provision or fitting a full maintenance hangar was another significant size (and initial cost) driver. Anyone familiar with my 2018 publication will realise this is not a good approach to ESSD. It was too narrow a role definition and lacked proper a Requirements Elucidation approach (Andrews, 2011; 2018), hence the above (Table 1 in Andrews, 2022) comment about the end of the Cold War and then stretching the ship size for the numerically main class of frigate in the fleet, which is what the T23s became as the post concept design was developed (see Bryson, 1984, Thomas & Easton, 1992).

There were a series of bad decisions made at "high stakeholders" level that compounded the problems the class had operating in the RN post-Cold War:

- Short life was stated to be 18 years, rather than the usual 25 to 30 years per ship. This was said to avoid a midlife extension and any capability updates. Of course, that high level intent was not held to, but it did mean the design has been more costly to keep going due to its detailed design 'cut long -life corners" as the "short life" style choice was given to the shipbuilder as a cost saving feature.
- No Board Margins, which if provided would have enabled any major enhancements through the actual design life of the class to be more readily accommodated and the design intent better maintained.
- Ships of the class were designed (note the hull form was established from the concept design) for cold North Atlantic weather rather than tropical operations, which became the norm (resulting in it being difficult to then incorporate appropriate air conditioning).
- Mandated (by 2ndSL) to only have accommodation for crew of 150 ('to keep ship small and save on Manning costs'). This proved totally stupid as when the first ship of 18 finally went to sea, the full weapon fit could not be manned adequately and all the accommodation margins (i.e., Training and Advancement) were then lost from this large part of the fleet. These margins then had to be put back in the next escort design. That was the Type 45 Destroyer, which then had to have 50 rather than 25 T&A billets, demonstrating why individual ship design choices ought to be considered as part of the fleet impact and not "sub-optimised" by "stand-alone" IPTs. (Interestingly another low margin design of the same time was the US FFG7 Frigate class, which also had accommodation limited (by a senior "stakeholder") to 150 and then "couldn't fight all its weapons" – so the UK cost pressures and consequences were not unique and also those ship designers found difficult to refuse.)
- OA in GRIUK gap did not see the need for a medium calibre gun – but of course it was essential for a GP frigate post-Cold War and that further stretched the required upper deck length to accommodate a post-Falklands medium gun. That ought to have led to a deeper hull girder (as can be seen in the T23's relatively low freeboard amidships) but again good ship design practice was compromised on the short-sighted imperative of holding to the initial cost ceiling estimate.

I would not criticise other decisions which were made in the "political environment" that DKB shows to have been part of the RCNC NSD world since 1883 (Brown, 1983) and is also typical of the "Fuzzy nature" of NSD (see Keane (2018) and Andrews (2022)). I have in mind: the adoption of a hybrid hull girder structure for the T23s to meet a shipbuilder suggested cost saving measure and the rejection of the Cellularity innovation (Gates, 1986) that could have helped accommodate ever changing electronic fits. (I have argued previously (Andrews 1984) that had my architecturally led synthesis approach (Andrews, 1984; 2018) been adopted earlier then, that would have shown this innovation could actually save space (and hence even initial cost) that 'the classic numeric synthesis' could not readily demonstrate, when the T23's concept's style was being set. I also would not criticise the decision by the then Controller (still held sensibly by an Executive Branch admiral (Vice Admiral Reffell)) to introduce the early T23s into service without their (still being developed) new combat management system (CMS). This was an example of a major technology step, going from big computer rooms with "dumb/slave' Ops Room terminals, to modern linked intelligent desk top computers. (A similar step change lay behind much of the problems with the WE system in the Collins Class SSKs (Yule and Woolner, 2008).)

A similar set of 'political design decisions' could be argued for several of the actual NSD programmes in Table 1 of Andrews (2022), such as HMS OCEAN, the two LPD(R), RFA ARGUS, SSNOZ, FSC/Type 26.

5.2 WHAT IS THE ESSENCE OF NSD (THAT WE SEEM TO HAVE KEPT TO OURSELVES)?

Briefly:

(a) A better focus on ESSD, which was particularly argued in Section 3 of Andrews (2018) at a strategic level, and recognising the human nature of the fuzzy NSD process captured in Figure 1 (spelt out in Andrews and Andrews, 2021).

(b) The ESSD consists of three overlapping steps (see Section 4 of Andrews, 2018) the first of which (Exploration of the solution space) is caught in Figure 8 of Andrews (2018).

(c) For each design (exploratory) option, the ship design team need to decide on: the "ship style" (Andrews, 2018); the design novelty level (Table 3 of Andrews (2018)) to select the 'right' ship synthesis approach; the extent to which it is necessary in 'Requirements Elucidation' terms to create a 'balanced concept solution' (i.e., in S^5, combat system and (if novel) main machinery choices).

(d) Explore sufficiently in ESSD how much needs to be architecturally varied to see the sensitivity to each key substyle choices (see Table 1 of Andrews, 2018) and apparent constraints (Table 2 of Andrews, 2018).

All of these aspects need to be explored for their capability and Through Life (TL) cost impact for the range of possible acquisition strategies. This includes how to conduct subsequent design phases: feasibility (MoD purple "Assessment" term is "ship process blind"); contract definition (to provide the Build Specification and a contractual cost); Detailed design and construction; TL exercise of Design Authority (see Andrews (2022) and the impact on the emergent and changeable top-level requirements (to be agreed with the 'requirements owners' and other major stakeholders).

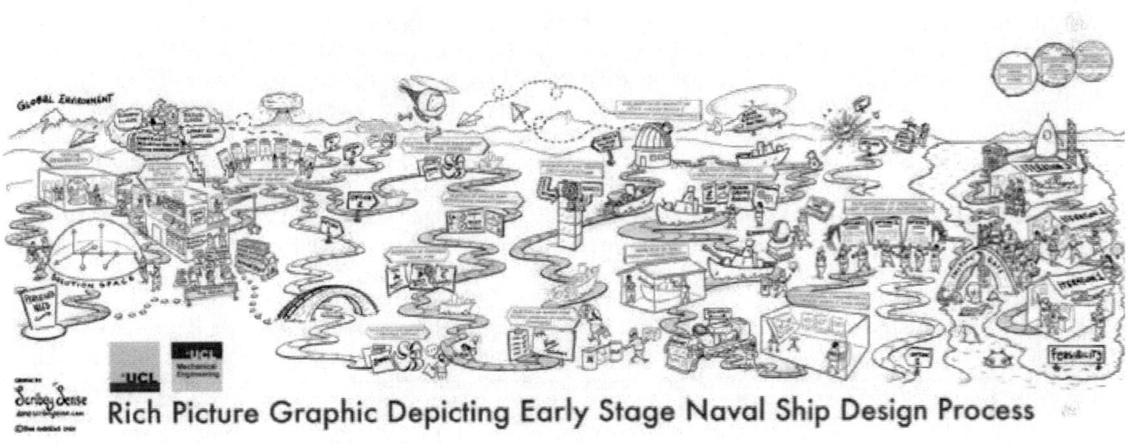

Figure 1: Rich Picture Graphic of a Typical Naval Ship Design Process (Andrews & Andrews 2021)

5.3 PRACTICING GOOD NAVAL SHIP DESIGN

All of the above leads to the following messages, that are necessary to cope with uncertainty in a geopolitical (questionably) sustainable and AI driven future:

- The essence of NSD is that warships should be designed for the unexpected (even the end of the Cold War showed a highly tuned (i.e., OA justified) design is quite inappropriate (and likely to turn out poor VFM. The latter requires to be properly costed – in terms of cost of ownership and not "Stakeholders" obsessed Unit Procurement Cost (UPC), which is a figment of government accounting rather than a good measure of value or even expenditure – *pace Oscar Wilde's jibe at Accountants.*
- A First-rate design (see Brown and Andrews (1982)) is invariably better VFM than a Second-rate one (albeit the latter might seem marginally cheaper in initial cost terms). Thus, sensible Board Margins (or in USN terms Advancement Margins (Gale, 1975)) have to be provided; and the ship's architecture designed from 'inside out' to be readily adaptable, to make the ship design is more resilient.
- Undertake a proper Concept Phase (Andrews, 2018) of Requirements Elucidating with the aim of solving the "wicked problem" (see also Kana et al., 2016) by the widest concept exploration of the solution space and questioning the balance of the emergent requirements, while noting this is hard to

do through a contractual design interface unless MoD exercises the 'control' spelt out by the former US Secretary of the Navy, (Winter, 2007).

- Stand up to senior stakeholders making "uninformed" detailed design inputs – at least argue for a concept design study to demonstrate the consequences of each of such "ideas". Taking Jackie Fisher's quote at the beginning of this paper, one could add (as he practiced): "Leave the ship design to the ship designers".

6. CONCLUSIONS ON WHY THE TASK OF NSD NEEDS TO BE MORE WIDELY UNDERSTOOD

The issue matters. Were historians (and their important readers) influenced by Simon's false view on such complex design processes; were designers often too obedient? (RCNC, like the Genie Maritime (now Corps de l'Armament) and USN NAs were trained and formed as a Civilian corps to able to stand up to Admirals, Permanent Secretaries and Politicians.) Haddon-Cave (2009) called for iconoclasts to resist bureaucracies. The RCNC started to lose its prime role from the late 1950s and into the 1960s when DNC as Head of RCNC stopped signing off every new design's Book of Calculations. These have now been lost with much else, such as "Ship's Covers" in dispersed CAD and database systems. Furthermore, there are the growing implications for design databases with AI and ML (Andrews, 2024).

My approach to ESSD and acknowledging the 'Fuzzy half' of NSD listed in Table 2 in Andrews (2024), better reflects what Concept Ship Design is and should be taught to NAs. Also, Marine Engineers and Combat Systems Engineers need to appreciate that the decision-making steps in the design of complex vessels (see Figure 4 of Andrews (2018)) are better described as the human designer journey of Figure 1. Furthermore, why isn't a simple tool, such as the "125 Exercise" that the Naval Staff Course attendees used to undertake, still used so they could appreciate the impact of requirements choices? Noting it was definitely not for them to believe that such a simple cost-capability model showed how a balanced ship synthesis is actually conducted for a multiplicity of types of ship options, as it could only show a simple sizing of a very un-innovative naval combatant. The need to understand a bit better the NSD process seems to have been lost in "rationalising" 'MoD Purple staff training'? Is this yet another issue to be sorted before we find ourselves at war?

NSD must be seen as a collaborative effort. That is the essence of Requirements Elucidation (Andrews 2018) and its subtlety is indicated by Figure 1 and needs to avoid being an exercise of mutual distrust between the ship designers and the naval staff plus other stakeholders and, apparently, the media/historians/the public. So those undertaking ship design (or less preferably, at least 'controlling' NSD (in Winter's (2007) terms) must be professional competent, seeing their primary role as designing vessels to win in conflict and ensure their safety (which means something different for ships of war than for merchant vessels).

CONCLUDING POINTS:

- Decision makers involved in NSD must listen to Naval Ship Designers who create ships and understand whole ship design, thereby ensuring ships are not just safe (Andrews, 2021) but also balanced and affordable (Andrews, 2022).
- Give technical authority to those whole ship design professionals who are competent. This should be a legal duty (as it is for Professional Engineers on the North American Continent).
- Ship and submarine safety must only be signed off by an actual designer. (See case of the then DG Ships (Admiral Spencer) and his sole naval architecturally qualified and experienced Project Director as outlined in Andrews (2021).
- Conduct ESSD properly as the key to NSD (see Andrews (2018), which the RINA Council (2022):

"considered to be essential reading for all naval architects and marine engineers, and not just those working in concept design."

So, I ask: Have you read it?

7. ACKNOWLEDGEMENTS

The views in this paper, when not referenced, are solely those of the author. The author would like to acknowledge Dr Hary Mukti for their assistance.

8. REFERENCES

1. ANDREWS, D J., 'Synthesis in Ship Design', *PhD Thesis, London University*, 1984.

2. ANDREWS, D J., 'Technology, Shipbuilding and Future Combat beyond 2020', *Chapter 15 in O'BRIEN*, 2001.

3. ANDREWS, D J., 'A 150 Years of Ship Design', *IJME, Vol. 152, Part A1, Discussion and Author's reply IJME Vol. 153 Part A1 2011*, 2010.

4. ANDREWS, D J., 'Marine Requirements Elucidation and the Nature of Preliminary Ship Design', *IJME Vol.153 Part A1*, 2011.

5. ANDREWS, D J., 'The View from Bath – a Naval Constructor's Perspective, Chapter 23 in From Dreadnought to Daring – Centenary of the Naval Review', *Edited by Captain Peter Hore*, January 2013, 2013.

6. ANDREWS, D J., 'The Sophistication of Early Stage Design for Complex Vessels', *Trans RINA IJME Special Edition*, Oct 2018.

7. ANDREWS, D J., 'Design Errors in Ship Design', *JMSE Special Edition on Maritime and Ship Design 2021, Vol. 9*, 34, 2021.

8. ANDREWS, D J., 'Babies, Bathwater and Balance – the Fuzzy half of Ship Design and Recognising its Importance', *SNAME Maritime Convention 2022, Houston, TX*, Sept 2022.

9. ANDREWS, D J., 'The Expanding Scope of Ship Design Practice', *Keynote for XVth IMDC, Amsterdam*, June 2024.

10. ANDREWS, D J., and ANDREWS, D A W., 'A New Way to Visualise the Design Process of Complex Vessels', *The Naval Architect*, Jan 2021.

11. ANDREWS, D., and BROWN, D K., 'Cheap Warships are not Simple', *Symposium on Ship Costs and Energy, New York*, Sept 1982, SNAME 1983. (Reprinted in Journal of Naval Engineering 1984), 1984.

12. BAKER, R., 'Ships of the Invasion Fleet', *Trans RINA, Vol. 99*, 1957.

13. BARNETT, C., 'The Swordbearers', *Hodder and Stoughton, Sevenoaks, Kent*, 1987.

14. BARNETT, C., 'Engage the Enemy more Closely: The Royal Navy in the Second World War', *Hodder and Stoughton, Sevenoaks, Kent*, 1991.

15. BELL, C M., 'Churchill and Sea Power', *OUP, Oxford*, 2013.

16. BETTS, C V., 'VANGUARD to DREADNOUGHT: 30 years of Naval Construction', *RINA, London*, 2018.

17. BROWN, D K., 'A Century of Naval Construction', *Conway Maritime Press, London*, 1983.

18. BROWN, D K., 'The Future British Surface Fleet: Options for Medium-Sized Navies', *Conway Maritime Press, London,* 1991.

19. BROWN, D K., 'History as a Design Tool', *Trans RINA, Vol.134*, 1993.

20. BROWN, D K., 'Sir Rowland Baker', *Warships*, 1995.

21. BROWN, D.K., 'The Grand Fleet: Warship Design and Development 1906-1922', *Chatham Publishing, London*, 1999.

22. BROWN, D K., 'Nelson to Vanguard - Warship Design and Development 1923-1945', *Seaforth Publishing, Barnsley, Yorkshire,* 2000.

23. BROWN, D K., 'The Way of the Ship in the Midst of the Sea: The Life and Work of William Froude', *Periscope Publishing, Penzance, Cornwall*, 2006.

24. BROWN, D K., and ANDREWS, D J., 'The Design of Cheap Warships', *Proc. of International Naval Technology Expo 80, Rotterdam, Netherlands*, 1980.

25. BROWN, D K. and MOORE, G., 'Rebuilding the Royal Navy: Warship Design since 1945', *Chatham Publishing, London*, 2003.

26. BRYSON, L S., 'The Procurement of a Warship', *Trans RINA, Vol.127*, 1984.

27. DANIEL, R J., 'The End of an Era: the Memoirs of a Naval Constructor', *Periscope Publishing, Penzance, Cornwall*, 2004.

28. DIMBLEBY, J., 'The Battle of the Atlantic; How the Allies Won the War', *Penguin Random House, London*, 2015.

29. ERICHSEN, S., 'Management of Marine Design', *Butterworths, London*, 1998.

30. FARRELL, K. et al., 'The DDH-280 Class Design', *Marine Technology*, Jan.1972.

31. FISHER, J., quoted by BROWN, D K., 'Future British Surface Fleet: Options for Medium-Sized Navies', *Conway Maritime Press, London*, 1991.

32. FORESTER, C S., 'The Ship', *Michael Joseph, London*, 1943.

33. GALE P., 'Margins in Naval Surface Ship Design', *US Naval Engineers Journal*, April 1975.

34. GANNON, M., 'Black May', *Aurum Press Limited, London*, 1998.

35. GATES, P J., 'Cellularity', *Trans RINA, Vol. 128*, 1986.

36. GATES, P J., 'Surface Warships: An Introduction to Design Principles', *Revised Edition, Brassey' Sea Power, Pavilion Books, London.* 1987 (Revised 2005), 2005.

37. GORDON, A., 'The Rules of the Game: Jutland and the British Naval Command', *John Murray, London*, 1986.

38. HADDON-CAVE, C., 'The Loss of RAF Nimrod XV230', *Final Report to UK Sec of State for Defence*, 28 Oct 2009.

39. HART DYKE, D., 'Four Weeks in May: The loss of HMS COVENTRY A Captain's Story', *Atlantic Books, London*, 2007.

40. HASTINGS, M., and JENKINS, S, 'The Battle for the Falklands', *Michael Joseph, London*, 1983.

41. HASTINGS, M., 'Operation Pedestal: The Fleet that Battled to Malta 1942', *William Collins, London*, 2021.

42. HILL-NORTON, P., and DEKKER, J., 'Sea Power', *Faber and Faber, London*, 1982.

43. KANA, A. et al., 'Why is Naval Design Decision-making so Difficult?', *RINA Warship 2016, June, Bath*, 2016.

44. KEANE, R G., Discussion to ANDREWS, 2018.

45. KEANE, R G., Discussion to ANDREWS, 2022.

46. KEEGAN, J., 'The Price of Admiralty', *Century Hutchinson, London*, 1988.

47. KENNEDY, P., 'Engineers of Victory', *Random House, New York*, 2013.

48. MAIER, M W., 'Architecting Principles for Systems-of-Systems', *Systems Engineering,* 1 (4), 1998.

49. MANLEY, D., 'The Loss of HMS SHEFFIELD: A technical Re-assessment', *RINA Warship 2015, Bath, UK*, June 2015.

50. MARWOOD, W J., and BAILEY, D., 'Design data for high speed displacement hulls of round bilge form', *National Physical Lab. Report No. 99, Teddington, UK*, 1969.

51. MAWDSLEY, E., 'The War for the Seas: A Maritime History of World War II', *Yale University Press, London*, 2019.

52. MEEK, M., 'There Go the Ships', *The Memoir Club, Spennymoor, Co Durham*, 2003.

53. MURRAY, O A R., 'The administration of a fighting service', *J. Public Admin*. July 1923.

54. O'BRIEN, P P. (Ed), 'Technology and Naval Combat in the Twentieth Century and Beyond', *Frank Cass Publishers, London*, 2001.

55. PRESTON, A., 'The World's Worst Warships', *Conway Maritime Press, London*, 2002.

56. PURVIS, K., 'RN Post-war Frigate and Destroyer Designs', *Trans RINA, Vol.116*, 1974.

57. RAWSON, K., 'Ever the Apprentice: the Memoirs of a Perpetual Learner', *The Memoir Club, Stanhope*, 2006.

58. RINA COUNCIL, '2019 Annual Report: The Royal Institution of Naval Architects', *RINA London*, Apr. 2020.

59. RODGER, N A M., 'The Price of Victory: A Naval History of Britain', *Penguin, London*, 2024.

60. RYDILL, L J., 'No, the future of Corps is not in Systems Engineering', *RCNC Journal, MoD, Bath*, 1978.

61. RYDILL, L J., Personal Communication to Author, 1984.

62. SIMON, H A., 'The Sciences of the Artificial', *M.I.T. Press, Cambridge, Mass*, 1969.

63. SIMON, H A., 'The Structure of Ill-structured Problems', *Artificial Intelligence, Vol.4*, 1973.

64. SPECTOR, R., 'At War At Sea: Sailors and Naval Warfare in the Twentieth Century', *Allen Lane, Penguin Press, London*, 2001.

65. THOMAS, T R., and EASTON, M S., 'The Type 23 Duke Class Frigate', *Trans RINA, Vol. 134*, 1992.

66. TILL, G., 'The Royal Navy in a New World 1990-2020', *Chapter 13 in O'BRIEN*, 2001.

67. UK NES 109, 'UK MoD Naval Engineering Standard - Surface Ship Stability Standards', *SSC, MoD, Abbey Wood, Bristol*, 1990.

68. USHER, P., and DOREY, A., 'A Family of Warship', *Trans RINA, Vol.123*, 1981.

69. Voltaire quoted in NOBLE, P., and TIPPING, M., 'Boundary Pusher: Brunel's Great Ships', *Marine Technology*, Jan 2022.

70. WARSHIP TECHNOLOGY, 'SASC leaders take aim at restructured LCS programme', March 2016.

71. WINTER, D C., 'Getting Shipbuilding Right', *US Naval Institute Proceedings*, June 2007.

8. APPENDIX A - NAVAL HISTORIANS' TYPICAL VIEWS OF NAVAL SHIP DESIGN

A.1 "THE PRICE OF ADMIRALTY"

This pithy title by the eminent military historian, John Keegan (1988) ought to say something about NSD but doesn't. Rather its most pertinent part is on the Battle of the Atlantic, which several key works on WWII acknowledge was the most important campaign of that nadir of modern civilisation. Keegan identifies that the Allies victory was due to greater resources and better organisation when compared to the German effort's lack of flexibility. Keegan's last chapter from a 1988 perspective focused on how the introduction of nuclear submarines changed the nature of the subsequent Cold War. However, Keegan didn't foresee the end of the Cold War – who did? Nor did he foresee the post-Cold War naval based power projection, exercised for the last three decades, particularly by the dominant US amphibious and carrier battlefleets. The sophisticated NSD developments on which this was all based should have been recognised.

A.2 "ENGINEERS OF VICTORY: The Problem Solvers who turned the Tide in the Second World War"

Paul Kennedy (2013) focuses on a few specific technological decisions in 1943/44 that swung things decisively in the Allies favour. Just considering his two maritime cases (i.e., "How to get Convoys safely across the Atlantic" and "How to Seize an Enemy-Held Shore"), again the Battle of the Atlantic was the crucial campaign and key to victory. (Even if Churchill was led astray by the RAF obsession that strategic bombing could win the war on its own and when a relatively small amount of reinforcing of Coastal Command's long range air cover could have shortened the war and directly saved many Merchant Navy lives.) Kennedy identifies much the same technologies and organisation outlined in in the more detailed Section 2.9, but not the effort put in by Admirals Backhouse, Henderson and Fraser (each in turn as Controller) supporting Goodall, as DNC from 1936 to 1944, and his RCNC team to restore the aging and under-resourced fleet for war and then enable the pursuit of 'sea power' (see Brown, 1983). The second of Kennedy's naval cases was D-Day where he focuses on the US Seabees (due to highly visible US Admiral Ben Morell) and General Percy Hobart's "Funnies" and not Rowland Baker's work for Mountbatten in Combined Operations (Baker, 1947) as well as other RCNC members on the Mulberry Harbour and PLUTO (Brown, 1983). Kennedy concludes with Churchill's summary that victory was due to "intelligent application of superior force" but no reference to DK Brown's substantial technical histories detailing the RCNC contribution (see Section 2.10 in contrast).

A.3 "AT WAR AT SEA: SAILORS AND NAVAL WARFARE IN THE TWENTIETH CENTURY"

American author (Spector, 2001) on 20th Century naval warfare with strong emphasis on WWII US Carriers: "Really battleworthy Capital Ships" (a view belied by DK Brown (2000) p.191: "the new capital ship, the carrier, was far more vulnerable … The battleship died because it had very little capability for damaging the enemy.") Some good statistics on the Battle of the Atlantic: 70% of merchant ships lost 1940-44 sank in under 15 minutes. 50% of 23 corvettes in under 10 minutes. Both statistics are sanguine messages for those obsessed with 'cheap commercial standards' and cheap cramped warships (see Type 22/Type 42 comparison in Andrews (2022).) Some interesting quotes from Spector that are worth debating further:

- "What appears most important is not the possession of the most advanced technology but in the skill in using the technology one has";
- "Navies appear conservative even reactionary" (belied by Spector's evidence) "navies ... are extremely adaptive" (a series of examples provided);
- Social structure of navies is characterised by: "The navy isn't what it used to be – and never was";
- "Civilian intervention is neither a necessary nor a sufficient condition for military adaption" (p.438)

(Does this apply to US and UK constructors or the wider ship design community, such as their political direction? Either way little actually provided by Spector on ship design compared to DK Brown's design analyses (1983, 2000), and no acknowledgement of designers, such as Goodall, Baker (in WWII) and for the USN: Keane et al., post WWII. (See major references in Andrews, 2022)

A.4 "THE WAR AT SEA: A MARITIME HISTORY OF WORLD WAR II"

Mawdsley's (2019), a British naval history of WWII bears comparison with Spector's American version with excellent reviews by other eminent historians: "rich in detail on tactics and technology that mattered", but technologically little on ship design and ship designers (no entry on Goodall as DNC). Some good strategic comments, such as those on the Battle of the Atlantic (BoA), which was 80% fought by RN and RCN) being more important than the war in the Pacific; the Mediterranean victory was essentially that of the RN, but at a high cost of men and ships and the scale of the fighting second only to BoA. All these historians seem to pull punches regarding the lack of decent carrier aircraft due pre-War control by the RAF or the RAF not providing enough long-range maritime patrol (for Coastal Command), both of which can be laid at the feet of the RAF command's obsession with strategic bombing?

A.5 "CHURCHILL AND SEA POWER"

Bell (2013) as a Canadian historian covers the whole of Churchill's involvement with the RN, with whom he enjoyed a longer and closer involvement than any other British politician, including of course his immense role for six years from the start of WWII. Clearly from a purely maritime perspective both over the "air gap" in the BoA and earlier lack of provision of decent carrier aircraft, Churchill's decisions were nuanced or before he was in his supreme "war lord" role, respectively. However, in his final (early 1950s) premiership he fought to sustain the big carrier capability as defence cuts bit ever deeper. Thus, Bell concludes that Churchill understood the navy's capabilities and limitations better than any other politician in the first half of the 20th Century: "The nation was fortunate he was so frequently and prominently involved in managing its naval affairs." It would have been better from a naval constructor's reading if Bell (like Item 2.10 due to Rodger (2024)) had read DK Brown's several key publications so he could have acknowledged the significant input to British sea power by both DNC's in WWI and II (Brown, 1983) and (undoubtably) Churchill's direct dealings with both these significant contributors to British 'Sea power'.

A.6 "ENGAGE THE ENEMY MORE CLOSELY: THE ROYAL NAVY IN THE SECOND WORLD WAR"

Barnett (1991), eminent historian and author of "The Swordbearers" (1966), which considered four key military leaders in WWI, including Jellicoe: "the one person who could have lost the war in one day" with Part 2 entitled "Sailor with a flawed cutlass" – 'though modern consensus is that the flaws lay in the rigidity of Edwardian class ridden society (Gordon, 1996) not the RCNC's design of the Grand Fleet (Brown, 1999). Barnett's detailed (900 pages) history starts with 1920s cuts and economic decline leaving the Navy seriously stretched as WWII started. Many official documents consulted but not Goodall's diaries (nor is Goodall mentioned) and no reference to Brown's technical histories on British NSD.

A.7 "THE BATTLE OF THE ATLANTIC: HOW THE ALLIES WON THE WAR"

Dimbleby (2015) produced the first of two popular histories considered here, this on the crucial Battle of the Atlantic, and with Item 2.9 a focus on the sailor experience, rather than more academic works and therefore possibly even more influential with 'wider stakeholders? No reference to Goodall or indeed A. W. Watson, the senior constructor responsible for many of the WWII escort designs (Brown, 1983). Thus, the ubiquitous Flower Class corvettes were conceived as coastal escorts but easily (and cheaply) produced in fishing boat yards ("perfectly safe but far too small to be effective in bad weather", such that sailors described them as "unmitigated hell" (Brown, 1983)). Dimbleby is good on the strategy (and Churchill delaying taking bombers from "Bomber" Harris to close the "air gap") and how the campaign against the U-boats got better, once the great admirals gained strategic command (e.g., Cunningham and Horton).

A.8 "OPERATIONAL PEDESTAL: THE FLEET THAT BATTLED TO MALTA 1942"

Hastings (2021) covers a specific crucial convoy supporting action 'to save Malta' given overwhelming Axes air control (and E boat/Italian MTBs, submarines and a significant Italian battle fleet in such a narrow stretch of water), when the tide had yet to turn in the Allies favour. Despite there being a sizeable RAF land-based force available they failed to support the naval force and even the most crucial 'straggler': the oil tanker Ohio 'that saved Malta'. Hastings says little on ship design but remarks twice on Churchill's view: "warships exist to fight, and if necessary to sink…".

A.9 "BLACK MAY: THE EPIC STORY OF THE ALLIES DEFEAT OF THE GERMAN U-BOATS IN MAY 1943"

American author's (Gannon, 1998) close-up look at the decisive convoy action that actually spelt out the U-boat defeat, such that Doenitz withdrew all 134 units then at sea (especially the 58 deployed in 4 wolf-packs across the NN Atlantic) following Convoy ONS5's defence sinking 6 U-boats for 13 merchant ships lost and additionally 7 more U-boats badly damaged and 11 more 'so badly handled' that they were immediately withdrawn. The RN led success was due to the escorts' "tenacity" using HF/DF to drive the U-boats to submerge and then using Asdic (now called Sonar), depth charges and the new "frightening" Hedgehog ahead mortars. But little is said about the (quickly designed) escort utility aircraft carriers' impact and also that of new, more seaworthy escort designs (than the early "cheap and nasty" Flower Class corvettes), which were coming into service by 1943 (See Brown, 2000).

A.10 "THE PRICE OF VICTORY"

Rodger's (2024) recent monumental history stands out and ought to be included, although not specifically limited to WWII (though he sees that as the culmination of the 1815 to 1945 RN dominance of the world's oceans). But also, Rodger recognises (unlike most of the above) the significant role the RCNC (with its strange British version of Colbert's Genie Maritime) played in achieving that "Price". This can be seen in that he references extensively more than a dozen of the technical books on the history of naval ship design authored by D K Brown. In particular, Rodger draws on "100 Years of Naval Construction" (Brown 1983) and on "Nelson to Vanguard" (Brown, 2000). Interestingly Rodger doesn't reference DKB's many RINA (technical) papers, which were largely on current ship design practice, so strictly post-WWII focused, but these drew extensively on the WWII experience. However, the two above DKB books are sufficient for Rodger to acknowledge the importance of naval ship design and the contribution by the RCNC (which for WWII is considered further in Section 3). The contribution of Sir Stanley Goodall, the Head of the RCNC and Director of Naval Construction from 1936 to 44, is fully acknowledged (unlike the other histories just considered before Item A.10). Thus, Rodger provides several direct quotes from the Goodall Diaries, including DKB analysis of the HUNT Class design error (summarised in Andrews (2021), however Rodger does not refer to DKB's (2000) detailed analysis of Britain's NSD WWII performance, which is summarised in Section 3 of the main text.

9. APPENDIX B – COMMON PROBLEMS IDENTIFIED WITH INTRODUCING NEW TECHNOLOGIES FROM DISCUSSIONS BETWEEN USN AND BRITISH NAVAL SHIP DESIGNERS (Page 116, Brown (1991))

- "Failure to make clear the vital contribution of (*ship*) technology to fighting value" (i.e., USN HME auxiliary systems)
- "Lack of proper investment appraisal" so 'cash-limited ships" (e.g., T23) prevent introduction
- "Insufficient R&D resources to complete and validate research" (see case of RV TRITON the 'prototype' Trimaran where the naval programme had no funding (Andrews, 2001)
- 'Insufficient time for development… if to properly integrate into a typical ship acquisition programme"
- "Lack of a coherent and integrated ship technology strategy" (Often budgets in conflict across bureaucratic boundaries.)
- "Lack of clear authority to introduce innovative technologies"
- "Wide-spread belief that ship technologies are fully developed." (A belief an investment appraisal would show to be fallacious.)
- "Contractual difficulties directing (once price agreed) that shipbuilder adopt new technology."
- "Basic resistance to new technology" – yet this can be battle winning. "

10. AUTHORS BIOGRAPHY

Professor David Andrews FREng, BSc(Eng), MSc, PhD, FRINA, FSNAME, FIMechE, FRSA, RCNC
David Andrews was made Professor of Engineering Design at University College London in September 2000, following his early retirement from a wide-ranging career in the UK Ministry of Defence, extensively involved in the design of a large number of naval vessels. He retired from the One Star post of Director of Frigates and Mine Countermeasures; prior to this he was the MoD Professor of Naval Architecture at UCL. Before that he was Assistant Director Future Design in the Future Projects (Naval) Directorate after being Warship Project

Manager for the procurement of the UK's Replacement Amphibious Shipping Programme. David Andrews is an acknowledged international authority on ship design methodology, with particular reference to naval ship design, including submarines and unconventional vessels, particularly Trimaran ship design. He has served RINA as Vice President, Editor-in-Chief and chaired several committees and many working groups and panels, as well as being on Council for twenty five years. From 1997 he edited all the International Marine Design Conferences State of the Art Reports on Marine Vehicle Design Methodology up to 2018 and from 2015 to 2022 was Chair of the IMDC International Committee. In 2000 he was elected a Fellow of the Royal Academy of Engineering for the developments he pioneered in ship design methods. His comprehensive and seminal paper on early-stage design of complex vessels was published by RINA in October 2018 and it was uniquely recommended for "Council Commendation" and that it also be issued to all young members of the Institution. In April 2020 the RINA Council approved his award of the William Froude Medal in recognition of his contribution to the field of ship design and having made a conspicuous contribution to naval architecture. On 29th October 2021 at the SNAME Maritime Convention in Providence, RI, he was awarded the David W Taylor Medal for his notable lifetime achievement in Naval Architecture and Marine Engineering.

SUPPORT TO CONCEPT ANALYSIS AND REQUIREMENT DEVELOPMENT THROUGH WARGAMING

Professor D Manley, CEng, FRINA, RCNC, and **Dr N Bradbeer**, CEng, MRINA, RCNC, University College London, UK

SUMMARY

Evaluation of potential solution spaces for novel concepts and the application case for new technologies can prove daunting, especially in the early stages of a project where up front funding for operations analysis is short. The paper discusses the application of manual wargaming techniques as an aid to this work, serving as a precursor to more focussed OA as a project develops, enabling an early evaluation of a "system of systems" approach and informing the development of system, platform and force requirements. The processes involved are illustrated by examples of recent work at UCL covering seabed security and the protection of critical undersea infrastructure, application of autonomous systems to anti submarine warfare in congested environments and early-stage consideration of alternative future fleet composition

1. INTRODUCTION - WHAT IS WARGAMING?

Professional wargaming refers to a systematic and organized approach to simulating and analysing military, political, economic, and other strategic situations within a controlled setting. This practice is frequently employed by military institutions, government agencies, and various organizations to formulate and assess strategies, train personnel, and investigate possible outcomes of intricate scenarios. The term "wargame" encompasses a broad array of activities, making it nearly impossible to pinpoint a single, definitive definition. However, Peter Perla, a prominent figure in contemporary professional wargaming, offers a definition that is widely acknowledged, if not universally accepted, in the field [1]:

> *"A wargame is a warfare model or simulation whose operation does not involve the activities of actual military forces, and whose sequence of events affects and is, in turn, affected by the decisions made by players representing the opposing sides."*

Professional wargaming generally encompasses several essential components:

- **Scenario Development:** Wargames initiate with the formulation of a comprehensive and credible scenario tailored to the specific issue or context being explored. These scenarios can involve military engagements, geopolitical tensions, disaster management, or business planning.

- **Participants:** A wargame typically includes a varied group of participants, such as subject matter experts, military personnel, policymakers, analysts, and other relevant stakeholders. These individuals assume different roles pertinent to the scenario, such as military leaders, diplomats, or executives.

- **Simulation Tools:** Wargames can utilize a range of tools and techniques, including tabletop exercises, computer simulations, and complex war rooms with models. The selection of tools is determined by the wargame's objectives and its level of complexity.

- **Rules and Adjudication:** Established rules and protocols govern the decision-making and actions taken during the simulation. Adjudicators or facilitators are responsible for ensuring adherence to these rules and guiding the game's progression in line with the scenario.

- **Analysis and After-Action Review:** A key aim of professional wargaming is to evaluate the outcomes and impacts of different actions and decisions. Following the conclusion of the wargame, participants typically partake in a thorough after-action review (AAR) to analyse the results, pinpoint lessons learned, and extract insights for practical application.

- **Training and Strategy Development:** Wargaming serves an important role in training military and civilian personnel in areas such as decision-making, crisis management, and strategy formulation, allowing them to hone their skills within a risk-free environment.

- **Policy and Strategy Development:** Wargaming can play a significant role in shaping policies, strategies, and plans by offering perspectives on potential challenges and opportunities presented in different scenarios. This enables decision-makers to refine their strategies and minimize risks

In summary, professional wargaming serves as an effective method for navigating complex and uncertain situations, promoting collaboration among stakeholders, and enhancing decision-making processes across various sectors beyond the military, including government, business, and academia.

The history of professional wargaming is rich and expansive, dating back centuries. In ancient times, leaders and military strategists utilized tabletop games and simulations to formulate strategies and instruct on tactics. An example of this is the Chinese game "Wei-Hai," which originates from the 5th century BC and featured pieces that represented military units.

In the early 19th century, the Prussian military created a wargame known as "Kriegsspiel," which translates to "war game" in German. This system employed intricate rules, maps, and miniatures to replicate combat scenarios and was specifically designed to educate Prussian officers in tactics and decision-making. The impact of Kriegsspiel extended to other military forces throughout Europe.

Naval Wargaming gained in prominence in the late 19th century, especially in the United Kingdom. The Royal Navy employed wargames to develop and test its naval strategies, developing critical insights which served them well during the Great War. The US Navy also similar practices; from an analysis of the battle of Jutland to a development of its own wargaming practices the US Navy wargamed extensively in the interwar period. Their main focus was operations facing the Imperial Japanese Navy in the Pacific. The series of games conducted were so extensive, analysed and documented that Admiral Chester Nimitz is quoted as saying:

> *'The war with Japan had been re-enacted in the game room here by so many people and in so many different ways that nothing that happened during the war was a surprise – absolutely nothing except the Kamikaze tactics towards the end of the war; we had not visualised those'.*

[2].

Wargaming also showed its worth during wartime. The Royal Navy's Western Approaches Tactical Unit, also known as WATU, was established in Liverpool in 1941. WATU was a training and research facility that initially trained warship captains in the skills needed to conduct anti submarine warfare in the North Atlantic, but rapidly grew in scope to become an ASW tactical analysis and development unit. WATU employed wargaming to develop and test innovative tactics, techniques, and technologies for countering U-boats in the defence convoys in the North Atlantic.

Figure 1: WRNS personnel conducting ASW training at WATU

WATU used realistic scenarios to train naval officers and escort ship crews, allowing them better understand U-boat tactics, and to develop tactics to counter them. The unit's work in improving convoy escort tactics and coordination, as well as developing new anti-submarine warfare tactics as u-boat tactics and platform capability developed, was crucial to the eventual Allied victory in the Battle of the Atlantic. WATU's work reduce the devastating impact of U-boat attacks on Allied convoys, ensuring the flow of vital supplies and reinforcements to Europe during the war.

Today, professional wargaming is used across a wide range of fields for training, strategy development, and policy analysis. It continues to evolve with advances in technology, incorporating computer modelling, virtual reality, and other innovative approaches. Professional wargaming has been demonstrated to be a valuable tool for decision-makers, helping them to anticipate challenges, test strategies, and develop effective responses to complex and dynamic situations. Its rich history reflects its enduring relevance in addressing a variety of strategic and operational challenges. With this in mind we have turned to its application in supporting engineering design and concept analysis/development

2. WARGAMING AS A CONCEPT ANALYSIS TOOL

Senior proponents of wargaming will stress that wargaming is not an operational analysis tool [3]. However, wargaming can be used as a precursor to OA work, allowing concepts to be explored and developed in a "safe to fail" environment and, in the case of manual wargaming[1] in particular, at relatively low cost. The Naval Architecture and Marine Engineering team at University College London (UCL) has used wargaming in this way Since 2012 to support students in their ship design activities [4, 5]. Wargaming is used in a number of ways:

- As a training aid – at the stage in their academic careers that students find themselves in the ship design exercise, students generally have a good understanding of the basics of ship design. However, they are usually unfamiliar with the ways that navies and warships operate, the interrelations between ship types and platform level combat systems. The UCL staff have developed series of wargames that can be used to develop naval domain awareness, and go further to introduce other capabilities such as maritime patrol aircraft, space and cyber assets). Other games illustrate the value of particular aspects of combat system design choice in close quarter engagements, the benefits or otherwise of detailed ship system layout and survivability or weight of fire and "depth" of missile silos is engagements with peer threat navies.
- As a simple "operational analysis" tool in the Ship Design Exercise. Recognising that wargaming is not OA, it is used as a rapid substitute in the UCL Ship Design Exercise (SDX) and the limitations of

[1] "Manual" wargaming is conducted using maps, dice or other randomising systems, counters and printed rules, as opposed to wargames conducted on computers (although some manual wargames use computers for record keeping and other management aspects

its use in this role are explained to the students . UCL's range of games are used to allow the students to explore aspects of their designs, for example the composition of air groups in an aircraft carrier or the design of offensive and defensive weapon systems and the impact of signature control. Some design projects may require bespoke games to be constructed by the academic staff to analyse niche areas of warfare and design. A recent example is UCL's "Cobalt Rocks" seabed warfare game which was developed to support a design project that was centred on the protection of Critical Undersea Infrastructure (CUI) [5]. Cobalt Rocks will be discussed later in this paper.

- Peer-on-peer performance assessment. A highlight of the SDX is Tactics Night, a large operational wargame in which the student designs are tested against opponents based on real world threats and/or successful student designs taken from previous year's exercises.

UCL's games were designed primarily for use in an educational environment, but their use has expanded into other training sectors such as team and domain training in the UK MOD and defence contractors in the EU, whilst the game focussing on close quarter combat has been used as the basis for a training game at the Royal Navy's Operational Advantage Centre. However, their use has also expanded to cover applications to "real world" projects and concepts. This aspect will be discussed in the rest of this paper.

3. APPLICATION TO CONCEPT ANALYSIS

The idea of using wargaming to support the warship design process stemmed from its use to illustrate use cases in NATO's Specialist Team on Mission Modularity (ST/MM). In June 2014 operational level wargames were conducted in the UK comparing the performance of traditional fleets with fleets comprising significant modular capability. Members of the NATO Specialist Team on Naval Ship Systems Engineering (ST/NSSE) took part in the event. On completion ST/NSSE raised the possibility of using wargaming as a formalised element of the ship design process, applied to concept evaluation and assessment and to support the development of platform and system requirements. Wargaming was included as part of NSSE's programme of work as efforts to support Mission Modularity wound down in 2018. Plans were raised to conduct a campaign of example games, but these were significantly delayed by the covid epidemic (wargaming is an inherently "social" activity and gatherings of players were not allowed.

Planning began in earnest in late 2022 with the aim to conduct the campaign in Summer 2023. The purpose of the campaign was to demonstrate the utility of wargaming, and hence a generic and hypothetical setting would have sufficed. However, it was felt that the use of a current area of interest to NATO would better serve as an illustration as the outcomes and insights could be better situated by seniors in NATO and nations currently engaged in national programmes. Discussions between UK MOD, NATO and UCL staff rapidly homed in on the use of UCL's wargaming capability as an ideal route to conduct the games.

4. THE NATO ASW BARRIER WARGAME

The selection of the ASW barrier project pointed towards UCL's "A Balanced Fleet" game system. This has been developed incrementally since 2012 and is used extensively in the hip Design Exercise. Its main obvious feature are detailed ship mats that describe the vessel, its systems, speed, weapon systems and a geographic representation of key equipments and systems as they are installed in the ship. Performance characteristics for weapon and sensor systems were originally included on the ship mats but this approach is being replaced by system capability cards, allowing options for weapons and sensors to be rapidly inserted.

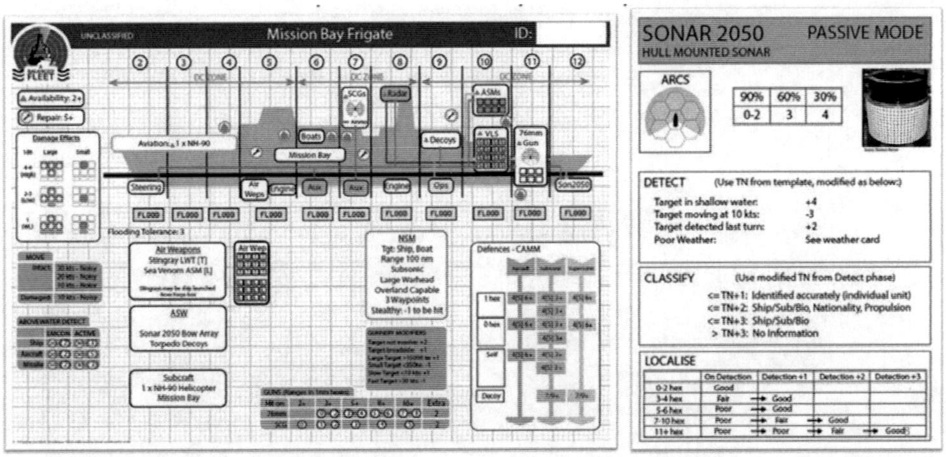

Figure 2: Example ship mas and capability cards used in ABF

Up until 2023 its use had concentrated mainly on above water engagements; consideration of ASW meant that the core ABF system required additional development to cover the capabilities explored in the proposed wargame series. As well as ASW the games would feature offboard systems which had the potential to be affected by weather, hence consideration of offboard systems and their characteristics such as environmental and performance limitations, performance characteristics, and launch and recovery was also required. To focus the direction and output from the games ST/NSSE set a research question:

"To explore the use of offboard autonomous systems for the defence of an amphibious task group against enemy submarines during the deployment, sustainment and recovery phases of an operation."

It was decided that the campaign would step through the various stages of a humanitarian evacuation conducted in the face of an uncertain and potentially hostile threat nation During the wargaming exercise the performance of an "offboard enabled" future force would be compared with the performance of traditional ASW assets. Several scenarios would be played. After each scenario the effectiveness of the ASW barrier would be evaluated and changes to the composition of the offboard autonomous systems mix would be proposed, in order to improve barrier effectiveness (without adding a huge amount of extra cost).

The wargame series was conducted as follows:

- A training scenario on day 1 to bring participants up to speed with the game mechanics and to iron out any issues,
- An initial play through of the scenario using existing ASW platforms and tactics, and,
- At least two games featuring task groups where those traditional forces were replaced with vessels and capabilities representative of platform/force design using Mission Modularity principles.

The detailed overall campaign setting was as follows. The setting for the wargame campaign were the fictional countries of Florin and Guilder, the former beset by civil war and natural disaster, the latter with territorial aims on Florin whilst also supporting rebel factions in the civil war. Blue[2]was to conduct a non-combatant evacuation operation (NEO) from Florin, Red (Guilder) sought to frustrate the evacuation, primarily through the use of small conventional submarines.

[2] For simplicity, "friendly" forces conducting the NEO were referred to as Blue, their opposition as Red.

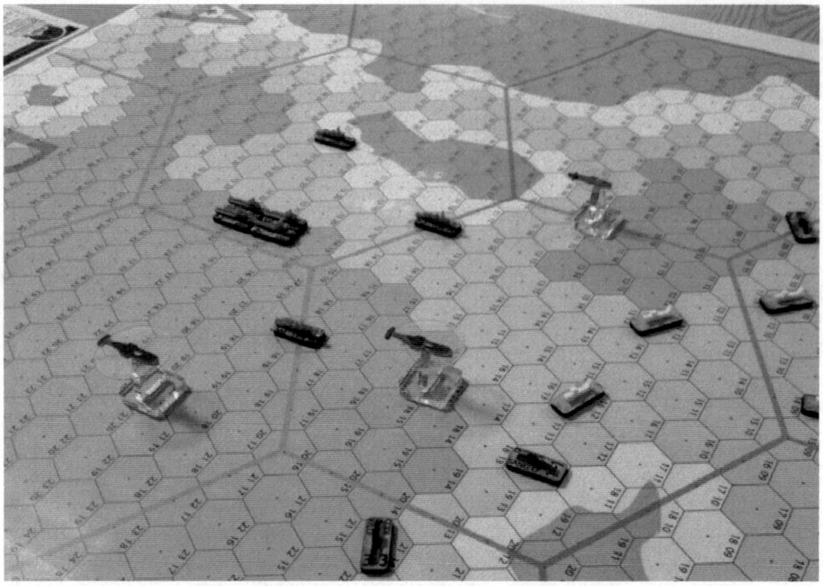

Figure 3: Playing area and pieces for the NATO ASW Wargame

4.1 GAME EXECUTION

The campaign was divided into a series of scenarios that were designed to test different elements of the concept force. Five games plus a training game were developed:

- Game 0 – training game for all participants
- Game 1 – Approach phase- Traditional fleet vs. Emplaced SSKs.
- Game 2 – Evacuation Phase - Conventional Force as ASW Barrier, NEO underway
- Game 3 – Evacuation Phase - Modular Force as ASW Barrier, NEO underway
- Game 4 – Evacuation Phase, Modular Force as ASW Barrier (player-selected force mix), NEO underway
- Game 5 - Exfiltration phase, HVUs (high-value units: the large commercial vessels) depart the OA.

It was found that players quickly became familiar with the rules and game mechanics. This allowed for each game to be run twice simultaneously, to increase learning benefit and generation of insights from the games.

Participants were divided into three teams, or cells. The competing teams were the Red and Blue cells, game control was executed by the White Cell. The game was run "double blind", whereby Blue and Red Cells were located in separate rooms with their own maps and counters; the only forces deployed on their maps were their own and any opposing forces that they had detected. White Cell was based in a third room with a map that featured counters denoting the location of all forces involved. A fourth team, the Green Cell, covered commercial and neutral shipping; this team was managed as an adjunct to the White Cell in this game.

Participants in the White Cell comprised the following:

Game Controller: a single person, an experienced wargamer, they acted as the game "umpire", controlling the flow of the game, managing game progression, "injects[3]", environment and overall game play. One of the Game Controller's key roles was to resolve situations that the game rules do not explicitly cover. An important activity is to determine whether a particular result should be reversed or ignored[4].

---

[3] Injects are specific actions outside the control of the player teams that are "injected" into the game to provide additional events that needed to be managed.
[4] If an unexpected or extreme event occurs that would end a game prematurely the Game Controller may reverse that decision to allow game play, learning and analysis to continue. The situation is recorded and

Adjudicators : The adjudicators (also known as 'runners') were the primary liaison between the White Cell and the Red and Blue Cells. At least one adjudicator was dedicated to each of Red and Blue. The adjudicators received orders and other transmissions from their respective team, brought these to the White Cell for adjudication and resolution, and returned to their teams with results and other relevant information. The adjudicators usually conducted die rolling on behalf their respective teams, for example, to determine if a Blue ASW frigate detects a Red submarine.

Red and Blue Cell participants comprised two to three players dedicated solely to determining the actions of their side in the game.

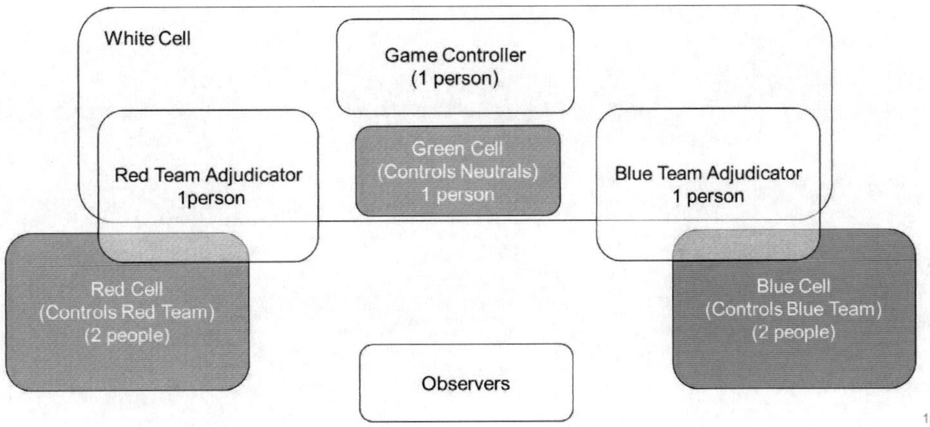

Figure 4: ASW Game Team/Cell Organisation

4.2 ORDER OF PLAY

Each game comprised the following phases:

Scenario Briefing – An overall brief given in plenary to all participants on the scenario to be played, followed by individual briefings to each team. Weather conditions for the next 7 days were determined by the White Cell, and forecasts for the next 3 days were briefed to Red and Blue.

Initial Planning Phase – Blue made their initial plans such as ship, helicopter and UxV patrol lines etc. taking into account the duty cycle, or roulement for the various platforms involved. Aspects to consider included launch and recovery times, transit times and time for rearming/refuelling/recharging. Red planned their submarine ingress and egress routes and other activities (in some scenarios Red wished to land special forces at observation points, or make use of covert teams in commercial vessels).

4.3 SCENARIO EXECUTION

The game comprised a number of turns, each turn comprising the following phases :

- Teams write orders and submit to their adjudicators
- Adjudicators brief Game Controller on their teams actions
- White cell determines results of encounters, detections, attacks and vessel damage (this may involve adjudicators returning to the teams for in-turn briefings, decision making and reporting back to the White Cell, particularly during detection and attack)

---

included in the end of game analysis. For example, in one game covered here a key Red submarine was sunk at the outset of the game. Its loss would have made the rest of the game largely redundant, hence the submarine was reinstated and it was noted that early detection of submarines at long range could considerably ease the burden on the ASW team.

- Adjudicators record results of actions in the turn, and return to their teams to read out results of which their team is aware. Care needs to be taken at this point as the adjudicators for each team are likely to know the locations and actions of BOTH teams; extreme discretion and confidentiality is required.

Game turns continue until either side achieves its victory conditions or the Game Controller judges that further play is nugatory.

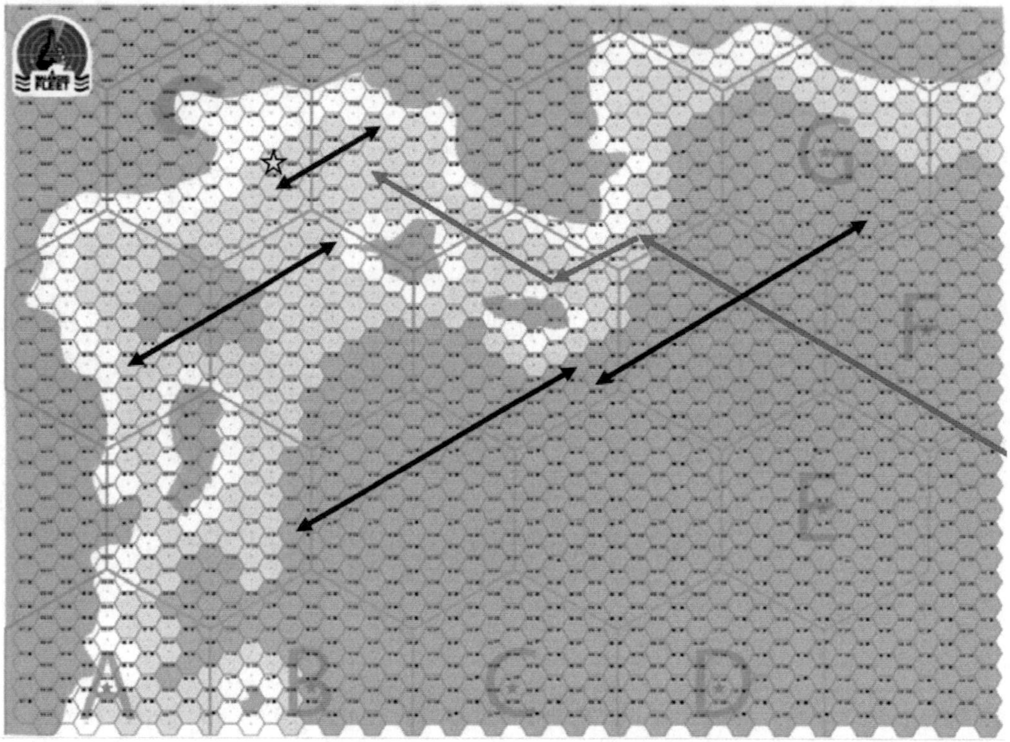

Figure 5: Wargame playing area showing anti submarine patrol routes (blue), submarine approach routes (red) and the protected anchorage (yellow star)

4.4 PLAYING THROUGH THE CAMPAIGN

The wargaming exercise was conducted over three days, each day comprising two sessions, one morning, one afternoon. Games were fought to a conclusion, with all participants responsible for recording their actions, the rationale behind those actions, and observations on their own aspects of the game (both in terms of the game events and also the game execution). At the conclusion of each game the players were brought back together in plenary to discuss the conduct and outcomes of the game, and to identify and record any learning points that arose. This discussion generated the insights both into the utility of the game to facilitate analysis of the concepts employed, as well as the efficacy or otherwise of those concepts.

Following each game, the participants (stand fast the Game Controllers) were encouraged to move between teams and cells. For example a participant may have played in the Red Team for game 1, moved to the Blue Team for game 2, thence the Red Cell for game 3 and the Blue Cell for game 4.

Each day concluded with an additional longer "hot wash" session to capture higher level observations and to allow modifications to the game plan for the following day. This gave an opportunity for new capabilities to be proposed, considered and developed, and for the game design team to generate new game material (usually in the form of new capability cards)to reflect these developments. For example, Blue felt that a deployable seabed acoustic monitoring system would be of significant benefit. The likely characteristics of such a system were discussed and agreed, including input from relevant subject matter experts. The design team determined the characteristics of the system in game terms and created new game components overnight for use in the games the following day.

Figure 6: Playing area, ship and capability cards

4.5 CAPABILITIES EXPLORED

Blue's initial forces were, as discussed, based on existing conventional ASW capabilities such as specialist frigates using hull mounted and towed array sonars, with torpedoes delivered by helicopter or (in extremis) the frigates themselves. After the first game was concluded Blue's force was replaced with modular frigates, essentially UxV "motherships" but with significant anti air and anti surface capabilities [5]. Throughout the series of games the offboard capabilities available to the Blue Team were reviewed and enhanced where capability gaps were identified. The range of planned enhancements included:

- Uncrewed Surface Vehicles (USV) using thin line towed arrays for submarine detection. These had the benefit of speed and the opportunity for constant communications, but were considered to be sea state limited, easily detectable and relatively vulnerable to attack. The Red team used periods of bad weather to challenge these systems with some success.
- Uncrewed Underwater Vehicles (UUV) using similar thin line towed arrays. These craft had the benefit of operating in the same medium as the target, were quiet, less likely to be adversely affected by bad weather and hard to detect but were slow, and in, the absence of a subsea communications network, required to operate on planned routes with pre-plotted reporting cycles
- Uncrewed Aerial Vehicles (UAV) were used to deploy sonobuoys and lightweight torpedoes. Payload considerations, in particular when carrying torpedoes, would limit their speed and endurance so the Blue teams invariably chose to operate them in concert with traditional ASW helicopters, a mix that proved to be quite effective.

As the games progressed, capability gaps were identified and proposals made for new systems to fill those. These included:

- Seabed communications infrastructure – this was discussed but was not deployed in the games. Had this happened the utility of the UUVs would have been enhanced, and whilst this would

[5] In the end these capabilities were, apart from one game where the use of submarine launched anti ship missiles were a feature, generally not exercised. The emphasis was clearly on ASW

have given a clear benefit in the defensive situation of the HVUs at anchor it would have been difficult to deploy to support open ocean ASW.

- Deployable passive sonar system – a "spoke and hub" array of seabed sensors with wired links to an active central communications buoy that allowed real time communications and data sharing with controlling sites in ships or ashore. This created a near-impenetrable barrier for the defended HVUs but it took time to deploy and recover, meaning that in the exfiltration phase of the operation Blue ships involved in recovery would have exposed the themselves to the submarine threat. As a result most of these assets had to be abandoned.
- USV-based torpedo defence system. The scenario required the ASW barrier and the frigates sustaining it to be operating at some distance from the anchorage in which the HVUs were operating. This meant that a hostile submarine could penetrate the barrier and achieve a firing solution, even at long range, against the noisy HVUs allowing them to launch a spread of homing torpedoes that were then unlikely to be countered. A torpedo hard kill system had already been included as an option for the modular frigates and it was suggested that the same system cold be deployed on USVs to protect the anchorage. This was developed as a capability enhancement and employed in the games on Day 3, where it was successful in providing an effective defence.

4.6 CONCLUSIONS FROM THE ASW GAMES

At the conclusion of the campaign the participants engaged in a final washup to discuss their thoughts, observations and conclusions as to the effectiveness of the game, and to expand on learning points and observations concerning the ASW barrier project. This discussion extended into a following day involving the members of ST/NSSE who were present. The main conclusions were compiled in a post-game report [6]. A summary of key points is as follows:

4.6 (a) Concept And Capability Assessment

1. The games revealed capability benefits, development paths and demonstration of concepts that are in line with "real world" trajectories, despite the event being run using only unclassified open source data.
2. SMEs engaged constructively and remained involved throughout. There was a concern that SMEs would focus on elements of limited realism arising from open source information, but this did not happen.
3. Wargaming acted as "Design validation" for some of the advanced platform, system and operating concepts under consideration.
4. Wargaming provides a framework that assists designers and others in knowing what questions to ask. It provides a "birds eye view" of a problem or capability gap before launching in to more extensive analysis.
5. Wargaming can be used to investigate a broad range of options and then focus in on options that show the greater potential benefit. This can focus limited resource in follow-on operational analysis work.
6. The agility of the process was well demonstrated. New elements were introduced and tested quickly, whilst capability effectiveness can be varied easily for sensitivity analysis. This was aided by SME involvement, with experts giving their opinions on capability performance (based on their experience) that could rapidly be turned into more realistic game mechanics, units etc.
7. The process allowed the effect of disruptive technologies to be investigated quickly.
8. Wargaming can identify areas that appear promising but which deliver little benefit. Within the campaign the use of UUVs was seen as of little benefit – but the reason why (and potential future mitigation paths) were also identified.
9. The ability to quickly consider and create future capabilities brings with it the ability to quickly consider how they may be countered. It allows for a virtual "arms race".
10. Understanding of the context within which a capability or concept exists allows the interfaces and interactions with other capabilities to be explored. It can uncover (higher order) effects that are not obvious from simply discussing the concept in a traditional environment.
11. It can provide assistance in informing decisions, forming a framework for explaining why a concept may or may not be worth pursuing.

4.6 (b) Game Execution

1. Post game washups after each game are essential in understanding interaction between red/blue actions. Capture insights/feedback whilst the debrief is happening.
2. White cell insights come from an overarching perspective. The adjudicators are critical, and should be writing down their thoughts as they happen. Consider having at least one dedicated rapporteur for a game, and preferably one for each of the Red, Blue and White Cells/Teams.
3. Consider where outcomes were "lucky" or were robust. These should be noted as insights. It may be beneficial to reset after particularly lucky (either good or bad) events.
4. Importance of establishing clarity in desired outcomes, capability needs, etc. Everyone should be aware of the aims and objectives of the game. That should not prevent insights and observations of other aspects being recorded.
5. Losing is not a bad thing, it is the mechanism by which you learn. Wargames should be a "safe place to fail". If a player feels bad at losing then that is really a good thing, they achieved good investment in their role and have probably gathered a learning insight that will stick..
6. The game was relatively quick to construct (speed depends on pre-existing material and the basis UCL ABF system was already established). This game took roughly 1 month of effort to prepare. The game design process established that a new ASW variant could be up and running in a week or so with dedicated effort from the game design team..
7. Iteration – and hence speed – is critical for this kind of game to succeed. But speed should not be a goal in itself if it risks the loss of insight and learning opportunity
8. Repetition – players may tend to overcompensate in some areas, but may then iterate down to a sensible medium. Benefit of overall experience shows.
9. Player engagement is essential. Players should be kept occupied, but take care to balance workload, especially in Red and Blue teams.
10. Managing player numbers – too many players in a team results in "decision paralysis"
11. Rotation of players through the various cells worked well, different insights obtained from different perspectives were valuable, especially in later games where players had been in different roles.
12. Running multiple games allowed for the concept to be quite well explored. Had this exercise been a real concept analysis task rather than a "proof of principle" we would probably want more iterations, e.g. variation on modular frigate mission bay size. Hence more time would be needed – perhaps a week – with options for more parallel games.
13. "Misson Creep" is a risk, hence the game controllers need to be strict on the scope for the range of studies / variables. Be careful not to stifle innovation though.
14. Be wary of the "Psychology of Red", suicidal tendencies and "last turn" frenzy. Unexpected game endings should mitigate.
15. Perceived attractiveness of sides was apparent, Red was thought of by Blue players s having "fun stuff to play with", although Red players sometimes felt they has little to do.
16. The game was developed and run successfully at UNCLASS. There is an obvious ability to run at higher classification for specific studies within suitable environments.
17. Including engagement (besides detection, tracking, classification and identification) is essential to player involvement and engagement and would increase the generation of learning, discussion points and insight into force capabilities.
18. A future iteration could have two teams each playing Red and Blue at the same time – this would help balance out workloads, may mitigate "Psychology of Red" issues, and give everyone insights into both sides.

4.6 (c) Training And Other Benefits

1. The games demonstrated a clear training and familiarisation benefit across non-engineering personnel. It was felt that participation in events such as this, either as analysis events or training events would be of significant benefit to those outside engineering and operations. For example, commercial/finance personnel would gain insights into the impact of decisions in their areas, and gain an understanding as to why "military" requirements are important.
2. The games gave non-specialists in ASW a significantly better appreciation of tactics, procedures, and systems involved in this warfare area. For specialists in other areas it provided a good vehicle to understand design and operations in context – how their specialist area feeds in to the greater whole

(e.g. survivability, noise, manoeuvrability, acceleration, flight deck ops, lean crewing, automated systems).

3. The games showed the benefit of engaging seniors / decision makers in physical games. Wargames promote interaction, personal investment and an opportunity for seniors to gain wider insights into their projects and capability areas.

4.6 (d) Conclusions Drawn By The Nato Team

ST/NSSE members were involved in various roles throughout the campaign. They observed how stakeholders and system designers exchanged information. The stakeholders are subject matter experts: persons who have experience with anti-submarine warfare, and who have a unique view on what is needed for the new system of offboard maritime unmanned systems to become successful. In general terms it was observed how the system designers learned important lessons on how systems are deployed and operated in the predefined operational setting. The subject matter experts learned about the potential capabilities and limitations of new technology. It was concluded that the information exchange was important and relevant, of critical importance to designing an effective new anti-submarine warfare capability. In summary, both the test and the test results are considered to be a great success. So ST-NSSE is going to include wargaming as an essential domain-specific activity in the new naval ship design NATO standard.

ST/NSSE's primary objective was to demonstrate and explore the role of wargaming to support design and concept analysis. It is felt that this was achieved. The games demonstrated (within the unclassified restrictions):

- the potential effectiveness of an ASW barrier system using offboard assets compared with a traditional arrangement,
- the benefits and drawbacks of various elements of such a system,
- capability gaps in the proposed concepts and potential solutions,
- wargaming is a very powerful method for discovering which components and characteristics of a system composed of many components, such as the deployable ASW barrier system, are actually essential,
- wargaming includes many relevant factors such as the complexity of sensor and weapon modelling, environmental effects, system capabilities and limitations, probability modelling, rules of engagement, etc.
- wargaming is an activity that needs support from experienced wargamers (e.g., from UCL) to make sure that the learning points, including those with negative experience on what actually does not work, are relevant,
- wargaming and modelling and simulation (M&S) serve different purposes: wargaming looks at the whole picture and therefore cannot be used for studying many variations of the operational problem (it would take far too much time); M&S can be used to study an operational problem in great detail with many variations, with a much smaller problem scope compared with wargaming,
- the results of M&S can be used to develop the wargaming model.

It was considered by the game design team that the event successfully demonstrated the benefit and worth of using wargaming as a concept analysis tool.

5. SEABED INFRASTRUCTURE PROTECTION – "COBALT ROCKS"

As well as developments of ABF, UCL also develops bespoke games to cover unconventional areas of interest. An example of this from 2022-23 was "Cobalt Rocks" which was showcased at the Connections 2023 conference [7]. One of the subjects for that year's Ship Design Exercise was a seabed operations ship tasked with (amongst other duties) the protection of seabed infrastructure – pipelines, power and internet cables. None of the design team had experience in this type of operation and so it was decided to explore this through the medium of wargaming.

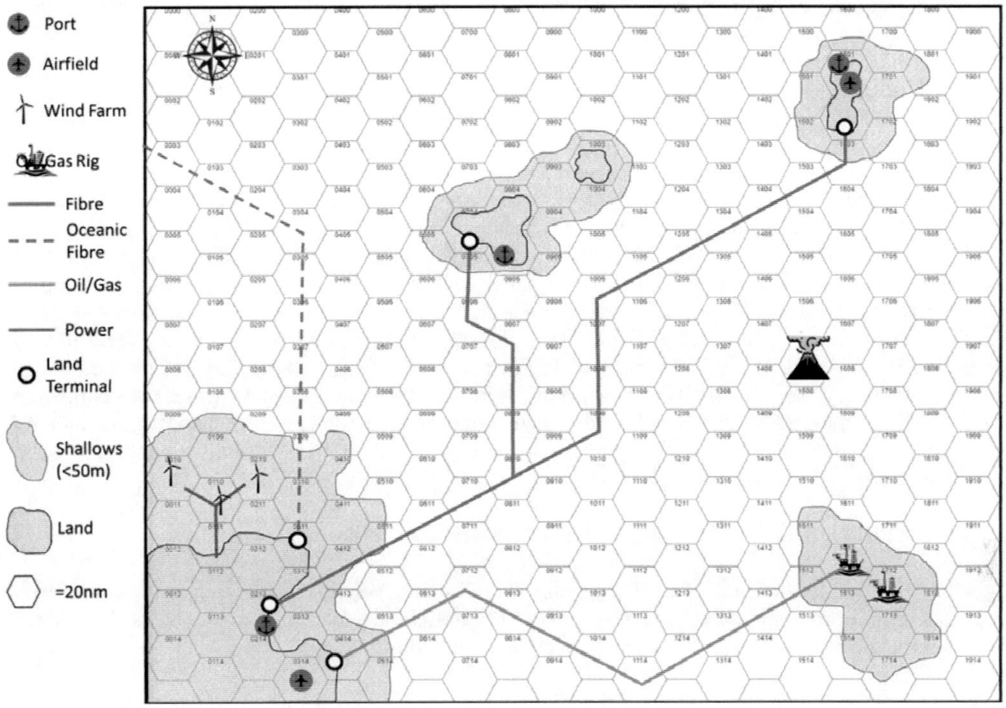

Figure 7: The generic environment for "Cobalt Rocks", featuring communications cables, pipelines, gas production platforms, wind farms and electrical interconnectors

5.1 THE COBALT ROCKS DESIGN AND EXECUTION PROCESS

The initial phase of the game covered the development of attack and defence strategies for critical undersea infrastructure (CUI). The design team (Blue) researched CUI and proposed what they thought would be a reasonable national CUI infrastructure within which their ship would operate, and techniques for CUI surveillance and protection. Simultaneously the Red Team was considering methods by which they would covertly survey and interfere with / destroy CUI. Roles were then reverse; each team then considered their opponent's conclusions and developed approaches by which they could counter the defence/attack strategies. Based on the outcome from this phase, Blue and Red developed their finalised plans, which in the case of blue included capabilities and equipments to be embarked in the ship they were designing, as well as the characteristics of the ship itself. The results were reviewed and moderated by the White Cell to ensure no "capability inflation" was taking place.

5.1 (a) Mission Planning And Adjudication

In the Mission Planning Phase, each team devised an operational plan, assuming their desired capabilities were available and then developing a mission plan covering 2-3 weeks of operations. Each plan would detail ship and air surveillance, deployment and routes of UXVs, detailed attack plans (with timings) for Red. Plans were then submitted to the White Cell for adjudication in the next phase, which was the main wargame itself.

Adjudication Phase essentially comprised an "events based" wargame. The White Team members compared each team's plan and looked for overlaps, chiefly the first point at which a team would become aware of their opponents' activities. At this point the detecting team was advised as to what they had observed and were allowed to modify their plan to suit. Only when both sides became aware of the other's presence would the game transit to a more traditional turn-based game. Thus it was possible for one side to be completely unaware that they had been detected by the opposition. For example, in the first game played, Blue's search patterns proved to be completely ineffective, allowing a Red vessel to deploy UUVs that surveyed and tapped into an underseas cable without detection. In another game, Red's attack on a pipeline that involved placing delayed action explosives was discovered by Blue who trailed the attacking UUV and then disarmed the charges once

Red's UUV had moved off, all without Red being aware that they had been detected. In both cases defeat for the losing team came as a complete surprise.

5.1 (b) Evidence Gathering

An essential element of the process is recording not just what the players chose to do, but also why. This applies throughout the various phases of the game. To facilitate this, turn record sheets, capability development records and other documents include a "why" element that captures the players thought processes and rationales. This then forms an important basis for post game analysis, in particular in providing evidence as to why a particular capability decision was made. For example, an early iteration of the design team's ship did not feature aviation facilities other than a small flight deck where light helicopters could land and embark personnel and small store. The rationale was that the ship was engaged in seabed operations and so aviation was not needed. Key events in a game required the players to react rapidly to a developing incident some 200 miles from the ship which, with a top speed of 15 knots, meant a steaming time of over 310 hours. In the post game capability development discussion the players introduced a flight deck and organic UAV capability, with the UV able to deploy a small UUV. Whilst this increased the cost of the vessel the deciding rationale was that it allowed reaction times to be reduced from 3-4 hours to 15 minutes or less depending on the exact choice of UAV embarked.

5.2 TYPICAL GAME OUTCOMES AND LEARNING POINTS

Cobalt Rocks was run several times to support the ship design development exercise, and several times since as an academic exercise and in anger as part of NATO and UK defence contractor studies. In general, each game has identified most if not all of the following learning points and outcomes.

Protection of CUI falls into the "really difficult" category of maritime operations. The most effective methods of protection involve near constant surveillance of the seabed assets to be protected, coupled with rapid identification of potential threats. Those threats should then be put under constant close and overt observation since this removes the cloak of anonymity and plausible deniability that an adversary may have.

Rapidity of response to a developing situation, particularly at range, makes a high speed intervention capability of significant benefit. The Blue players identified the need for organic UAVs for local reconnaissance, and a long range, high speed UAV capable of deploying a UUV and acoustic communications buoy which then remained on station to provide real time intelligence back to the ship or other command station

Where the ship is operating within a national defence capability, and particularly where this includes civilian authorities a unified command structure or highly effective interfaces are needed to ensure that intelligence is shared so that it can be acted upon promptly

Effective communications with deployed subsea assets at range is also essential to allow timely exploitation of information. In several games, UUVs were able to detect threat craft but were unable to pass that information on without coming to periscope depth. The players postulated communications nodes built in to pipelines and cables to allow this to happen.

It was notable that many of the learning points identified above were areas of development that were observed in various CYI related elements of the NATO "REPMUS" trials, and the CUI defensive activities of the NATO Joint Expeditionary Force (JEF) in the Baltic, both in the latter half of 2023. Whilst it is not suggested that the outcome from the games influenced these events it was gratifying to see that the students' wargaming activities drew the same conclusions as those in the real world.

It should also be noted that many of the insights drawn from the various games played in terms of game design, execution, ;earning and training opportunities are the same or similar to those drawn from the NATO ASW study and there is no need to repeat them here. It does however indicate a consistency in lessons and insights that can be drawn from wargaming.

5.3 CURRENT AND FUTURE USE OF COBALT ROCKS

At the time of writing, Cobalt Rocks is being used as the basis for a series of NATO wargames testing the "Baltic Sentry" mission to protect subsea infrastructure in the Baltic and to consider potential uncrewed options that may form part of NATO's "Task Force X". It is being evaluated for use in the US and other allied nations and has been used by a UK defence contractor to evaluate the use of medium and large uncrewed vessels in the CUI protection role. Information gathering on the subject of CUI attack and defence, as well as the wider nature of commercial and military seabed operations, has allowed additional detail and capability to be included in the game both in terms of offensive and defensive capabilities and the mechanics of commercial seabed operations and infrastructure.

6. CONCLUSION

These two studies have successfully demonstrated the ability of wargaming to be used as a concept analysis tool across a range of maritime areas and activities. The speed with which a series of wargames can be developed has been shown, as has the ability to take up rapid changes to the games and their constituent elements to take advantage of developing ideas and concepts within the wargaming event. Finally, the extension of the approach from a NATO study and a game supporting student studies into the protection of critical undersea infrastructure into real-world applications in significant naval and maritime security projects has been shown.

7. REFERENCES

1. PERLA, P, 'The Art of Wargaming', *History of Wargaming Project*, 2011.

2. MOD Wargaming Handbook, *UK MOD*, August 2017.

3. DOWNES-MARTIN, S., 'Advanced Topics in Wargaming', *Connections US, National Defense University Washington DC*, 2023.

4. BRADBEER, N., 'Naval Wargaming as a Teaching Tool for Warship Design Engineers', *IMDC*, 2022.

5. MANLEY, D., 'Wargaming as a Design and Concept Analysis Tool at UCL', *DSET*, 2023.

6. MANLEY, D., 'NATO UXV ASW Barrier Wargaming - Game Report', *NATO ST/NSSE*, 2023.

7. MANLEY, D., 'Wargaming Seabed Operations in a Platform Design Context', *Connections US, National Defense University Washington DC*, 2023.

8. AUTHOR BIOGRAPHIES

David Manley is the current MOD Professor of Naval Architecture at University College London, He is the MOD Senior Fellow for Maritime Vulnerability Reduction, DE&S Head of Specialism for Platform Survivability and Technical Authority for Maritime Survivability. He leads and supports survivability development through membership of various NATO, ABCANZ and TTCP working groups and committees. David has been a wargamer since childhood, and has been active in professional wargaming since 2012. He is the designer of "Cobalt Rocks", the first open source wargame covering contemporary offensive and defensive seabed operations. He is a Fellow of the Royal Institution of Naval Architects and a Constructor Captain in the Royal Corps of Naval Constructors.

Nick Bradbeer is the MoD Associate Professor of Naval Architecture at University College London, where he teaches submarine design and provides operations analysis and wargaming support to the MSc capstone ship design exercise. He has been developing engineering-focussed analytical wargames in support of warship design since 2016, with a research interest in how wargames can most effectively contribute to the procurement process. He is a Member of the Royal Institution of Naval Architects and a Constructor Commander in the Royal Corps of Naval Constructors.

LARGE UNCREWED SURFACE VESSELS – BALANCE OF SURVIVABILITY

D Patten, BMT, UK, and **C Clark,** Survivability Consulting Limited, UK

SUMMARY

This paper explores the requirements, opportunities and constraints space for the survivability characteristics of Large Uncrewed Surface Vessels (LUSVs). To achieve this, the paper has been broken into 2 distinct parts; the first explores how squadron survivability could and should be achieved by individual LUSV platforms through a combination of susceptibility reduction, vulnerability reduction and recoverability measures, using qualitative and quantitative approaches including Purple Fire vulnerability analysis. The second part explores how survivable a military system using LUSVs as the primary assets would need to be to achieve a high probability of mission success in an enduring conflict using tools such as the CONEMP, wargaming, and simple engagement models.

Acronyms and abbreviations

Acronym	Definition
AAW	Anti-Air Warfare
ASW	Anti-Submarine Warfare
CESM	Communication Electronic Support Measures
CODAD	Combined Diesel and Diesel
CONEMP	Concept of Employment
EMCON	Emissions Control
ESM	Electronic Support Measures
IR	InfraRed
LPD	Low Probability of Detection
LUSV	Large Uncrewed Surface Vessel
MPA	Maritime Patrol Aircraft
PECD	Probability of Escaping Critical Damage
PEH	Probability of Escaping Hit
P_{Surv}	Probability of Survival
RCS	Radar Cross Section
RESM	Radar Electronic Support Measures
RF	Radio Frequency
RINA	Royal Institution of Naval Architects
RoE	Rules of Engagement
SA	Situational Awareness
SATCOM	Satellite Communications
SSN	Nuclear Powered Attack Submarine
TAS	Towed Array Sonar
UAS	Uncrewed Air System
UCL	University College Londin
UUV	Uncrewed Underwater Vessel

1. INTRODUCTION

USVs are increasingly prevalent in naval fleets, primarily in the workboat size range but Large USVs (LUSVs) are now starting to emerge, for example; the USA's SeaHunter and the Chinese ORCA-A. As these platforms transition from basic role capabilities to becoming an indispensable part of fleets, survivability – or the ability to operate in a hostile threat environment and achieve military effectiveness, becomes ever more important.

As a starting point it is worth remembering why navies exist in the first place. In the view of the author, navies exist to wield soft and when needed, hard power. The two are heavily intertwined, especially when the means of hard power is not credible, as summed up by Joseph Nye: *Soft power without hard power is often ineffective. The ability to attract and persuade must be backed by the ability to coerce if necessary [1].* The point being that if a LUSV based capability is not credible or perceived to be so, its ability to be used as a tool in those soft or hard power aims is diminished. Survivability is an essential thread of capability and credibility, a heavily armed vessel with no survivability may 'go down fighting', but it will nonetheless go down.

LUSVs were chosen as the subject for this paper as it is clear that the utility afforded by them (along with their air and sub-surface counterparts) is likely to be revolutionary, in a similar way to Dreadnought battleships and aircraft carriers in their time. For this study a nominal LUSV concept has been developed with particulars as detailed in Table 1.

Table 1. Concept LUSV particulars

Length	65m
Beam	6.5m
Depth	7.5m
Draught	2.75m
Payload	100Te
Displacement	650Te
Speed	20kts
Range	~3000nm

A length of 65m was chosen because it is the point at which the platform has capacity to meet the range, payload and seakeeping qualities required of the vignettes as well as being the size at which vulnerability reduction may enable some reasonably large threat weapons to be survived if the platform is hit. The potential utility of LUSVs is vast [2], but in this paper we have considered primarily defensive roles as these are likely to be the initial roles carried out by LUSVs due to the need for a 'human in the loop' for offensive operations. This study sits alongside research BMT are currently undertaking into medium and large USVs through the USV ModUS project, which incorporates workstreams for ship design, autonomy, modularity and mission systems.

2. PLATFORM SURVIVABILITY

2.1 OVERVIEW

In many respects a LUSV is no different from a similarly sized conventional craft and is subject to the same physical design constraints, though in others the lack of persons on-board will be a defining difference, this is particularly true of the recoverability aspects. Each of the three elements of survivability (susceptibility, vulnerability and recoverability) are considered in the following sections.

2.2 SUSCEPTIBILITY REDUCTION

Susceptibility reduction is the ability to operate in a hostile environment without being hit by a weapon, this is especially important for intrinsically vulnerable platforms such as smaller submarines and surface craft against disproportionate threats. For a platform with a high degree of autonomy the primary challenge will be in tactical decision making and execution, including considerations over Rules of Engagement (RoE) when faced with extremely complex situations. Susceptibility is measured through PEH (Probability of Escaping Hit) and encompasses many layers contributing to the overall capability. In this paper, the layers of interest to ship design are considered, these are platform signatures (the means by which the platform can be detected, classified or tracked by its interaction with the external environment) and self defence systems.

The key signatures of importance for multi-role surface platforms are radar return, Infrared (IR), acoustic, electromagnetic and wake fields.

2.2 (a) Signatures

Radar is the most prevalent and most effective detection method and emanates from two sources; radar reflections determined by Radar Cross Section (RCS), and self-emissions. For the former, to avoid detection altogether requires very low radar returns in the direction of a transmitting radar. Typically, power received by a capable radar can be interpreted as a signal in the femtowatt range, by using the simplified radar equation [3] at distances of around 100km this would require a RCS in the order of 10^{-3} square metres. The design parameters needed to achieve this level of return are well understood but remain challenging in practice. A design indicative of the measures needed to achieve this is the Swedish Visby class corvette, which happens to have a similar capability to the platform types considered in this paper.

The second source of radar emissions, from the platform's own electromagnetic transmissions such as radars and communications antennae, are a very challenging design aspect for uncrewed vessels. This is due to their heavy reliance on communications systems, a large source of self-emission. In order to achieve a signature low enough to avoid detection would require the use of receive only (Emissions Control/EMCON) or Low Probability of Detection (LPD) communications systems in either RF [4] or optical wavelengths [5].

Infrared (IR) signature control may be easier with a LUSV as compared to crewed vessels, which generate a large proportion of their signature from the need to heat or cool accommodation spaces in response to ambient temperature changes, no such requirement exists for a LUSV. However, they will still need to manage other hot compartments such as machinery spaces and areas with large thermal inertia such as fuel and ballast tanks as well as solar heating. Another key source of IR emissions is from machinery exhausts, which is typically managed using devices to directly cool the plume or by having the option to exhaust close to or under the waterline, there is no difference between a crewed or uncrewed platform in this regard. There is however an opportunity in this area for LUSVs, where alternative fuels such as methanol and ammonia could be used, which have lower temperature burn characteristics [6] and may provide lower IR signatures. BMT intend to carry out further research in this area.

Acoustic signature targets are challenging to manage through design due to the complexity of structural and machinery interactions. The most effective way to manage machinery noise is with large machinery rafts coupled with controllable machinery, for example using variable frequency drives. The other large noise source comes from fluid borne noise, this is typically managed through managing flow rates, valve placement and pipework layouts. The only difference between a crewed and uncrewed vessel in this area is the elimination of noisy systems associated with accommodation spaces, which is likely to improve the signature albeitby a small amount.

2.2 (b) Self Defence Systems

Self defence encompasses two broad capabilities; soft-kill and hard-kill. Soft-kill systems use many techniques to prevent a weapon from hitting the platform either above or underwater but all are heavily dependent on platform signatures being low enough that the soft-kill system can create a more attractive target. Introducing soft-kill systems to uncrewed platforms should be relatively straightforward from a ship design standpoint. Hard-kill systems conversely are likely to be more challenging as they would almost certainly need to be operated without a human in the loop to be effective, especially in a communication denied environment. Physical platform integration of hard-kill systems including laser and Radio Frequency (RF) Directed Energy Weapons, missile systems and gun based weapon systems is no different than the challenge of doing so on a crewed platform and may in fact be simpler due to the removal of constraints around accommodation adjacency.

2.3 VULNERABILITY REDUCTION

The setting of vulnerability requirements, measuring of designs against them and the mitigation of vulnerabilities is a mandated and well-established part of UK naval design. Purple Fire is SCL's vulnerability/lethality assessment tool and has been used in the design process of Type 26, Type 31, Tide Class and Fleet Solid Support, as well as vulnerability reviews across the whole Royal Navy and Royal Fleet Auxiliary surface fleets. An overview of the tool and assessment process is available in previous RINA Warship papers [7,8,9].

The vulnerability reduction process is well understood for crewed vessels but the methods can be equally applied to a LUSV. Design philosophies that should be followed to achieve a reduced level of vulnerability include:

- Separation of redundancy within propulsion, generation and steering systems;
- Zoning of key service systems;
- Shock mounting of key systems;
- Protection of high value compartments, magazines and machinery spaces;
- Strengthening of key bulkheads to protect compartments or zones.

These could potentially be more effective for LUSVs as the human element doesn't have to be accounted for. The design can be assessed against various threat weapons and damage mechanisms such as blast, fragmentation, flooding, shock and whipping. By assessing damage from all potential points at which a weapon can hit a ship, an average vulnerability value for each role a design can undertake is established. Here the initial LUSV design is assessed, then the results are used as a base to compare against an improved design that accommodates vulnerability reduction methods.

2.3 (a) LUSV Vulnerability Model

The outline layout of an example 65m LUSV is shown below in Figure 1.

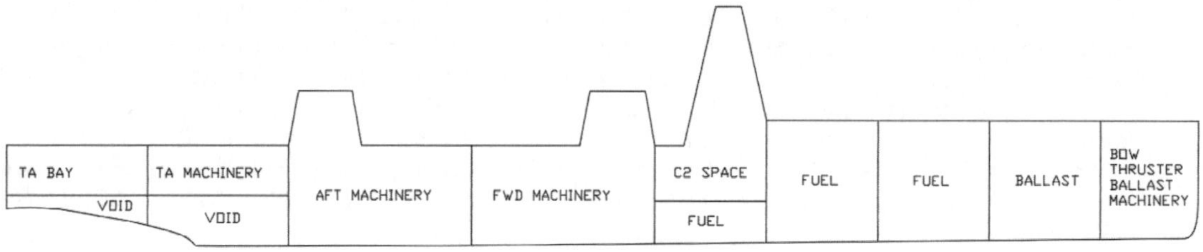

Figure 1: Outline USV layout

To define a functional Purple Fire model of the LUSV it was specified that:

- Each machinery space should contain a diesel generator and a switchboard to distribute power;
- The propulsion equipment should consist of a CODAD arrangement with a single medium speed diesel and reduction gearbox for each of two shafts;
- The steering gear compartments should have separation port and starboard of the Towed Array bay;
- There are two redundant server racks which require power and cooling, located in the C2 space;
- The sensor fit is two each of Radars, SATCOM and Electro Optic Devices.

Note that the model does not include cabling or other system detail, commensurate with an early stage design. The resulting Purple Fire model is shown in Figures 2 and 3.

Figure 2: Initial Purple Fire USV model

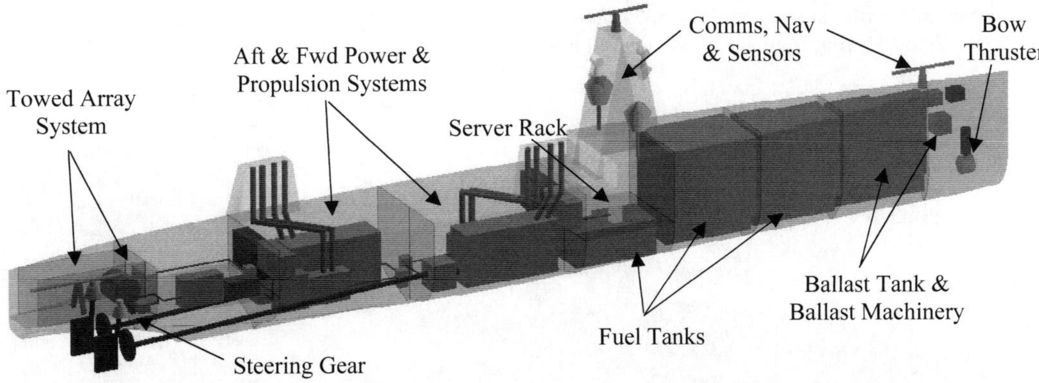

Figure 3: Purple Fire USV model system layout

2.3 (b) LUSV Vulnerability Assessment & Reduction

The initial LUSV model was assessed against small, medium and large sea-skimming missile threats hitting from port or starboard. From each side, a grid of potential impact points was generated. Results have been quoted in terms of Probability of Escaping Critical Damage (PECD) for each primary system (e.g. ballast, propulsion, communications) and a combination of the primary systems that reflect the overall ability of the LUSV to undertake its potential roles such as anti-submarine/mine operations, intelligence gathering etc. PECD is the average probability, over all hit points considered, that the system will remain available after a single threat detonation. Note that Purple Fire can also consider various multi-hit scenarios, but these are not assessed here.

Figure 4 gives an example of a grid of points at which the threat will interact with the target.

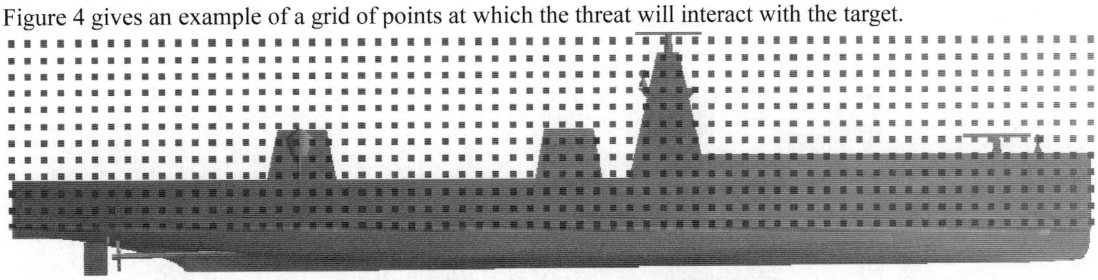

Figure 4: Threat grid used against LUSV

Table 2 shows the PECD of the main systems of the initial LUSV design against each threat, averaged over all attack angles and hit points used.

Table 2. PECD of initial LUSV design

	Small	**Medium**	**Large**
Ballast System	88	77	79
Minimum Mobility	93	86	80
Towed Array	82	51	42
Communications	82	58	48
Awareness	88	65	52
Undertake Role	**63**	**22**	**15**

The PECD of the "Undertake Role" function is quite low for the initial layout. The lack of separation between the propulsion spaces or the fuel tanks is a major source of vulnerability. This can be improved by rearranging the compartment layout and improving system redundancy. Figure 5 shows the LUSV model after vulnerability mitigations have been taken into consideration, with the main focus being separation of redundancy in propulsion and power generation.

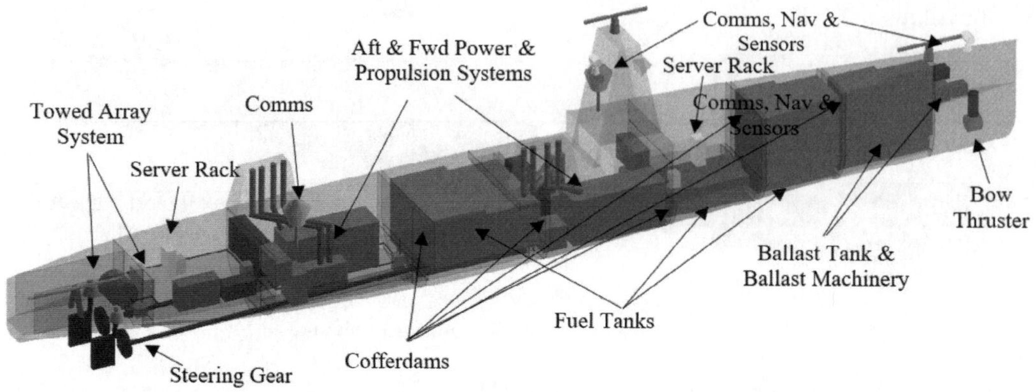

Figure 5: Updated Purple Fire USV system layout

The propulsion systems have been separated by moving one of the fuel tanks in between machinery spaces. 300mm cofferdams have also been added to all fuel tanks as a means to limit spillage into the compartments at the cost of reducing the fuel capacity of the ship. Separation of C2 redundancy has been improved by duplicating and moving servers apart as well as increasing distance between sensors and communication devices.

Figure 6 shows the PECD of the "Undertake Role" function for the updated LUSV model compared with the initial version.

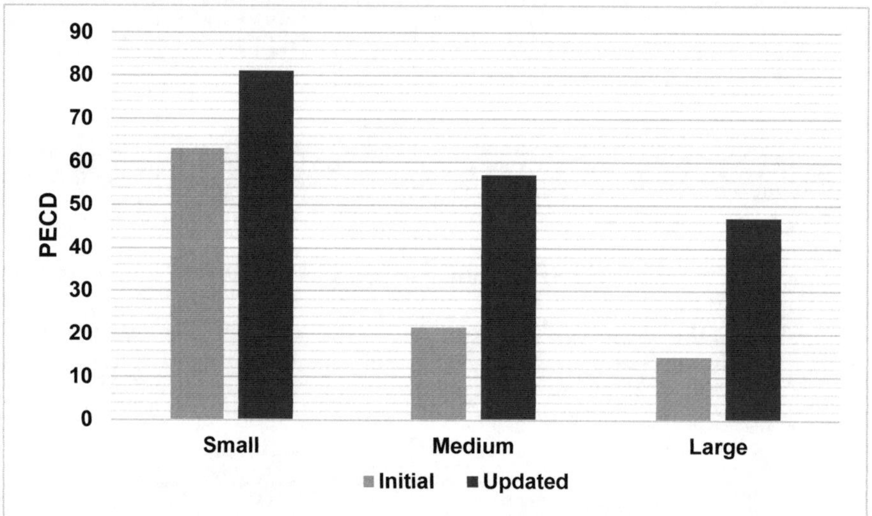

Figure 6: Updated USV compared against initial LUSV

There has been an improvement in the PECD of all of the ship's systems, which combine to give substantial improvements in the PECD for the overall ability of the ship to undertake its role. The use of redundancy in systems such as 'Communications' has made it much harder for one attack to kill the equipments that have been duplicated and positioned at different areas of the LUSV. The 'Minimum Mobility' function has become invulnerable to single hits by having the propulsion and generation systems separated further.

2.3 (c) Vulnerability Process For A LUSV

Whilst there may be differences in the survivability balance (between susceptibility, vulnerability and recoverability) in the requirements setting process for an uncrewed versus a crewed ship, existing processes and tools can be used to measure, identify and mitigate vulnerability. Indeed, removing the need for a design to support a crew removes many constraints. It is likely that lower vulnerability can be achieved overall for uncrewed ships of a similar size.

2.4 RECOVERABILITY

Whilst it has been shown that vulnerability may be optimised for the platform size, it should be recognised that without effective recoverability actions including containment and effective prosecution of fire and flood, a mission kill on the platform is all but inevitable and vulnerability must therefore be considered as intertwined with recoverability. This is the area in which the largest changes from a crewed platform to an uncrewed platform are necessitated because recoverability is currently conducted as a heavily human centric activity. Whilst some of the recoverability aspects can readily be automated, there exists a large challenge to several aspects particularly in situational awareness, containment and repair activities. By way of example, an attack that has resulted in internal blast and fragment damage to the LUSV and its systems and how the sub-disciplines of recoverability is considered has been presented below:

Situational Awareness:
On a crewed platform Situational Awareness (SA) is performed using a combination of search parties and sensors however, in a weapon event the local sensors and their cables are likely to be damaged and without significant hardening they are likely to be unusable or unreliable in the very place they are most needed.

Management:
Management of an internal battle may in fact be easier for an autonomous platform as it is not prone to common human errors, especially those due to complexity, stress, fatigue and exhaustion.

Containment:
Containment is arguably the biggest challenge with an autonomous platform for the simple reason that any system local to the weapon event is likely to be damaged. Therefore, the platform must achieve containment in one of 2 ways, either by utilising a system remote to the weapon event or by using passive containment such as blast and fragment resistant structure that will not be damaged in the event. At present, there are no autonomous systems in open source material that can achieve containment of a compartment/zone other than electrical and fluid isolating systems. Therefore the current generation of LUSVs would be forced to choose passive zonal containment through hardened structures capable of withstanding blast, fragmentation, fire and flood, which is possible but likely a significant cost driver.

Prosecution:
If containment can be achieved, the next challenge is prosecution in the form of extinguishing fires and removing flood water in a compromised compartment. Flood waters can be removed by eductor systems, just as for crewed vessels, but they will need to be located such as to be protected from weapon damage, an option for this could be integration into the double bottom where the system would be protected by layers of structure. Prosecution of fires will remain a challenge as fixed firefighting systems are generally vulnerable to damage. Crewed vessels have manual hoses for this eventuality but this is not currently an option for LUSVs and the most viable option at present is to simply flood the compartment via a protected watering/dewatering system.

Recovery:
Recovery of damaged systems is again a challenge for an uncrewed platform with the only viable strategy currently being the redundancy of systems and smart re-routing of fluid and electrical systems, which is likely a sub-optimal use of funds.

2.4 (a) Recoverability Summary

Recoverability for Large USVs is likely the most challenging, financially burdensome and least rewarding means of survivability, but without it any vulnerability reduction strategy is likely to fail due to the inevitable spread of damage over time.

3. DETERMINING THE REQUIRED LEVEL OF SURVIVABILITY

The starting point for determining the overall level of survivability required of a LUSV is to first assess what missions they are suited to and how they would undertake the task, for this the CONEMP is used. BMT are currently undertaking R&D in this space and have produced a CONEMP [10] suitable for LUSVs undertaking roles including; Anti-Submarine Warfare (ASW), intelligence gathering, seabed warfare and air-defence operations.

The roles identified are all defensive in nature and have threat environments ranging from permissive/low through to high in the case of air-defence. Given that the LUSV is likely to have modular mission payloads, the inherent characteristics of the parent platform must enable operations in a high threat environment.

On the assumption that the capability, as provided by multiple distributed LUSVs, must survive within this high threat environment, the question as to whether it can survive becomes one of attrition and probability. As the system components are attritted, the probability of survival diminishes and at a certain point becomes unacceptably low. This may vary depending on the role and the criticality of an individual unit to the overall capability. An example of this is the ASW role where a group of USVs are tasked with creating an ASW barrier, as given the probability of detecting an adversary submarine using sonars diminishes with range, the loss of a single unit would mean that the range needed to be covered by the rest of the squadron increases thereby reducing the probability of successfully deterring the enemy submarine.

Where a squadron of LUSVs is used to provide 'coverage' to a task there may be some overlap which can be used to maintain full capability, even with some attrition, this essentially provides system survivability through quantity rather than at the individual platform level, which is a useful strategy for smaller platforms where vulnerability reduction is not a viable approach.

3.1 CONEMP

The CONEMP for a LUSV based system is likely to rely on a mix of crewed and uncrewed platforms, with direction from the command vessel:

"Distribution serves two distinct purposes. First, a more routinely distributed force enables increased persistence, and greater and more lasting effect. This ensures the Maritime Force is routinely orientated to deliver effects below the threshold of conflict, and when the full array of Maritime Task Group capabilities is not required. Secondly, distribution enhances lethality and survivability. Taken in concert with a system of systems approach, the Maritime Force will more routinely exploit cooperative sensor-decider-effector systems to increase the range (and frontage) at which adversary threats can be held at risk or targeted. Distribution is used to maximise effect, to extend lethality and increase survivability. Distribution enhances the range, reach and effect of the Maritime Force."

The CONEMP describes 4 primary roles in which the Large USV can be used to gain an advantage, as previously detailed. The key benefits identified are:

- Endurance and persistence;
- Distribution of assets;
- Extended coverage;
- Scalability.

Survivability is important in this context as whilst a distributed system may inherently be more survivable, if each individual unit is not also survivable then the rate of attrition will be high, undermining the argument that survivability can be achieved with quantity and distribution.

3.2 WARGAMING

To aid the development of the survivability aspects, wargaming was used to explore the impacts of certain variables on the effectiveness of the LUSV based system. Wargaming was conducted using the game Cobalt Rocks developed by David Manley, Professor of Naval Architecture at UCL. The game is designed around seabed operations but has mechanics that can be used for ASW and Intelligence gathering missions. The Air Defence vignette was not wargamed as this is out with the game mechanics but inferences can be drawn from the other games.

The game is designed around a hexagonal (hex) grid with a nominal size of each hex of 20 nautical miles and a grid of 14 x 24 hexes or 280 x 480 nautical miles, which allows for dispersed operations. The game is played by two teams, red and blue. Red's objectives are offensive in nature whereas blue is the defending team. Each team has assets at their disposal with which to complete their objectives. The assets range from small UUVs to large Warships and Maritime Patrol Aircraft (MPA), each has their own sensors and effectors as well as attributes used in detection phases of the game. The probabilistic elements of the game are determined with the use of a D10 die. The game rules have set probabilities for a large number of activities, and these will be discussed further as each vignette is discussed in detail.

Seabed Warfare:

The first vignette gamed was the one most closely aligned to the game's ruleset – seabed warfare. Blue's objective is to prevent reds UUVs and submarines from gathering intelligence or compromising undersea infrastructure such as data and power cables, sensor networks as well as hydrocarbon pipelines. The primary detection method employed for underwater warfare is the Towed Array Sonar (TAS). The rules covering the use of passive TASs are fairly straightforward. There is a high probability of detection where the target and search platform share the same hex, moderate if in adjacent hexes and low if 2 hexes apart. These base rules are further modified by characteristics such as littoral environments, acoustic stealth and platform speed. A LUSV has a base probability of detecting a submarine within the immediate area of 7 hexes of 50%, or 20% in the area determined by a distance of 2 hexes, giving a total detection area of 19 hexes, as indicated in Figure 7 below. This demonstrates the potential effectiveness of a distributed system, in probabilistic terms. In this scenario in particular, the area of operations is also geographically constrained by the physical location of the underwater infrastructure, effectively reducing the game area and therefore making detection easier by a well distributed LUSV squadron.

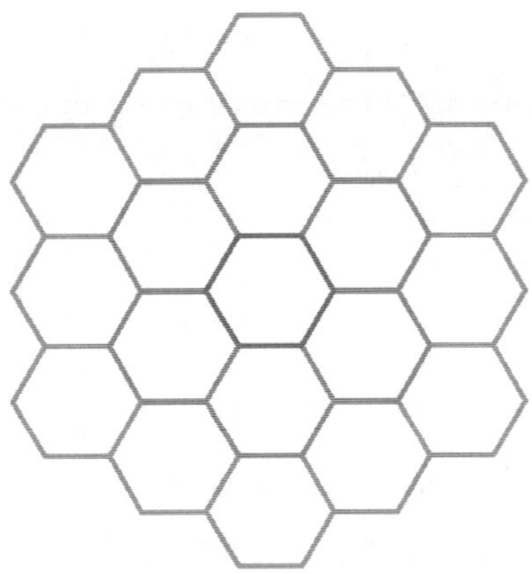

Figure 7: TAS detection probabilities: Green = 20%, Orange = 50%, Red = 80% (total area = 32%)

The other aspect to consider is the precision of the detection, though generally speaking once an underwater asset has been detected localisation and classification of it using air assets is relatively straightforward. If air assets are not available, LUSVs are able to 'go active' with their sonar, which gives the position of the LUSV away to the red assets in a relatively large area but crucially enables the LUSV to continuously track the red submarine and share information with the rest of the blue assets. This would ordinarily be a huge risk for a crewed platform, but a Commander may be better able to tolerate the risk of a LUSV being attacked.

Therefore, if red's submarine or UUV assets are detected the presence of blue's platforms will have a clear deterrent effect as red cannot defend itself against multiple blue platforms, crewed or uncrewed, due to the risk they could be armed with depth charges or torpedoes thereby forcing red to retreat. This shows a clear benefit of the distributed capability offered by LUSVs, something that would be prohibitively expensive to replicate using crewed vessels with a similar capability.

Anti-Submarine Warfare

In the ASW vignette the game environment is much less constrained, effectively open ocean bounded on either side by land, representative of a trans-oceanic bottleneck such as the Iceland-UK gap. In this scenario, the mathematics of detection become more favourable to red. The interesting dynamic that was observed in this scenario was that if red had good situational awareness of blue's surface asset location and the ability to communicate that to underwater units, it was easier for red to avoid detection, especially when blue uses predictable search paths.

In order to prevent red from having good situational awareness of blue's USV assets, a sensitivity study was undertaken that changed the surface rules involving the use of Radar and ESM in blue's favour to model the effect of having a very low Radar Cross Section and being able to operate at EMCON or by using Low

Probability of Detection comms. This enabled blue's LUSVs to remain undetected by red's surface assets, increasing the chance that blue could detect a red SSN.

This modification was rendered less effective when red was able to use a Maritime Patrol Aircraft (MPA) due to the coverage that the MPA capability provides. In future the author would like to re-run certain scenarios with the modified MPA rules to model the effect of LUSVs with low above water signatures as this would likely have a significant impact on the outcome of ASW vignettes. It is clear that this scenario shows the benefit of LUSVs as detection platforms due to the increased coverage provided by multiple platforms.

3.3 INTELLIGENCE GATHERING

In this scenario, it is the red team that uses LUSVs as the primary asset for collecting information on blue's fixed infrastructure using a variety of sensors including; optical surveillance and electromagnetic spectrum surveillance using Electronic Support Measures (ESM). This scenario yielded little in the way of information pertinent to this paper as red was able to achieve the objective with ease and it does not feasibly lead to a hostile engagement unless there are particularly trigger happy players. Blue may have greater success if it is able to deploy its own LUSVs in a sub-warfare manner, for example through jamming, cyber or nuisance tactics.

3.4 WARGAME SUMMARY

Whilst derived from a wargame with simplified encounters and engagements, the lessons that can be learned are generally valid as they are based on real-world constraints therefore while the actual variables involved may not be fully accurate, the methods of detection and engagement are. The key points raised by the games were that underwater warfare is incredibly challenging due to the vastness of oceans and limited detection ranges but also that LUSVs have key benefits in increasing the probability of threat platform detection due to their increased numbers and persistence. All of this relies on prior intelligence to work, if you don't know that the red team want to attack that particular area then you can't defend it with LUSVs, but this is also true of any warfare scenario. Also worthy of note is that there seems to be a more permissive RoE when an uncrewed vessel is considered, but this is likely to be an artefact of it being observed in a game. Anecdotally there are examples in the real world where this could have contributed, including the downing of a US RQ4 Global Hawk UAS by Iran in 2019, there has been much speculation over whether this would have occurred if the aircraft was crewed.

3.5 SYSTEM SURVIVABILITY SUMMARY

In order to achieve a desirable military outcome in the face of a capable adversary, attrition rates must be such that the capability remains effective until the end of the conflict. Even with a fleet of hundreds of LUSVs, a non-survivable system could be diminished to a non-credible capability within a matter of dozens of discrete engagements. If an individual platform has a vignette probability of survival (P_{Surv}) of a seemingly high 80% at the design threat level, the average number of engagements before suffering a mission kill would be just 5. On the other hand, if the P_{surv} were to be increased to 98% the average number of engagements required for a mission kill becomes 50. Whilst this is simplistic and does not account for attrition or damage in previous attacks, it raises a valid point. It is not inconceivable that in the era of robotic warfare a high value platform such as a LUSV may come under repeated attack in short order and therefore the author suggests a platform mission survivability of greater than 95% is essential to ensure an enduring system.

A high P_{Surv} can only be achieved with a very high PEH or a combination of a high PEH and a high PECD, as shown by the green cells in Table 3 below, noting that in this simplified example, PECD must also to consider secondary effects that may cause a mission kill some time later than the instantaneous damage.

It is also important to state that where multiple combinations of PECD and PEH give a similar overall P_{surv}, the value achieved with a higher PEH is always more desirable as repairs are not needed and full capability is retained, excluding the need to reload spent decoys and weapons.

Table 3. P_{surv} as a function of PEH and PECD

		PEH											
		0.1	0.2	0.3	0.4	0.5	0.6	0.7	0.8	0.9	0.95	0.98	0.99
PECD	0.1	0.190	0.280	0.370	0.460	0.550	0.640	0.730	0.820	0.910	0.955	0.982	0.991
	0.2	0.280	0.360	0.440	0.520	0.600	0.680	0.760	0.840	0.920	0.960	0.984	0.992
	0.3	0.370	0.440	0.510	0.580	0.650	0.720	0.790	0.860	0.930	0.965	0.986	0.993
	0.4	0.460	0.520	0.580	0.640	0.700	0.760	0.820	0.880	0.940	0.970	0.988	0.994
	0.5	0.550	0.600	0.650	0.700	0.750	0.800	0.850	0.900	0.950	0.975	0.990	0.995
	0.6	0.640	0.680	0.720	0.760	0.800	0.840	0.880	0.920	0.960	0.980	0.992	0.996
	0.7	0.730	0.760	0.790	0.820	0.850	0.880	0.910	0.940	0.970	0.985	0.994	0.997
	0.8	0.820	0.840	0.860	0.880	0.900	0.920	0.940	0.960	0.980	0.990	0.996	0.998
	0.9	0.910	0.920	0.930	0.940	0.950	0.960	0.970	0.980	0.990	0.995	0.998	0.999
	0.95	0.955	0.960	0.965	0.970	0.975	0.980	0.985	0.990	0.995	0.998	0.999	1.000
	0.98	0.982	0.984	0.986	0.988	0.990	0.992	0.994	0.996	0.998	0.999	1.000	1.000
	0.99	0.991	0.992	0.993	0.994	0.995	0.996	0.997	0.998	0.999	1.000	1.000	1.000

7. CONCLUSIONS

Through the activities carried out and discussed in this paper, it has been shown that in many respects LUSVs are no different to their conventional crewed counterparts and that furthermore they are likely to become essential assets. As such, the survivability of these assets must be considered and if high survivability is achieved LUSVs can be employed as valuable force multipliers.

The different roles undertaken by LUSVs will demand tuneable levels of survivability, with permissive operations below the level of conflict requiring no survivability measures other than a baseline resiliency compared to high threat roles such as Anti-Submarine and Anti-Air Warfare operations against a capable adversary, which will require the platform to be able to survive multiple encounters. It is also inferred that the level of survivability should not necessarily be dictated by the platform cost but by the value of the system as a whole. While the platform itself may be cheaper than a crewed alternative, the goal is not low-cost but to deliver the greatest effect for the budget and this is best achieved by having a system that would not need to be replaced at a high rate through an enduring conflict.

The challenge with achieving survivable LUSV platforms are two-fold. The first is a function of size as although large for a USV, 65m is a relatively small warship and all but the smallest threat weapons will overmatch the platform.

The second challenge is in achieving an effective replacement of the human activities in 3 primary areas:

- Pre-engagement tactical Susceptibility Reduction;
- Mid-engagement (weapon en-route) actions;
- Post-engagement actions, RoE & Lethality or recoverability.

From this we can conclude that the balance of survivability for LUSVs should be delivered through a layered approach with greater emphasis on susceptibility reduction, with vulnerability reduction and recoverability being limited to measures against proportionate threats. A reliance on susceptibility reduction characteristics over the ability to survive a successful weapon engagement may permit LUSVs to be smaller and cheaper, which in turn may allow susceptibility to be decreased further.

The particular measures that will need to be adopted by a platform will be determined from the role, for example a low magnetic and acoustic signature will be critical for a vessel conducting mine-warfare, whereas a vessel conducting air defence will need low above-water signatures as well as self-defence measures.

8. FURTHER WORK

This study has demonstrated clearly that LUSVs that are required to operate in hostile threat environments must be proportionately survivable to the role being undertaken. In the case of Anti-Submarine and Air-Defence operations, the LUSVs will be essential assets and therefore further work will concentrate on these particular roles. The aim of a future study will be to develop a concept platform capable of undertaking these roles as well as the survivability characteristics required, particularly in the susceptibility reduction space.

9. ACKNOWLEDGEMENTS

The authors would like to thank Professor David Manley for allowing them to use his wargame for the study and Ms Lucy Collins for providing her editorial skills to this paper.

10. REFERENCES

1. Soft Power: The Evolution of Concept; Nye, Joseph, 2021.

2. SAVITZ., et al., U.S. Navy Employment Options for Unmanned Surface Vehicles (USVs), 2013.

3. Introduction to Radar Systems: The Radar Equation, MIT Lincoln Laboratory, 2018.

4. WEEKS, G., TOWNSEND, J., and FREEBERSYSER, J., 'A method and metric for quantitatively defining low probability of detection', IEEE Military Communications Conference (1998), pp. 821–826.

5. WEISMAN, et al, 'Analysis of the low-probability-of-detection characteristics of ultraviolet communications, 2020.

6. YUSUF, T., et al, 'The Effect of Methanol-Diesel Blended Ratio On CI Engine performance', *University of Southern Queensland*, 2013.

7. Combat Safety – The Role of Vulnerability Assessment and Purple Fire in Meeting the Duty of Care, *RINA Warship 2024*, 2024.

8. Submarine Vulnerability Reduction WYVERN and Purple Fire, *RINA Warship 2023*, 2023.

9. Purple Fire Survivability & Lethality Assessment Tool, *RINA Warship 2018*, 2018.

10. Large USV CONEMPT, *BMT*, 2024.

9. AUTHOR BIOGRAPHIES

Daniel Patten currently holds the position of Senior Naval Architect at BMT. He is responsible for managing the survivability capability within the Naval Architecture department. The author's previous experience includes roles as the lead survivability engineer for multiple Royal Navy ship procurement projects and previously as above water vulnerability reduction lead engineer within the UK Naval Authority and Technology Group.

Cameron Clark holds the current position of Survivability Engineer at Survivability Consulting Limited, SCL. The Author's previous experience comprises nine months of working for SCL where he has completed vulnerability analyses of the Royal Navy Fleet, modelled new ship designs and conducted research into methods of improving vulnerability analysis for above water detonations.

CAN CURRENT AND FUTURE DIGITAL TOOLS SUPPORT EFFECTIVE AND EFFICIENT EVALUATION OF SUBMARINE CREWING?

H Morley and **J Green,** BAE Systems Submarines, UK

SUMMARY

Despite recent advances in autonomy and artificial intelligence, the human remains an integral part of any maritime platform capability. It is therefore critical to account for people related characteristics associated with performance and the safety case throughout the design lifecycle. Digital tools have been developed to support naval domain experts and stakeholders understanding of crew role, size and performance, for example: The Canadian Simulation for Crew Optimisation and Risk Evaluation (SCORE) tool and the Complement Generation Tool (CGT). However with digital tools now enabling at sea data collection this opens up opportunities to develop predictive human performance models which support efficient and valid ways for identifying performance issues with crewing roles, numbers, work system and watch schedules. For example psychophysiological measures can now be collected that provide data on crew fatigue, stress levels or cognitive load. Human data technologies, bio-mathematical modelling and associated methods of monitoring cognitive and physical performance can be used to provide evidence based crewing models. This paper presents a practical commentary on the potential value these tool offer, helping to support a focused approach to the development and uptake of these digital tools.

ACRONYMS

BCR	Basic Crewing Requirements
CGT	Complement Generation Tool
HF	Human Factors
HFI	Human Factors Integration
HP	Human Performance
LFE	Learning From Experience
MOD	Ministry of Defence
N/A	Not Applicable
SCORE	Simulation for Crew Optimisation and Risk Evaluation
SQEP	Suitably Qualified and Experienced Personnel
TAD	Target Audience Description
UK	United Kingdom

1. INTRODUCTION

The development of a maritime platform capability follows a four stage design systems engineering lifecycle: 1) Specification 2) Design 3) Development, Integration, Testing and Acceptance 4) In-Service [1, 2], and the human element of the capability needs to be considered at all stages of this lifecycle including activities associated with the generation and validation of the platform crewing solution [Figure 1].

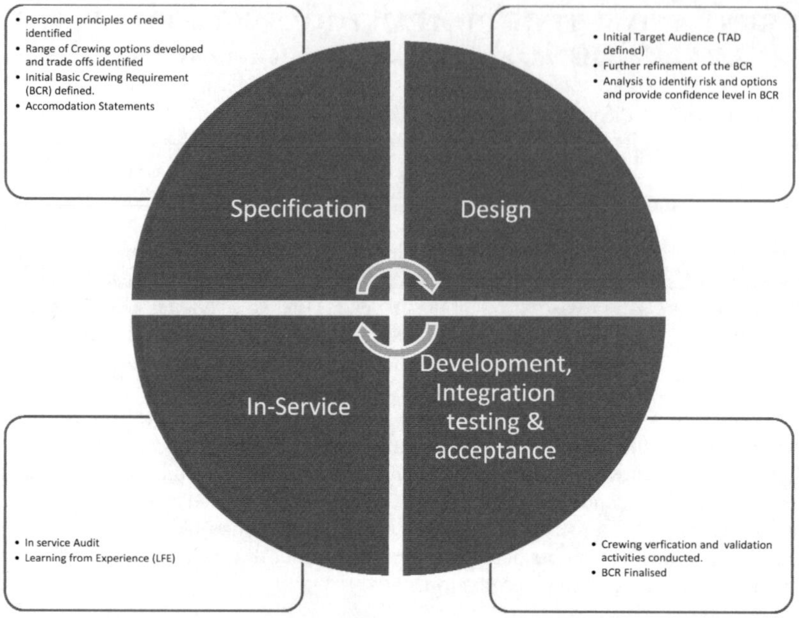

Figure 1: Four Stage Design System Engineering Lifecycle [1]

Human Factors Integration (HFI) is a management process for identifying, tracking and resolving human related issues to ensure a balanced development of both technological and human aspects of the capability [1, 2]. Human Factors (HF) is the technical discipline and profession concerned with the understanding of interactions between human and other elements of the system. HF focuses on delivering usable, safe, high performing and cost effective systems and the HFI process helps to ensure that a human centred process is maintained throughout the design lifecycle.

Generating the overall crew size and composition, and evaluating how equipment and technology capability changes impact on crewing roles, numbers and performance is a key HFI activity that needs to occur throughout the acquisition process, including early stages where the capabilities and limitations of systems and equipment can be unknown or subject to change. Digital tools have been developed to support naval domain experts and stakeholders understanding of crew role, size and performance and so this paper provides:

- An overview of approaches used in the generation and validation of the crew size and composition,
- A commentary on the key challenges faced with the use of digital tools for the use in crewing generation and validation,
- A way forward for the use of digital tool to support crewing generation and validation.

2. GENERATION AND VALIDATION OF MARITIME PLATFORM CREWING

2.1 CURRENT APPROACHES TO THE GENERATION AND VALIDATION OF CREWING

Through life personnel and training costs form a large proportion of defence expenditure often outweighing equipment costs, so there is a pressure to reduce overall numbers. Early identification of any opportunities to reduce crewing numbers is crucial if the navies are to align crewing systems, recruitment, organisational structure and training with the platform and support solution. With the emphasis on reducing numbers on board maritime platforms and overall through life personnel costs, understanding the risks of crewing changes including impact on habitability, human performance, safety, recruitment and retention is a key focus for HFI to ensure that there is a focus on optimisation on crew not just reducing numbers [3]. The complex nature of crewing on-board maritime platforms means that whole platform analysis has to be undertaken to understand if the crewing solution is optimised for not just cost, but performance and safety. For example it may be identified that a combat system technology removes an operator from an Operations/Control room, but this does not mean the removal of a person from the overall crew as they will have other roles for different states, watches and evolutions. Although the emphasis is on looking to reduce overall crew numbers crewing generation and analysis is needed not just to identify and understand areas of reduction but also to support:

- Habitability requirements and decisions e.g. number of bunks, sizes of catering areas etc. This is essential not just to determine specific requirements but also in order to establish the overall size of the Submarine platform since the areas like the accommodation and messes form a substantial percentage of the overall arrangement.
- The comparative analysis of proposed crewing solutions throughout the design lifecycle. This needs to take account of automation and new technology and how it interacts with crew size.
- Identification of gaps in roles, numbers in terms of operating & maintaining systems/equipment and issues of human performance (e.g. fatigue, cognitive load) for proposed crewing solutions, including the watch systems.

If conducted as an evidence based and integrated part of the design process the generation and validation of the crewing will save costs and potentially delays in terms of getting a platform into service. Early estimates of the size and composition of a maritime platform's crew are compiled within the Basic Crewing Requirement (BCR). The BCR should be used as part of the HFI process as the link between the equipment and system human operational and maintenance needs with personnel policy, standards and guidance. However since there are many uncertainties that will impact on the crewing option digital tools have been developed to support naval domain experts and stakeholders to generate and validate the BCR by providing analysis of crew role, size and performance, for example: the Canadian Simulation for Crew Optimisation and Risk Evaluation (SCORE) tool [4], and the Complement Generation Tool (CGT) [5]. It is worth noting that there is no agreed set of tools for maritime platform crew generation and validation, instead good practice has been developed to support BCR development, including information on the strengths and weaknesses of tools available [6].

The current digital tools available to support crewing generation and validation vary in terms of whether they take a 'Bottom up' approach or a 'Top down' approach to the generation of crewing options. Table 1 provides a summary of the typical input for the two different approaches. The key difference between the 'Bottom up' and 'Top Down' approaches is that the 'Bottom up' utilises current crewing structure and numbers, and amends according to differences in technology, policies and procedures. Whereas the 'Top down' approach starts by considering the mission and scenario demands and attempts not to be restricted by current policy, roles and qualifications. 'Bottom up' approach is by far the easiest approach to use, as means that the analyst can pull through current maritime platform crewing information. However unless the assumptions are reviewed and challenged this may mean that the crewing option generated is based on current practices and ways of working. The 'Top down' approach is thought to support a more innovative crew generation, but may go against current policy and personnel organisation for example it may result in a small crew number, but the captain may be assigned to husbandry activities. In practice a combination of both 'top down' and 'bottom up' approaches are taken, and that the generation and validation become mutually dependent with a progressive approach to the development of a final BCR.

Typically crew solution option validation and comparison is conducted by exploring a scenario and looking at crew usage in terms of an evolution or watch. This can be done by digital analysis, and both 'Bottom up' and 'Top down' digitals tools support crew validation by looking at crew utilisation. However table top exercises are also used as the table top activity supports a more interactive and collaborative approach to review and validation. These table top exercises usually use print outs of the layout of the platform and pins/disc to represent the crew. Both the table top approaches and digital tools can support the identification of gaps/omissions in the crew composition being analysed, but only the digital tools can be used to understand high operator/maintainer utilisation and provide objective data for crewing comparison. Therefore good practice for the validation of crewing options would involve both the use of digital tools and table top exercises [6].

Table 1. Typical Inputs for 'Bottom Up' and 'Top Down' Crewing Digital Tools

Inputs for 'Bottom up' Crewing Digital Tools	Inputs for 'Top Down' Crewing Digital Tools
Starting Point - a Watch and Station Bill from comparative platform	**Starting point** - a functional breakdown based on assumptions regarding technology, systems and equipment
Roles - Given technologies, policies and procedures	N/A
Scenarios – Missions, evolutions	**Scenarios** – a set of concurrent activities.
Crew – Number, occupations, ranks	**Crew** – number
Evolution demands – roles, hours and qualifications	**Evolution demands** – hours
Maintenance demands – roles, hours and qualifications	**Maintenance demands** - hours
Watchkeeping demands – roles, hours and qualifications	**Watchkeeping demands** – hours

2.1 CREWING VALIDATION and HUMAN PERFORMANCE

One of the key considerations when determining crew sizes is whether the crew option generated can maintain and sustain high levels of performance [7]. However as figure 2 summarises the current crew generation and validation digital tools available focus on utilisation; identifying crewing options and roles with high/low utilisation, and gaps in terms of personnel. It is the human performance (HP) models that will provide the information as to where there are potential performance issues by identifying crewing options and roles where there are high/low workload and fatigue. Therefore validation of crewing solutions to identify if it can maintain and sustain high levels of performance should conducted by using human performance models. That way there is understanding not just of crew/role utilisation and gaps but areas of potential performance risk due to fatigue and workload.

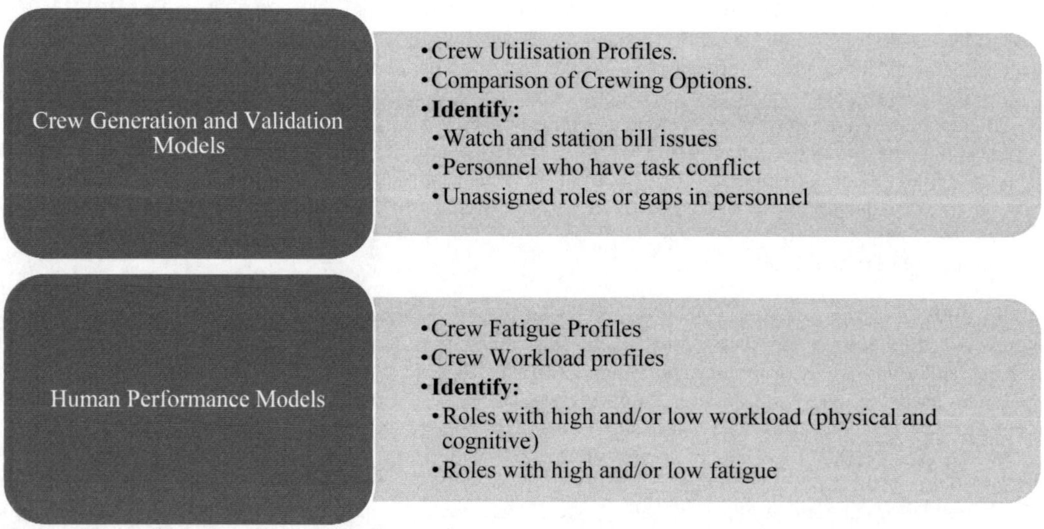

Figure 2: Output of Crew Generation and Validation Models and Human Performance Models

The issue with the HP models is that to have predictive validity there is a need to collect HP data, which requires a high level of time and effort for maritime platforms which tend to have large crews. However with the increase in the wearable technologies to collect HP data at sea this opens up opportunities to collect the data to support the development of high validity predictive human movement and performance modelling tools. For example psychophysiological measures can now be collected to provide data on crew fatigue, stress levels or cognitive load [8]. This is an exciting opportunity evantually enabling the creation of digital human/crew shadows, where simulations can replicate what happens in the real world but not when it is actually happening [8]. At the moment organisations tend to use separate models for crew generation and validation and HP, but ideally a digital tool that combines the predicted work schedule for crewing options into maritime platform work and sleep models would be used so that seamlessly able to look at crew utilisation, gaps, fatigue and performance in one tool [7].

3. CHALLENGES WITH USING CREWING DIGITAL TOOLS

There are digital tools already available to support the generation and validation of crewing options [4, 5 & 7], however there are challenges associated with the uptake and development of these including: 1) No agreed tool set for the generation and validation of crewing options for maritime platforms 2) Restrictions in sharing of scenario, functional and task analysis data 3) Conducting crewing analysis in the early stages of the lifecycle 4) Governance of personal data and 5) Time and effort associated with updating and managing large and complex data sets within digital tools [summarised in figure 3]. Sections 3.1 to 3.5 discuss these challenges with suggestions for mitigations.

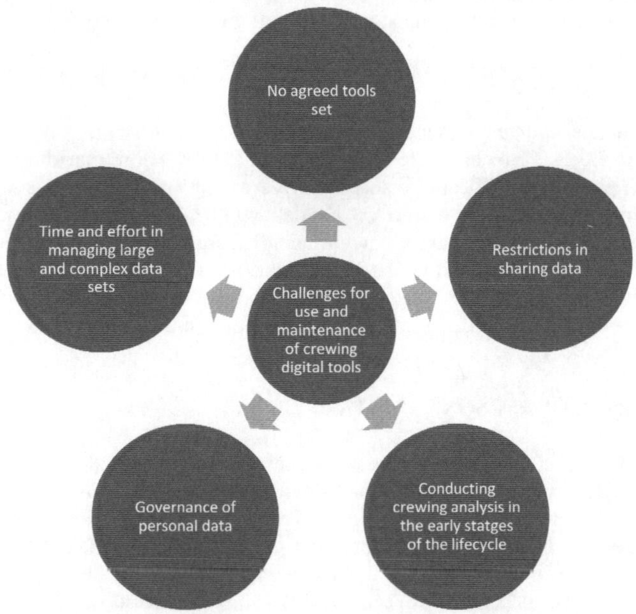

Figure 3: Challenges associated with the use and uptake of digital tools to support Crewing Evaluations

3.1 TIME and EFFORT to MODIFY LARGE and COMPLEX CREW MODELS

Crewing analysis is highly iterative, and there can be changes for many reasons [7]. For example Personnel Policy can mean changes to roles and structure and crewing assumptions; technology development may mean radical changes to the functions and tasks involved. This means that the crewing models need to be updated and run frequently, and the modification of large and complex models can take time and effort. If the models have complex data collection needs and the appropriate suitably qualified and experienced personnel (SQEP) are not available then this may mean they are not able to be updated in time to support programme decision making. Therefore there has also got to be a balanced consideration to the academic rigour of crewing generation and validation digital tools and the fact that they are to be used on an engineering programme. This means when down selecting a crewing generation and validation toolset the review is conducted not just by academics but those personnel who have the practical knowledge of engineering programmes and how personnel programmes are run in the owning organisation.

3.2 MAINTAINING A CREWING DIGITAL TOOLSET

There is not an agreed crewing generation and validation toolset maritime platforms rather a programme/organisation specific use of tools, and even within organisations different digital tools will be used. Since the crewing generation and validation is conducted by different organisations using different tools this means crewing technologies can come and go. The result of this is there is not an ongoing development of the tools or the knowledge and skills needed to use them, and sharing of the data from them may not be optimal due to differences in format and content. An enduring and centralised crewing technology and data capability would address this challenge, as would having an agreed toolset for the generation and validation of crewing for maritime platforms.

3.3 SHARING OF SCENARIO, FUNCTIONAL and TASK ANALYSIS DATA

Scenario, functional and task data is developed for every crewing activity for a maritime platform, but the sharing of the data between programmes can be difficult due to: 1) legal, security and commercial restrictions, 2) different formats and contents of the scenario, functional and task analysis data 3) the leaving of staff who were managing the crew generation and validation element of a maritime platform programme. Ideally owners of the data are identified at the specification stage of the programme who are able to ensure the appropriate permission for data in place that that support sharing of data across programmes. The use of a common digital toolset would also help to address this challenge, as this would mean that at the specification stage a clear need would be to pull through and re-use the data stored within digital tools for equivalent programmes.

3.4 CONDUCTING CREWING ANALYSIS IN EARLY STAGES OF LIFECYCLE

To evaluate the demands and tasks placed on operators and maintainers requires understanding of the equipment and systems, yet the technologies being considered for future maritime platforms may not have been used in other ships within the fleets, so access to relevant and detailed data is not necessarily available. This is also an issue in terms of human data as the HP data will be collected from currently serving personnel, and this may mean HP models and digital shadows are not representative of the knowledge, skills and abilities of personnel who will crew future maritime platforms. This is a particularly difficult challenge to address, but it does mean that there have to be varied stakeholders and experts involved who can challenge and debate the validity of data and assumptions. It is also important that the crewing digital tool can be kept, shared and used to inform simulation.

3.5 GOVERNANCE OF PERSONAL DATA

Since HP models will require data from serving personnel the governance of this data needs to be considered and when working on defence maritime platforms the permissions and controls are complex, and become more so when working on an international scale. Those involved in the collection and management of HP data will need training so that data is handled with respect to current data protection legislation and organisations will need the appropriate processes and procedures in place to ensure that there is the appropriate control over who sees the data and how the data is passed between networks and organisations. Consideration of this should be at the specification stage of the lifecycle as it may delay and even prevent the re-use and sharing of data if the appropriate consent has not been sought from participants.

4. CONCLUSIONS

There is an emphasis on trying to understand if technologies and policies can reduce the crewing numbers on maritime platforms. This has resulted in many of the current crewing generation and validation digital tools focusing on the identification of gaps in roles, numbers in terms of operating & maintaining systems/equipment rather than issues of human performance (e.g. fatigue, cognitive load) for proposed crewing solutions, including the watch systems. Part of the reason for this is due to the difficulty in collection and maintaining a data set for maritime platforms which tend to have large crews compared to Air and land platforms. The development of wearable technologies now offers an opportunity to collect and maintain a data set that can be used for maritime HP digital tools. Although there is an increased opportunity to develop and use HP models for crewing generation and validation there are challenges that need to be addressed, particularly in the managing, updating and sharing of complex data sets containing personal data. However it is likely that having an enduring and centralised crewing technology **and** data capability for maritime platform crewing would help organisations to address many of the challenges associated with uptake and use of digital tools for crewing analysis and to ensure that the appropriate governance and processes are in place for the collections, storage and sharing of HP data.

5. ACKNOWLEDGEMENTS

Would like to thank our colleagues at BAE Systems who have been involved in crewing generation and development for many year and provided documents and comments to support this paper including Alan Felstead, Chris Vincent, Vicki Wall and Charlie Mckee.

6. REFERENCES

1. JSP 912 Human Factors Integration (HFI) for Defence Systems. Part 1 Directive. V3.0 MOD. March 2024.

2. JSP 912 Human Factors Integration (HFI) for Defence Systems. Part 2 Guidance. V3.0 MOD. March 2024.

3. Maritime Acquisition Publication (MAP) 01 – 011 Maritime HFI Technical Guide. Official 2018.

4. CHOW, R., WANG, W., LAMB, M., COATES, C., PERLIN, M., and MCKAY, P., 'Simulation for Crew optimisation and Risk Analysis (SCORE)', In *TTCP HUM TP-18 Workshop on Crewing for Maritime Warfare, Portsdown West, UK*, May 2014.

5. WANG, W., 'Functionality and Usability Assessment of the Complement Generation Tool (CGT): A Preliminary Review', *TM 2011-057, Defence R&D Canada – Toronto*, April 2011.

6. COOPER-CHAPMAN, C., 'Identifying Good Practice in BCR Generation', *Naval Design Partnership (NDP) Presentation at the BAE HF Maritime SIG. Official*. Nov 2022.

7. CHOW, R., CDR MURKE, R., and LT CDR WITSKE, D., 'A Systems Approach to Naval Crewing Analysis Coping with Complexity', *DRDC- RDDC Report* P 117. 2015.

8. HAN, S., LI, F., LEE, C-H., WANG, T., and DIACONEASA, M A., 'Mirror the mind of the crew: maritime risk analysis with explicit cognitive processes in human digital twins', *Advanced Engineering Informatics Vol.62 Part B*. 2024.

7. AUTHOR BIOGRAPHIES

Harriet Morley holds the current position of Human Factors Technologist at BAE Systems Submarines. The Author is responsible for technical supervision of Research and Technology looking at both tools and models to support HFI Management Process and for HF applied activities including the generation and validation of crewing options. The Author's previous experience includes supporting the crew generation and validation for various MOD maritime programmes including Astute, T45, The Aircraft Carrier and T26, and SSNA. She was also the Chair and the UK representative on a five nation's panel and providing the UK input into several crewing digital tool Research and Development activities.

Jonathan Green holds the current position of Principal Operability Engineer at BAE Systems Submarines. The Author is responsible for ensuring that submarine design considers the role of the operator; and that systems, equipment and compartments can be operated and accessed by all personnel within the timescales required for both routine and emergency scenarios. His particular area of responsibility is for the Reactor Compartment and Manoeuvring Room Module on the next-generation SSNA platforms. The Author's previous experience includes ten years' service as a Royal Navy Marine Engineering Officer (Submariner), incorporating three deterrent patrols conducted onboard Vanguard-class submarines, two years spent in a project support capacity with the Submarine Delivery Agency and one year helping to generate the crew of an Astute-class submarine in build.

ADAPTING COMMERCIAL HYBRID-ELECTRIC PROPULSION SOLUTIONS FOR SMALLER NAVAL VESSELS: CHALLENGES AND OPPORTUNITIES

IS Makkar and **P Singh**, Indian Navy, India,
N Francis and **S Anish,** Cochin Shipyard Limited, India

SUMMARY

This study explores the potential of hybrid and electric propulsion systems for naval vessels, with a focus on the evolving landscape of the Indian maritime ship-building sector. As global naval operations increasingly prioritise sustainability, energy efficiency, and reduced environmental impact, the role of advanced propulsion technologies, particularly hybrid and electric systems, has become more critical. This study discusses how commercial hybrid propulsion solutions can be adapted and scaled for naval applications, drawing from India's experience in the commercial maritime sector, particularly through Cochin Shipyard Limited (CSL), a premiere ship builder for the nation, spearheading this green revolution. It examines CSL's work on electric-powered ferries, support vessels, and their pioneering efforts in all-electric naval vessel design, alongside global trends in hybrid electric systems for both commercial and naval ships.

This study emphasises the growing relevance and challenges of hybrid propulsion in small naval platforms, such as Yard Crafts, Auxiliaries, small naval ships, and other support crafts. These vessels, with unique operation-specific dynamic operating regimes and sizes for maximum propulsion speeds, spend the majority of their service life in low-part-load efficiency regimes, resulting in a significant loss in endurance. Hence, these platforms are ideal candidates for hybrid propulsion, as they typically operate in low-speed, quiet regimes and have substantial hotel load demands, making hybrid solutions efficient and effective alternatives. The key benefits highlighted include a reduction in emissions (~10 -15 %), maintenance requirements (CAPEX), increase in operational availability, and the potential for total ownership cost (TCO) reduction with design flexibility after a comprehensive operational exploitation review. Additionally, these solutions offer increased endurance, as they enable more flexible operational profiles and reduced reliance on fuel consumption (~10% reduction).

Beyond the environmental and operational advantages, this paper also outlines the strategic opportunity for the global maritime industry to leverage growing expertise in hybrid electric propulsion in India. The nation is uniquely positioned to provide cost-effective, sustainable, and high-performance propulsion systems for a range of naval applications, based on the expertise gained in hybrid solutions by Indian shipyards and niche capabilities developed by a range of Indian industries.

KEYWORDS

Hybrid-electric propulsion, naval vessels, auxiliary crafts, emission reduction, green shipping, battery technology, shore infrastructure, total cost of ownership, Indian maritime industry.

NOMENCLATURE

ASTDS	Approved Standard Tug Design and Specifications
D/E	Diesel Engine
DG	Diesel Generator
Fi-Fi	Fire Fighting
GCTP	Green Crafts Transition Program (India)
GHG	Green-House Gas
GoI	Government of India
GTG	Gas Turbine Generator
ICE	Internal Combustion Engine
IMO	International Maritime Organisation
INECR	Indian Navy Environment Conservation Roadmap
IOCL	Indian Oil Cooperation Limited
IPA	Indian Ports Association
LNG	Liquified Natural Gas
LPG	Liquified Petroleum Gas
MAKV	Maritime Amrit Kaal Vision 2047 (India)
MCS	Megawatt Charging System

MIV	Maritime India Vision 2030 (India)
MNRE	Ministry of New and Renewable Energy, Govt. of India
MoD	Ministry of Defence, Govt. of India
MoPSW	Ministry of Ports, Shipping and Waterways, Govt. of India
NCoEGPS	National Centre of Excellence on Green Ports and Shipping (India)
PRV	Polar Research Vessel
TCO	Total Cost of Ownership
TERI	The Energy and Resources Institute (India)

1. INTRODUCTION

The integration of hybrid and electric propulsion systems in naval vessels represents a significant advancement in maritime technology, offering numerous advantages such as improved fuel efficiency, reduced emissions, and enhanced design/operational flexibility. The Indian Navy plays a pivotal role in defending the nation's maritime interests and ensuring security across vast, strategically vital coastlines. As the focus is on modernising naval capabilities, the adoption of these systems is becoming increasingly relevant. With global attention turning towards sustainability and eco-conscious development, navies worldwide are being encouraged to reduce their environmental impact. The idea of a "Green Navy" has emerged in this context - one that integrates environmentally responsible practices into its operations without compromising combat readiness. A major development in the Indian ecosystem is the adoption of such 'green technologies' for in-house designed and built commercial ships. The shipbuilding industry in India, including the major public sector and private shipyards, is currently actively involved in this transition towards global sustainability targets. This study examines how global navies can pursue a sustainable future by adapting commercial green technology-based strengths while continuing to fulfil their critical national security mission. Multiple short- and long-term technologies, including design, propulsion systems, fuel systems, and operational practices, are available for this transition. The analysis primarily focuses on smaller warships and auxiliary yard crafts, which constitute nearly 35%-40% of the global naval fleet, to maintain or enhance operational readiness. The key benefits highlighted include a reduction in emissions, maintenance requirements (TCO and CAPEX), increase in operational availability, and the potential for total ownership reduction with design flexibility after a comprehensive operational exploitation review.

1.1 BALANCING ENVIRONMENTAL PERFORMANCE WITH OPERATIONAL READINESS

Naval shipbuilders are under growing pressure to align with emission reduction targets established by the International Maritime Organization (IMO) and broader national climate commitments (Figure 1 and 2). In response, there has been a marked shift toward integrated power and propulsion solutions that prioritise the environmental performance without compromising the operational effectiveness of military platforms. Unlike commercial vessels, naval ships must contend with highly dynamic mission profiles, where the power demand can shift dramatically within seconds and not within minutes. Therefore, any emission-conscious solution must ensure rapid power availability, redundancy, and resilience under combat and emergency conditions. This balancing act drives innovation in hybrid electric and integrated electric propulsion systems, energy storage integration for peak-load handling, and advanced power management systems. The steps taken are also to optimise fuel use and reduce emissions during variable-speed or low-load operations. The ability to reduce fuel consumption and greenhouse gas emissions is a key performance metric for future naval platforms. These developments reflect a broader trend toward green naval capabilities, in which sustainability and combat performance are designed to coexist and not compete. India's national commitment to achieve net-zero carbon emissions by 2070, in line with global environmental sustainability requirements, places shared responsibility across all sectors, including defence [1].

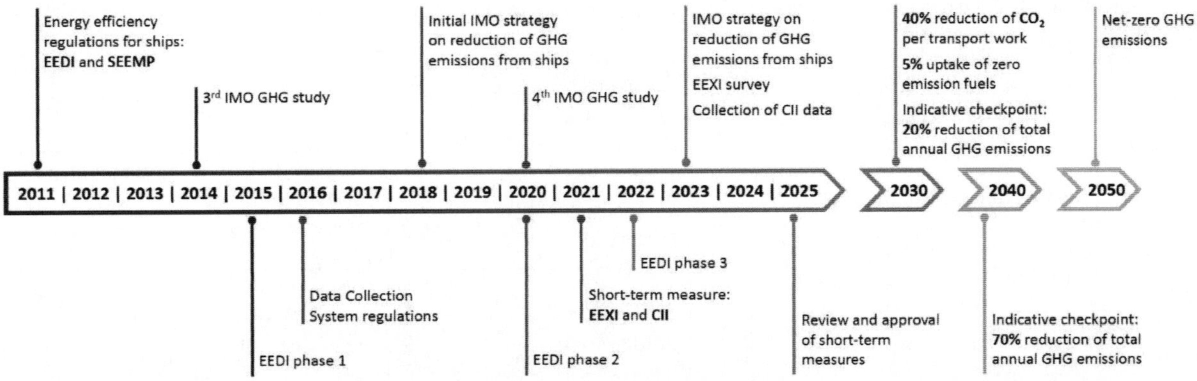

Figure 1: Chronology of the most significant measures implemented and an overview of the IMO's plan for GHG emission reduction by 2050 [2].

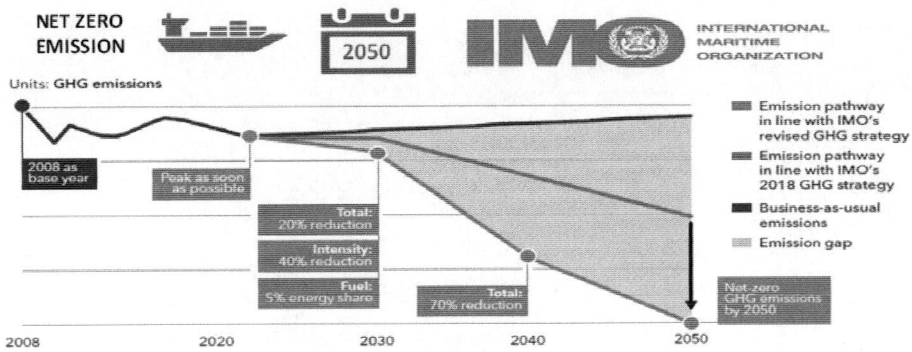

Figure 2: Emission Pathway for Maritime Industry [3].

1.2 INSTITUTIONAL COMMITMENT AND GREEN INITIATIVES

1.2 (a) Future Indian Ship Building/Green Shipping Industry Policies

India's shipbuilding sector is growing, with major ports expanding their capacity and implementing modernisation initiatives. India aims to play a significant role in the global shipbuilding industry by adopting sustainable practices. By focusing on sustainability and modernisation, India is poised to achieve its vision of becoming a leading shipbuilding nation, positively impacting its economy and maritime industry. The primary initiatives under the directives of the Government of India (GoI) and the Ministry of Ports, Shipping & Waterways (MoPSW) are presented in Table 1.

It is important to acknowledge that these initiatives are primarily directed towards the commercial shipping industry. This expansion is also expected to enhance inland/coastal cargo transport on Indian-built vessels (Figure 3), which are similar to auxiliary crafts/small naval vessels. While the intention is to increase the contribution of waterways from the current 8% with high dependence on roadways (Waterways shipping in Europe, USA: > 45%), this certainly remains a challenge with aging above 20 years for 50% of the current coastal fleet (Figure 4). Notwithstanding, with recent initiatives by the government towards shipbuilding and greener solutions, this also presents an opportunity for the global and Indian shipbuilding and ancillary industry [4].

Table 1. Government of India Initiatives for Shipbuilding and Green Maritime Transition [1] to [6].

Government Initiative	Ministry/ Organization	Year	Target Industry/ Vessel Type	Key Goals
Maritime India Vision (MIV) 2030	MoPSW	2021	Shipbuilding, Ports, Logistics, Inland Waterways	• 150+ initiatives to place India in top 10 maritime nations • Port modernization • Inland logistics
Harit Sagar (Green Port Guidelines)		2023	Shore Infrastructure at Major Ports	• Shore-to-ship power • Green hydrogen/ammonia hubs • Shore power by 2030
National Centre of Excellence for Green Ports and Shipping (NCoEGPS)	MoPSW & TERI	2023	Ports and Shipping Sector	• Promote sustainability through policy/ technology roadmaps • GHG emission reduction
National Green Hydrogen Mission	MNRE	2023	Maritime fuel supply chain, ports, and vessels	• Develop India as a global hub for green hydrogen production, storage and utilization • Support decarbonization of ports and adoption of hydrogen fuel in ships
Green Tug Transition Programme (GTTP)	MoPSW & IPA	2024	Harbour Tugs at Major Ports	• Convert all tugs to green fuels by 2040 • Reduce emissions by 30% by 2030 and 70% by 2047
Maritime Amrit Kaal Vision (MAKV) 2047	MoPSW	2024	Shipbuilding, Ship Repair, Transshipment Ports	• Place India in top 05 shipbuilding nations by 2047 • 4 million GRT capacity • Carbon neutrality
Harit Nauka		2024	Inland Waterway Vessels	• Promote green vessels • Enhance energy efficiency Adopt low or zero-emission technologies
Green Crafts Transition Programme (GCTP)	MoPSW & IPA	2025	Port Crafts (e.g., pilot boats, mooring boats)	• Promote eco-friendly operations • Retrofit port crafts with green technologies

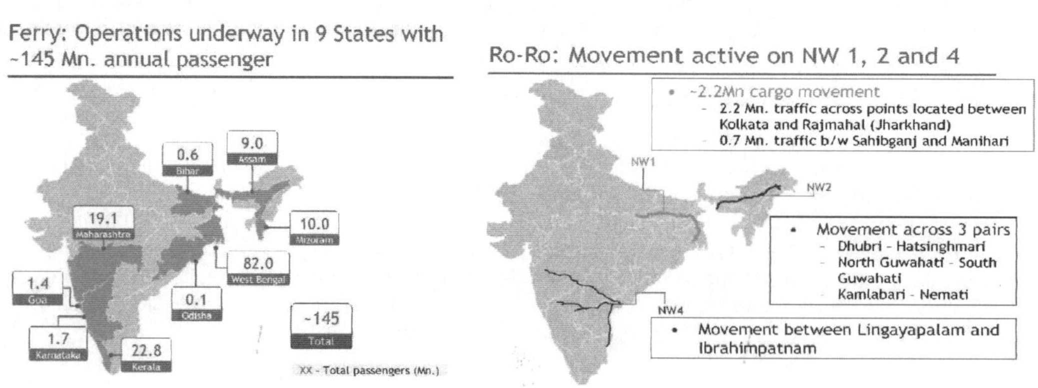

Figure 3: Inland Vessels Green Transition Plan (India): '*Harit Nauka*' [6].

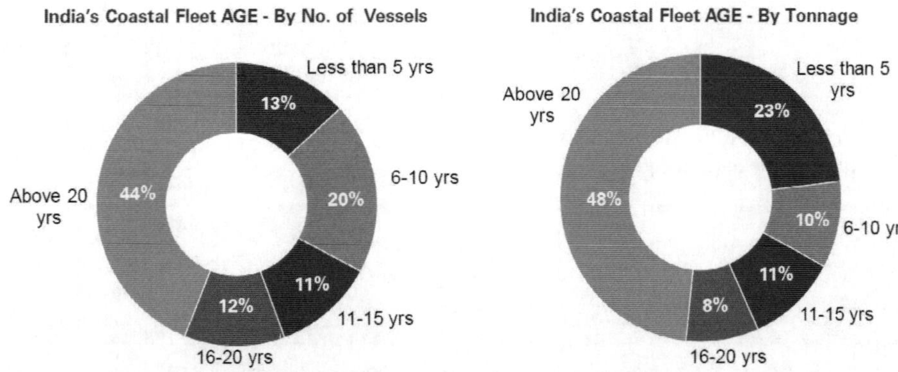

Figure 4: India's Coastal Fleet: Age Profile [4].

1.2 (b) Indian Naval Initiatives

The Indian Navy's commitment to sustainability was guided by the Indian Navy Environment Conservation Roadmap (INECR) launched in 2014. A 'Green Cell' oversees initiatives in alternative fuels, solar energy, waste management, and energy conservation. The Navy aims to reduce its environmental impact while maintaining operational readiness. In collaboration with Indian Oil Cooperation Ltd. (IOCL), the fuel specifications were revised to reduce emissions and maintenance requirements. The Navy emphasises ocean ecology protection and has implemented MARPOL regulations, equipping ships with pollution-control equipment. Energy efficiency is prioritised through the adoption of solid-state lighting and increased use of renewable energy. The Navy focuses on reducing power consumption and integrating environmental conservation into its operations. By adopting green practices and adhering to environmental standards, the Indian Navy demonstrates its commitment to ecological stewardship and sustainability, aligning defence preparedness with environmental responsibility. These efforts contribute to the nation's sustainability targets and showcase the Navy's proactive approach to environmental conservation.

2. TRANSITION OPTIONS FOR GREENER MARITIME PROPULSION

In light of global sustainability commitments, the accelerated adoption of alternative propulsion methodologies for future ships is crucial, particularly given the ageing fleet and stringent emission norms. Encouragingly, there is growing organizational support and increased interest from shipowners, shipyards, and ancillary industries in implementing these changes.

2.1 MARINE DECARBONISATION OPTIONS

Globally, several alternative options for this transition have been identified (Table 2 and Figure 5 [7], [8]. The table also highlights the Technology Readiness Level (TRL) for these options both worldwide and in India. This study primarily examines the option of propulsion methodology modification, which has demonstrated a significant impact on sustainable vessel performance, particularly in the commercial maritime industry for smaller naval vessels such as Auxiliary/Yard Crafts and Patrol Vessels (Table 2). The subsequent sections provide a rationale for identifying hybrid electric propulsion as an ideal candidate for early adoption in the target vessel category. The discussion further elaborates on the benefits and challenges associated with the adoption of this technology by naval vessels. An essential component of this transition is the presence of indigenous ancillary industries and shore infrastructure. Consequently, a thorough analysis was performed, and the results of potential emission savings and recommendations for future auxiliary naval fleets are detailed in the following sections.

Table 2. Technology readiness levels for key marine decarbonisation and propulsion innovations.

Alternate Options: Technology Area	Example Use Case	Global Technology Readiness Level (TRL)	Indian Technology Readiness Level (TRL)
Hydrodynamic Optimization	Hull form optimisation for reduced drag	TRL 8-9 (Widely implemented in new builds and retrofits)	TRL 7-8 (Tested in naval prototypes and commercial applications)
Advanced Propulsion Methodologies	Hybrid-electric propulsion, Shaft generators etc.	TRL 7-9 (Deployed in multiple green/hybrid ships)	TRL 7-8(Hybrid/electric/ Solar powered ships & ferries; limited appln in naval vessels)
Alternative Propulsors	Advanced propulsors- Pods, Waterjets, Azimuth thrusters etc.	TRL 7-8 (Operational in commercial vessels and fast crafts)	TRL 5-6 (Limited use in harbour crafts and research platforms)
Diverse Fuel Options	LNG, Hydrogen, Biofuels, Fuel Cells etc.	TRL 6-8 (LNG widely used, dual fuel, pilot projects in hydrogen and ammonia)	TRL 5-6 (LNG / Dual fuel tested; Hydrogen/Ammonia under trials/studies)
Operational Methodology Modifications	Speed profile optimisation and power management	TRL 8-9 (Adopted across fleets for efficiency and emission control)	TRL 7-8 (Applied in navy and commercial optimization strategies)
Shore Facilities	Onshore charging stations for electric ferries and cold ironing for ports	TRL 8-9 (Infrastructure in place at leading ports)	TRL 6-7 (Infrastructure in limited Indian ports/waterway routes and demonstration underway at selected ports)

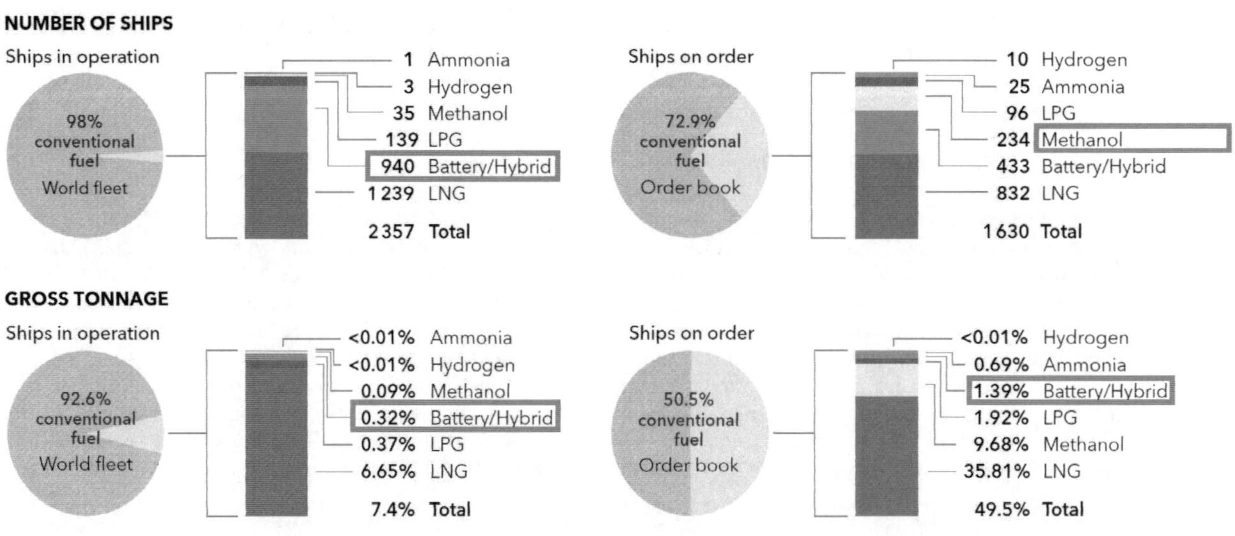

Figure 5: Alternative fuel uptake in the world fleet (based on the no. of ships and gross tonnage) [7].

2.2 GLOBAL UPTAKE: HYBRID ELECTRIC PROPULSION

Maritime hybrid electric propulsion systems are emerging as promising technologies for ships that offer a balance between environmental sustainability and operational efficiency. The statistics of the DNV class reports are shown in Figure. 6, indicating a high number of battery hybrids in operation (940 out of 2357 alternate fuel vessels) and on order (433 out of 1630 future ships), underscores the growing adoption of these systems. Using LNG or LPG for naval vessels poses challenges in terms of storage space, combat survivability, propulsion complexity, limited refuelling infrastructure, and increased thermal/signature risks, making them unsuitable for naval platforms [7], [8].

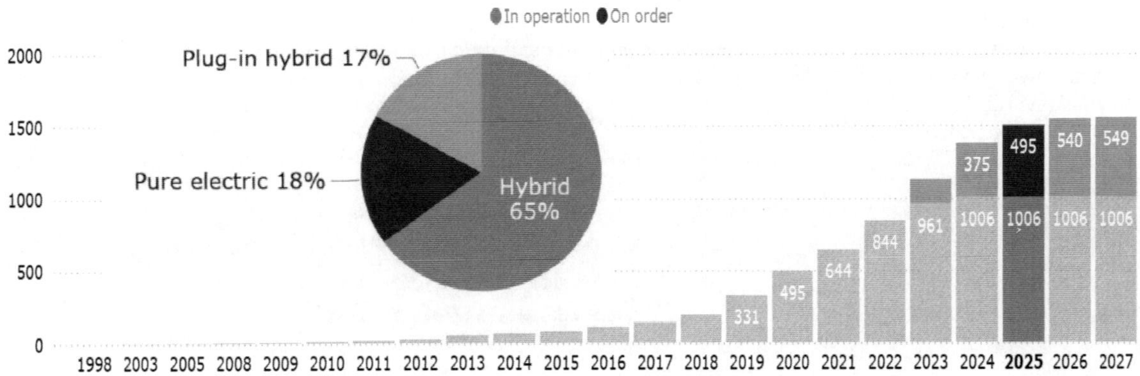

Figure 6: Battery Ships by Application. Source: Alternative fuel insights [7].

The choice between direct diesel generator-driven plants, battery options, and completely green sources, such as wind and solar, depends on factors such as vessel type, operational profile, and route characteristics. Although battery hybrids are gaining popularity owing to their immediate environmental benefits and operational flexibility, completely green sources are still in the developmental stages for large-scale maritime applications. The optimal solution often involves a combination of these technologies tailored to specific vessel requirements and operational conditions [3], [4], [9].

Hybrid electric propulsion systems for ships have various configurations, each with unique advantages. Integrated Full Electric Propulsion (IFEP) systems offer flexibility in vessel layout and power distribution, thereby enhancing efficiency and adaptability. Hybrid systems with batteries provide the benefits of energy storage, allowing for peak shaving and improved fuel efficiency. Plug-in charging systems with smaller batteries offer a compromise between electric operation and conventional propulsion, and are suitable for vessels with predictable routes and frequent port calls. A comprehensive examination of the available configurations, along with their respective advantages and disadvantages, has been conducted by multiple researchers [10]–[16], including a few recent publications by Sha et al. [3]. and Radica et al [4]. This examination encompasses the components of hybrid electric systems and the principles guiding their selection. Furthermore, TCO calculations for sample vessels akin to auxiliary naval crafts were performed by Sandith [9]. Currently, multiple options for hybrid propulsion are available, which need to be selected, modelled, and integrated along with the system components, considering the operating profile of the vessel. A comparative analysis of the various propulsion types is presented in Table 3 for ease of reference, based on the insights from the current study and the findings of Radica et al. [2].

Table 3. Comparison of the Propulsion System Configurations.

Propulsion Type	Emission Reduction	Maintenance Required	System Stability	Costs	Noise Reduction
Mechanical propulsion	✓	✓	✓✓✓✓✓	✓	✓✓✓
Electric propulsion	✓✓✓	✓✓✓	✓✓✓✓	✓✓	✓✓✓✓✓
Hybrid propulsion	✓✓✓✓	✓✓✓	✓✓✓✓	✓✓	✓✓✓✓
Hybrid propulsion with hybrid energy sources	✓✓✓✓✓	✓✓✓✓✓	✓✓✓✓	✓✓	✓✓✓✓✓
Electric propulsion, DC hybrid energy source	✓✓✓✓	✓✓✓✓	✓✓✓	✓✓✓	✓✓✓✓✓

2.3 HYBRID ELECTRIC PROPULSION SYSTEMS: DESIGN FLEXIBILITY

One of the primary reasons for this is that variable-speed-controlled electric motors exhibit high efficiency (approximately 95 %) across a range of 5% to 100% of the rated power. In contrast, combustion engines achieve an optimal efficiency of approximately 85–90% of the rated power, with a significant decline at lower load operations, exceeding the losses associated with electrical transmission. In a simplified manner, the electric propulsion system divides the power plant into several smaller units, rather than relying on a single large prime mover. This configuration allows for the starting and stopping of generator sets in response to varying propulsion and power demands, thereby ensuring that generator engines operate near their optimal performance point, as illustrated in Figure 7 [17]. The electrical losses at full power are compensated for by the increased efficiency during the part-load operation of the vessel. Thus, this fundamental principle is valid. However, during the 1990s, 2000s, and the 2010s, numerous variations and additional arguments have emerged regarding the benefits of electric propulsion.

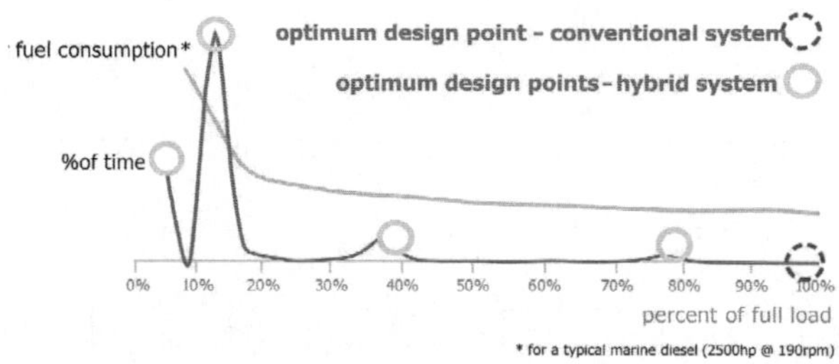

Figure 7: Hybrid Electric Propulsion Systems: Design Flexibility [17].

2.3 (a) Advantages of Improved Hybrid and Electric Propulsion Systems

Hybrid and electric propulsion systems present considerable advantages in marine applications, as extensively documented in the open-source literature. The appeal of hybrid electric propulsion lies in its technical maturity, simplicity, and adaptability. The primary advantages of this method are listed below.

- Flexibility to meet varying power requirements in different operational modes.
- Improved fuel economy by distributing propulsion load across multiple power sources.
- Lower maintenance costs owing to reduced engine runtime.
- Enhanced workplace health with reduced noise and hygiene of machinery space.
- Reduced emissions through optimised engine efficiency, meeting emerging environmental standards.
- Suitability for new construction and retrofitting of existing vessels.
- Particularly beneficial for vessels with frequent idle periods or fluctuating power requirements.

2.4 SHORE INFRASTRUCTURE

The availability of charging infrastructure and shore power connections is essential for effective implementation of battery-powered vessels. Recent regulations advocate for increased shore power availability in ports to mitigate emissions. Currently, charging facilities are predominantly located in Northern Europe, with some present in North America and East Asia [6]. The low Technology Readiness Levels (TRL) in India, as compared to global standards, are highlighted in Table 2, necessitating concerted efforts towards infrastructure enhancement to achieve future sustainability goals in alignment with planned ships. The Government of India (GoI) has shown significant interest in this endeavour under the Harit Sagar initiative and MAKV 2047 [1]. There is a pressing need for dedicated high-power charging infrastructure to support larger battery capacities.

2.4 (a) Selection of Shore Infrastructure

The determination of shore infrastructure upgrades, which influence the cost and efficiency of the system as well as the total cost of ownership (TCO) of vessels, is primarily based on the following factors:

- Operational Requirements: Charging time availability, battery type, operational profile of vessels, and cost.

- Shore Charging/Grid Type: Alternating Current (AC) or Direct Current (DC)
- Incoming Supply Type and Cost: Renewable or non-renewable sources.
- Cable Management Facilities: Options include manual, automated, tower, and induction types.
- Cable Connection: Combined Charging System (CCS) or Megawatt Charging System (MCS).
- Communication and Control Protocols: Battery Management System (BMS) and Power Management System (PMS).
- Supply Chain Availability/reliability and safety: Considerations for new installations and maintenance.

2.5 UNIQUENESS OF NAVAL VESSELS

Although naval vessels are typically not bound by international environmental regulations such as those outlined by the International Maritime Organization (IMO) or the MARPOL Convention, this exemption does not diminish the Navy's obligation to contribute to national climate goals. A gradual but firm move toward lowering its environmental footprint is both timely and necessary. The Indian Navy's foremost obligation is to maintain maritime security and protect national interests at sea along a coastline of more than 7500 km, as a well-balanced, future-proof, credible, and cohesive force. Although eco-friendly operations are becoming increasingly necessary, they must not come at the cost of operational strength or strategic agility. Hence, any shift towards greener technologies or sustainable procedures must be carefully calibrated to ensure that mission effectiveness is preserved. Accordingly, the selection of platforms and solutions for this transition should ensure long-term sustainability, unhindered operational readiness, and a reduction in the total cost of ownership (TCO = CAPEX + OPEX) [9].

3. LESSONS FROM COMMERCIAL INDUSTRY

3.1 MAJOR GLOBAL STAKEHOLDERS

Several key global players are at the forefront of the transition to hybrid electric propulsion systems by introducing innovative technologies and systems for maritime applications. Table 4 lists notable firms/shipyards in the maritime hybrid electric propulsion ecosystem. Among the Classification Societies, DNV is at the forefront of providing services for hybrid, electric, and green shipping requirements, with the majority of global ship orders [7]. While these major companies lead the charge, smaller firms and startups also make notable contributions, especially in niche areas, such as battery technology and energy management systems. Moreover, the entire maritime industry is being pushed towards more sustainable practices due to regulatory demands and environmental concerns, which may foster increased collaboration and technological progress across various sectors such as automobile, aviation, and land-based electric systems. This dynamic environment suggests that a range of stakeholders may continue to evolve as new players enter the market and existing companies expand their offerings.

Table 4. Notable Firms/Shipyards in Global Maritime Hybrid-Electric/Electric ecosystem.

Company	Country	Key Hybrid-Electric Systems/ Services Offered
ABB Marine & Ports	Switzerland	Electric propulsion systems, energy storage solutions, DC grid systems, Azipod propulsion, energy management systems, automation
Kongsberg Maritime	Norway	Integrated hybrid propulsion systems, energy storage, energy/power management, automation, propulsion control
GE Vernova (Marine)	USA	Electric propulsion systems, power conversion, energy storage, Marine propulsion motors, generators, power electronics
Siemens Energy Marine	Germany	Electric propulsion, energy storage, power distribution systems, motors, power management
Wärtsilä	Finland	Hybrid propulsion systems, energy storage, power management, hybrid power modules, energy management systems, motors
Rolls-Royce Power Systems (MTU)	UK	Hybrid propulsion systems, power management, energy storage, hybrid power packs, energy management systems
VARD Electro	Norway	Electric and hybrid system design and integration, hybrid propulsion integration, energy management systems
MAN Energy Solutions	Germany	Hybrid propulsion systems, energy storage, power management
Berg Propulsion	Sweden	Hybrid propulsion systems, energy storage, power/energy management, hybrid thrusters
AKA Canada	Canada	Hybrid propulsion systems, energy storage, power management
Leonardo DRS	USA	Electric propulsion systems, hybrid-electric and power conversion systems, energy storage, power management, DC/AC converters, propulsion modules
Corvus Energy	Norway	Energy storage systems, battery solutions, marine battery systems, energy management systems
Navalt Solar & Electric Boats	India	Ship design and construction, custom hybrid solutions, hybrid propulsion integration, energy management systems, battery systems,

Company	Country	Key Hybrid-Electric Systems/ Services Offered
Cochin Shipyard Advanced Systems (CSAS)	India	Ship design and construction, custom hybrid solutions, energy management systems, hybrid propulsion integration

3.2 MAJOR STAKEHOLDERS IN GREEN INITIATIVES: INDIAN SHIPBUILDING INDUSTRY

The Indian shipbuilding industry is shifting towards green technologies, focusing on alternative fuels and electric propulsion to reduce greenhouse-gas emissions. Public sector shipyards, such as Cochin Shipyard Limited (CSL), Mazagon Dock Shipbuilders Limited (MDL), and Garden Reach Shipbuilders and Engineers Limited (GRSE), are pioneering hybrid and electric propulsion technologies backed by government policies promoting indigenous shipbuilding. Private shipbuilders such as Chowgule & Company Shipbuilders Pvt. Ltd. and Navalt Solar & Electric Boats also deliver hybrid and fully green solutions, driving the trend towards sustainable green shipping. These shipbuilders are catering to both Indian and global maritime industries with a combination of in-house and foreign designs (Table 5, Figure 8 – Details from official shipyard/design house websites and press releases). The industry is also exploring e-fuels and battery-electric propulsion, aiming to reduce net fuel consumption and emissions, and aligning with India's commitment to decarbonisation and sustainable maritime practices. The specific projects undertaken by the CSL are listed in Table 6, which primarily provides data for the extant study.

Table 5. Major Indian Shipyards and Design Houses: Green/Sustainable Vessel Development.

Shipyard/Company	Sector/ Ownership	Green/ Sustainable Vessel Details	
		Constructed or in progress	Design Origin
Shipyards (India) in Green/ Sustainable/Hybrid Vessel Development			
Cochin Shipyard Ltd, Kochi	Government/ Min of Shipping	Electric Ferries*, Hydrogen Powered Ferry*, Zero Emission Autonomous Electric Barge, Zero Emission Feeder Container Vessel, CSOV, SOV, PSV, Passenger Ships, Green Tug*	*In-house & Owner
Garden Reach Shipbuilders and Engineers Ltd (GRSE), Kolkata	Government/ Min of Defence	Electric Passenger Ferry*, Hybrid MPV, PRV	*In-house & Owner
Mazagon Dock Ltd, Mumbai	Government/ Min of Defence	Hybrid MPV, Hydrogen Powered pilot Ferry*	In-house & Owner
Chowgule and Company Pvt Ltd, Ratnagiri	Private	Electric hybrid Plug-in General Cargo, Diesel Electric Short Sea Vessels & GC Ships.	Owner
Navalt Solar & Electric Boats, Kochi	Private	Solar Powered All-Electric Ferries & Workboats	In-house
Shoft Shipyard Ltd, Gujrat	Private	Electric Work Boat	In-house
Ship Design Firms (India) in Green/Sustainable/Hybrid Vessel Development			

Smart Engineering & Design Solutions (SEDS), Kochi	Private	Passenger vessels, MPV, Short-Sea Cargo Carrier, Electric Ferry, Ro-Ro ships	Owner/ SEDS
Vedam Design, Mumbai	Private	Green Hybrid Tug, CNG Tanker, Electric Ferry	Owner/ Vedam

Notes:

- The above list is dynamic with increasing interest in hybrid, electric propulsion and green shipping observed in shipbuilding sector - accelerated by recent government policies and global requirements.
- Traditional shipyards involved in conventional vessels are entering hybrid, sustainable and green shipping ecosystem.
- Design firms have requisite experience to bridge the requirements between foreign/Indian Owner and shipbuilder for green/sustainable vessel development.

Kochi Metro Ferry, CSL

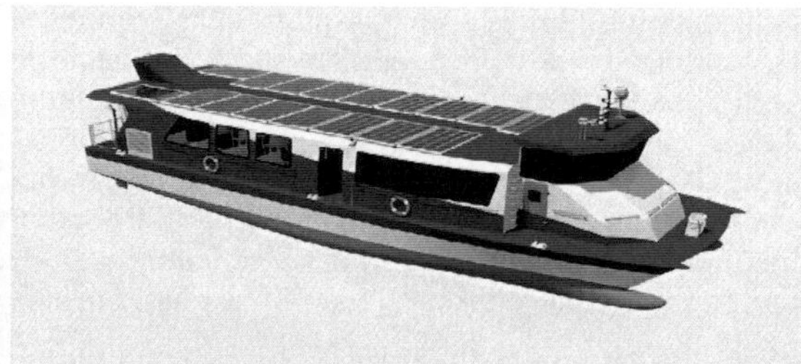

Electric Passenger Ferry, GRSE Ltd

Aditya Solar -Electric Ferry, Navalt

Baracuda Solar -Electric Boat, Navalt

MV Electramar, Chowgule and Company Pvt Ltd

Samskip Sea Shuttle, CSL

Figure 8: Some notable Green/Electric/Hybrid-electric Vessels built in India

Table 6. List of Green/Hybrid Vessels from M/s Cochin Shipyard Limited.

Vessel	Type	Ship Design	Green/ Hybrid Integrator	Class	Green/ Hybrid Tech	Remarks
Kochi Metro	Ferry	Owner (KMRL) & CSL (Indigenous)	Siemens	IR Class	All Electric	19/23 delivered
Hydrogen Boat	Ferry	CSL (Indigenous)	CSAS (In-house)	IR Class	All Electric: Fuel Cell+ Battery	01 delivered
Hybrid Electric Catamaran Vessel (HECV)	Ferry	CSL (Indigenous)	CSAS (In-house)	IR Class	All Electric	02/08 delivered
Green Tug	Tug	Robbert Allen & CSL (Indigenous)	-	IR Class	Plug-in Hybrid/All-electric	Preliminary Design completed
Asko Autonomous Barge	Barge	From Owner (Norway)	Kongsberg Maritime	DNV	All electric	02/02 delivered
Samskip Sea Shuttle	Feeder Container Vessel	From Owner (Netherlands)	ABB	DNV	Hybrid Electric - Fuel Cell + Battery	01 Construction in progress
CSOV	Offshore Supply Vessel	From Owner (Netherlands)	Kongsberg	DNV	Hybrid electric + Methanol ready	02 vessels construction in progress
SOV	Offshore Supply Vessel	From Owner (UK)	Vard Electro	DNV	Hybrid Electric & Alternate Fuel	04 vessels construction in progress
PSV	Platform Supply Vessel	From Owner (India/UK) & CSL	Konsberg/Vard	DNV	Diesel electric	10 vessels delivered
Passenger Ship	500 PAX Passenger	SEDS(India) & CSL	Siemens	ABS	Diesel electric	02 delivered

3.3 CASE STUDIES: GREEN/HYBRID VESSELS

A typical naval auxiliary fleet consists of various operation-oriented barges, such as fuel barges, ammunition barges, and sullage barges, as well as tugs, workboats, and ferries, all aligned with the requirements of fleet and shore establishments. To further explore green and hybrid electric designs for the potential identification of suitable candidates (for hybrid elects/green propulsion) among smaller naval vessels and auxiliary crafts, the authors examined completed and ongoing projects at Cochin Shipyard Ltd., Kochi. Subsequently, a proposal for a hypothetical small naval ship with an assumed dynamic speed profile to evaluate hybrid electric propulsion was presented, along with indicative advantages for a typical naval fleet. The following key projects at the CSL were considered.

- Case Study 1: Hybrid Electric Ferry, CSL Water Metro Ferry.
- Case Study 2: Green Ferry, CSL Hydrogen Boat.
- Case Study 3: Green Tug.

3.3 (a) Case Study 1: Hybrid Electric Ferry, CSL Water Metro Ferry

The Kochi Water Metro project, initiated by Kochi Metro Rail Limited (KMRL) and the Kerala Government, aims to connect ten islands through 15 routes spanning 78 km (in National Waterway 3). It will be the world's second-largest water transport system after Venice, utilising 78 hybrid electric ferries (Phase 1:19/23 boats delivered–being built at CSL). The custom-built low-emission project was a maiden electric vessel design for the Shipyard. The Kochi Water Metro aims to integrate rail and water transport, enhance connectivity, and improve the socioeconomic status of Kochi residents.. The project's total investment is Rs 819 crore (USD 100 million), with Rs 579 crore (85 million Euros) funded by Kreditanstalt für Wiederaufbau (KfW) through long-term loans. The vessels were twin-screw Aluminium catamarans with LTO batteries supplied by Echandia (Table 7, Figures. 9 and 10). Fast charging stations can charge batteries within 15 min. The PM propulsion motor, DC switchboard, and electric propulsion system integration were performed by Siemens [6], [18].

Although these vessels are relatively small compared to the larger commercial ships typically listed in Shipyard's order books, the project offered numerous valuable insights that were essential for advancing the yard's and the nation's green initiatives. These insights include the creation of domestic IR Class regulations for electrically powered ships, introduction of the first DNV Class approved LTO battery of this magnitude in the maritime sector, optimisation of the propulsion system and total cost of ownership, and identification of challenges related to the supply chain localization/indigenization of future electric or hybrid propulsion systems.

One main obstacle in designing boats for these waterways is ensuring navigation at low depths and through narrow passages while protecting the shores. In addition, boats must achieve an adequate speed to compete with road transportation. Maintenance poses another challenge in operating boats in backwaters. A daily upkeep is necessary to ensure that boats remain safe and operational. The project's estimated internal rate of return is 1.42%. The annual GHG emissions due to net near-zero emissions and post usage of shore charging from renewable energy sources is nearly 10,300 MT/year.

This model can be replicated in other Indian cities with inland waterways. Furthermore, this design is recommended for naval auxiliary ferries. However, route and operational cycle studies are critical prior to replication directly/with minor modifications.

Table 7. All-Electric Kochi Metro Ferry Details

General		**Propulsion System**	
Hull Material	Aluminium	Type	Hybrid Twin Screw
Superstructure Material	FRP	Diesel Generators	2 x 40 kW
Vessel Type	Passenger Ferry	Propellers	2 x FPP
Classification	Dual Class (DNV & IRS)	**Battery**	
DNV	+1A, IN (1,2) Metro Ferry	Type	LTO
	Battery Power)	Capacity	~120 kWh
IRS	+IWL, +IY, Zone 2, Ferry	**Passengers**	
Dimensions		Passengers	100
Length OA	24.8 m	Wheelchairs	2
Beam	6.4 m	Crew	4
Depth	1.7 m	**Emission Reduction**	
Draft	0.9m	GHG Savings	10,300 MT/year for the fleet
Performance		**Suitability for Naval/Coast Guard Auxiliary Craft**	
Design Speed (Hybrid Mode)	10 knots	***High***	
Service Speed (Electric Mode)	8 knots	*Minor modifications – Battery Type/Sizing, Partial Solar, Speed/Ops exploitation regime*	
Service Speed (Diesel)	8 knots		

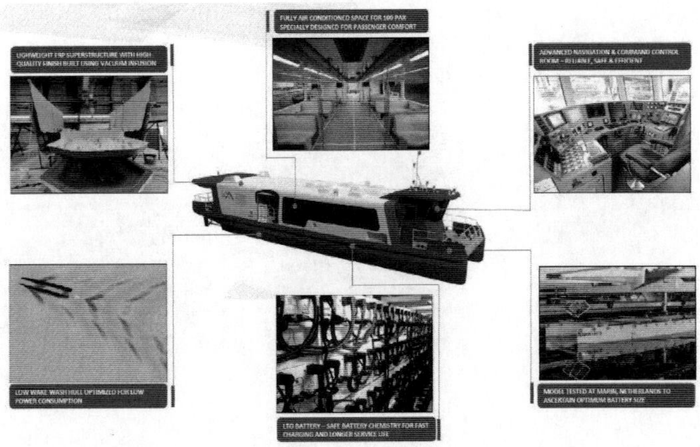

Figure 9: CSL Water Metro: Key Features.

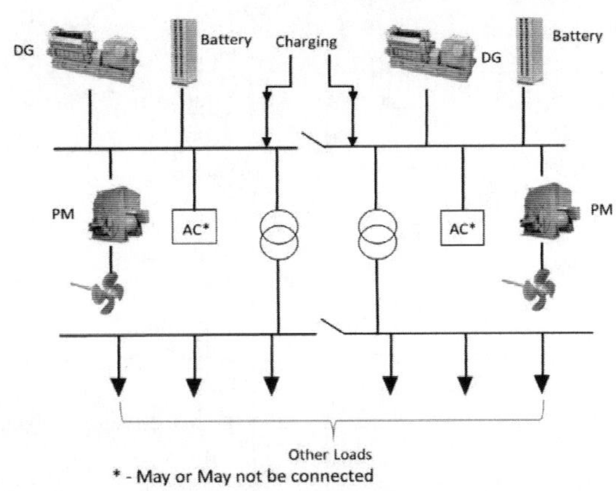

Figure 10: CSL Water Metro: Power Management Architecture.

3.3 (b) Case Study 2: Green Ferry, CSL Hydrogen Boat

In alignment with the Government of India's green hydrogen initiative and the 'Harit Nauka' mission, CSL has pioneered the creation of the country's first entirely indigenous hydrogen fuel cell catamaran ferry vessel as a pilot project in 2024. The zero-emission 100-passenger Hydrogen Fuel Cell Passenger Catamaran Vessel was designed and developed by CSL in partnership with KPIT, Pune, a leader in automotive industry innovations for the Inland Waterways Association of India (IWAI). Following pilot testing, the vessel is planned for deployment in Uttar Pradesh, specifically in Varanasi and the National Waterway - 3.

The project achieved significant indigenisation of the propulsion system, including the fuel cell and the integration of electrical systems. CSL's in-house services division, Cochin Shipyard Strategic & Advanced Solutions (CSAS), manages the crucial task of electrical system integration. The project also advanced the establishment of localised supply chains and offered valuable insights into the challenges of the hydrogen shore infrastructure. Notably, the electrical battery was designed to complement the hydrogen fuel stack in the green mode, utilising LFP technology for safer operation (Table 8, Figure 11 to 13).

In addition, inland waterway regulations for electric operations were facilitated, supporting subsequent electric ferry projects at the shipyard. This initiative promotes the use of hydrogen in marine applications, and the early adoption of hydrogen fuel cell technology in the marine sector will provide a global competitive edge, aligning with the industry's sustainable green energy goals for net zero emissions.

In its current form, the project may not immediately be suitable for naval implementation. However, in the long term, the stabilisation of hydrogen transport and shore infrastructure issues presents a promising solution.

Figure 11: CSL Hydrogen Ferry.

Table 8. Hydrogen Fuel Ferry Details, CSL.

General	
Hull Material	Aluminium
Superstructure Material	FRP
Vessel Type	Passenger Ferry
Classification	IRS
IRS	IRS Swastika IWL, Zone 2, Ferry Swastika IY, Battery Prop, FC (Main) – Hydrogen
Dimensions	
Length OA	24.8 m approx.
Beam	6.4 m approx.
Depth	1.7 m approx.
Draft	0.9 m (at full load)
Performance	
Design Speed (Fuel Cell + Battery)	6.5 knots
Endurance	Abt 8 trips (One trip: 1h operation +15 mins break)

Propulsion System	
Type	Twin electric motor propulsion
Fuel Cell	2 x 25 kW
Propellers	2 x FPP
Battery	
Type	LFP
Capacity	36 kWh
Passengers	
Passengers	100
Emission Reduction	
GHG Savings	244 kg CO_2
Suitability for Naval/Coast Guard Auxiliary Craft	
Medium	
Shore Infrastructure limitations for naval applications	

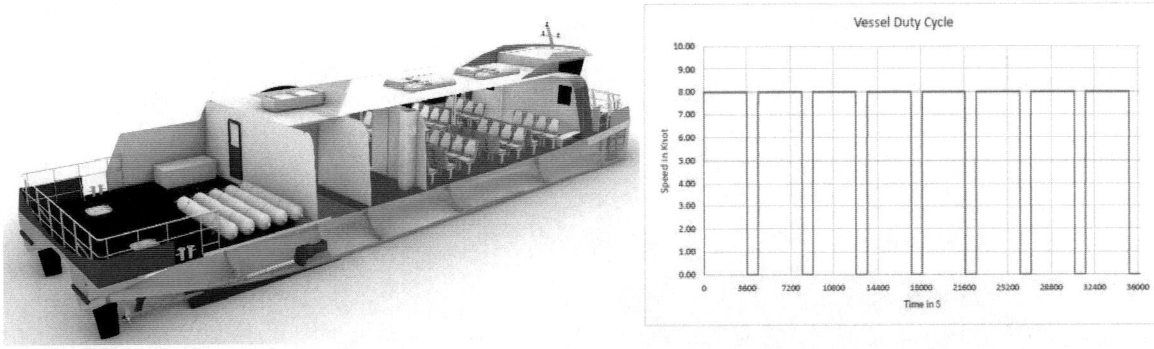

Figure 12: CSL Hydrogen Boat: Operational Profile

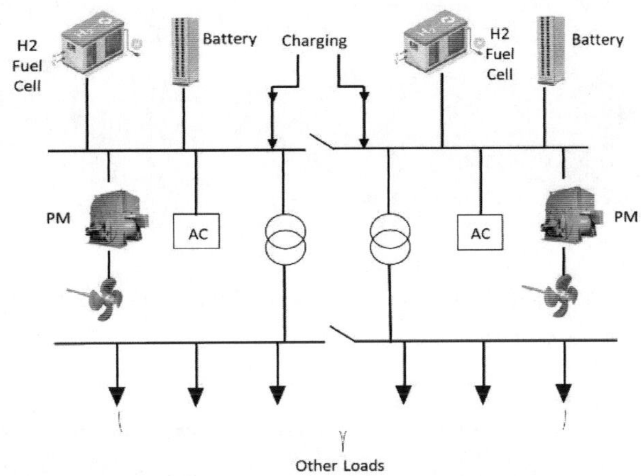

Figure 13: CSL Hydrogen Ferry: Energy Sources and Consumers (High Indigenisation).

3.3 (c) Case Study 3: Green Tug

Tugs are identified as a significant category of coastal vessels in India, contributing to the largest fleet among the vessel types. In October 2024, the Government of India introduced the Green Tug Transition Program (GTTP) as a strategic initiative to decarbonise the maritime sector. This program is designed to replace diesel-powered harbour tugs at major Indian ports with zero-emission green tugs utilising advanced technologies and cleaner fuels, such as battery-electric, hybrid systems, methanol, and green hydrogen. The first phase of the GTTP, spanning 1 October 2024 to 31 December 2027 will involve four major ports: Jawaharlal Nehru Port Authority, Deendayal Port Authority, Paradip Port Authority, and V.O. Chidambaranar Port Authority, each committed to acquiring or chartering a minimum of two green tugs. The initiative anticipates an investment of INR 1000 Crores, fostering domestic innovation and manufacturing under the 'Make in India' initiative, and generating employment opportunities in shipbuilding and design sectors.

By 2040, the program envisions a complete transition to green tugs at Indian Major Ports, with new tugs constructed post-2033 required to comply with ASTDS-GTTP standards. The GTTP is aligned with Maritime India Vision 2030 and Maritime Amrit Kaal Vision 2047, which aim to enhance sustainability, reduce carbon emissions, and integrate renewable energy into port operations. This programme underscores India's dedication to environmental sustainability and innovation within the maritime industry.

The primary configuration under consideration for this transition was initiated by the CSL (Figure. 14). The shipyard has undertaken detailed studies on four possible variants, as shown in Table 9 (including 20 years TCO, battery change at 10 years). These hybrid electric variants of tugs are suitable for naval tug implementation, considering their lower utilisation than that of commercial port tugs.

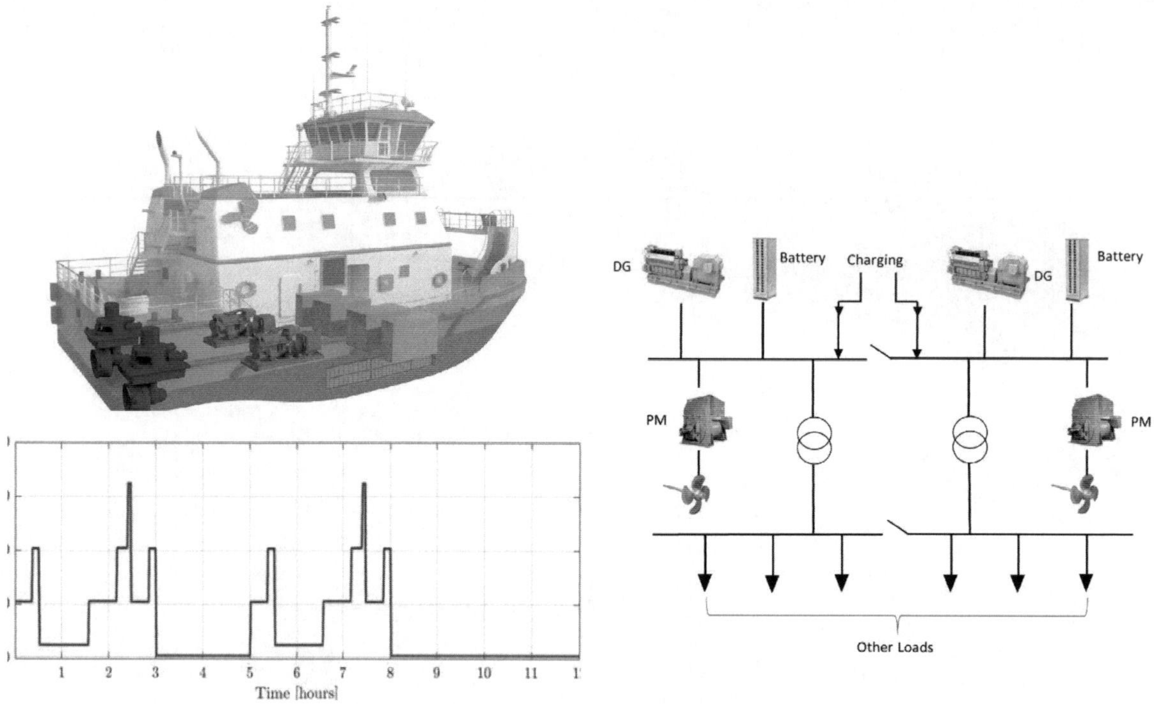

Figure 14: Green Tug Variant, Typical Operating Profile and Propulsion Layout Option (H1 Variant).

Table 9. Green Tug Variants, CSL.

Tug Variant	Propulsion Configuration	Emission Reduction	Remarks
M1 – Diesel Mechanical (Baseline Configuration)	2 x Main Engines + 2 x Gensets – No Batteries	-	Baseline configuration
H1 – Diesel Hybrid Configuration	Main Engines with Electric Boost to Props + 1 x Genset + Small Battery Storage	~20%	• Cost competitive with 20% CAPEX Subsidy. • Uncertain if sufficiently green to be eligible to obtain the subsidy
E1 – Battery Electric Configuration	Large Battery Storage (2 operating cycles on single charge) + 2 x Gensets for range extension & Fi-Fi	~65%	Cost competitive with 20% CAPEX Subsidy
E2 – Battery Electric Configuration	Very Large Battery Storage (>2 operating cycles on single charge) + 2 x Gensets for range extension & Fi-Fi	Zero Emissions	Cost competitive with 30% CAPEX Subsidy
Suitability for Naval/Coast Guard Auxiliary Craft:			
High			
Shore Infrastructure limitations for naval applications			

4. MAJOR FINDINGS: ADOPTION OF HYBRID-ELECTRIC DESIGNS FOR NAVAL AUXILARY CRAFTS AND SMALL NAVAL VESSELS

4.1 PROPOSED LAYOUT: HYBRID ELECTRIC SMALL NAVAL VESSEL

Drawing from the case studies in the previous section and a literature survey, a hypothetical configuration for a small naval vessel with a dynamic operational profile was conceptualised, as depicted in Figure 15. The prevailing design philosophy emphasises maximising speed performance, which necessitates appropriately larger propulsion systems. Currently, propulsion systems and diesel generators (DGs) for auxiliary loads operate at partial loads, which are notably fuel-inefficient and characterised by high specific fuel consumption (SFC) points. Various hybrid propulsion options are viable for these small vessels, ranging from all-electric systems to DGs with either direct motor/propulsor coupling or indirect configurations. In the indicative propulsion layout (Figure 16), the shock standards for the electric propulsion plant can be maintained as commercial/class standards for cost/weight reduction, and the conventional ICE system can be designed for shock standards to ensure survivability. The hybrid electric configuration is likely to provide significant GHG reduction (10-15%) along with fuel consumption optimisation (10%). The system affords a naval platform with a longer operational cycle and shorter maintenance duration, significantly enhancing the operational availability of the platform. However, a comprehensive understanding of operational practices is essential because these electric propulsion components are highly sensitive to speed variations and may result in an underpowered or overpowered system if not aligned with specific naval requirements. Notably, a configuration developed by a particular navy or coast guard may not be directly applicable to another naval fleet.

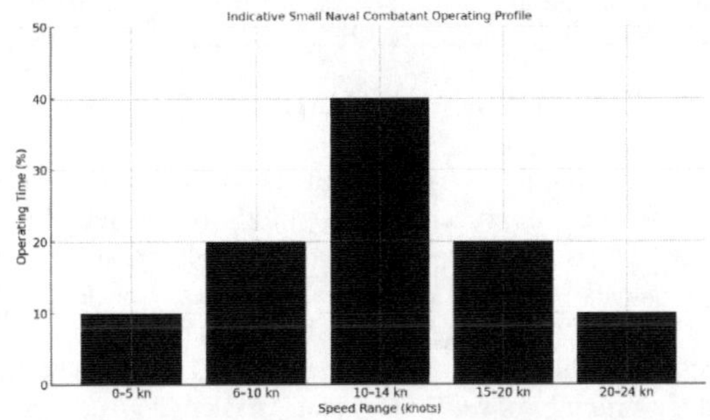

Figure 15: Hybrid Electric Small Naval Vessel: Indicative Operating Profile.

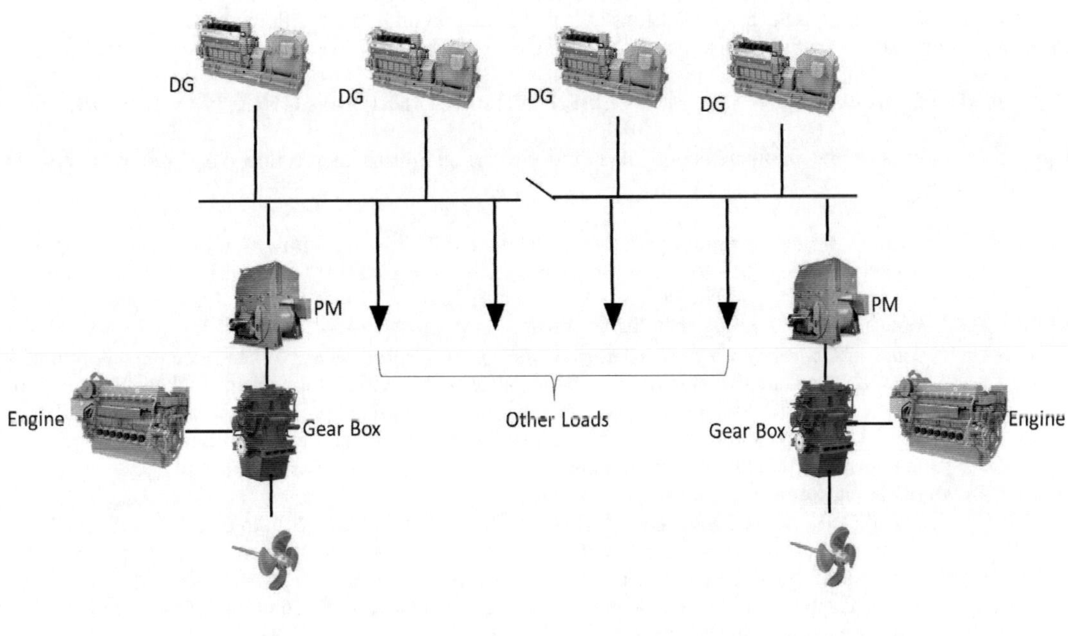

Figure 16: Hybrid electric small naval vessel: propulsion layout option.

4.2 IDENTIFICATION OF RELEVANT NAVAL VESSEL TYPES

Based on the discussions in Sections 3 and 4.1, and the identified advantages, the list of vessels recommended for green/hybrid transition along with indicative configurations are tabulated below:

Table 13. Recommended configurations: Naval auxiliary crafts and small naval vessels.

Vessel Type	Propulsion Type	Emission Reduction
Ferry	Battery-electric (Plug-in)	✓✓✓✓✓
Barge	Diesel-electric hybrid/ Battery hybrid/ Hydrogen fuel cell	✓✓✓✓✓
Tug	Diesel-electric hybrid/ Battery hybrid/ Full battery-electric	✓✓✓✓
Small Naval Vessel	Diesel-electric hybrid	✓✓✓

4.3 INDICATIVE SAVINGS FOR A TYPICAL NAVAL FLEET

The implementation of hybrid electric propulsion systems has been linked to a reduction in GHG emissions by approximately 10-15%, contingent upon the specific configurations utilised. Furthermore, optimising prime movers in accordance with the operational profile can result in a 10% decrease in fuel consumption. Flexibility in design and decreased maintenance costs present additional advantages. For smaller naval vessels and auxiliary crafts, a notable benefit is a substantial increase in operational cycles and enhanced availability. Overall, the total cost of ownership (TCO) of these vessels can be significantly reduced under both new construction and post-retrofit scenarios. However, it is crucial to recognise that challenges such as the development of shore infrastructure, technology obsolescence, and supply chain management continue to present concerns. Notwithstanding, the adoption of hybrid electric options by leading navies on more than 1000 smaller/auxiliary naval crafts can contribute to near doubling of the projected future hybrid electric vessels globally, along with associated emission reduction. Such contributions by naval forces, in alignment with the IMO objectives, would significantly support the maritime industry's efforts to achieve reduced emission targets.

4.4 FINDINGS TOWARDS NAVAL IMPLEMENTATION: HYBRID-ELECTRIC PROPULSION.

The specific findings of the implementation of hybrid electric platforms in meeting naval requirements are enumerated below:

- *Total Cost of Ownership Reduction*: A detailed TCO analysis with a conducive procurement strategy is essential for these ships instead of the traditional CAPEX-based selection for conventional ships. The design must be configured such that the benefits accrued in OPEX outweigh the CAPEX expenditure increase.
- *Study of Operational Profile & Review of Owner's Requirements*: When selecting an optimal power and propulsion system for a naval platform, it is critical for decision makers to begin with a thorough understanding of the vessel's intended operational profile. This forms the foundation for designing a system that is both technically appropriate and mission effective. Therefore, it is imperative that the owner's requirements be practical and close to the actual operating profile of the platform.
- *Naval Operational Uniqueness*: Naval operations often introduce unique demands that differ significantly from commercial applications, including variable mission roles, tactical mobility, survivability, and energy-intensive systems such as weapons, sensors, and communications. However, these will have to be met with judicious selection of a combination of propulsion and power-generation machinery.
- *Emission Reduction*: The selection of hybrid electric and battery-powered vessels is likely to

reduce life-cycle emissions by nearly 10 -15 %.

- *Fuel Optimisation*: Reduction in non-renewable energy consumption by nearly 10 % is achievable by implementing a simplistic hybrid-electric propulsion system.
- *Enhanced Operational Availability*: Significant operational period extension is feasible by reducing the off-design running hours of the ICE with an associated reduction in refit cycle/duration. This, coupled with the TCO reduction, may facilitate defence budget optimisation, which is critical for any global navy.
- *Multiple Technology Options*: Varied Hybrid-electric configurations/architectures are feasible (Grid, Battery, Motor, PMS, BMS, EES, charging methodology, etc.). Suitability must be assessed based on the operational profiles.

4.5 CHALLENGES FOR NAVAL IMPLEMENTATION: HYBRID-ELECTRIC PROPULSION.

The study highlights the following obstacles to implementing hybrid electric propulsion in naval ships/yard crafts:

- Operational practices/training for varied hybrid/battery-electric options.
- Ambitious user requirements for auxiliary crafts/smaller naval vessels.
- Shore Infrastructure and energy considerations
- Heavy reliance on non-renewable energy sources with an established supply infrastructure.
- Technological obsolescence. E.g. Battery, Motor, Fuel technology evolution and availability.
- Certification of electrical system components: Class/international regulatory framework with evolving technology and considering naval requirements.
- Limited technology expertise with smaller shipyards.
- Battery disposal and recycling.
- Global supply chain availability, including the local indigenisation component for continued availability.

4.6 RECOMMENDATIONS FOR NAVAL IMPLEMENTATION: HYBRID-ELECTRIC PROPULSION.

Based on the analysis of successful commercial projects worldwide and considering the expanding Indian ecosystem alongside the case studies presented in the preceding section, it is recommended that future small naval ships and auxiliary crafts adopt hybrid-electric and electric propulsion systems. These designs, which share characteristics with their commercially successful counterparts, present a viable option for a swift transition to a more sustainable propulsion strategy that offers assured environmental benefits. The specific recommendations for this transition are as follows:

- Collaboration between engineering and operational stakeholders for robust solutions:
  Rationalise speed requirements for auxiliary crafts/smaller naval vessels to optimise hybrid design efficiency and develop clear operating regimes (including in-service data exchange and continued analysis).
- Consider the operational profile, lifecycle, durability, and total cost of ownership (TCO) in the propulsion selection and procurement strategy.
- Address safety and operational risks in hybrid systems.
- Minimise the electrical conversion steps to improve the overall system efficiency.
- Ensure system flexibility and scalability for future upgrades.
- Pursue hull form optimisation, propulsion efficiency improvement devices, and hybrid electric propulsion methodology for higher effectiveness.
- Adapt mature commercial maritime technologies/platforms for naval applications.
- Review existing auxiliary crafts for retrofitting potential and extended service life.
- Concurrently, shore infrastructure has been developed to support green vessel induction.
- Crew training programs for energy-efficient operational practices.
- Encouraging indigenisation initiative/subsidies in the green shipping sector.
- Partner with global navies for sustainable cost-effective ship development and cross-pollination of technological advancements towards unified IMO goals.
- Invest in research and development for advanced electrical power technologies, including inter-sectoral interactions (automobile, railways, land-based plants, aviation, etc.).
- Sustainable institutional recycling policies.

5. CONCLUSIONS

The study concludes that hybrid solutions present a promising avenue for decarbonisation in the shipping industry. The integration of batteries with electric propulsion systems or standalone hybrid-electric propulsion configurations can significantly reduce energy consumption and emissions. The primary advantage of this transition is its ease of adoption and effectiveness in the near future. With the global and Indian commercial success of this hybrid electric transition, this study identifies potential classes of ship candidates for adoption within naval and coast guard fleets, specifically auxiliary crafts and smaller naval vessels. Given the emission reductions identified and reported in similar commercial vessels, the potential for overall emission reduction without compromising operational efficiency and with improved total cost of ownership (TCO) presents a compelling case for the adoption of these technologies in future global naval fleets. Most importantly, these options facilitate higher operational cycles, increased platform availability, and associated optimisation of the defence budget. With a favourable landscape for shipbuilding, a renewed focus on green shipping solutions and the saturation of available construction slots among global shipbuilding leaders, India is steadily advancing towards becoming a hub for sustainable, cost-effective ship designs and construction for the global market.

6. ACKNOWLEDGEMENTS

The authors express their sincere gratitude to Cochin Shipyard Limited for offering invaluable insights into critical green and hybrid electric shipbuilding projects. We also extend our heartfelt thanks to Professor O.P. Sha from the Department of Ocean Engineering and Naval Architecture at IIT Kharagpur, as well as Cmde. Vineeth Aerat (Retd), GM (Design) at Garden Reach Shipbuilders & Engineers (GRSE), Mr. Vivek Ramnath, CTO at Smart Engineering & Design Solutions (SEDS), Mr. Shaji P Janardhanan, CGM (Water Transport) at Kochi Metro Rail Ltd, and Mr. Sandith Thandasherry, CEO at Navalt, for generously sharing their expertise and experience. Their contributions significantly enriched the quality and perspective of this study.

7. DISCLAIMER

The views and opinions expressed in this publication are solely those of the authors and do not necessarily reflect the official positions, decisions, or policies of their affiliated institutions, including the Indian Navy and Cochin Shipyard Limited, India. Additionally, the data presented on indicative small naval platforms are intended solely to illustrate the potential advantages of the proposed technology, and are not representative of any existing or operational naval platforms.

8. REFERENCES

1. MOPS&W, 'Maritime Amrit Kaal Vision 2047', *Ministry of Ports, Shipping and Waterways, Govt of India*, 2023.

2. RADICA, G., VIDOVIĆ, T., ŠIMUNOVIĆ, J., and JURIĆ, Z., 'Overview of Hybrid Marine Energy System Configurations and System Component Modeling Approaches', *Energies*, 2025.

3. SHA, O P., SREEDHARN, R., and NAGARAJAN, V., 'Hybrid electric propulsion for GHG mitigation: Opportunities and challenges', *FROST International Conference on Green Technologies in Coastal and Inland Water Transportation, Bhubhaneshwar*, 2024.

4. VERMA, A., 'Leveraging defence ship building to catalyse India's shipbuilding industry', *KPMG India Publications*, 2020.

5. MOPS&W, 'Maritime India Vision 2030', *Ministry of Ports, Shipping and Waterways, Govt of India*, 2021.

6. DNV, 'Indian Coastal Green Shipping Programme 2024', *DNV Maritime Publications*, 2024.

7. DNV, 'Maritime Forecast to 2050', *DNV Maritime Publications*, 2024.

8. DNV, '2022 Energy Transition Outlook', *DNV Maritime Publications*, 2022.

9. THANDASHERRY, S., '100 Passenger Ferry Boats from Wooden/Steel Diesel to GRP Solar-Electric', *ResearchGate Preprint*, 2025.

10. PESTANA, H., 'Future trends of electrical propulsion and implications to ship design', *Maritime Technology and Engineering*, 2014.

11. ALEXANDER, D., 'Hybrid Electric Drive for Naval Combatants, *Proceedings of the IEEE*, 2015.

12. HANSEN, J F., and WENDT, F., 'History and State of the Art in Commercial Electric Ship Propulsion, Integrated Power Systems, and Future Trends', *Proceedings of the IEEE*, 2015.

13. SKJONG, E., VOLDEN, R., RODSKAR, E., MOLINAS, M., JOHANSEN, T A., and CUNNINGHAM, J., 'Past, present, and future challenges of the marine vessel's electrical power system', *IEEE Transactions on Transportation Electrification*, 2016.

14. KALIKATZARAKIS, M., GEERTSMA, R. D., BOONEN, E. J., VISSER, K., and NEGENBORN, R R., 'Ship energy management for hybrid propulsion and power supply with shore charging', *Control Engineering Practice*, 2018.

15. BALSAMO, F., CAPASSO, C., LAURIA, D., and VENERI, O., 'Optimal design and energy management of hybrid storage systems for marine propulsion applications', *Applied Energy*, 2020.

16. VERMA, J., and KUMAR, D., 'Recent developments in energy storage systems for marine environment', *Materials Advances*, 2021.

17. AKA Hybrid Propulsion – AKA energy system. (n.d.), from https://www.aka-group.com/marine-power/aka-hybrid-propulsion/.

18. MOEFCC, 'Kochi Water Metro Project EIA', *Ministry of Environment, Forest and Climate Change*, 2018.

10. AUTHOR BIOGRAPHIES

Commander (Dr.) Ishaq S. Makkar is a Senior Naval Architect and has done his B.Tech in Naval Architecture & Ship Building from Cochin University of Science and Technology and his M.Tech in Ocean Engineering & Naval Architecture from IIT Kharagpur. He also holds a doctoral degree from IIT Delhi in Experimental and Computational Fluid Dynamics. An alumnus of the prestigious Technical Staff Course for Tri-Service Officers at the Military Institute of Technology, Pune, has over 19 years of experience in warship operations, construction, repairs, inspections, and training. He has served as an adjunct faculty in the Applied Mechanics Department at IIT Delhi, where he taught courses on warship propulsion/design and submarine design. As a qualified Dock Master, he has led more than 500 docking operations for Indian naval ships and submarines. He has also contributed to the state-of-the-art Scorpene submarine project at Mazagon Dock Limited, Mumbai, in collaboration with the DCNS, France, and Navantia, Spain. Currently, he is the Senior Naval Construction Overseer of the Warship Overseeing Team (Kochi), Cochin Shipyard Limited. As the owner representative of the Indian Navy, he leads the overseeing team for the hull department on projects involving future anti-submarine warfare shallow water crafts and new-generation missile vessels. He is actively involved in ship hydrodynamics, aerodynamics, model testing, liaising with classification societies and maritime robotics.

Commander Prashant Singh is a distinguished Senior Naval Architect with over 19 years of specialised experience in ship design, construction, repair, and training. He holds a Bachelor of Technology degree from the Cochin University of Science and Technology and a Master of Technology degree from the prestigious Indian Institute of Technology, Kharagpur. Currently, serving as the head of the Forward Design Group at the Warship Design Bureau, he spearheads the development of next-generation warship designs for the Indian Navy. His leadership extended to pioneering futuristic naval platforms that defined India's maritime capabilities for years to come. He has played a pivotal role in the design and development of advanced stealth frigates and corvettes for the Indian Navy, significantly contributing to India's indigenous shipbuilding capabilities. His expertise encompasses design optimisation strategies aimed at maximising operational effectiveness while implementing streamlined process flows. His professional focus centres on enhancing platform performance through innovative design solutions, ensuring that Indian naval vessels maintain their technological edge in modern maritime operations.

Nibin Francis holds a B.Tech in Electrical and Electronics Engineering from RIT Kottayam and brings over 18 years of experience in ship design. He currently serves as an Assistant General Manager, heading the Basic Design Group at Cochin Shipyard Limited, where he leads the development of cutting-edge designs for both commercial and defence vessels. Notably, he led the design team behind India's first indigenously developed hydrogen fuel cell-powered vessel, and was actively involved in the Kochi Water Metro electric ferry. His professional interests lie in green and energy-optimised marine designs, with a focus on alternative fuels, such as hydrogen and battery technologies.

Anish S is the Deputy General Manager at Cochin Shipyard Limited's Strategic and Advanced Solutions (C-SAS) division. He has led system integration efforts for CSL's green and hybrid-electric vessel programs, including inland and coastal ferries. He plays a key role in India's transition to eco-friendly maritime solutions, with more than 15 years of expertise in marine electronics and sustainable propulsion. Mr. Anish is actively involved in advancing the electric and hybrid propulsion technologies through indigenous shipbuilding initiatives.

Warship 2025: The Future Fleet: Smart Technology, Sustainability and Autonomy
16th – 17th June 2025
Glasgow, UK

VALIDATING OPERATIONAL SCENARIOS THROUGH SIMULATION

J Orlowski[1], **K Droste** and **B Mes**, Damen Naval, Netherlands, **A, Kana** Delft University of Technology, Netherlands

SUMMARY

This paper presents an approach for integrating naval operations simulations into the early stages of vessel design, providing a means to evaluate the operational capabilities of proposed designs. The primary objective is to support designers in making informed decisions during the initial design phases, where insights into the potential operational impacts of design choices are often limited. To establish the approach, a study was conducted to examine current norms, regulations, and available simulation software options, identifying key factors that must be considered when applying the proposed method within the design process. The method was then formulated and tested through a case study related to USV operations, providing simulation results for both the overall duration of the operation and the effectiveness of potential systems. The findings highlight how simulation-based evaluations can enhance the design process by supporting informed decision-making and ensuring that vessel capabilities align with operational requirements.

1. INTRODUCTION

Ships are widely recognized as some of the largest and most intricate man-made systems, characterized by their complex structures and the numerous regulations and requirements they must adhere to. The ship design process reflects this complexity, integrating multiple considerations and requiring expertise from various disciplines. Among the different stages, the concept design phase is considered one of the most critical. At this stage, the requirements and mission objectives elucidated by stakeholders are translated into the specific capabilities the vessel must possess [1].

During the concept phase, designers often make decisions that have a lasting impact on the entire project. These decisions not only influence costs and restrict design flexibility but also shape subsequent choices later in the process [2], [3]. However, at these early stages, designers often lack sufficient knowledge about the implications of their decisions and their potential consequences [4] (see Fig. 1). This limited insight can reduce the project's overall effectiveness and may lead to costly modifications in later phases.

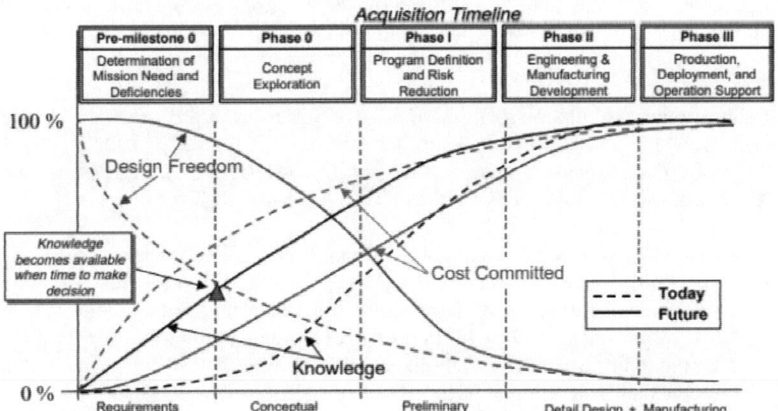

Figure 1: The relationship of design freedom, knowledge and cost committed. [4].

As a result, the need for a new design method has emerged. Recently, many companies and organizations recognized the advantages of Model-Based Systems Engineering (MBSE) and Mission Engineering (ME) [5]. While these approaches offer valuable benefits, they do not fully resolve the underlying challenges. Traditional design methods often rely heavily on the experience of designers, which can be limiting—especially for innovative solutions. Expertise is typically specialized, accessible primarily to seasoned professionals, and less applicable to

[1] Work performed while at Delft University of Technology

novel designs, which are essential for companies looking to stay competitive and develop long-term solutions or national defense organizations operating in an evolving global threat environment.

This issue is particularly evident in modern businesses, which must continuously refine their processes to deliver high-quality products and maintain their market position. Therefore, the research gap addressed by this project lies primarily within the concept design domain. The objective is to enhance the understanding of potential solutions, enabling better-informed decision-making before critical commitments are made.

The aim of this project is to enhance the assessment of capabilities during the conceptual design stage by integrating military operations simulations with a system model of the operational requirements for future military platforms. This approach seeks to improve the understanding of a vessel's required capabilities early in the design process by providing a tool to explore the 'performance space' before the full system composition is finalized [6]. Additionally, the outcomes of this process will serve as an evaluation for operational needs analysis, which is currently based primarily on the expertise and experience of specialists.

2. REQUIREMENTS AND REGULATIONS

Given the primary focus on developing a method for real-time applications, it is essential to closely examine current norms and regulations to ensure that the proposed solutions align with established recommendations and restrictions.

2.1 NORMS AND REGULATIONS

In recent years, legislative efforts have focused on maximizing the potential of modeling and simulation (M&S) within the North Atlantic Treaty Organization (NATO) to enhance operational efficiency and cost-effectiveness [7]. These initiatives resulted in the development of the Modelling and Simulation Strategic & Implementation Plans for the Alliance, which serve as key implementation documents and establish the foundation for future M&S advancements in 2010.

With the introduction of High-Level Architecture (HLA) in evaluation and simulation systems, the STANAG 4603 standardization agreement was issued. This document's purpose is to provide system-level interoperability between command and staff, tactical, and individual modeling, and simulation systems. All participating nations are required to incorporate specific standards into their processes, as outlined in various norms published on the subject [8]. The outlined standards establish guidelines for the simulation's adherence, encompassing aspects such as model construction, behavior, interactions, and specific connections among different simulation elements. Primarily, these norms address the structure of the simulation software and the interactions among its various components.

On the other hand, some popular simulation software options may presently conform to the Distributed Interactive Simulation (DIS) norms, which serve as precursors to the current HLA regulations. Originally designed to meet the requirements of the Defense Advanced Research Project Agency (DARPA), these norms were subsequently incorporated into the NATO standardization agreement STANAG 4482 [9].

In the context of this research, it was determined that non-compliance with contemporary norms would not be a significant concern. This is because the project's primary focus—method development—remains unaffected by HLA and DIS standards, which primarily influence simulation software creation rather than its application. Additionally, given NATO's prior acceptance of DIS norms, client opposition to this approach is expected to be minimal. However, it is important to note that compliance may become crucial when integrating the developed method into the company's existing design practices.

2.2 METHOD REQUIREMENTS

Selecting the appropriate simulation software is a critical factor in determining a project's success. The right choice enables the team to concentrate on the development while ensuring the simulation provides the necessary level of detail. Notably, most advancements in military simulation are primarily tailored for operational evaluations. However, since this research focuses on assessing the initial design phase, it is essential to identify software that, while originally may be intended for different applications, can still effectively meet the project's specific requirements. The following descriptions are based on insights gained from studying various solutions available on the market, with particular emphasis on AFSIM [10], GCAM [11], STORM [12], JFOrCE [13], JWARS [14], and CPE [15], [16].

It is crucial to select a simulation that is appropriate for evaluating the capabilities of a single unit within a naval environment. The chosen software for the project should facilitate the simulation of naval operations for both individual vessels and groups of vessels. Moreover, to ensure accurate evaluations, the simulation environment must account for the key factors that influence vessel capabilities and operations.

Certain projects may also require the modeling of additional critical elements, such as rules of engagement orders, which can significantly affect the outcomes. The vessel's behavior must align with predefined operational instructions from the client—in this case, NATO. This behavior should be reproducible with a representative level of detail. Additionally, to enhance understanding of the events during the simulation and facilitate communication with stakeholders, the software should feature a clear, user-friendly graphical user interface (GUI).

Based on research findings and general requirements, several key capabilities must be met. These include the ability to clearly describe the vessel, its systems, and subsystems for simulation purposes, while omitting details that are not yet available during the preliminary design stages. Furthermore, the ability to accurately recreate specific events or operations using predefined vessels based on previously prepared input data is essential. Finally, the simulation should allow for the evaluation of missions and present the outcomes in a clear and comprehensible manner.

Furthermore, when selecting the appropriate simulation software for evaluation, potential compatibility with external coding scripts should be considered. As demonstrated in existing solutions [17], the use of Python scripting is particularly beneficial for efficiently reformulating and analyzing simulation results (output processing). This feature would significantly reduce the duration of the overall process.

Several issues have been identified in existing simulation environments for similar applications. First, their focus on campaign-level evaluations limits detailed assessments of individual units. Second, key operational factors are often omitted, reducing simulation accuracy. Third, most simulation software requires highly detailed input data—such as precise vessel dimensions and onboard system placement—which is often unavailable in early design stages, making initial evaluations challenging. Addressing these issues would require additional effort and potentially significant source code modifications. Additionally, while agent-based simulations are useful for assessing communication channels and situational awareness in campaigns, these functionalities are unnecessary for this project, as they would not be fully utilized.

3. METHODOLOGY

This section outlines the primary method for using simulation software to evaluate vessel capabilities during the conceptual stage of ship design. The process is divided into three main stages: data gathering and input, simulation setup, and output data export and handling. This comprehensive approach involves assessing available data, defining the methods to be employed, and considering various influencing factors.

3.1 INPUT TFORMULATION

The first step involves clearly identifying the data required for the process. This includes determining which data is essential for creating the simulation and pinpointing the sources responsible for providing this information. At the beginning of the process, the input data is predetermined by the simulation software's requirements. However, it can be organized into four distinct datasets, each responsible for different aspects of the process.

First, the primary input concerning the vessel should be described. This input must adequately detail the current (or predicted) capabilities of the vessel, including systems onboard, fuel storage capacity, main propulsion characteristics, and basic design features. Key characteristics should include information such as general dimensions (approximated), crew size, and vessel class, while the systems list must precisely specify the allocated systems and their capabilities or performance under certain conditions.

The second input dataset focuses on the scenario description used for the evaluation. This dataset should contain all relevant information about the tactical situation, including its precise location, mission objectives, and other factors defining the scenario. Given the complexity of different scenarios, this dataset is expected to be the most time-consuming to create, as the number of descriptions varies based on the simulation's complexity.

Next, a third input dataset governs the behavior of assets within the simulated scenario. This includes the instructions for the evaluated vessels and opposing forces, created based on reputable military sources like NATO handbooks, and should cover elements such as Rules of Engagement (RoE) or Emissions Control (EMCON). This dataset should be developed in close collaboration with stakeholders to ensure that the tactics used are as realistic as possible.

Lastly, the final dataset provides any additional information used during the simulation process. The specific structure of this dataset remains relatively flexible, depending on factors such as the desired outcomes (performance measures), capabilities to be tested, and any additional variables introduced. This dataset should be carefully reviewed, as some simulation software may require the desired output composition to be determined upfront. The relevant information would then be included within this dataset.

During the research phase of presented project, key stakeholders responsible for contributing to the datasets were identified. The design team holds primary responsibility for providing the initial dataset related to vessel capabilities, including system predictions and design decisions. The "scenario" dataset, which defines operational scenarios will be developed in collaboration with Operational Analysts. These analysts create the Operational Capability Description (OCD), the primary source of information for simulation scenarios. Additionally, the dataset concerning vessel behavior will be developed with input from Naval Operations Experts. Their insights into naval operations and commander perspectives will enhance the realism of the vessel's behavior within the simulation.

To streamline and enhance the process, predefined document drafts will be used to collect the necessary dataset information. These drafts will standardize data requirements and reduce the level of simulation-specific knowledge needed from participants. While implementation may vary across companies, the core principle remains the same: using structured input dataset drafts to simplify participation in the evaluation process. Additionally, this approach aims to improve traceability and efficiency throughout the evaluation.

3.2 IMPLEMENTATION OF DATA AND SIMULATION SETUP

The process of implementing the data established in the previous section into the simulation software requires the creation of a special 'connection.' The data must be translated into a format compatible with the software to define the scenario, allocate the vessel with its properties and characteristics, and implement the intended behavior of the assets within the simulation.

This 'connection' can be realized through two distinct approaches, which can be categorized based on the level of proficiency required from the user. The first approach involves the addition of a dedicated team member responsible for manually inputting the datasets into the simulation software and creating the scenario. This process would primarily be conducted using the built-in GUI which allows for the manual entry of the predetermined data.

The second approach involves automating the process of exporting input data into the simulation software using a specially created function or script. This script would handle the correct utilization of the input datasets while simultaneously generating the scenario. The exact method used would depend on the chosen simulation software. Some simulation tools may even offer an integrated scripting language designed for such tasks.

During this step the base simulation file—the scenario that consolidates all predefined elements—will be directly used for evaluation. Once the input datasets are formulated and exported into the simulation software, the system must integrate all created elements into the main scenario. This scenario should then be saved and submitted for independent evaluation.

Initially, members of the design team responsible for formulating the input data will review the scenario to ensure it accurately represents all necessary factors for assessing the vessel's capabilities. Additionally, if needed, a separate evaluation should be conducted by a simulation specialist who was not involved in the scenario's creation. This independent review helps verify the scenario's accuracy and reliability.

The culmination of this part of the process is reached when the previously prepared and evaluated scenario is used to assess the desired capabilities of the designed vessel. The specific method of final evaluation may vary in detail, but the core principles should remain unchanged. The evaluation should be carried out through multiple iterations of the simulation, with each scenario starting anew and the results of one iteration not influencing the others. This approach guarantees that each simulation run is independent and yields valuable outcomes that can inform decision-making processes.

3.4 EXPORT OF THE RESULTS

The concluding step of the method focuses on gathering, presenting, and evaluating the data obtained from the simulation. The entire process of data collection must span all steps of the method. The specific data to be collected needs to be determined and implemented through appropriate datasets. Subsequently, an action must be implemented within the simulation software to save all relevant data from the simulation into the desired file format.
Saved data may include all parameters processed by the software, such as vessel positions, vessel status, or reports of actions. The precise composition of saved information may depend on investigated case. For example, it could include detection cycles for prediction of sensor composition onboard. Moreover, some simulation software can provide output files used to recreate the conducted scenarios in 3D environments, effectively creating visualizations for simulated processes.

At the end of the process, the results would need to be exported from the simulation software into specific files and then integrated into a predefined draft. This draft would serve as the official result of the complete evaluation process. It is believed that incorporating the final draft with the results into the design process documentation would greatly improve traceability and provide justification for decisions made.

The process of extracting the data obtained from the evaluation could also be conducted as described for the initial data export into the simulation software. This is anticipated to be done by a "simulation engineer" initially, while the complete method is still being researched and, with time, it may evolve into a fully automated process, using additional or integrated scripting languages.

The described process is designed to ensure that the designer gains a comprehensive understanding of the data obtained from the evaluation. It allows the design team to explore design space and identify potential solutions to design challenges without committing to decisions that may have significant implications in later project stages. Additionally, utilizing simulations offers the opportunity to evaluate the same design across a range of operational scenarios that the vessel is expected to perform. This is particularly valuable, as predicting the design's effectiveness under off-design conditions can help ensure the vessel meets its objectives throughout its entire lifecycle. Moreover, by integrating the results into the appropriate documentation, they can be effectively communicated to stakeholders, providing further context and justifications for the decisions made.

The complete process described above was also recreated using a systems engineering tool called Capella MBSE software. This step was undertaken to gain additional insights into the relationships between the different steps in the method and to help identify the participating entities and specific information that should be provided throughout the process. A part of the model, which showcases the relationship between the design team, the method, and the simulation software, is provided below (see Fig: 2).

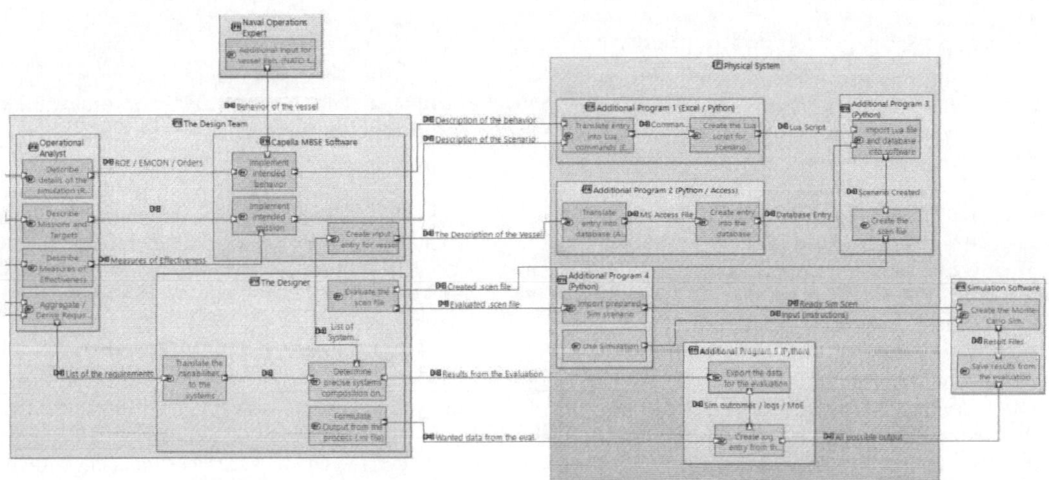

Figure 2: The system model used to visualize the relations within the method. Created with Capella MBSE Software. [18].

4. CASE STUDY EVALUATION

The primary objective of this case study is to demonstrate how the method works and its potential application within the design process, aiming to improve understanding of the design space by evaluating operational capabilities during the conceptual stages. This evaluation was conducted using the implementation method outlined in the previous chapter.

4.1 INTENDED OPERATIONS

The first simulated task focuses on conducting a patrol operation on critical underwater infrastructure (CUI) within the Dutch Economic Zone. Protecting CUI has become increasingly important due to recent sabotage attempts by unidentified actors. Modern naval doctrine recommends deployment of Unmanned Surface Vehicles (USVs) to handle patrolling, investigation, and prevention is similar operations.

The scenario begins with an intelligence report on a suspicious vessel behaviour near critical underwater infrastructure in English Channel. To prevent potential sabotage, two USVs are dispatched from a nearby naval base to monitor the activities of the suspected enemy vessel and conduct inspections of the underwater structures for any signs of explosive materials.

The precise location of the Opposing Force (OPFOR) vessel is known from the outset, and the USVs depart from the Den Helder naval base to patrol the area near the suspected contact. Unbeknownst to the USVs, the OPFOR vessel has already placed at least one explosive device on the underwater structure. The designated USV must navigate to the operational area, identified by the OPFOR vessel's position. Once there, it must detect any potential threats and report its findings back to base. The operational area is located at the northern entrance of the English Channel, where underwater cables extend from 99 nautical miles (Zeus Cable) to 66 nautical miles (Farland North Cable) from the starting position.

The second task scenario involves a helicopter, carrying sensitive equipment or data, crashing in the North Sea. Due to the nature of the equipment in the wreckage and the potential interest from opposing parties, a rapid response is crucial. The USV must quickly travel to the crash site, located 70 nautical miles from the starting location, locate the wreckage, and deploy an Unmanned Underwater Vehicle (UUV). The UUV's objective is to either retrieve the contents or destroy them. Timeliness and precision are critical to prevent hostile forces from obtaining the valuable materials.

Both tasks necessitate the USV's ability to conduct underwater search and neutralization operations. While these capabilities are initially structured for naval applications, the solutions identified could have broader applications in civilian markets. It's important to emphasize that the method presented here can be freely applied within both civil and naval realms of ship design.

4.2 EXAMINATED VESSEL DESIGNS

For evaluation purposes, several configurations of the USV were developed. The primary differences among these configurations relate to the number of UUVs onboard and key performance characteristics of the USV, such as speed and fuel consumption.

To establish the design baseline, the Damen 5009 Patrol Vessel was selected, under the assumption that it would be equipped with autonomous systems. Operational equipment would be integrated in a modular fashion. Where data were unavailable, values were extrapolated based on similarly sized USVs, primarily referencing the USV Nomad. Initially, it was assumed that the Kongsberg HUGIN would serve as the UUV, forming the basis for the underwater capabilities in the presented designs.

In the first configuration, the USV is equipped with a single UUV deployment system on the deck. The USV itself does not carry any onboard sensors for underwater search; its role is limited to transiting to the designated area and deploying the UUV. The second configuration also deploys a single UUV but adds a hull-mounted sonar system (TSM 2022 Mk3, as listed in the CPE Database), enabling the USV to conduct searches independently of the UUV.

Due to the high cost of the HUGIN system, the third configuration retains a single UUV deployment capability but replaces the HUGIN with the more cost-effective Mk 18 Mod 2 Kingfish. This change is considered as a potential measure to reduce overall project costs.

The performance indicators and capabilities for each design are summarized in Table 1. As shown, range and speed vary across configurations, primarily due to changes in onboard system weight, which directly affect vessel performance.

Table 1.The performance data and capabilities of evaluated designs.

Configuration Number	Speed (knots)	Range (nm)	Capabilities
1	18	273	1 UUV (HUGIN)
2	15	236	1 UUV (HUGIN), TSM 2022 Mk3
3	18	273	1 UUV (Kingfish)

4.3 RESULTS FROM THE EVALUATION

The following evaluation was conducted using the Command: Professional Edition software [15]. Each operation was repeated 30 times for each vessel design. For the first operation, the average compilation time was 23 minutes, while for the second operation, each evaluation concluded in 6.5 minutes. The results from the evaluations are presented below. These results include information on the time of arrival at the designated area of operation, the time taken to locate the target, the success rate in target identification, and the overall time to complete the operation.

First configuration exhibited an average operation time of 11 hours and 46 minutes (see Tab.2), with a remarkably low standard deviation of only 30 minutes. There were two instances where the vessel failed to detect the target in the final area of search. These results give indication as to what may be expected from this composition and may be used as benchmark for other configurations.

Table 2. Results obtained from evaluation of first vessel configuration obtained during CUI operation.

First Operation / First Configuration: Vessel 1 x2 / Hugin UUV		
	From Scenario Start:	From Previous Area:
Average time of arrival to Area 1:	05:52:49	05:52:49
Average time of arrival to Area 2:	04:01:10	04:01:10
Average time of arrival to Area 3:	08:46:49	00:49:21
	Average Time:	Rate of Success:
Target detection in Area 1:	00:13:51	100 %
Target detection in Area 2:	00:54:52	100 %
Target detection in Area 3:	01.28.55	93.3 %
Operation Completion Time:	11:46:49	

In the case of the second configuration (see Tab.3), the average operation time was 11 hours and 18 minutes, showing an improvement of almost half an hour compared to the first configuration. Moreover, the standard deviation for the recorded results was only 4 seconds, indicating high consistency. The reduction in total operation time can be attributed to the inclusion of hull-mounted sonar, which contributed to 74% of the total detections.

Table 3. Results obtained from evaluation of second vessel configuration obtained during CUI operation.

First Operation / Second Configuration: Vessel 2 (sonar) x2 / Hugin UUV		
	From Scenario Start:	From Previous Area:
Average time of arrival to Area 1:	06:51:27	06:51:27
Average time of arrival to Area 2:	04:44:07	04:44:07
Average time of arrival to Area 3:	08:18:32	00:55:33
	Average Time:	Rate of Success:
Target detection in Area 1:	00:02:56	100 %
Target detection in Area 2:	00:04:22	100 %
Target detection in Area 3:	00:24:10	96.67 %
Operation Completion Time:	11:18:33	

Finally, the third configuration, proved to offer a slightly lower success rate of detection with a shorter time to detect the target compared to the configuration based around Hugin UUV (see Tab.4). However, due to the smaller range of the sensors onboard the UUV, it required more time to fully examine the area of interest. This directly influenced the overall operation time, which was hour longer than the overall time for the first configuration.

Additionally, the standard deviation was equal to 1 hour and 42 minutes, which is not consistent enough for similar operations.

Table 4. Results obtained from evaluation of third vessel configuration obtained during CUI operation.

First Operation / Third Configuration: Vessel 1 x2 / Kingfish UUV		
	From Scenario Start:	From Previous Area:
Average time of arrival to Area 1:	05:42:30	05:42:30
Average time of arrival to Area 2:	04:01:27	04:01:27
Average time of arrival to Area 3:	09:23:15	00:29:08
	Average Time:	Rate of Success:
Target detection in Area 1:	00:08:09	100 %
Target detection in Area 2:	00:17:46	96.67 %
Target detection in Area 3:	00:46:36	96.67 %
Operation Completion Time:	12:53:15	

In summary, the second configuration—featuring a hull-mounted sonar and a single UUV—proved to be the most effective overall. However, if stakeholder priorities lean toward cost efficiency while allowing for slightly longer operation times, alternative configurations may be considered. For example, the first configuration exhibited an average mission duration approximately 30 minutes longer (which corresponds to its standard deviation), while maintaining a high detection success rate, identifying only one fewer target than the best-performing design.

For the helicopter operation scenario, the vessel configurations remained unchanged. In this case, the simulation assumed that a single vessel should be capable of fulfilling the primary objective using a deployable unmanned underwater vehicle. Additional attention was given to fuel efficiency during this scenario to assess the feasibility of reallocating the vessel from a different mission. The results from this final evaluation are presented in the tables below.

In the first tested configuration of the second scenario, the average time between UUV deployment in the search area and initial target contact was just 26 seconds (see Table 5). This rapid detection is likely attributable to the capabilities of the UUV system, with the primary detection tool identified as the generic echo sounder used by the HUGIN. Moreover, the total mission time was 5 hours and 20 minutes, with a remarkably low standard deviation of just 21 seconds, demonstrating the configuration's consistent and reliable operational performance.

Table 5. Results obtained from evaluation of first vessel configuration obtained during second scenario.

Second Operation / First Configuration: Vessel 1 / Hugin UUV	
Time to enter the area of operation:	04:14:55
Time to locate the wreck:	04:15:21
Rate of success of the detection:	100 %
Amount of fuel consumed by the USV:	4259.21 kg
Amount of fuel needed for the transit:	2901 kg

Second tested configuration, outfitted with sonar, located the target 5 minutes after entering the area of operation. However, due to its slower transit speed, the overall operation time increased compared to the first test results, with an average value of 6 hours and 16 minutes (see Tab.6). Furthermore, it's worth noting that the USV would still need to deploy the unmanned underwater vehicle to closely inspect the target and fulfill the mission. The standard deviation for this scenario was also considerably low, at only 6 minutes and 52 seconds.

Table 6. Results obtained from evaluation of second vessel configuration obtained during second scenario.

Second Operation / Second Configuration: Vessel 2 (sonar) / Hugin UUV	
Time to enter the area of operation:	05:01:00
Time to locate the wreck:	05:06:15
Rate of success of the detection:	100 %
Amount of fuel consumed by the USV:	4283.39 kg
Amount of fuel needed for the transit:	3016.6 kg

Lastly, the configuration utilizing a smaller and more affordable UUV was tested. Although the results show that this configuration can complete the mission in a similar average time to its counterpart (1 minute faster than the first USV with the Hugin UUV), the success rate for this design is significantly lower at only 73% (see Tab. 7). The standard deviation was 5 minutes and 54 seconds, with the shortest time to conduct the operation being 5 hours and 17 minutes, and the longest time at 5 hours and 46 minutes. While not measured in the simulation, this difference in detection effectiveness may lead to an increase in mission duration, as more time would need to be spent scanning the entire area of operation.

Table 7. Results obtained from evaluation of third vessel configuration obtained during second scenario.

Second Operation / Third Configuration: Vessel 1 / Kingfish UUV	
Time to enter the area of operation:	04:03:55
Time to locate the wreck:	04:13:17
Rate of success of the detection:	73.3 %
Amount of fuel consumed by the USV:	4259.21 kg
Amount of fuel needed for the transit:	2901 kg

The overall differences in the second scenario were less pronounced compared to the first, mainly due to its lower complexity and shorter simulated time. It can be concluded that the first tested configuration, which relied solely on the Hugin UUV for underwater search operations, demonstrated the highest capability and achieved the best results. However, it's important to note that the second configuration, equipped with sonar and the Hugin UUV, could achieve similar or slightly lower time-to-detection values if the UUV were actively involved in searching for the wreckage. The primary distinction between these two configurations lies in the transfer time, which is notably longer in the second case.

5. DISCUSSION

Throughout presented research it became evident that introducing simulation-based evaluation into ship design practices remains relatively uncommon. However, the obtained results demonstrate its potential to enhance understanding of operational capabilities, particularly beneficial for tackling novel and complex design challenges where experience in assessing capabilities is limited.

A significant advantage of the presented approach for operational capability evaluation is its ability to initiate evaluations at the conceptual design stages without needing detailed design information. This capability provides early insights into whether initial concepts are viable and identifies potential drawbacks of specific design configurations.

The versatility of the method allows for analysis of desired properties across various design possibilities, pinpointing strengths and areas for improvement in each configuration. Many simulation tools also support additional customization and expansion through integrated scripting languages. As highlighted, potential implementation of scripting languages can significantly enhance simulation flexibility. Future research should explore further scripting methods and languages to expand evaluation capabilities.

Additionally, future research is recommended to further enhance the method by integrating external software for supporting calculations, incorporating wargaming techniques, and analysing potential calculation models within various simulation software options. This would help identify the most critical areas to focus on for future evaluations.

It should be noted that the evaluation of early-stage designs through simulation may already be implemented by certain companies within the industry. This was recognized during the research and is also highlighted by Mission Engineering needs. However, the author was unable to find any publications on this specific topic or a detailed description of the process in publicly available sources.

6. CONCLUSIONS

The method presented in this paper demonstrates promising potential for evaluating the operational capability of an initial design without requiring specific data typically obtained in later project stages. Furthermore, integrating this approach into the design process may enhance several aspects of project development, including a deeper understanding of design considerations, improved recognition of trade-offs, and strengthened communication channels among stakeholders.

This work provides future users with a structured method that enables them to conduct initial research on the most suitable tools for their specific evaluations. Due to its unified nature, the method can function as a standalone evaluation process, as described in the publication, or be integrated into broader frameworks, such as the Mission Engineering Approach. While composed method effectively addresses the research gap identified at the project's outset, it remains in the early stages of development. Additionally, several areas for improvement have been identified, underscoring that the full potential of this approach has yet to be realized.

7. ACKNOWLEDGMENTS

This work was performed as a part of the MSc thesis of the first author, which was conducted at TU Delft in collaboration with Damen Naval Company.

8. REFERENCES

1. CUI, W., FU, S., and HU, Z., 'Encyclopedia of Ocean Engineering', *Springer Singapore*, 2022.

2. ZWAGINGA, J., STROO, K., and KANA, A., 'Exploring market uncertainty in early ship design', *International Journal of Naval Architecture and Ocean Engineering, Volume 13, pages 352–366*, 2021.

3. BERNSTEIN, J., 'Design Methods in the Aerospace Industry: Looking for Evidence of Set-Based', *MSc Thesis. Massachusetts Institute of Technology*, 1998.

4. MAVRIS, D., and DELAURENTIS, D., 'Methodology for examining the simultaneous impact of requirements, vehicle characteristics, and technologies on military aircraft design', 2000.

5. Office of the Under Secretary of Defense for Research and Engineering, 'Department of Defense Mission Engineering Guide', 2023.

6. DUCHATEAU, E., 'Interactive evolutionary concept exploration in preliminary ship design', 2016.

7. North Atlantic Council, 'NATO modelling and simulation master plan', 2010.

8. North Atlantic Treaty Organization,'STANAG 4603: Modelling and simulation architecture standards for technical interoperability: High level architecture (HLA)', *IEEE*, 2015.

9. FULLFORD, D., 'Distributed interactive simulation: it's past, present, and future', *Proceedings Winter Simulation Conference*, IEEE, 1996

10. CLIVE, P., JOHNSON, J., MOSS, M., ZEH, J., BIRKMIRE, B., and HODSON, D., 'Advanced framework for simulation, integration and modeling (AFSIM)', 2015.

11. CAUNCE, J. and SEARLE, G., 'Introduction to the general campaign analysis model (GCAM) version 3.3', 2009.

12. SEYMOUR, C N., 'Capturing the full potential of the synthetic theater operations research model (STORM)', 2014.

13. AU, T A., HOEK, P J., and LO, E H S., 'Combat Analysis of Joint Force Options using Agent-Based Simulation', *2018 Military Communications and Information Systems Conference MilCIS*, 2018.

14. MAXWELL, D T., An overview of the joint warfare system, 2000.

15. Matrix Pro Sims, 'Command professional edition: 23 nations, 150+ orgs, 3000+ users', *Article on official publishers' website (command.matrixgames.com)* ,2023.

16. Matrix Pro Sims, "Game Manual for Command Modern Operations", 2023

17. HAGLEY, S., LO, E., HOEK, P., DONOHOO, A., and PEACOCK, J., 'Command professional edition: Initial insights of a commercial military operational modelling and simulation tool', *24th International Congress on Modelling and Simulation, pages 869–875*, 2021.

18. ROQUES, P., 'Systems Architecture Modeling with the Arcadia Method', *Elsevier*, 2018.

9. AUTHOR BIOGRAPHIES

Jakub Orlowski is currently a Process Engineer at Damen Naval. Originally from Poland, he earned a BSc in Ocean Engineering from Gdańsk University of Technology in 2022. In 2024, he obtained an MSc in Maritime Technology from Delft University of Technology, specializing in Ship Design and Structural Engineering.

Austin A. Kana is an Associate Professor of Early Stage Complex Ship Design in the Department of Maritime and Transport Technology at Delft University of Technology. His research involves developing new methods and tools to support early-stage complex ship design problems under uncertainty. He received his PhD in Naval Architecture and Marine Engineering from the University of Michigan in 2016.

Koen Droste is process architect in the process development and innovation department at DAMEN Naval. After working on a PhD at Delft university on the topic process driven ship arrangement design. At DAMEN Naval he has been involved in various developments and research related to MBSE, System Modelling, architectures, and design processes.

Björn Mes is the Process Development and Innovation Manager at Damen Naval. He is responsible for identifying and implementing innovations that enhance ship design and production processes. He graduated in 2002 with a European Media Master of Arts (EMMA) degree from the University of Portsmouth, specializing in Digital Media Design.

MULTI-ATTRIBUTE DECISION MAKING FOR THE DESIGN OF A NAVAL SURFACE COMBATANT

L Nugent, P Louvros, and **E Boulougouris,** University of Strathclyde, UK

NOMENCLATURE

AAW-	Anti-Air Warfare
ASW-	Anti-Submarine Warfare
ASUW-	Anti-Surface Warfare
Cb-	Block Coefficient
CFD-	Computational Fluid Dynamics
dB-	Decibels
GPFF-	General Purpose Frigate
GM-	Metacentric Height
GZ-	Righting Lever
kN-	Kilo-Newtons
Kts-	Knots
MOD-	Ministry of Defence
m-	Meters
m^3-	Cubic Metres
m.deg-	Metre Degree
mm-	Millimetres
MADM-	Multi-Attribute Decision Making
MCDM-	Multi-Criteria Decision Making
MHz-	Mega-Hertz
NATO-	North Atlantic Treaty Organisation
nmi-	Nautical Miles
RCS-	Radar Cross Section
SBD-	Set Based Design
t-	Tonnes
TOPSIS-	Technique for Order Preference by Similarity to Ideal Solution
URANS-	Unsteady Reynolds-Averaged Navier-Stokes
U.S.-	United States

1. INTRODUCTION

The modern surface warship is a complex vessel that is often designed for a multi-purpose role in a strike group as opposed to a single objective design. Traditionally, a naval vessel would be designed with a single mission role in mind, such as Anti-Air Warfare or Anti-Submarine Warfare. However, in modern naval warfare, there is a need for surface combatants to deliver traditional roles as well as being capable of providing humanitarian aid, fulfilling anti-piracy roles and conducting enhanced surveillance missions. This requirement for a multi-purpose vessel typically results in a completely new vessel being designed (Lye, H., 2021). Therefore, the goal nowadays is to achieve a comprehensive design across many aspects of its operation so that it can effectively execute any role assigned to it as efficiently as a single-purpose vessel would.

This approach is expensive and time-consuming (McKinsey & Company, 2024). Creating a new class of vessel can take decades, costing billions of pounds, with bitter recent experience as many projects run over budget and are significantly delayed (U.S. Government Accountability Office, 2025). When compared to similar sectors, naval shipbuilding is significantly outpaced. When looking at the design of merchant ships, such as containerships and tankers, many variations of design are produced with significant changes being implemented slowly (Jacobson, 2023). It can be said that this is due to a more iterative design approach used and larger classes of vessels being produced. While naval ships are significantly more complex than many of their merchant counterparts, the use of decision-making methodologies can aid designers in making changes to aspects of existing designs to fulfil mission objectives while reducing the time and cost to design the next generation of naval combatant. Design aspects include but are not limited to performance aspects, susceptibility and vulnerability.

This paper aims to demonstrate the effectiveness of multi-attribute decision-making processes in altering an existing hull, designed by the author, designated as the 'Type X', for a multi-purpose role within a strike group. Designed as an Anti-Air Warfare Frigate with a length of 143.8 m, beam of 16.5 m, and draft of 5.55 m, a displacement of 6004 Tonnes and powered for a maximum sustained speed of 30 knots and a range of 6000 nmi, it is designed for AAW missions. Multiple attribute design is the approach that evaluates and selects the best alternative from a range of results using real-world scenarios. complexities and considerations.

2. BACKGROUND

The design of naval surface combatants is an inherently complex and a multidisciplinary challenge, requiring the simultaneous consideration of performance, cost, survivability, and mission effectiveness. Traditional design approaches, which are based on sequential and discipline-specific methods, are increasingly being replaced. These new integrative frameworks support early-stage trade-off analysis and decision-making under uncertainty.

Multi-objective optimisation (MOO) plays a central role in modern conceptual ship design, offering a structured means of managing competing objectives. In naval contexts, typical trade-offs include endurance versus speed, survivability versus displacement, and cost. MOO enables the identification of Pareto-optimal solutions, which results in a large solution set. This requires further refinement and selection based on stakeholder priorities to identify the most suitable design priorities. Sen, J. and Yang, J. (1998) outline the use of Pareto Optimality within Multi-Criteria Decision Making (MCDM) to assess an attribute across various operating conditions. During the Pareto method a feasible boundary called the Pareto Front, it is beyond this front that the "best" point will be situated. Sen, J. and Yang, J. (1998) additionally introduce MCDM as the strategy that provides the "best" solution when considering multiple design criteria. This is initiated firstly by understanding the types of attributes the designer is dealing with, quantitative and qualitative. For example, in the context of selecting a ship, quantitative attributes would be its speed, deadweight, range and fuel efficiency. Qualitative attributes would be aspects such as seakeeping ability, manoeuvrability, adaptability, and crew comfort. Subjective attributes can be graded on a simple scale using a hierarchical approach. This combines subjective and objective assessments with a decision-making process used allows for design decisions to be chosen confidently.

A common foundation in this research domain is the parametric design approach, wherein a baseline concept is incrementally modified through changes to key parameters. While this allows focused exploration near a known feasible point, it risks missing more diverse or disruptive alternatives and can embed early design biases, which may limit innovative outcomes. early design biases. In contrast, this work adopts a Set-Based Design (SBD) methodology which begins with broad sets of feasible design alternatives and progressively narrows the field through iterative constraint satisfaction and performance filtering.

Erikstad, S. O. and Lagemann, B. (2022) discuss set-based design strategies and their implementation particularly in studies at the University of Michigan. The method uses a broad set of parameters that are open throughout most of the design process which allows for trade-offs to be made, reducing the set, and resulting in an optimised design and increased confidence. This method was proven to give preferable results when compared to the more traditional Spiral Method. This is also confirmed in Singer, D. J. et. al. (2009) where it is shown the method was used in other industries such as aviation and automotive manufacturing.

Set-Based Design (SBD) represents a significant paradigm shift in engineering design. It is particularly relevant for developing complex and safety-critical systems, such as naval surface combatants. Andrews (2024) reflects on this shift within the broader context of evolving ship design methodology, arguing that the increasing complexity of naval platforms and the adoption of holistic systems-level thinking necessitate flexible frameworks like SBD that better align with modern acquisition and capability requirements.

Van Houten et al. (2024) further advance this methodology by introducing a quantitative framework to assess how design spaces can become fragile, or vulnerable, when knowledge or requirements change. Their probabilistic strategy supports the core tenet of SBD: deferring decisions until supported by sufficiently comprehensive information. Their work shows that while designing the DDG(X) program, their work shows that while SBD guides the principle, and point-based designs serve as knowledge vehicles for early prototyping, integration testing, and cross-disciplinary coordination within the broader set-based structure. This supports ability assessments, reinforcing the argument that SBD serves as a viable scaffold that integrates performance, survivability, and layout evaluations in the conceptual phase. Building upon these advances, this paper applies SBD as a foundational design strategy for early-stage naval combatant development. It also embeds Multi-Attribute Decision Making (MADM) as a complementary decision-support tool for the structured down-selection of candidate concepts.

Papanikolaou (2022) introduces a holistic approach to ship design. This approach underscores the integration of multidisciplinary optimization and advanced digital tools to address the challenges of modern maritime engineering. This methodology was developed through the HOLISHIP project and leverages integrated design software platforms to facilitate the exploration of extensive design spaces. By incorporating virtual reality testing and enabling distributed, multi-site collaboration, the approach supports the advancement of novel maritime concepts aligned with low-emission strategic objectives. The holistic framework not only advances the ship design procedure but also guarantees that various design aspects are considered smoothly at early stages, promoting rationally optimized designs through state-of-the-art tools and virtual prototyping.

The use of the holistic methodology to inform design decisions within a SBD, maintains design flexibility and encourages innovation by deferring early commitments and allowing multiple alternatives to evolve concurrently. This is particularly advantageous in complex systems such as naval surface combatants, where operational requirements, stakeholder inputs, and technical constraints may evolve throughout the design process. Through this integration of SBD and MADM, the research positions itself within the evolving landscape of holistic naval design. It offers a flexible yet rigorous methodology for conceptual decision-making in high-stakes engineering domains.

Traditionally, a naval vessel is designed with a single mission objective in mind (Bottero, 2024). In the modern navy, there exists an evident requirement for adaptability. Therefore, a streamlined decision-making approach should be applied during the conceptual design stage to consider the needs of a modern naval vessel, such as hydrodynamic performance, seakeeping, vulnerability and susceptibility in long-range detection.

3. METHODOLOGY

An assessment of the hierarchical, multiple-attribute decision-making process will be conducted making decisions across the discussed design parameters. Using a similar method to that set out by Walker (2016), the conclusion is expected to be a vessel design that is a measurable improvement on the basis vessel with trade-offs that allow the ship to operate in its designed roles with a better performance envelope. The subsequent stages will lay out the philosophy, reasoning and analytical steps conducted to design the Type X Neo.

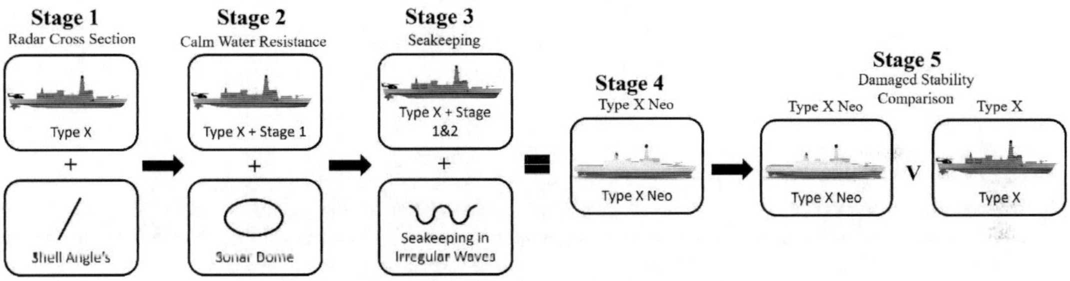

Figure 1: Methodology Flowchart.

3.1 BASIS VESSEL

The basis vessel for this case study is the Type X frigate. The Type X is an AAW frigate (Table 1).

Table 1. Type X Main Particulars.

Type X Main Particulars	
Speed (kts)	30
Length between perpendiculars (m)	138.2
Length Overall (m)	143.8
Beam (m)	16.526
Depth (m)	10.47
Draft (m)	5.55
Lightship (t)	4193.45
Displacement (t)	6002
C_B	0.463

The Type X is designed to operate Vertical Launch Missile Systems with accompanying Sampson and L band radar arrays. In addition to this, it has an Mk45 MOD Turret Gun on the forward quarter for surface-to-surface combat as well as two phalanx systems, one on either side of the vessel for close-range threat elimination. The Type X utilises an aviation hangar capable of handling Augusta-Westland Merin and Lynx helicopters with the flight deck large enough for chinook operations. The helicopters carried are considered the main method of hunting and attacking submerged threats.

3.2 NEW VESSEL REQUIREMENTS

It is crucial to first identify the requirements that the vessel be expected to satisfy and that the requirements and be separated into mission requirements and survivability requirements. Mission requirements being items like speed, range, armament stc. While survivability requirements would be items like subdivisions, machinary arrangements and the likes.

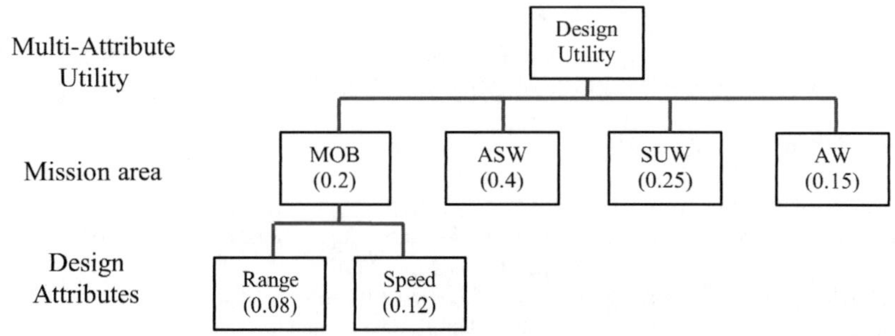

Figure 2: MAUT Structure.

The vessel will have three main operational requirements, the ability to conduct Anti-Submarine, Anti-Air and Anti-Surface warfare. Each of these abilities have their own attributes that define the role and that make a surface vessel effective in these areas.

3.2 (a) Anti-Air Warfare

Air superiority and surveillance is critical for the effectiveness in any mission for the individual vessel as well as for a strike group. With this, the vessels capability in this capacity comes down to the equipment installed on board and the vessels' ability to remain undetectable- or partially masked- to enemy sensors. In most cases of modern AAW vessels, the platform is designed around a specific sensor array such as the Royal Navy's Type 45 destroyer (Birkler, J. et. al. 2002). This method results in the vessel being optimised for AAW operation, however, does not leave a balanced design when considering multi-objective roles. Design aspects such as vessel speed for combat area relocation along with extended endurance is crucial as the threat of aircraft at sea is ever present with the inclusion of aircraft carriers in the battlespace resulting in operability spanning thousands of nautical miles, out of reach of any allied ports. Methods to deceive radar are often adopted to help the vessel stay undetectable to aircraft or propelled weapons such as surface-to-surface and Air-to-Surface missiles. The greater the time undetected, the greater the advantage that the vessel has over the enemy aircraft. This deception is often done using the vessels' own structure, creating shapes that deflects radar, Kang, et al. (2017). Perhaps the most effective deterrent a surface vessel can possess is the array of weaponry on board, specialised for use against aircraft such as Surface-to Air Missiles (SAMs) such as Sea Sparrow missiles; Close-in-Weapon Systems (CIWS) such as Phalanx guns; and Electronic Warfare (EW) systems like radar and communication jammers. These weapons combined with the passive methods of radar evasion make the surface combatant an intimidating adversary.

3.2 (b) Anti-Submarine Warfare

For surface vessels, it is often the case that submerged submarines have the advantage of being able to identify the enemy vessel before they themselves can be detected by the surface vessel (Hudson., 2020). Using passive sonar, the hunter can remain silent while listening to the marine environment where sensors on board the vessel analyse the incoming noise and are able to detect vessels or submarines and provide the weapons control with firing solutions to attack the hunted enemy. With this the vessel with the more effective sonar capability has the advantage and in live fire situations are more likely to survive.

While the sonar equipment itself is important to this objective there are other aspects of the vessel's design that affect the effectiveness of ASW. Propulsion choice is the largest decision in making an effective ASW vessel. Direct propulsion such as shaft and propeller can generate vast amounts of noise through machinery operation, shaft vibration and cavitation. Options such as electric motor propulsion including azimuth podded units means that the arrangement of machinery can allow for vibration mitigation such as shock absorbers and anti-vibration mounts and compartmentation of machinery spaces. Control over vessel speed is also critical. Electric propulsion allows for finer control of the vessels speed when compared to conventional propulsion arrangements. This is critical for the capability to sweep sea areas with sonar arrays to detect submerged targets. With this sonar placement is important also with options being at the bow, integrated with a bulbous bow, midship and towed sonar array. Each arrangement has their benefits however for the purpose of this study I will only consider the impact of these arrangements on the resistance of the vessel.

3.2 (c) Anti-Surface Warfare

Of all roles adopted by any surface vessel, the most traditional role is conducting surface-to-surface operations. In the modern navy this role is commonly seen as anti-piracy and anti-drug trade roles. However, the vessel must be equipped for close quarters combat with similar sized vessels. For this purpose, like those previously, the sensor and targeting array is crucial to the vessel's success in surface combat. As well as this, mobility, speed, and endurance are hugely important attributes that determine the vessels effectiveness in ASUW operations. The greatest passive asset to any surface vessel is again, its ability to stay hidden from the enemy for the longest possible time. For the modern naval combatant this is best done using the vessels shape. A small radar cross section area means that at long ranges the vessel can be masked or appear smaller than it is, often mistaken for merchant vessels. The vessels speed and endurance are also key in this capacity. The ability to proceed and reposition at high speeds can be crucial in many surface missions as mentioned above.

3.3 HOLISTIC DESIGN CHANGES

When considering the Type X as a base from which to redesign a multi-role vessel, a holistic view must be adopted. As discussed through the methodology, the focus of the project will be on aspects relating to the hull, superstructure, fighting and hunting capability and shipboard operations. As well as this design decisions will be made based on philosophical reasoning to make the basis design more robust and ready for enhanced design and optimisation as a multi-role vessel. Propulsion is one such aspect which will be altered in this way. The Type X was designed with twin azimuth podded propulsion units, to offer manoeuvrability in close quarter and shallow water scenarios. For a multi-role purpose vessel these are not high priority considerations and as such the propulsion choice must be reviewed.

Considerations for the Type X Neo's propulsion system must be redundancy and serviceability and repair ability at its homeport, at sea and in foreign allied ports. While podded units offer great manoeuvrability among other benefits, they also give susceptibility in combat through their profile and positioning. As well as this their level of redundancy is lower than what would be desirable on a naval vessel. For example, if one unit is damaged by shock or direct impact there is a significant chance that propulsion and directional control of the unit is lost. In this situation the vessel would be restricted in its ability to manoeuvre and thus be an effective warship in combat. If both units were damaged, then the vessel would be immobile and susceptible to attack without the ability to evade and potentially retaliate. Further to this, when deployed the vessel will have to rely on using allied ports for maintenance and repairs after considerable damage. As such it is more feasible and favourable to use components that are commonly used on allied vessels. An example of this is common parts and weapons used and shared by NATO navies. These are common parts that if required can be used by several types and class of vessel. Relating to propulsion, a conventional arrangement is more favourable to this effect as many components of the driveline are transferrable from ship to ship and navy to navy allowing for a greater scope for repair on deployment allowing the vessel to operate at its designed lethality.

Considering these points, the Type X Neo should be designed with a conventional shaft and rudder propulsion and manoeuvring arrangement. Owing to the size, purpose, and power requirement a twin-screw arrangement is adopted. As the hull form was designed particularly for podded units, the aft end will be altered to accommodate the new arrangement. Changes to the hull include the addition of a skeg structure and deepening of the aft end, both augmentations are being philosophically made, with the aim to provide additional buoyancy in the aft end which should aid the vessels response in irregular waves, a theory that will be assessed and analysed in subsequent sections (Lee, C., Lee, S., 2020).

3.4 RADAR CROSS-SECTION ANALYSIS

When considering the detectability, the greatest threat to the vessel often comes from the air or on the surface. One of the most reliable tools for an enemy hunter is radar. Radar is a tried and tested method of scanning large areas of seascape to detect combatants and give the hunter the upper hand in combat and surveillance (Defense Advancement., 2024). Knott et al.(2004) outline radar fundamentals by explaining that radar functions as both a transmitter and receiver are effective from the same point. The return signals are influenced by the geometry and material properties of the target. Khan et al. (2012) discuss RCS reduction methods into passive methods, such as design and material choices, and active methods that disguise the vessel's true RCS signature. Yang and Guo (2008) explore shape optimization to deflect radar signals away from the receiver. Cerruti et al. (2015) expands this work by developing a simulation tool, integrating 3D numerical RCS analysis with graphical outputs, thereby facilitating efficient iterative design processes for design engineers.

To decrease the chance of detection, as discussed by Knott et al. (2004), the use of composite materials and rubbers can help with minimising return signals from radar however this study assesses the angle of the surfaces exposed to atmosphere that could reflect radar signature. Basic radar theory dictates that a return signal will be returned at the greatest strength if the surface targeted is normal to the source. For the case of a vessel at sea normal can be taken as vertical to the sea surface.

The Type X was designed with radar signature in mind with bulkheads offset from the vertical normal by 20° without any radar cross-section analysis being done at the time. As well as the radar reflectance itself, when considering the angle of the shell and external bulkheads, the designer must consider the impact on the internal volume available to fit crucial systems, and compartments and fulfil reserve buoyancy requirements. With this, the hierarchical weighting can be determined, with the benefit of reduced radar signature being the highest priority attribute with internal volume being second priority.

For the overall design and optimisation of RCS, three angles were evaluated in an isolated specimen test to analyse the radar signature reflected to the source (MDPI., 2024). These were 20°, 30°, and 40°. Using the RCS package in Paramarine by QINETIQ the radar reflectance was first tested using a flat plate placed at 15 km from the source with and emission elevation of 4.019 km, this results in an emission azimuth of 15°, which can be considered a real-world estimate of a radar sweep from an enemy aircraft. The radar specifications used (Table 2) will remain constant through all tests, allowing for a fair benchmark that all specimens will be assessed on.

Table 2. Radar Test Parameters.

Radar Parameters	
Locator Height (km)	4.019
Locator Distance (km)	15
Range Gate (km)	30
Radar Frequency (MHz)	1200
Radar Wavelength (mm)	249.83

3.5 CALM WATER RESISTANCE

When analysing the hull design of the basis vessel it is crucial to understand which areas will yield the greatest results from hydrodynamic optimisation. Gui et al. (2002) study the forces, moments, and wave patterns on naval surface combatants in regular head waves, emphasizing measurement accuracy and uncertainty to provide foundational insights into hydrodynamic interactions. Sharma and Sha (2005) focus on integrating a bulbous bow with a sonar dome to optimize hull efficiency, illustrating how this integration achieves a balance between form and performance without compromising functionality. Wu et al. (2017) apply the Neumann-Michell Theory for multi-objective hull form optimisation. This combined theoretical with computational analysis methods to compare competing designs. For this case study, the addition of an integrated sonar in the hull of the vessel to expand its submarine warfare capacity is explored. As discussed previously, common choices are bow sonar domes, mid-ship sonar domes or no sonar dome- along with the use of a towed sonar array.

Sharma, R. and Sha, O. (2005), argue that in the case of towed sonar arrays, noise and turbulence from the vessel is minimised or even eliminated- depending on the length and depth of tow- something that cannot be achieved by integrated sonar domes. However, a major drawback of such a system is that when under tow, the vessel must operate between a speed envelope typically between 3 and 20 knots to maintain the sonar unit's buoyancy, greater than twenty knots and there is a significant risk of the tow cable snapping due to the increased drag. A towed array therefore limits the operating capacity of the vessel to react to an everchanging battlespace. If the vessel had to relocate at full speed or react to an enemy unit, the vessel would firstly have to retrieve the sonar unit, which depending on the length of tow, could result in a sizable delay and leave the vessel vulnerable.

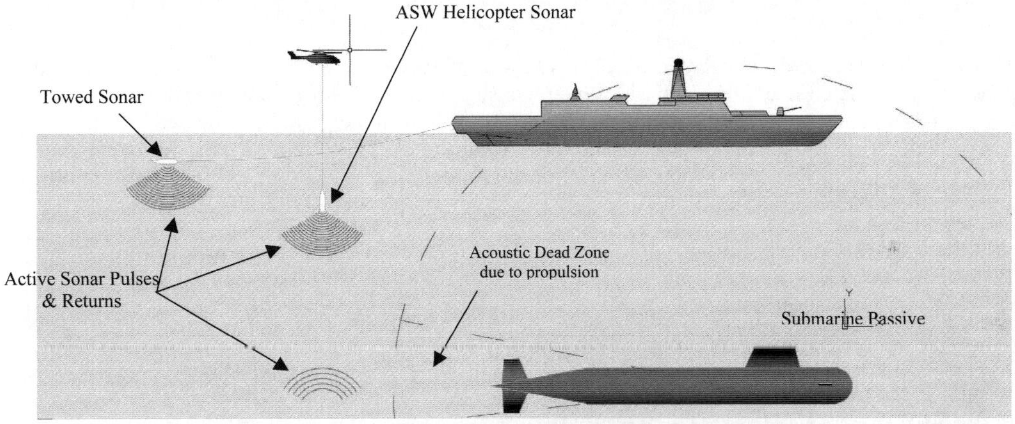

Figure 3: Example of Alternative Sonar Utilisation.

To overcome this shortcoming, the Type X Neo will be fitted with a hull-mounted sonar. There are alternative options to fit the sonar on the vessels hull with the two main considerations being radiated noise from the vessels propulsion and machinery, and the hydrodynamic effect that the inclusion of a sonar dome would have on the ship's resistance- an increase in resistance would result in an increase of power required, which in turn increases the emitted noise. A lower resistance would have the opposite effect. With a lower resistance- depending on the sonar equipment- the vessel can expand operational capabilities to sweep larger area for submerged targets by scanning at a higher speed at a lower power and thus, a lower mechanical noise emission.

Although noise emission plays a large role in the capability of sonar on the vessel, I will not be analysing noise radiation in this case study, instead focusing on the resistance of the vessel in calm water. Given this, I will assess the calm water resistance of three different arrangements to carry forward to the final design of the Type X Neo.

- Bare hull- Towed array.
- Hull mounted sonar dome.
- Bow mounted sonar dome.

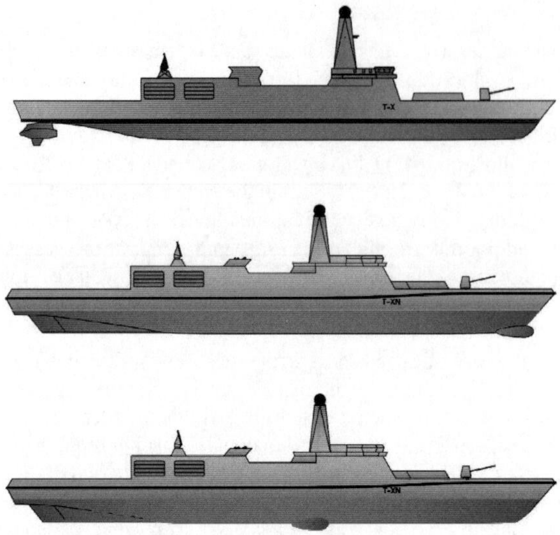

Basis Type X

Type X Neo with Bow Sonar and Skeg

Type X Neo with Midship Sonar and Skeg

Figure 4: Comparison of CFD Models.

These will be evaluated in Star CCM+ CFD software at speeds of 5 knots, 15 knots and 30 knots. The parameters for the CFD experiments will remain constant across all three simulations to give a benchmark from which to compare the three arrangements. The hierarchical weighting can be set with the benefit of reduced calm water resistance being the highest priority attribute with additional mechanical noise interference being the second consideration although this will be based on a more philosophical consideration instead of a numerically based determination.

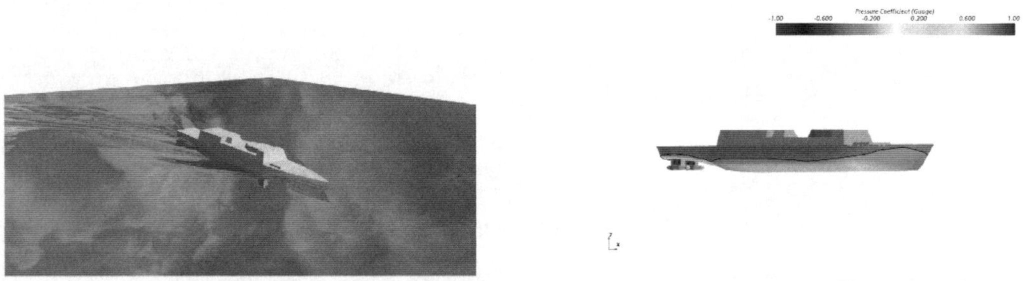

Figure 5: CFD Rendered View (Left), Pressure Coefficient (Right).

3.6 SEAKEEPING IN IRREGULAR WAVES

An integral part of the modern surface platform is the ability to conduct aviation operations from the aft flight deck. Most commonly flight operations are conducted by helicopters. Operations can include ship-to-ship transfers, anti-piracy operations, anti-drug operations, anti-submarine combat and ship-to-shore transfers of personnel or stores. With technological advances, drones are being introduced to the naval combat theatre to conduct these operations which limits the human factor through automation throughout the operation (Risk Intelligence., 2023). These operations, manned or unmanned, are commonly referred to as 'sorties'. regardless of helicopter or drone, the aircraft requires a steady deck to safely land on and be secured to upon landing.

Seakeeping has been a long-standing focus in naval research, as seen in Kehoe (1973) and Wilson et al. (2006). Kehoe compares American and Soviet destroyers with the focus being on how factors such as form, length, displacement, and appendages- including bulbous bows, bilge keels, and fin stabilisers- influence their motion and operability. In Kehoe's analysis, Soviet designs, with standardised fin stabilisers and better bow protection, demonstrated improved seakeeping and reduced deck wetness compared to their U.S. counterparts. Wilson et al. (2006) advance this study. Their method, however, achieves approximately 1% uncertainty in roll decay predictions – outperforming traditional techniques. While accurate the URANS method requires significantly more computational resources. For this study, seakeeping will be assessed using the more efficient strip theory method. Collectively, research from Olsen (1978), Kiss (1990), and Wilson et al. (2006) charts the evolution of seakeeping analysis from empirical to high-fidelity numerical approaches.

It is the aim of this aspect of the case study not to directly assess the accelerations and motions of the vessel with a specific assessment criterion in mind, but instead to compare the basis vessel to the new vessel. The key aim is to determine which has more favourable seakeeping characteristics in the way of motions and accelerations observed on the flight deck of the vessel. The sea states that the vessel will be based on the operation locations of the vessel. These are the Black Sea, the Aegean Sea, and the Mediterranean Sea. For each location, three sea states and three speeds of headway will be assessed resulting in nine sets of data from which to assess the vessel's response to irregular waves and trend in motion at various headings relative to the oncoming waves.

3.7 DAMAGE STABILITY AND VULNERABILITY COMPARISON

In terms of damaged stability, improvement of a specific aspect is not the aim of this project. Instead, the two vessels will be compared against each other, and philosophical reasoning will be used to determine whether to design changes made will impact the damaged stability of the vessel in a positive or negative manner.

Boulougouris et al. (2017) argue that survivability today is influenced more by sea state than by damage distribution, criticising the implementation of a fixed significant wave height. Instead proposing that the prevailing sea state should reflect the vessel's operational area. Only the German Navy, they note, uses probabilistic assessments – and only when a ship fails to meet initial criteria. The author expands on work by Boulougouris and Papanikolaou (2004, 2013), which offers a more realistic way to assess survivability and vulnerability for sea-going combatants. Van't Veer (2003) points out that survivability often ends with crew evacuation, even though complex compartmentation – common in naval ships – means that flooding behaviour requires a deeper analysis. His work uses the FREDYN time-domain flooding model to simulate how water spreads through a damaged ship in a high level of detail.

In reference to the survivability of the vessel, a comparison will be made between to two vessels in terms of the probabilistic damage that could be inflicted to both platforms and discuss how the design changes made throughout the case study could positively or negatively impact the survivability of the vessel. Although these will be philosophically based arguments, they will still be acknowledged in the decision-making process. While the investigative research into vessel survivability and recoverability is still an unknown area in the public eye, general naval architecture design principles can still inform the designer as to the direction of favourable designs. To do this I will use the data gathered in previous sections to comment on both vessel's survivability and suggest further analysis and improvements in further iterations of design.

3.8 DECISION-MAKING PROCESS

The process of decision-making in this report relies on two methods, Hierarchical Decision-making process- to highlight the priorities of the project as a whole and the TOPSIS technique to identify the best option in design that is closest to the ideal solution.

3.8 (a) Hierarchical MADM

A Hierarchical Multiple Attribute Decision-Making process will be used through the case study as previously outlined. A numerical weighting normalised to 10 is given to the assessed attributes to reflect the needs and requirements of the new vessel. This method is deemed appropriate as philosophically, hierarchical decision-making methods place emphasis on structure, accountability, and coordination at the heart of the decision-making process. These attributes are key when designing a naval platform which will involve hundreds, thousands of designers focusing on different areas of the vessel often with varying priorities. For time and cost-efficient ship design and construction, the process must be unified so that all requirements of the project are known, understood, and appropriately captured from the preliminary design stage. Once the operational requirements are known, variables can be identified that will change between designs and their impact will be assessed. They will be assessed in the Design Vector Map Matrix:

	Attributes						
	ASW		AAW		ASuW		
	Endurance	Sonar Search Speed	Radar Cross Section	Internal Volume	Accelerations	Roll RAO	
Propulsion Arrangement	7	8	0	4	0	0	19
Subsea Hunt Capability	7	9	0	0	0	0	16
Surface Search Capability	9	0	3	0	8	7	27
Air warfare Capability	0	0	9	8	0	0	17
Damaged Capability	0	0	0	8	3	5	16
	23	17	12	20	11	12	

Figure 6: Design Vector Map Matrix.

Mission areas identified above are Mobility, ASW, AAW and ASUW. The design variables are ranked with by their criticality to the mission area as shown above. This will allow for ranking of the importance of variables.

3.8 (b) TOPSIS MADM

The TOPSIS method is based on the theory discovered by Hwang and Yoon (1981), that the alternative chosen should be the solution with the shortest path to the ideal solution. Like many MADM theories, this method requires a decision-making matrix. These matrices are created with the attribute being assessed for each stage and uses weighting normalised to 1.00 to help determine the ranking of solutions. The method of this can be seen below.

$$\overline{X}_{ij} = \frac{X_{ij}}{\sqrt{\sum_{i=1}^{n} X_{ij}^2}} \quad (1)$$

$$V_{ij} = \bar{X}_{ij} \times W_j \quad (2)$$

$$S_i^+ = \left[\sum_{j=1}^{m} \left(V_{ij} - V_j^+\right)^2\right]^{0.5} \quad (3)$$

$$S_i^- = \left[\sum_{j=1}^{m} \left(V_{ij} - V_j^-\right)^2\right]^{0.5} \quad (4)$$

$$P_i = \frac{S_i^-}{S_i^+ + S_i^-} \quad (5)$$

4. CASE STUDY

4.1 RADAR CROSS SECTION

A sample simulation was conducted on a set of varying surface angles of 20, 30 and 40 degrees (Table 3). As expected, from the reflected radar signature it is evident that at angles deviating from the vertical normal, radar reflectance diminishes. The more 'angular' the vessel the lower the radar signature. In the image below this effect can be seen. Colours closer to red indicate a high radar return signal while green shows a lower return signal.

Table 3. RCS Baseline Test.

20 Degree		
	Volume (m3)	Return Signal
	2222	18.07
% Diff.	0%	0%

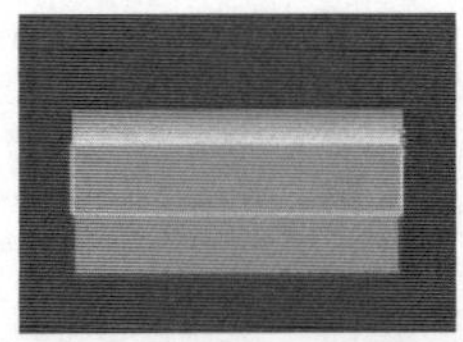

30 Degree		
	Volume (m3)	Return Signal
	2030	15.39
% Diff.	-9%	-17%

40 Degree		
	Volume (m3)	Return Signal
	1795	14.28
% Diff.	-24%	-23%

Analysis of this sample simulations was conducted to show the reduction of signal return as well as the volumetric changes in the geometry, from this it can be determined that the preferable option would be bulkheads angled at 30°, with a reduction of RCS of 17% while only a 9% reduction of volume. Although 40° yields the greatest reduction in RCS it also greatly reduces the internal volume of the space. This reduction would mean that the vessel would have to increase in size to accommodate all necessary compartments thus inevitably increasing RCS. As such, full analysis shall be conducted between the basis ship and the new 30° variant. To confirm this a TOPSIS decision-making method (Table 4) was adopted as per the steps detailed in section 3.8 (b).

Table 4. RCS Decision Making Matrix.

weighting	0.4	0.6	Normalised Matrix		Weighted Matrix		Euclidean Matrix Ranking			Rank
	Volume	RCS	Volume	RCS	Volume	RCS	Si+	Si-	Pl	
20°	2222	18.07	0.634076	0.65234	0.25363	0.39141	0.3754	0.049	0.115	2
30°	2030	15.39	0.579286	0.02664	0.231714	0.01599	0.0219	0.376	0.945	1
40°	1795	14.28	0.512226	0.51552	0.20489	0.30931	0.2973	0.082	0.216	3

The results from the above TOPSIS analysis show that 30° angle offset for the bulkheads yields the best results to be considered and carried forward in the design process.

The design and optimisation of RCS were conducted using the RCS package in Paramarine by QINETIQ. The Radar emission source was set as previously discussed, such that a range of 15 km is maintained with the midship of the vessel being analysed. Radar return signatures were calculated and recorded at 1° intervals through 360o. the result of this simulation is a polar graph and 'heat map' of radar hot spots that show the areas of the ship that reflect to the source or amplify return signals. The same procedure is conducted at angles of 10o- to simulate mid-altitude aircraft; 5°- to simulate low-altitude aircraft such as helicopters: and ~0°- to simulate other surface ships.

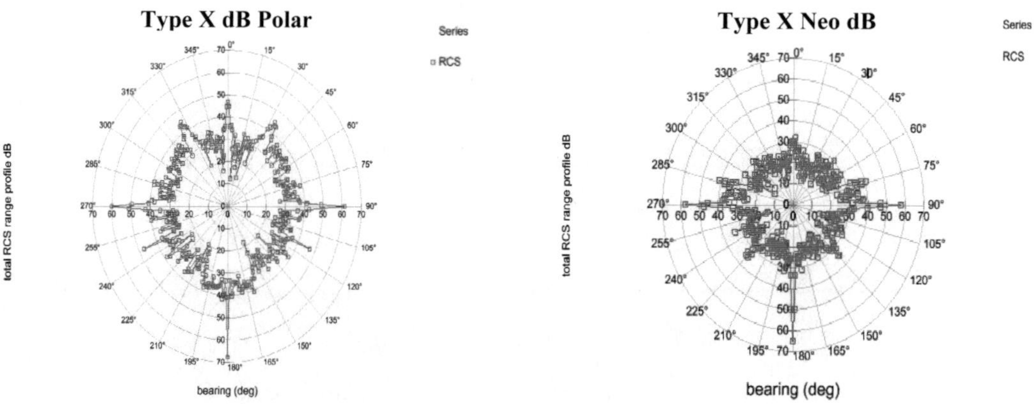

Figure 7: Type X (Left), Type X Neo (Right) dB Polar Graphs.

Table 5. Numerical Comparison of RCS.

Type X		Type X Neo		
Bearing (Deg.)	Total RCS Images dB	Bearing (Deg.)	Total RCS Images dB	Diff. Tot. RCS Image
0	40.139	0	23.692	-69%
45	27.577	45	20.913	-32%
90	60.793	90	57.636	-5%
135	32.392	135	35.122	+8%
180	66.881	180	64.827	-3%
			Average Difference	-20%

The graphical and numerical (Table 5) representation shows that, as expected from the sample simulation, the Neo design with bulkhead angles at 30° rather than 20° shows a significant reduction in RCS through all degrees of simulation. This can be visualised further with the following heat map comparisons.

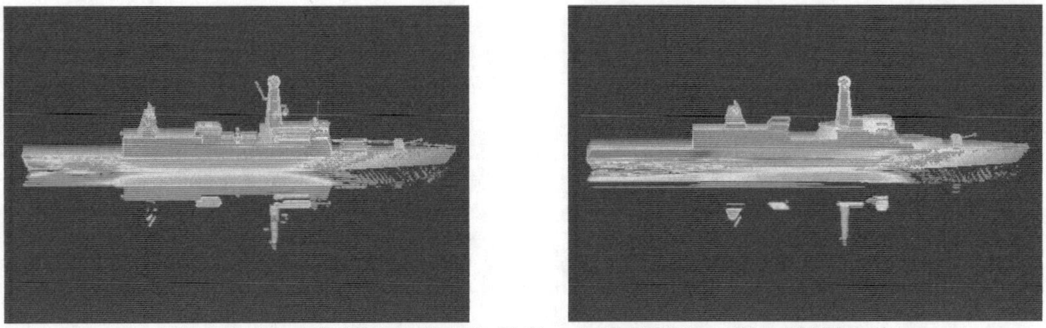

Figure 8: Type X (Right), Neo (Left) Broadside RCS Heat Map 15° Elevation.

Initially, a clear comparison can be made with regards to the heat maps shown above. The Neo design with surfaces at 30°, clearly deflects radar return signatures away from the source minimising reflected signature observed by the enemy. Looking closer at the features of the vessel one can see that the surfaces above the main deck benefit most from these alterations, signature on the hull is still the strongest observed however, signal reflection from the water's surface has decreased sizably. This trend may change as analysis is conducted with the source at different angles to the vessel.

Figure 9: Type X (Right), Neo (Left) Bow RCS Heat Map 15° Elevation.

In the same manner, when targeted bow-on the 30° variant of design displays a significant improvement. Where before, the vertical launch system, the forward end of the superstructure and the wheelhouse were the areas or greatest return signal, now in the Neo these areas are neutral- equally as immersive, to a significantly lesser extent.

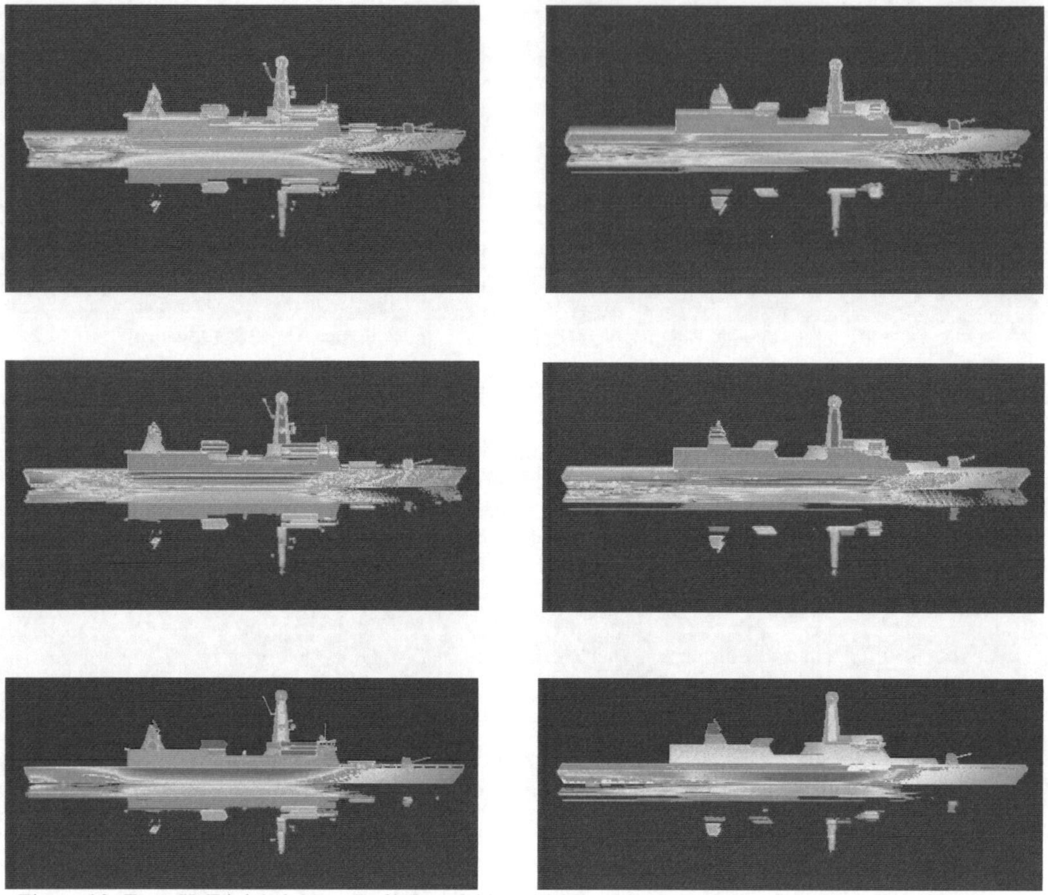

Figure 10: Type X (Right) & Neo (Left) Broadside RCS Heat Map 10°, 5°, 0° (Top to Bottom) Elevation.

From the figures shown above and results below (Table 6), the trend seen previously carries through and a reduction in RCS is consistent in the Neo design through all tested angles. An interesting observation is that as the radar emission source changes angle, it becomes normal to some surfaces of the vessel. This is the reason it is observed that some surfaces of the vessel have lesser signal return at 10° and 0° than it does at 5°, an example being the Sampson radar mast just forward of midships. While this is unfavourable it is unrealistic to have a geometric arrangement of surfaces on the vessel that will have a low signature at all angles relative to the emission source. Instead, the designer must choose a solution that best fits all scenarios.

Table 6. RCS Simulation Results.

Bearing (Deg.)	15° Azimuth		10° Azimuth		5° Azimuth		~0° Azimuth	
	Type X	Type X Neo	Type X	Type X Neo	Type X	Type X Neo	Type X	Type X Neo
	Total RCS Images (dB)		Total RCS Images (dB)		Total RCS Images (dB)		Total RCS Images (dB)	
0	40.139	23.692	44.838	22.554	52.353	34.127	54.458	26.12
45	27.577	20.913	32.267	30.48	37.146	31.174	27.748	27.571
90	60.793	57.636	62.234	57.717	57.939	56.654	64.106	61.27
135	32.392	35.122	29.388	31.472	21.505	31.193	35.614	27.44
180	66.881	64.827	67.578	65.021	56.054	60.316	63.99	63.75

4.2 CALM WATER RESISTANCE ANALYSIS

Analysis of the vessels' calm water resistance was conducted on Siemens Star CCM+. As stated before, the testing parameters for all three experiments were kept constant to achieve a fair benchmark to assess against.

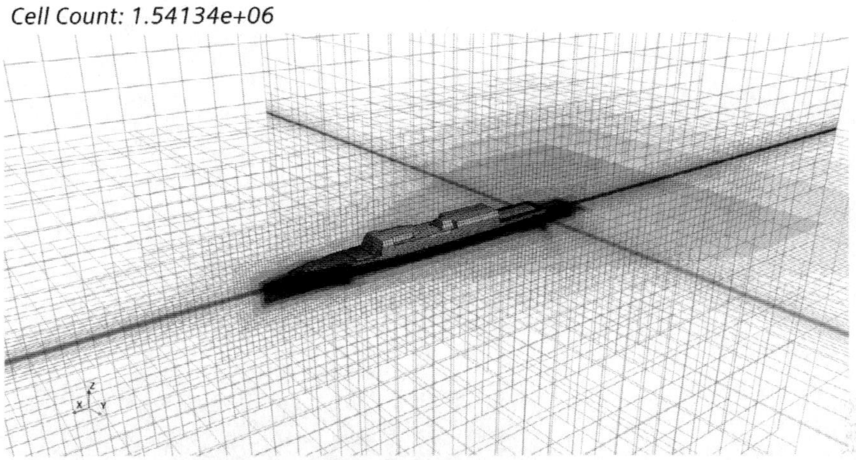

Figure 11: CFD Example Mesh.

The mesh visualised above is for a half-hull condition to increase efficiency with simulation times. While a larger cell count would yield a higher fidelity result, as stated previously the aim of this simulation is to present a comparison between competing designs. A finalised resistance will not be required at this stage however indications of trend will be used to make the decision to carry the design forward.

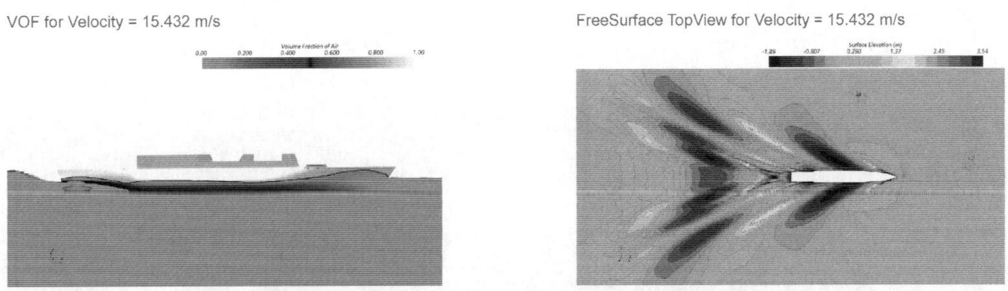

Figure 12: Type X CFD Graphical Outputs.

Figure 13: Type X Rendered Graphic.

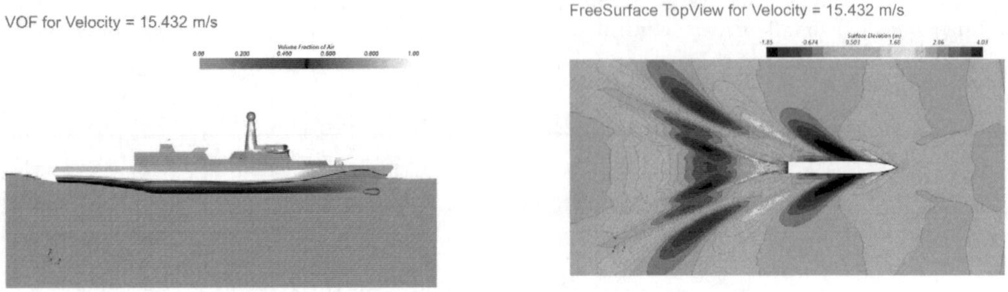

Figure 14: Type X Neo with Bow Sonar CFD Graphical Outputs.

Figure 15: Type X Neo with Bow Sonar Rendered Graphic.

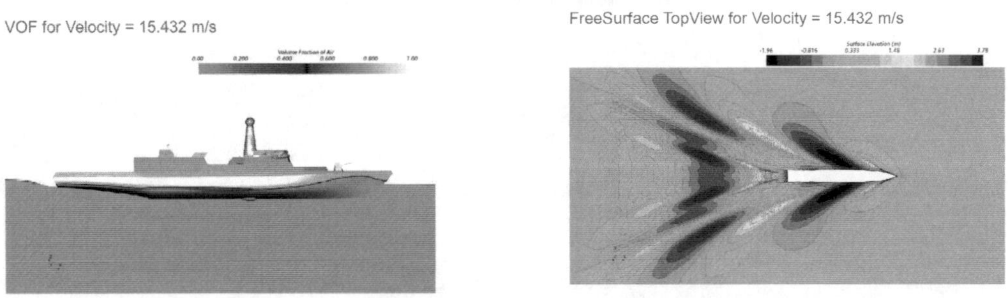

Figure 16: Type X Neo with Midship Sonar CFD Graphical Outputs.

Figure 17: Type X Neo with Midship Sonar Rendered Graphic.

From the graphical representations above, looking at the VOF representations for each velocity and each iteration, there does not seem to be any great reduction aft end wake when comparing podded arrangements to skeg configuration. The same can be said for the bow wave. The introduction of the bow sonar dome does not have an easily visible impact on the bow wave of the vessel. The difference is apparent when observing the wake field. Above 2.572 m/s it is there is an accentuated wake at the aft end of the Type X, indicated by the dark red shading. Contrasting this the Type X Neo variants have a less proponent wake in the aft area indicating the omission of the podded units streamlines flow in the aft end and reduces resistance. This can be confirmed by looking at the data retrieved from the simulation (Table 7 & 8).

Table 7. Full Hull Resistance Comparison.

Full Hull Resistance (kN)			
Speed (Knots)	Type X	Type X Neo (Bow Sonar)	Type X Neo (Midship Sonar)
5	23.69	21.31	20.57
17.5	348.71	269.98	273.08
30	1600.64	1483.67	1498.71

Table 8. % Resistance Reduction.

Original Resistance (kN)		Reduction in Full Hull Resistance	
Speed (Knots)	Type X	Type X Neo (Bow Sonar)	Type X Neo (Midship Sonar)
5	23.69	10%	13%
17.5	348.71	23%	22%
30	1600.64	7%	6%
	Avg. Reduction	13%	14%

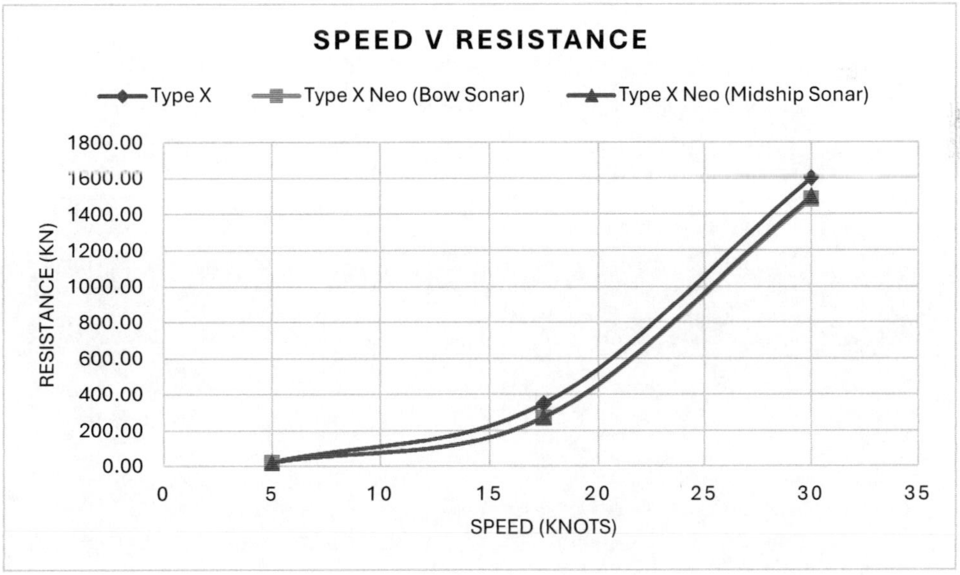

Figure 18: Speed v Resistance Graph.

To confirm the correct selection to carry forward in the design process, a TOPSIS decision-making method (Table 9 & 10) was adopted as per the steps detailed in section 3.8 (b).

Table 9. Resistance Weighting and Normalised Matrix.

Weighting	0.3	0.4	0.3	Normalised Matrix		
	5 Kts (kN)	17.5 Kts (kN)	30 Kts (kN)	5 Kts	17.5 Kts	30 Kts
None	23.69	348.71	1600.64	0.625	0.672	0.605
Bow Sonar	21.31	269.98	1483.67	0.562	0.520	0.0003
Midship Sonar	20.57	273.08	1498.71	0.542	0.526	0.566

Table 10. Resistance Weighted and Euclidean Matrix.

	Weighted Matrix			Euclidean Matrix Ranking			Rank
	5 Kts	17.5 Kts	30 Kts	Si+	Si-	Pi	
None	0.187	0.269	0.181	0.181	0.066	0.266	2
Bow Sonar	0.169	0.208	0.00009	0.064	0.181	0.740	1
Midship Sonar	0.163	0.211	0.170	0.181	0.012	0.061	3

The results from the resistance TOPSIS analysis shows that the Bow Sonar configuration yields the best results to be considered and carried forward in the design process.

4.3 SEAKEEPING IN IRREGULAR WAVES

As discussed previously, on deployment a naval combatant often relies on its on-board aviation platform to conduct missions from anti-drug operations to simply conducting ship-to-ship store transfers. As such, a combatant's effectiveness could be compromised when weather and environmental conditions make it difficult for helicopters to take-off and land. One crucial parameter that determines if a helicopter can conduct operations is the acceleration of the flight deck. Analysis of the accelerations of the Type X and Type X Neo hull configurations was conducted on Ship X to determine if the difference in propulsion arrangements was beneficial or detrimental to the vessels' ability to conduct flight operations. The analysis was then cross-evaluated using Maxsurf motions which then confirmed the anticipated trend in motions.

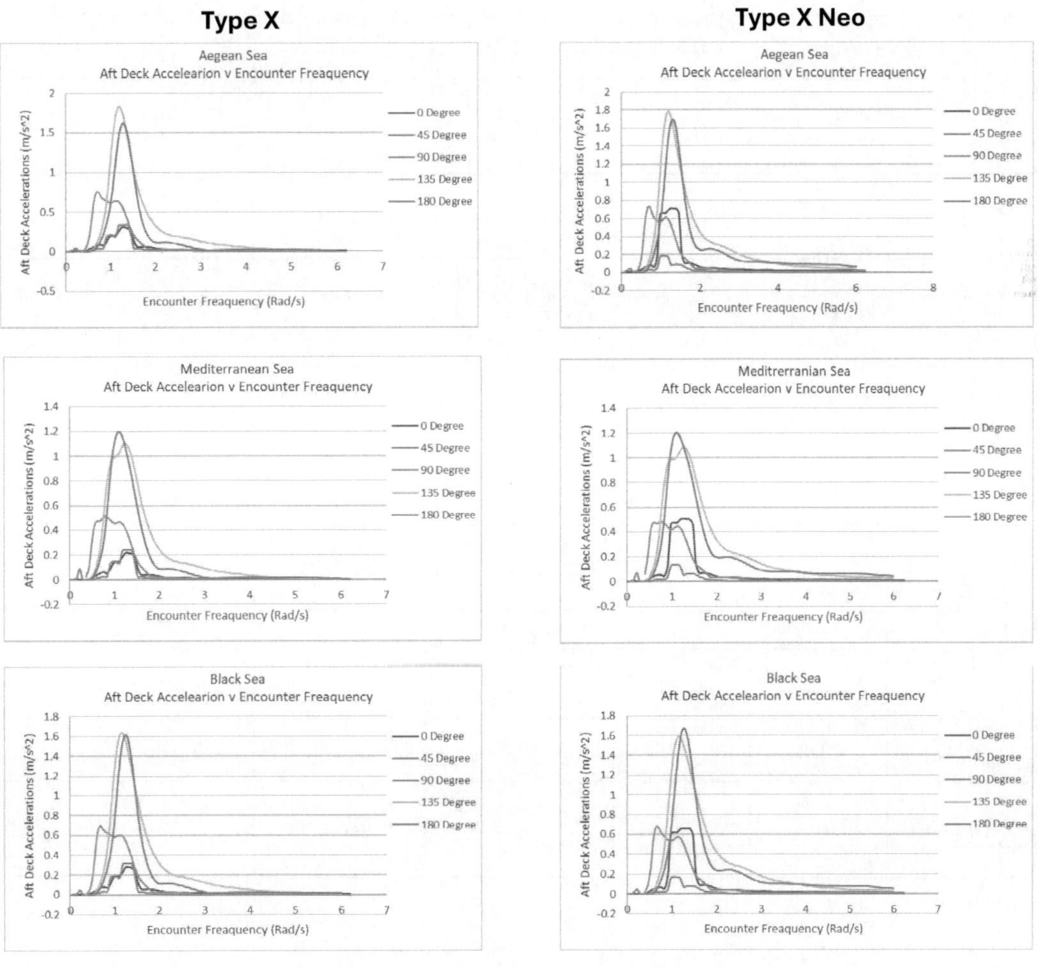

Figure 19: Type X &Type X Neo Flight Deck Accelerations.

The graphs above highlight that regardless of location and sea-state, the acceleration on the aft end is greatest at 180° & 135°, when the vessel is in head seas & waves approaching from the forward quarters. In the aft areas, while there is a reduction in accelerations for the Type X Neo with waves approaching the aft quarter or 45° there is also a sizeable increase in accelerations when waves approach from the stern or 0°. These trends are consistent through all sea-states which indicates that it is an inherent form-driven attribute (Table 11-14).

Table 11. Seakeeping Matrix.

weighting	0.12	0.12	0.12	0.12	0.12	0.25	0.15
	Accel . 0°	Accel . 45°	Accel . 90°	Accel . 135°	Accel . 180°	Redundancy	Manoeuvrability
Type X with Azipod	0.268	0.298	0.654	1.522	1.472	3	8
Type X with propeller & Rudder	0.626	0.167	0.632	1.485	1.517	7	5

Table 12. Seakeeping Normalised Matrix

Normalised Matrix						
Accel. 0°	Accel. 45°	Accel. 90°	Accel. 135°	Accel. 180°	Redundancy	Manoeuvrability
0.394	0.872	0.719	0.716	0.696	0.394	0.848
0.919	0.489	0.695	0.698	0.718	0.919	0.53

Table 13. Seakeeping Weighted Matrix.

Weighted Matrix						
Accel. 0°	Accel. 45°	Accel. 90°	Accel. 135°	Accel. 180°	Redundancy	Manoeuvrability
0.047	0.105	0.086	0.086	0.084	0.098	0.127
0.11	0.059	0.083	0.084	0.086	0.23	0.079

Table 14. Seakeeping Euclidean Matrix with Matrix/

Euclidean Matrix Ranking			
Si+	Si-	Pi	Rank
0.139	0.079	0.362	2
0.079	0.139	0.638	1

From the above, the skeg, propeller and rudder arrangement shall be carried through to the final design consolidation.

4.4 DAMAGE STABILITY AND SURVIVABILITY ANALYSIS

At this stage, a full, consolidated model with all the design features from the previous stage is produced. This is the first time that a full model is presented to serve as a direct comparison to the basis vessel. When considering probabilistic damage, unlike merchant vessels, the length of damage does not necessarily need to be a critical factor. This is due to longitudinal watertight subdivisions as well as conventional transvers subdivisions. Expanding this line of thinking introduces the thought of the vessel having a large angle of heel due to loss of compartments on one side of the vessel. The design changes made through this case study alone are not thought to have a notable impact on the damaged stability criteria of the vessel. The inclusion of a chine on the upper hull of the vessel initially was a concern as it was thought to reduce internal volume of the vessel and thus the reserve buoyancy of the vessel. However, regardless of chine, upon further investigation it is evident that at the angles of heel caused by damage the total loss of the vessel would be inevitable. This situation would mean that the freeboard deck would be awash, and points of down flooding would be breeched or severely compromised. To improve the vessels damaged stability criteria the free board was increased by ~1.3 m. this increase in freeboard would extend the extent of which the vessel could be damaged and remain operable. This also increases the internal volume to offset losses from the increased angle of the bulkheads. Summary of the stability criteria (Table 15 & 16) of both designs can be seen below.

Note: Deck A denotes weather deck

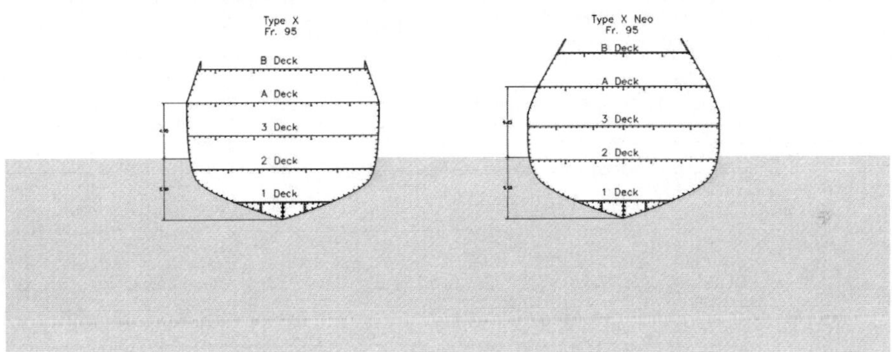

Figure 20: Freeboard Comparison of Type X and Type X Neo.

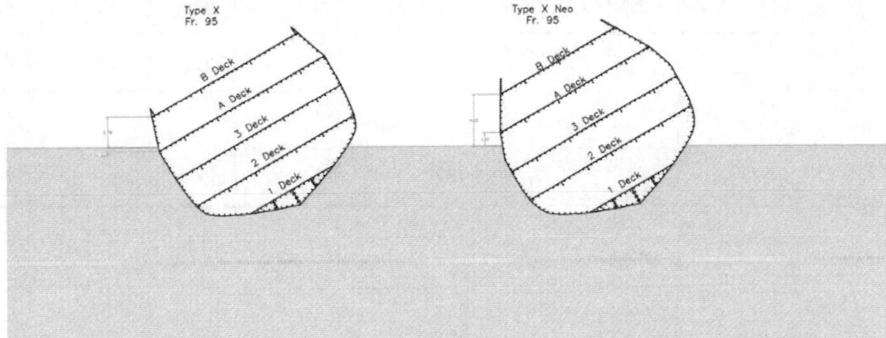

Figure 21: Freeboard Comparison of Type X and Type X Neo at 30° Heel.

Note: Deck A denotes weather deck.

Table 15. Type X NES-109 Stability Criteria.

Code	Criteria	Value	Units	Actual	Status	Margin %
Royal Navy; DS 02-109(NES109) Part 1	1.2.1.b: Area 0 to 30	4.584	m.deg	17.357	Pass	278.64
Royal Navy; DS 02-109(NES109) Part 1	1.2.1.b: Area 0 to 40	7.62	m.deg	28.92	Pass	279.53
Royal Navy; DS 02-109(NES109) Part 1	1.2.1.b: Area 30 to 40	2.75	m.deg	11.563	Pass	320.48
Royal Navy; DS 02-109(NES109) Part 1	1.2.1.b: Maximum GZ	0.3	m	1.216	Pass	305.33
Royal Navy; DS 02-109(NES109) Part 1	1.2.1.b: Angle of maximum GZ	30	deg	40.9	Pass	36.36
Royal Navy; DS 02-109(NES109) Part 1	1.2.1.b: GM fluid at equilibrium	0.3	m	2.444	Pass	714.67
Royal Navy; DS 02-109(NES109) Part 1	1.2.1.b: Range of positive stability	70	deg	78.4	Pass	12.03
Royal Navy; DS 02-109(NES109) Part 1	1.2.2: Stability in Beam Winds				Pass	
	Angle of steady heel shall not be greater than (<=)	30	deg	0.6	Pass	97.93
	Area1 / Area2 shall not be less than (>=)	140	%	440.88	Pass	214.91
	GZ (equilibrium) / GZ (max) shall not be greater than (<=)	60	%	2.14	Pass	96.43

Table 16. Type X Neo NES-109 Stability Criteria.

Code	Criteria	Value	Units	Actual	Status	Margin %
Royal Navy; DS 02-109(NES109) Part 1	1.2.1.b: Area 0 to 30	4.584	m.deg	19.2859	Pass	320.72
Royal Navy; DS 02-109(NES109) Part 1	1.2.1.b: Area 0 to 40	7.62	m.deg	31.451	Pass	312.74
Royal Navy; DS 02-109(NES109) Part 1	1.2.1.b: Area 30 to 40	2.75	m.deg	12.1651	Pass	342.37
Royal Navy; DS 02-109(NES109) Part 1	1.2.1.b: Maximum GZ	0.3	m	1.349	Pass	349.67
Royal Navy; DS 02-109(NES109) Part 1	1.2.1.b: Angle of maximum GZ	30	deg	49.1	Pass	63.64
Royal Navy; DS 02-109(NES109) Part 1	1.2.1.b: GM fluid at equilibrium	0.3	m	2.823	Pass	841
Royal Navy; DS 02-109(NES109) Part 1	1.2.1.b: Range of positive stability	70	deg	90.8	Pass	29.7
Royal Navy; DS 02-109(NES109) Part 1	1.2.2: Stability in Beam Winds				Pass	
	Angle of steady heel shall not be greater than (<=)	30	deg	0.5	Pass	98.23
	Area1 / Area2 shall not be less than (>=)	140	%	482.57	Pass	244.69
	GZ (equilibrium) / GZ (max) shall not be greater than (<=)	60	%	1.93	Pass	96.78

5. DISCUSSION

5.1 RESULTS

A summary of the decision-making results, as found through the TOPSIS method is presented below (Table 17-20).

Table 17. Summary of RCS Results.

Radar Cross Section						
	VOLUME	RCS	Si+	Si-	Pi	Rank
20°	0.254	0.391	0.375	0.049	0.115	2
30°	0.232	0.016	0.022	0.376	0.945	1
40°	0.205	0.309	0.297	0.082	0.216	3

Table 18. Summary of Resistance Results.

Resistance							
	Resistance (5 Kts)	Resistance (17.5 Kts)	Resistance (30 Kts)	Si+	Si-	Pi	Rank
None	0.187	0.269	0.181	0.181	0.066	0.266	2
Bow Sonar	0.169	0.208	0	0.064	0.181	0.74	1
Midship Sonar	0.163	0.211	0.17	0.181	0.012	0.061	3

Table 19. Summary of Seakeeping Results.

Seakeeping							
	Accel 0°	Accel 45°	Accel . 90°	Accel . 135°	Accel . 180°	Redundancy	Manoeuvrability
Type X with Azipod	0.047	0.105	0.086	0.086	0.084	0.098	0.127
Type X with prop & Rudder	0.11	0.059	0.083	0.084	0.086	0.23	0.079

Table 20. Summary of Seakeeping Results Continued.

Si+	Si-	Pi	Rank
0.139	0.079	0.362	2
0.079	0.139	0.638	1

5.2 MADM ANALYSIS

The decision-making results produced can be verified by looking at the overall improvement from the adopted changes driven by the decision-making process. Decisions made by a single designer can often be subjective and can easily miss the overall aim, secondary and tertiary considerations.

5.2 (a) Radar Cross Section

Starting with the RCS analysis, the TOPSIS method has determined that the preferable option is 30° angled bulkheads. To confirm that this is the correct choice, the average RCS return through 360° was obtained from the basis vessel and the new, 30° design on the Type X Neo. Both sets of results were obtained at the assessed 15°, 10°, 5° & ~0° (Table 21).

Table 21. RCS Summary of Results.

All Values in dB	Type X	Type X Neo
360° Avg. @ 15°	45.56	40.44
360° Avg. @ 10°	47.26	41.45
360° Avg. @ 5°	45.00	42.69
360° Avg. @ ~0°	49.18	41.23
Average RCS Return	46.75	41.45

The data shown in the table above illustrates that at all evaluated points, the Type X Neo performs better than the basis vessel, consistently displaying a smaller RCS return signal. This shows that the TOPSIS MADM chooses the objectively preferential option based on the weighted value selected in the initial stage of the calculation. This is affirmed by the Type X Neo's 11% decrease in radar signature return.

5.2 (b) Calm Water Resistance Analysis

For the calm water resistance simulations for the vessel, there were three options to decide from. The TOPSIS method has determined that the preferable option is a bow-mounted sonar dome. Three speeds were tested that represented 5 knots- for sonar hunting, 17.5 knots for cruising speed and 30 knots for rapid transits and redeployment. Priority and weighting of 40% was given to the resistance at 17.5 knots as this is the condition that the vessel will mostly operate in. The other speeds were given 30% each (Table 22).

Table 22. Resistance Summary of Results.

Original Resistance (kN)		Reduction in Full Hull Resistance	
Speed (Knots)	Type X	Type X Neo (Bow Sonar)	Type X Neo (Midship Sonar)
5	23.69	10%	13%
17.5	348.71	23%	22%
30	1600.64	7%	6%
	Avg. Reduction	13%	14%

It can be seen in the results that overall, the midship-mounted sonar dome has a lower resistance with an average reduction of 14% while the bow mounted variant has an average reduction of 13%. Whilst the results are close, at first instance the designer would select the midship sonar as the choice to carry forward as on average, it has less resistance. However, this would not be reflective of the operating conditions of the vessel. The TOPSIS method takes into consideration the operating conditions and tailors the decision-making process accordingly to reflect the needs of the platform.

5.2 (c) Seakeeping in Irregular Waves

The vessel was assessed in terms of seakeeping in irregular waves to analyse the motions of the aft deck in varying sea states and headings relative to the waves. In this scenario there were only two designs to choose from and the brief was simple in the sense of the criteria was for the lowest accelerations experienced on the aft flight deck.as well as the numerical values being assessed, two propulsion arrangements were being assessed that primarily changed the arrangements at the aft end of the vessel. Acceleration at the polar cardinal points were each weighted as being equally as important at 12% each totalling a 60% total. Propulsion redundancy is weighted at 25% and manoeuvrability 15% (Table 23).

Table 23. Seakeeping Summary of Results.

weighting	0.12	0.12	0.12	0.12	0.12
	Accelerations 0°	Accelerations 45°	Accelerations 90°	Accelerations 135°	Accelerations 180°
Type X with Azimuth Propulsion	0.268	0.298	0.654	1.522	1.472
Type X with prop & Rudder	0.626	0.167	0.632	1.485	1.517

The accelerations alone are similar however based on the increase at 0°, the designer would be more likely to select the Type X with azimuth-podded propulsion. It is only when redundancy and manoeuvrability are considered that the designer will get the full picture and be presented with the correct selection of Type X with Propeller and Rudder arrangements.

5.2 DISCUSSION

Throughout all stages of this case study there have been decisions to be made that when considered as the single attribute would be an objective decision easily made by a single designer. However, when designing a warship there are no one-dimension problems. Every aspect has the objective subject as well as the subjective requirements of the client. As proven above these subjective requirements are best captured and analysed in a MADM such as the TOPSIS or Hierarchical methods. Optimisation of the Type X design was carried out over all attributes discussed. An overall RCS reduction of 20% was achieved with a maximum of 69% reduction when observed at a relative polar angle of 0o. In terms of calm water resistance, after the utilisation of the TOPSIS method, a total resistance reduction of 13% was achieved while achieving the secondary objectives of reducing noise interference with the sonar itself. Seakeeping analysis had shown an improvement and expansion of sortie capability of only~1.3%, a negligible amount however when TOPSIS method was applied and consideration for redundancy and manoeuvrability in reference to the propulsion arrangements, it was found that the Type X Neo was the favourable option. Damage stability was analysed philosophically, discussing the possible impacts that the subsequent design changes would have on the vessel in a damaged state. The discussion behind these design decisions was based on typical stability criteria set out by Royal Navy standards NES-109. It was found that the Neo design had maximum GZ angle reductions ~27%, the range of positive stability increased by ~17% while heeling due to wind was the same with less than 0.5% reduction. The consolidated design, with the improvements stated, is proof that the combination of Hierarchical and TOPSIS methods is beneficial to the design process.

6. CONCLUSION AND FURTHER RESEARCH

This case study has explored the intricacies and difficulties of the design process when optimising a basis vessel to suit the needs of the modern naval fleet. The objectives set out in the opening chapters were executed as detailed in subsequent sections. A hierarchical decision-making process was utilised to evaluate the requirements of the vessel dictated by current and relative modern naval needs. Using a set of industry standard design and decision-making methodologies, specific aspects of the ship were optimised and analysed with the results being implemented in a TOPSIS MADM calculation to aid the designer in making the correct decisions both objectively and subjectively. The designs were then combined to form the final design of the Type X Neo with key areas assessed against the basis, Type X frigate.

Further investigation could also be undertaken to determine the role and integration of computational automation and optimisation within this complex design space. Alongside this, an in-depth assessment of the damaged stability could be conducted in conjunction with optimisation of internal arrangements. When considering the assessed attributes, more accuracy could be achieved when assessing seakeeping in irregular waves with the use of CFD to eliminate uncertainties due to uncoupling effects, particularly in head seas. In conjunction with this, analysis could span further than motions at the aft end of the vessel by investigating the motions, immersion and possible slamming of appendages particularly, the bow sonar dome.

In Conclusion, the decision methods used were confirmed to be effective when using objective and subjective attributes, where investigating one attribute itself the designer may come to one decision however when the subjective is quantified and implemented in the MADM process the decision can be made more clearly and with greater confidence. This shows that MADM can be used on an existing design to improve and alter the basis vessel for the needs of the navy without the need of a completely new design. In this project the Hierarchical and TOPSIS methods were proven to be accurate however further research would be beneficial in areas relating to MADM being used within a team of design engineers, where subjective weighting may vary.

7. REFERENCES

1. BIRKLER, J., et. Al, 'The Royal Navy's New-Generation Type 45 Destroyer: Acquisition Options and Implications', *Rand Corporation, Rand Europe 2002*, 2002.

2. BOTTERO, M., and GUALENI, P., 'Systems Engineering for Naval Ship Design Evolution', *Journal of Marine Science and Engineering.* 2024; 12(2):210. https://doi.org/10.3390/jmse12020210, 2024.

3. BOULOUGOURIS, E., and PAPANIKOLAOU, A., 'Optimisation of the Survivability of Naval Ships by Genetic Algorithms', 2004.

4. CERRUTI, M., PASTORINO, M., RANDAZZO, A., PERRA, F., and GUAGNANO, A., 'A Radar Cross Section and Radar Performance Evaluation Tool for The Early-Stage Ship Design (ESSD) Phase'. Oceans 2015 - Genova, 2015.

5. ERIKSTAD, S O., and LAGEMANN, B., 'Design Methodology State-of-the-Art Report' *SNAME IMDC 2022-301*, 2022.

6. GUI, L., et al, 'Forces, moment and wave pattern for surface combatant in regular head waves', *Part II. Measurement results and discussions. Experiments in Fluids*, 2002.

7. HOUTEN, J V., KANA, A., SINGER, D., and COLLETTE, M., 'Early Risk Quantification Strategy for Design Space Reduction Decisions in Set-Based Design', *International Marine Design Conference*. https://doi.org/10.59490/imdc.2024.852, 2024.

8. JACOBSON, D., 'The Evolution of Container Ship Design and Its Impact on Global Trade'. *Available at: https://www.universalcargo.com/the-evolution-of-container-ship-design-and-its-impact-on-global-trade/.* Accessed 04/05/2025, 2023.

9. KANG, H.-J., YANG, H.-K., JO, M.-C. and KIM, K.-H. (2017) 'Feasibility study of corner reflector for radar countermeasures and deception for conventional forces', *Journal of the Korean Society of Marine Engineering*, 41(2), pp. 171–175. https://doi.org/10.5916/jkosme.2017.41.2.171

10. KEHOE, J W., 'Destroyer Seakeeping U.S. and U.S.S.R'. *U.S. Naval Institute*, 1973.

11. KHAN, J., DUAN, W., and SHERBAZ, S., 'Radar Cross Section Prediction and Reduction for Naval Ships'. *Journal of Marine Science and Application*, 2012.

12. KNOTT, E F., SHAEFFER, J F., and TULEY, M T., 'Radar Cross Section (2nd Edition)'. *SciTech Publishing*, 2004.

13. LYE, H., 'Littoral Strike ship: Prevail Partners refines Multi-Role Vessel' *Naval Technology*. Available at: https://www.naval-technology.com/features/littoral-strike-ship-prevail-partners-refines-multi-role-vessel/?cf-view. Accessed 04/05/2025, 2021.

14. McKinsey & Company. 'How the increase in military spending is shaping global navies', Available at: *https://www.mckinsey.com/industries/aerospace-and-defense/our-insights/shifting-currents-the-role-of-shipbuilders-in-enhancing-naval-fleets*. Accessed 04/05/2025, 2024.

15. PAPANIKOLAOU, A., and BOULOUGOURIS, E., 'Risk-based design of naval combatants'. *Ocean Engineering*, 2013.

16. PAPANIKOLAOU, A D., 'Holistic Approach to Ship Design', *Journal of Marine Science and Engineering*. 2022; 10(11):1717. https://doi.org/10.3390/jmse10111717

17. SEN, J., and YANG, J., 'Multiple Criteria Decision Support in Engineering Design', *London: Springer*, 1998.

18. SHARMA, R., and SHA, O., 'Hydrodynamic Design of Integrated Bulbous Bow sonar Dome for Naval Ships', *Defence Science Journal*, 2005.

19. SINGER, D J., DOERRY, N., BUCKLEY, M E., 'What is Set-Based Design', *American Society of Naval Engineers*, 2009.

20. U.S. Government Accountability Office, 'Navy Shipbuilding: Enduring Challenges Call for Systemic Change'. Available at: https://www.gao.gov/products/gao-25-108225. Accessed 04/05/2025, 2025.

21. VAN'T VEER, R., 'Studying Survivability with FREDYN' MARIN', *MARIN Report 80*, 2003.

22. WALKER, J C., 'Multi-Attribute Tradespace Exploration for US Navy Surface Ship Survivability: A Framework for Balancing Capability, Survivability and Affordability', *Massachusetts Institute of Technology*, 2016.

23. WILSON, R., CARRICA, P M., and STERN, F., 'Unsteady RANS Method for Ship Motions with Application to Roll for A Surface Combatant', *Computers & Fluids*, 2006.

24. WU, J., et al, 'Neumann-Michell theory-based multi-objective optimization of hull form for a naval surface combatant', *Applied Ocean Research*, 2017.

25. YANG, D., and GUO, F J., 'Ship appearance optimal design on RCS reduction using response surface method and genetic algorithms', *Journal of Shanghai Jiaotong University (Science),* 2008.

8. AUTHORS BIOGRAPHY

Liam Nugent is a Masters student in Naval Architecture and Marine Engineering at the University of Strathclyde. His current research explores the application of multi-attribute decision-making techniques in the design of naval surface combatants. He has a strong interest in holistic ship design, parametric modelling, and set-based design methodologies. He has also contributed to projects such as the preliminary design of a sustainable tugboat design, focusing on systems and regulatory compliance.

TECHNOLOGY INNOVATION FOR DEFENCE-PLATFORM ENERGY-EFFICIENCY PROJECT (TIDE): ENERGY-EFFICIENT DESIGN APPROACHES FOR NEXT-GENERATION WARSHIPS

M T Ozdenoglu, D Sin, S T Tudora, N Townsend, T Tezdogan, M Kurt, J Banks, D Hudson, A J Murphy, and **S R Turnock,** University of Southampton, UK

SUMMARY

This paper contributes to the Technology Innovation for Defence-platform Energy-efficiency (TIDE) project, a collaborative initiative between BAE Systems and the Universities of Southampton and Strathclyde, aimed at enhancing the energy performance of future naval platforms. Within the hydrodynamic research stream, this study presents experimental and numerical investigations focused on reducing propulsion energy demands. Key contributions include stern foil experiments for wave energy interaction, resistance-based analysis supporting alternative RPM control strategies, and a voyage energy simulation framework using Simulink to evaluate full-mission power demand. Combined, these efforts support TIDE's overarching objectives of improving ship-level energy efficiency by 5–25% and reducing through-life energy costs through integrated design and operational solutions. The results highlight the potential of combining low-order models, hybrid propulsion analysis, and flow control technologies to support energy-aware control and planning strategies. While ongoing, this work establishes a foundation for next-generation warship design aligned with sustainability and operational resilience goals.

NOMENCLATURE

[Symbol]	[Definition] [(Unit)]
C_B	Block coefficient
C_M	Midship section coefficient
C_P	Prismatic coefficient
C_W	Wave-making resistance coefficient
CAPEX	Capital Expenditure
CFD	Computational Fluid Dynamics
D	Propeller diameter (m)
EEOI	Energy Efficiency Operational Indicator (g CO_2/ton-nm)
EMS	Energy Management System
ESS	Energy Storage System
F_x	Longitudinal (thrust) force on the foil (N)
F_z	Vertical (lift) force on the foil (N)
Fr	Froude number
J	Advance coefficient
KT	Thrust coefficient
P	Pressure (N m^{-2})
P_{prop}	Propulsive power (kW)
$P_{nonprop}$	Non-propulsive load (kW)
$PE_{delivered}$	Delivered electric power to propulsors (kW)
R_A	Air resistance (N)
R_T	Total ship resistance (N)
Re	Reynolds number
RPM	Revolutions Per Minute
SOC	State of Charge (of battery/ESS)
SEPM	Ship Energy Performance Model (%)
T	Thrust (N)
$T_{command}$	Commanded thrust (N)
TIDE	Technology Innovation for Defence-platform Energy-efficiency
OPEX	Operational Expenditure
V_a	Advance speed at propeller (m s^{-1})
u	Kinematic viscosity (N s m^{-2})
r	Density of water (kg m^{-3})

1. INTRODUCTION

The maritime industry, including naval defence platforms, faces significant pressure to reduce greenhouse gas emissions and operational costs in response to increasingly stringent environmental regulations and evolving fuel economy standards [1]. As future fuels, likely renewable or alternative energy sources, become necessary, operational expenses associated with naval vessels are anticipated to rise substantially due to higher fuel costs and infrastructure changes [2]. Consequently, achieving substantial energy efficiency improvements is no longer an option but a necessity for sustainable operations.

Historically, naval vessels have prioritised performance, speed, and capability over energy efficiency, leading to substantial energy consumption and environmental impacts [3]. However, recent shifts in global policy and rising operational expenses have necessitated reconsidering naval platform designs and operations. Recognising these critical challenges, the Technology Innovation for Defence-platform Energy Efficiency (TIDE) project, initiated in 2023, represents a strategic collaboration among BAE Systems, the University of Strathclyde, and the University of Southampton. The primary goal of the TIDE project is to significantly enhance energy efficiency across naval platforms by investigating and developing innovative technological solutions with clearly defined efficiency targets.

The TIDE project comprises three primary research areas: hydrodynamic-based energy efficiency improvements, innovative waste heat recovery (WHR) and energy storage (ES) systems, and a comprehensive decision-modelling framework to facilitate optimal technology integration and informed decision-making. Within hydrodynamic research, particular attention is given to advanced hull form optimisation, air lubrication technologies, and improved rudder and propeller systems, collectively aiming for up to a 30% reduction in hydrodynamic energy demand. Experimental and computational studies, such as Computational Fluid Dynamics (CFD) simulations and scaled model testing, are integral components of this research, providing the necessary evidence base and predictive capability for future naval designs [4].

Complementary to hydrodynamic advancements, the WHR and ES components seek to identify viable technology combinations that enhance overall system efficiency by at least 15%. These technological innovations require careful evaluation and informed decision-making, facilitated through a robust decision-modelling framework that balances initial investment with long-term operational efficiency [5].

This paper presents the holistic approach and key technical progress achieved within the TIDE project, providing insight into innovative strategies for energy reduction in next-generation naval defence platforms. The outcomes aim to advance technological understanding and guide naval procurement towards more sustainable, cost-effective operational practices.

2. LITERATURE REVIEW

2.1 WHAT CONTROLS THE ENERGY USE OF A SHIP?

Numerous interrelated design and operational variables govern the energy consumption of a ship. Operational profiles, such as calm water versus realistic sea conditions, significantly affect energy consumption through additional resistance experienced in waves [6]. Critical ship design features include hull forms, appendages, propulsion systems, and coatings, each having a pronounced impact on resistance and, consequently, on power requirements [6]. Hull form optimisation through careful consideration of the bow shape, stern geometry, and bulbous bow configurations directly influences hydrodynamic efficiency and fuel consumption. Appendages such as rudders, propellers, and additional energy-saving devices (ESDs) play crucial roles in minimising power needs through enhanced hydrodynamic performance [6]. Surface coatings designed to reduce skin friction resistance have also demonstrated significant energy-saving potential [7] [8].

2.2 CONVENTIONAL SHIP DESIGN VS LEAST THROUGH-LIFE ENERGY APPROACH

Traditionally, ship designs have been oriented towards achieving specified design speeds and operational performance, occasionally at the expense of long-term energy efficiency [6]. The conventional design approach often prioritises the minimisation of initial capital expenditures (CAPEX) such as hull construction and machinery costs without adequately considering through-life operational expenditures (OPEX), including fuel consumption, maintenance, and crew-related costs [9]. CAPEX refers to the upfront investment in physical assets, while OPEX covers the ongoing costs of operating a vessel throughout its service life [10]. In contrast, the least through-life energy design philosophy emphasises a comprehensive life-cycle analysis of energy consumption, promoting higher initial investments that yield substantial long-term energy savings and reduced OPEX. Key strategies in this approach include hull form optimisation, propeller enhancements, and the integration of energy-saving devices. Molland *et al.* [6] highlight various methods to reduce propulsive power and emissions, such as improving hull efficiency and adopting advanced propulsion technologies. Additionally, Tillig *et al.* [11] presented a generic energy systems model for evaluating the performance of energy-saving technologies in early-stage design.

2.3 CHANGES WHEN PRIORITISING LEAST THROUGH-LIFE ENERGY

Adopting a least through-life energy perspective significantly redefines ship design and operation priorities. Instead of focusing solely on immediate metrics such as speed or capacity, this approach emphasises optimisation of hull form, propulsion systems, energy-saving technologies, and external energy-harvesting structure aimed at long-term efficiency [6]. Examples include stern foil integration, RPM control strategies, and tailored operational profiles that reduce fuel use over the vessel's lifetime.

However, designing for a broad operational profile often involves multi-objective optimisation, which can increase initial costs due to more complex requirements [12]. These costs necessitate customer support for higher CAPEX, justified by long-term reductions in OPEX and enhanced lifecycle performance [11].

This approach also requires a shift in customer expectations, moving away from minimising upfront costs to valuing long-term efficiency, operational flexibility, and sustainability across the vessel's service life [13].

2.3.1 Energy-Saving Foil Systems and Their Effect on Delivered Power

Recent model tests have shown that wave-assisted foils can significantly reduce delivered power. Bowker and Townsend [13] demonstrated that a spring-loaded bow foil reduced delivered power by up to 50% in regular head waves and 12% in irregular seas by generating thrust and reducing pitch and heave motions. Their work also introduced key performance indicators, such as flapping phase, wave phasing, and feathering coefficient, relevant to foil optimisation. These findings are particularly appropriate for stern-mounted foils, where similar mechanisms can enhance propulsive efficiency.

2.3.2 Alternative RPM Control Strategies

Alternative RPM control strategies enable modulation of propeller speed in response to changing wave conditions, helping reduce delivered power and improve through-life energy efficiency. These approaches align thrust output with resistance variations, avoiding unnecessary energy losses associated with constant speed operation.

Recent studies have introduced nonlinear model predictive control (MPC) frameworks that incorporate environmental loads such as wave-induced forces [15]. These demonstrate that optimising RPM within a path-planning or tracking algorithm can enhance manoeuvrability and reduce voyage time and energy demand, particularly under regular wave conditions. Additionally, they support spatial and temporal reformulations of ship dynamics for better prediction and control.

The integration of RPM modulation within marine craft control systems has also been emphasised in marine hydrodynamics literature [16]. Building on these principles, this study implements a wave-aware RPM control scheme and evaluates its performance in regular wave scenarios, focusing on its contribution to delivered power reduction and propulsion efficiency.

2.3.3 Operational Demand Modelling through Voyage Simulation

The voyage demand simulation is an important computational approach to estimate the operational profile and onboard power demand, which is essential to understand the energy efficiency over the vessel's lifetime. By implementing the Ship Energy Performance Model (SEPM) that integrates the mathematical model of each subsystem of a vessel, the approach can provide a detailed energy profile for evaluation. With the capability, the approach can also be applied to investigate hybrid propulsion for modern vessels [17,18].

2.4 METHODS AVAILABLE FOR ENERGY OPTIMISATION

The optimisation of ship energy consumption employs various methodological approaches, primarily Computational Fluid Dynamics (CFD), Experimental Fluid Dynamics (EFD), and analytical evaluations. CFD methods enable detailed analyses of flow around hulls and propellers, providing insights into minimising resistance and optimising propulsion efficiency [6]. EFD, including towing tank experiments, Particle Image Velocimetry (PIV), and direct force measurement techniques, validate computational models and provides critical empirical data necessary for refining and confirming efficiency gains from various design modifications [6].

2.5 OPERATIONAL PROFILE OPTIMISATION

Optimisation of the operational profile, encompassing design speed, voyage speed patterns, and corresponding energy usage profiles, is vital to achieving minimum fuel consumption throughout a ship's operational lifecycle [6]. This approach carefully considers voyage management, weather routing, and operational speed adjustments. Adjusting operational trim, maintaining optimal hull and propeller surface conditions, and effectively deploying energy-saving devices collectively significantly reduce overall resistance and energy consumption [6].

2.6 SERVICE SPEED VS MAXIMUM SPEED

The balance between service speed and maximum achievable speed is central to operational efficiency. Technologies such as foils, stern flaps, wind-assisted propulsion, and air lubrication are increasingly used to reduce drag and improve performance without significantly increasing installed power [6].

Defining an optimal service speed, however, requires a mission-based analysis that accounts for how much time the vessel is expected to operate at various speeds. Incorporating real-world fleet data enables the development of an accurate operational profile, which is essential for effective ship design and optimisation [13].

These strategies support the holistic, data-informed design philosophy of the TIDE project, aimed at delivering energy-efficient and sustainable naval platforms.

3. METHODOLOGY

3.1 FOIL AND WAVE

3.1.1 Stern Foil Design and Experimental Setup

To investigate the energy performance of a flappable stern foil in realistic operating conditions, a series of towing tank experiments were conducted at the University of Southampton as part of the TIDE project. A scale model representative of a frigate-class hull (approximately 4–6 meters in length) was fitted with a passive, spring-loaded stern foil and tested without a propeller in regular following seas at two Froude numbers (Fr = 0.23 and Fr = 0.38). Flapping foil systems have shown potential for thrust generation and motion control in waves [19].

The stern foil was designed using a symmetric NACA0012 profile. The foil pivoted at 15% of the chord and was supported by struts and the foil system was suspended from the transom via a 3-axis load cell. Stern foil properties are presented in Table 1. The stern foil configuration used in the experiments is illustrated in Figure 1, which shows both the CAD design of the flappable foil system (Figure 1a), and the physical setup installed on the frigate-scale model in the towing tank (Figure 1b).

A National Instruments myRIO-based data acquisition system was used to record foil forces and motions (F_x, F_z, pitch angle), while drag and ship motions were monitored via the carriage system. Experiments were conducted under multiple wave conditions (Range λ/L 1.82 to 0.65, and two different wave amplitudes. The corresponding wave steepness H/λ ranged from 0.008 to 0.046) to examine energy recovery potential and wave interaction effects. Recent studies have demonstrated the potential for energy harvesting using passive and active foils in marine environments [14].

Table 1. Stern Foil Properties.

Property	Symbol	Unit	Value	Rationale
Foil section	n/a	n/a	NACA 0012	Symmetric profile, widely studied
Foil Span	S	n/a	5c	
Foil Chord	C	n/a	c	
Foil Aspect Ratio	AR		5	
Foil Area	A	n/a	$5c^2$	
Foil Pivot Point	a	%	15	15% chord length, from leading edge
Foil Spring Stiffness	k_θ	Nm/rad	$\approx 1.5(k_2)$, $\approx 3(k_1)$	
Foil Depth	d	n/a	1.55c	Water plane to top of foil
Foil Position	L_f	n/a	0.9c	From stern to foil pivot point

* Note that c is the chord length and all other parameters are nondimensionalised with c.

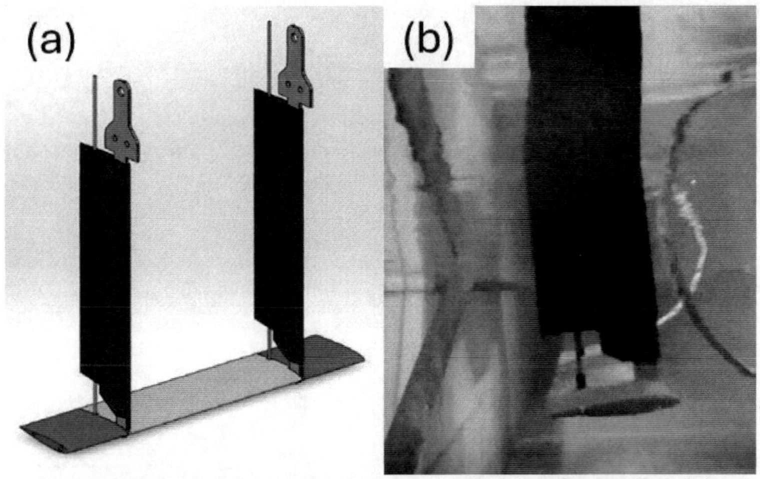

Figure 1: Stern foil design ((a) CAD model (b) Manufactured foil rig).

3.1.2 Experimental Campaign and Data Handling

Raw data were filtered using a fourth-order Butterworth low-pass filter (5 Hz cutoff) and synchronized using carriage pitch signals. Data were segmented into steady state runs and non-depersonalised for performance comparisons. Amplitude and mean values of thrust (F_x), lift (F_z), pitch amplitude, and drag were extracted using sine fitting techniques. Additionally, calm water runs were used as a baseline to evaluate wave-induced drag components and net performance changes due to the foil. The experimental setup for hull-propeller interaction is illustrated in Figure 2.

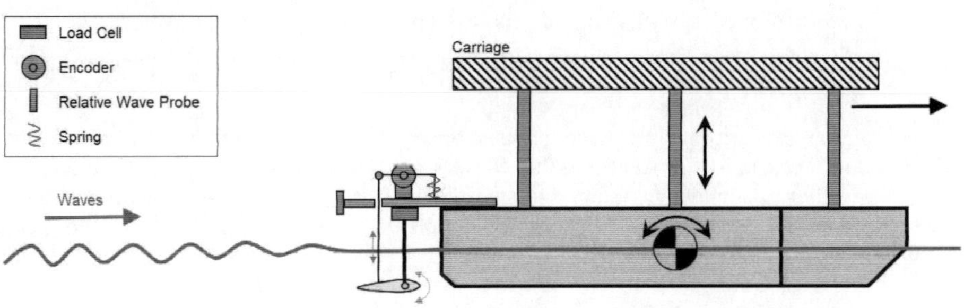

Figure 2: Experimental setup for the stern foil experiment.

3.1.3 Supporting Flow Field Visualization with PIV

To gain additional insight into the stern foil's effect on wake behaviour, a supporting set of stereo Particle Image Velocimetry (PIV) experiments were conducted using a KRISO Container Ship (KCS) model. Although the hull form differed from the frigate used in the main experiments, the KCS-based PIV results provided high-resolution 2D-3C velocity field data illustrating how stern foils influence flow separation, streamline behaviour, and wake structure.

This visualization complements the force-based evaluation by revealing flow alteration mechanisms that may underlie energy performance trends. Notably, the PIV images highlight the directional realignment of the wake and changes in velocity angle due to foil-induced effects. These findings are incorporated as qualitative validation of the hydrodynamic behaviour observed in the stern foil experiments.

3.2 ALTERNATIVE RPM CONTROL STRATEGY FOR ENERGY OPTIMISATION

To explore potential energy savings through propulsion control, a comparison was conducted between high-fidelity CFD data and two lower-order resistance prediction methods: the Holtrop-Mennen empirical model and Thin Ship Theory. Previous work comparing Holtrop and Thin-Ship Theory in frigate-scale designs has shown consistent divergence under high-speed conditions [20]. This evaluation aims to support the development of an alternative RPM control framework, allowing resistance-informed decisions without relying on computationally expensive simulations.

By comparing resistance curves generated from these three methods, operational deviations and prediction biases were identified. These insights serve as a basis for future control logic where engine input can be adjusted in response to estimated resistance characteristics.

This approach is conceptually illustrated in Figure 3, adapted from Sørensen [21], where RPM-based input control leads to higher power fluctuations, while torque-based control maintains steadier propeller dynamics and offers greater energy stability.

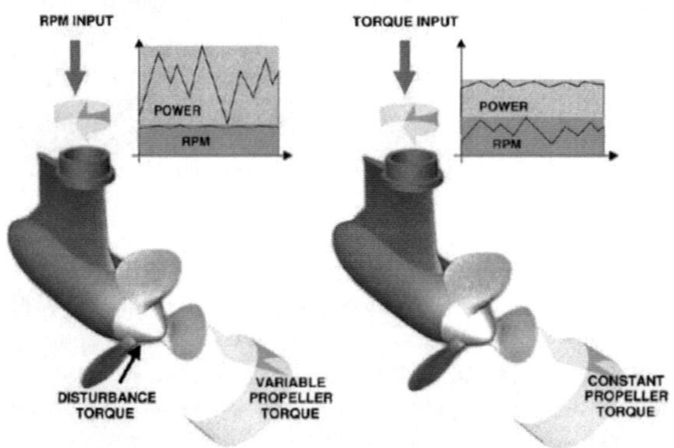

Figure 3: Comparison of RPM-based versus torque-based control approaches. Torque control maintains steadier propeller behaviour and enables improved power stability. Adapted [20].

3.3 VOYAGE DEMAND GENERATION FRAMEWORK

To enable system-level energy analysis under realistic operating conditions, a Ship Energy Performance Model (SEPM) was developed in MATLAB Simulink. Similar modelling frameworks have been used to simulate energy flows and hybrid propulsion system dynamics [22]. This model estimates energy demand for a given voyage by considering resistance components, propulsion system characteristics, and general vessel conditions.

The SEPM is structured into four primary subsystems:

- Resistance block, accounting for calm water, wave, air, and wind resistances.
- Propulsion system, including propulsor models and propeller performance curves.
- Power generation, simulating diesel generators and energy storage system (ESS) behaviour.

- Thrust and power management, using a rule-based EMS to control thrust distribution and energy flow.

Inputs such as sea state, vessel mass, set speed, and general control settings are used to drive the model dynamically. By solving the force balance and incorporating predefined external and vessel parameters, the SEPM provides output profiles for energy demand, thrust distribution, fuel consumption, and battery state-of-charge.

The framework is illustrated in Figure 4, showing the data flow and interconnection between modules. This approach supports fast simulation of full voyages with varying conditions and provides a flexible tool for evaluating operational energy efficiency, supporting mission planning and design trade-offs.

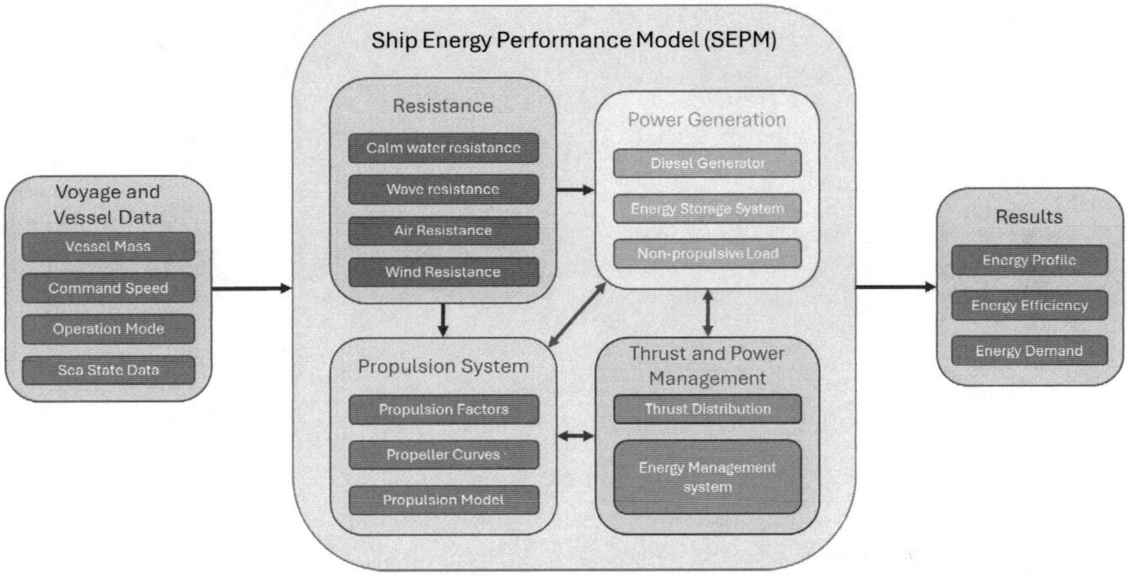

Figure 4: Ship Energy Performance Model (SEPM) framework for voyage demand generation. The system integrates resistance modelling, propulsion characteristics, power generation, and thrust/power management to simulate energy profiles based on voyage and environmental data.

4. PRELIMINARY RESULTS

4.1 STERN FOIL FORCES AND MOTION BEHAVIOUR

The performance of the passive stern foil was evaluated based on thrust force (F_x), vertical lift force (F_Z), and pitch motion across a range of wave conditions. As shown in Figure 5, the foil produced oscillatory thrust in the presence of waves. At zero forward speed (Figure 5a), the average thrust was directed astern, indicating drag dominance. However, once the vessel reached cruising speed (Figure 5b), F_Z shifted marginally positive at certain frequencies, indicating that the passive foil can generate net forward thrust through energy interaction with incident waves.

The vertical force component F_Z also showed a change in behaviour between the two operating conditions. While nearly symmetric around zero at rest, the onset of forward motion resulted in consistently positive F_Z, suggesting an increase in vertical lift. This behaviour may be attributed to changes in local flow conditions and angle of attack (AoA) induced by hull interaction.

Additionally, foil pitch amplitude was significantly reduced from ~40-100% to at most 20% during forward motion, confirming that foil dynamics became more restrained, likely due to dominant flow from the hull stern. This raises important considerations for foil placement and design tuning, as minimal flapping under cruising conditions may limit energy extraction and the responses to local flow variations.

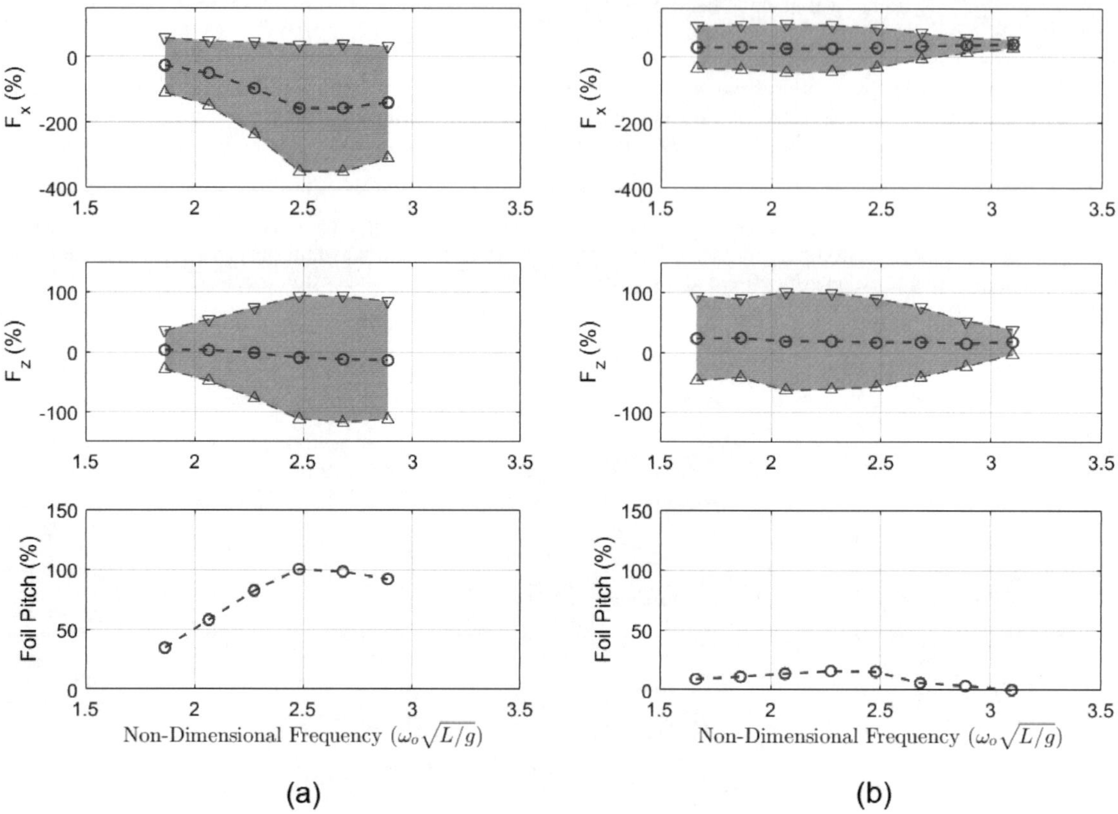

Figure 5: Non-dimensionalised foil responses (a) zero speed (b) with forward speed (Top: Fx, Middle: Fz, Bottom: Foil pitch) (Dimensionalised with respect to the maximum value) (mean and amplitudes)

4.1.1 Performance Ratio and Operational Implications

The performance of the foil was further characterized using the ratio of generated thrust (F_x) to total drag (F_D), as shown in Figure 6. This ratio ranged between 2.5–4% across tested conditions, which indicates the implementation of the stern foil consistently yields positive averaged thrust without reversal. The increasing trend of the averaged thrust suggests a potential performance threshold where the foil's contribution becomes more favourable. The observation aligns with existing literature highlighting improved effectiveness of such devices at elevated Froude numbers or under controlled actuation. These depicted the potential of utilising stern foil for vessel design in the future.

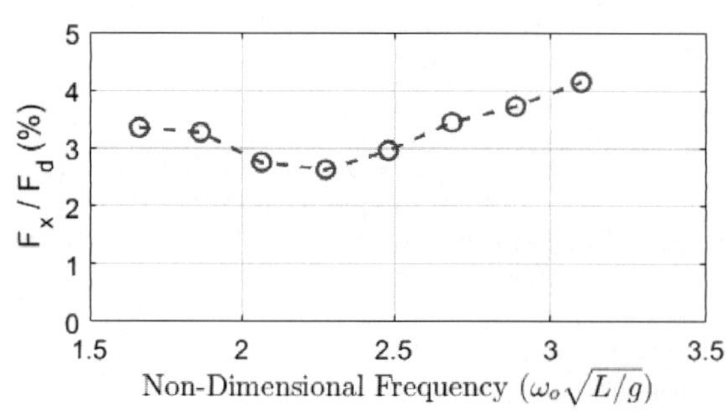

Figure 6: Ratio of foil thrust (F_x) to drag.

4.1.2 Supporting Flow Field Insights from PIV

To complement the force-based findings, flow visualization using stereo-PIV was conducted on the benchmark KCS model. Though geometrically different from the primary frigate model, these results revealed distinct alterations in wake structure and velocity angle due to the presence of the stern foil (Figure 7).

The foil introduced visible realignment of the flow field and changes in vorticity distribution. These qualitative observations reinforce the experimental evidence of stern flow alteration and energy-interaction potential, suggesting wake flow manipulation as a valid mechanism for future performance enhancement.

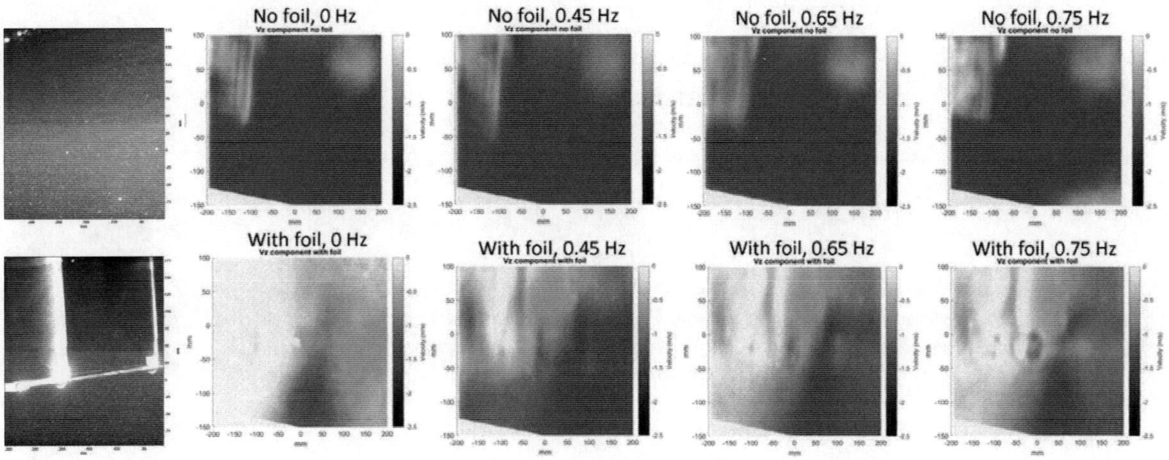

Figure 7: The V_z velocity distributions (view from camera 1) after processing for the model with stern foil at different wave conditions.

4.2 COMPARATIVE EVALUATION OF RESISTANCE PREDICTION METHODS FOR ENERGY CONTROL

To evaluate the viability of implementing an energy-aware RPM control approach, total resistance predictions from three methods—CFD (as a benchmark), Holtrop-Mennen (empirical), and Thin Ship Theory (analytical)—were compared. The resistance curves are shown in Figure 8, which highlights the deviations among the methods across varying Froude numbers.

While CFD provides the highest-fidelity prediction, the Holtrop method closely follows the CFD trend across most of the operational profile, confirming its potential as a reliable and computationally efficient tool for informing control strategies. Thin Ship Theory, although more variable due to its inherent assumptions, captures the general shape of the resistance curve and becomes more aligned with empirical methods when smoothed using a polynomial trendline.

These comparative trends are essential for constructing a resistance-informed RPM control framework. In such a system, the controller could use continuously updated empirical estimations (e.g., Holtrop-based or trendline-fitted thin-ship theory method outputs) to dynamically adjust engine input based on expected resistance profiles, improving propulsion efficiency without the need for continuous high-fidelity simulations.

The observed differences between the prediction methods emphasize the importance of choosing an appropriate model based on the control system's responsiveness and data availability, torque-based control systems may yield steadier and more energy-efficient performance, particularly when integrated with low-cost resistance predictors that inform propulsion loads.

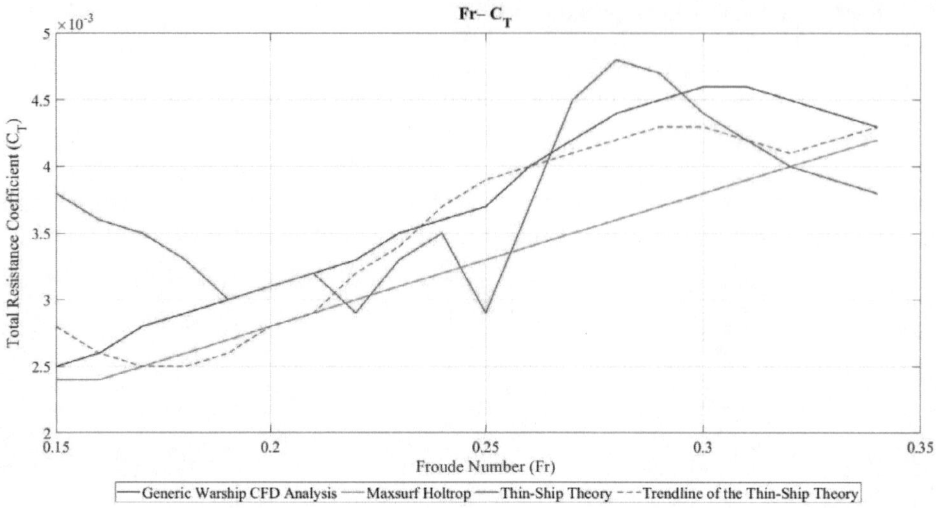

Figure 8: The resistance characteristics are presented in terms of non-dimensional parameters (Froude number vs total resistance coefficient, Ct) to allow for scale-independent comparison. This approach facilitates a clearer evaluation of hydrodynamic performance across different conditions and supports the subsequent analysis of RPM control effectiveness.

4.3 VOYAGE DEMAND GENERATION

The voyage demand simulation produced profiles of ship resistance, power usage, and energy-related system behaviour over a representative 32-hour route. The model, developed in Simulink, incorporates varying external and vessel-specific conditions. Specifically, the simulation captured the variation in total ship resistance due to environmental effects and vessel dynamics, as well as the corresponding propulsion power demand over time. It also tracked the state-of-charge (SOC) of the onboard energy storage system (ESS), along with the power split between the diesel generators and the ESS throughout the voyage. These outputs collectively demonstrate the model's capacity to replicate complex system interactions and validate energy performance during voyage execution.

To evaluate the impact of energy storage, a second simulation was conducted with battery integration under harsher winter sea conditions. As illustrated in Figure 9, the inclusion of batteries led to a noticeable reduction in diesel generator load between the 22nd and 28th hours. The battery buffered the increased power demand during this period, preventing activation of a third diesel generator.
The battery system was primarily charged during periods of low generator load, when excess power capacity was available. By charging the battery under these conditions, the diesel generators could operate closer to their optimal load range, thus reducing specific fuel consumption (SFC) and improving overall fuel efficiency. When demand increased, the battery supplied additional power, enabling the system to meet operational needs without engaging additional generators operating at low efficiency. After the battery is consumed, the system charged again in the optimal condition again and overall, the system could achieve better efficiency. Based on the simulation, the modelled hybrid power management strategy reduced total energy consumption by approximately 8%, demonstrating the potential of energy storage systems to enhance efficiency, reduce emissions, and lower operational costs.

The model's responsiveness and modularity support its use in operational planning, energy profile comparison, and power architecture assessment. Moreover, its rapid simulation capability (32-hour voyage computed in under 10 minutes) makes it suitable for scenario testing and design decision-making during early-stage development.

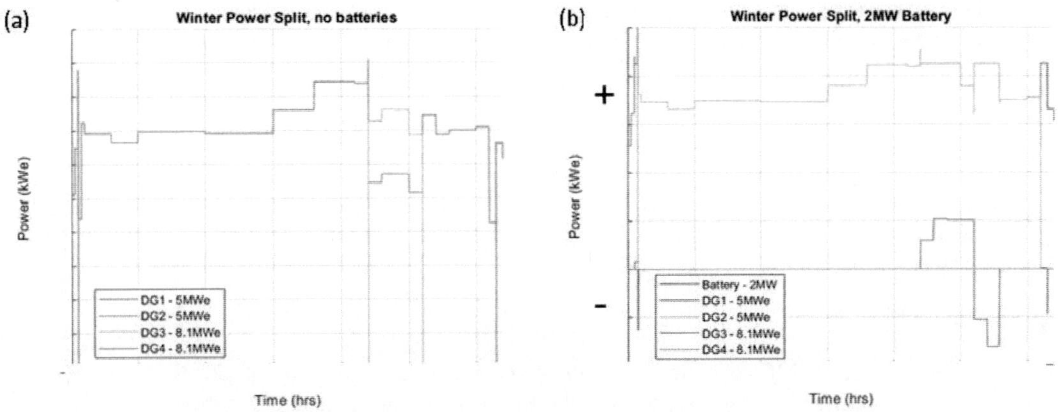

Figure 9: The power split (a) without and (b) with battery in winter.

5. CONCLUSIONS

This study presents a multi-perspective investigation into energy-saving strategies for future naval platforms, combining experimental hydrodynamic research, simulation-based control development, and system-level voyage demand modelling. By integrating contributions from stern foil force evaluations, resistance-informed RPM control methodologies, and dynamic energy profiling, the work provides a comprehensive foundation for developing more energy-efficient frigate designs.

The outcomes of each research stream directly support the broader goal of reducing through-life energy consumption in naval vessels:

The stern foil experiments confirmed measurable thrust contributions under wave conditions, while also identifying critical drag penalties—highlighting the need for further optimization of foil geometry and integration strategies.

The alternative RPM control framework, grounded in comparative resistance modelling, enables preliminary control logic to modulate engine output based on predictive hydrodynamic data, reducing unnecessary power expenditure.

The voyage demand generation model demonstrated the potential for detailed energy profiling over a range of representative conditions. The incorporation of energy storage systems yielded up to 8% fuel savings in high-demand conditions, reinforcing the value of hybrid solutions in mission planning and propulsion architecture design.

These findings contribute to three core industry insights:

Energy Efficiency Potential: The methodologies explored here—when fully implemented and extended—support the projected 5%–25% reductions in energy consumption through hydrodynamic and operational innovation.

Design and Operational Integration: The presented techniques, including resistance prediction models, foil-based flow manipulation, and voyage demand simulations, align with the trend of integrating technologies like air lubrication, controllable appendages, and optimized hull forms into a unified energy strategy.

Life-Cycle Cost Considerations: While technical feasibility is clear, wider adoption remains constrained by procurement strategies. Higher upfront costs for efficient technologies must be balanced against long-term fuel savings, endurance benefits, and operational flexibility. This work contributes quantitative evidence supporting such investment decisions.

While the presented results represent key milestones in the development of a next-generation energy-efficient frigate, the research is ongoing. Future work will expand the experimental dataset, refine control algorithms under seaway conditions, and further validate the voyage simulation framework under varied environmental scenarios. Collectively, these developments aim to support more sustainable naval operations and informed decision-making across both design and mission planning domains.

6. ACKNOWLEDGEMENTS

The research reported in this paper was funded by BAE Systems as part of the TIDE project. The authors would also like to thank the University of Strathclyde for their valuable contributions as part of the TIDE consortium. The authors also wish to thank the University of Southampton's Iridis High Performance Computing (HPC) facility for providing the computational resources necessary for the numerical simulations presented in this work.

7. REFERENCES

1. IMO, 'Fourth IMO GHG Study', *International Maritime Organization*, 2020.

2. BUHAUG, Ø., CORBETT, J J., and ENDRESEN, Ø., et al., 'Second IMO GHG Study 2009', *International Maritime Organization*, 2009.

3. KLEBANOFF, L E., PRATT, J W., and JOHNSON, T A., et al., 'Naval ship propulsion: challenges and opportunities', *Naval Engineers Journal*, 2017.

4. TURNOCK, S R., and WRIGHT, P N., 'Directly calculated hull-propeller interaction coefficients for the design of ship propellers', *Transactions of RINA*, 2000.

5. CALLEYA, J., PAWLING, R., GREIG, A., and BUCKNALL, R., 'Ship impact model for technical assessment and selection of carbon dioxide reducing technologies (CRTs)', *Ocean Engineering*, 2015.

6. MOLLAND, A F., HUDSON, D A., and TURNOCK, S R., 'Reductions in propulsive power and emissions', Chapter 17, *Ship Resistance and Propulsion: Practical Guidelines*, 2nd ed., Cambridge University Press, 2017.

7. CANDRIES, M., and ATLAR, M., 'On the drag and roughness characteristics of antifoulings', *Transactions of the Royal Institution of Naval Architects*, 2003.

8. DEMIREL, Y K., TURAN, O., and INCECIK, A., 'Experimental determination of the roughness functions of marine coatings', *Proceedings of the Shipping in Changing Climates Conference*, University of Strathclyde, Glasgow, 2015.

9. STOPFORD, M., 'Maritime Economics', 3rd ed., Routledge, 2009.

10. DNV, 'Energy Management Study for Shipping', Det Norske Veritas, 2013.

11. TILLIG, F., RINGSBERG, J W., MAO, W., and RAMNE, B., 'A generic energy systems model for efficient ship design and operation', *Proc. IMechE Part M: Journal of Engineering for the Maritime Environment*, 2016.

12. DEB, K., 'Multi-objective Optimization Using Evolutionary Algorithms', John Wiley & Sons, 2001.

13. DNV, 'Assessment of Ship Operational Profiles Based on AIS Data', Det Norske Veritas, 2020.

14. BOWKER, J A., and TOWNSEND, N C., 'Evaluation of bow foils on ship delivered power in waves using model tests'. *Applied Ocean Research*, *123*, p.103148, 2022.

15. FOSSEN, T I., 'Handbook of Marine Craft Hydrodynamics and Motion Control', *2nd ed., Wiley*, 2021.

16. KIM, D., LEE, S., and PARK, J., 'Time-optimal control of ship manoeuvring under wave loads', *Ocean Engineering*, Vol. 293, 2024.

17. BASSAM, A., 'Use of voyage simulation to investigate hybrid fuel cell systems for marine propulsion', *University of Southampton*, 2017.

18. McKINLAY, C., 'Dynamic energy system modelling to assess viable zero-emission shipping solutions', *University of Southampton*, 2023.

19. BELIBASSAKIS, K A., and FILIPPAS, E S., 'Hydrodynamic performance of foil systems in regular and irregular seas', *Ocean Engineering*, 2015.

20. ZHANG, Y., HE, X., and KINNAS, S A., 'Comparison of empirical and potential flow methods for resistance prediction in naval hulls', *Ocean Engineering*, 2020.

21. SØRENSEN, H C., 'Propulsion control strategies for marine applications', *Proceedings of the Marine Control Systems Workshop*, 2005.

22. STRANG, T., and STROBEL, T., 'Energy management strategies for hybrid marine power systems', *Proceedings of the International Conference on Maritime Energy Management*, 2019.

8. AUTHOR BIOGRAPHIES

Muhammed Talha Ozdenoglu holds the current position of PhD Researcher at the University of Southampton. The Author is responsible for conducting CFD and experimental analyses for energy-efficient propulsion and control strategies in maritime systems. The Author received his BSc degree in Naval Architecture and Marine Engineering from Istanbul Technical University.

Dixon T. Sin is Research Fellow at the University of Southampton, where he is responsible for Experimental Fluid Dynamics (EFD), Particle Image Velocimetry (PIV), and studies on fluid–structure interactions. His previous experience includes drag measurements on bioinspired surfaces, PIV analysis of near-wall flow fields, and slip-length quantification on functional coatings and engineered surfaces.

Stefan-Tiberiu Tudora is currently an undergraduate student at the University of Southampton. The author is studying for a MEng degree in Ship Science. The author's previous experience includes ship systems modelling and the development of simulation software tools for performance analysis.

Nicholas Townsend is an Associate Professor in Maritime Engineering at the University of Southampton. His research focuses on experimental hydrodynamics, with particular emphasis on marine renewable energy systems and autonomous maritime robotics. He has also supervised H2020 projects related to wave energy harvesting and the performance optimisation of marine systems.

Tahsin Tezdogan is an Associate Professor in Maritime Engineering at the University of Southampton. He received his PhD in 2015 from the University of Strathclyde on hydrodynamic analyses of mono- and multi-hull vessels. His research centres on marine hydrodynamics—particularly using CFD to investigate ship resistance, wave-induced added resistance, and vessel behaviour in shallow or restricted waters. He is also the Editor-in-Chief of Ocean Engineering.

Melike Kurt is a Lecturer (Assistant Professor) in Maritime Engineering at the University of Southampton. Her research focuses on fundamental fluid–structure interactions, bio-inspired design, and underwater robotics. She addresses key challenges in the design of maritime technologies operating in turbulent flow conditions, aiming to enhance manoeuvrability, improve hydrodynamic efficiency, and reduce noise. Dr Kurt often draws inspiration from nature, such as the coordinated movement of schooling fish or the unique surface textures of shark skin, including denticles and bristles, to develop innovative engineering solutions. Dr Kurt serves as a member of the EPSRC Engineering Early Career Forum. She is also the coordinator and admissions tutor of MSc Maritime Engineering programme at the University of Southampton.

Joseph Banks is an Associate Professor in Maritime Engineering at the University of Southampton. He leads experimental and computational research in fluid–structure interaction, sailing and wind-assisted vessels, and contributes to programme admissions. He serves on the ITTC's Wind-Assisted Ship Propulsion Committee and is on the editorial board of the *Journal of Sailing Technology*.

Dominic Hudson is Shell Professor of Ship Safety and Efficiency at the University of Southampton and leads the Centre for Maritime Futures. His research focuses on ship hydrodynamics, safety, and energy-efficient design. He advises Shell Shipping & Maritime on decarbonisation strategies and chairs the Maritime Engineering research group. He is co-author of the textbook *Ship Resistance and Propulsion*, widely used in academia and industry. and chaired several ITTC specialist committees (high-speed craft, seakeeping, ships in service, ships in operation at sea).

Alan J. Murphy is Professor in Maritime Engineering and Head of the Maritime Engineering Group at the University of Southampton. His research is centred on sustainable, zero-emission maritime systems, including carbon-free propulsion technologies and energy-efficient port operations, supported by real-world ship monitoring and autonomy. He serves as the UK standing committee member for PRADS and Editor-in-Chief of the *International Journal of Maritime Engineering.* He is a Fellow of the Royal Institution of Naval Architects and a Liveryman of the Worshipful Company of Shipwrights.

Stephen R. Turnock is Professor of Maritime Fluid Dynamics and Head of School of Engineering at the University of Southampton. He has led pioneering research in ship hydrodynamics, maritime robotics, and energy-efficient marine propulsion. He founded the Maritime Robotics Lab and the Performance Sports Engineering Lab. He also led the creation of the UK's largest university towing tank, a cornerstone of the Boldrewood Innovation Campus. Stephen has co-authored two landmark textbooks, *Ship Resistance and Propulsion* and *Marine Rudders and Control Surfaces*, and supervised over 50 PhD students worldwide. He has held key roles in ITTC and IEA Hydrogen Task 39 and continues to shape international maritime research and education. His contributions have recently been recognised with the William Froude Medal and the IMarEST Outstanding Contribution Award.

MODULAR NAVAL VESSELS AND THE IMPACT ON THE DESIGN AND EFFECTIVENESS

M L Stam, and **A A Kana,** Delft University of Technology, Netherlands, and **J E Streng**, Material and IT Command (COMMIT), Netherlands

SUMMARY

Modularity is often mentioned as a method to make a vessel's design flexible during its lifetime. However, literature often focuses on ways to apply modularity to naval vessels instead of evaluating the added value. This paper aims to build a framework to give valuable insights into the possible benefits and the technical impact of modularity. The framework consists of a Modular Function Deployment (MFD), Analytical Hierarchy Process (AHP), and Knowledge Based Engineering (KBE) model. The final effectiveness assessment is performed by application of the AHP again. A case study will be presented on large surface combatants of the Royal Netherlands Navy, of which the final effectiveness assessment results in a preference for a non-modular air surveillance radar and a modular Laser-Directed Energy Weapon. The results highlight the need for a structured approach to assess the added value of modularity in the design phase, as it does not always improve effectiveness.

1. INTRODUCTION

Designing naval vessels comes with its challenges. The first challenge is the unpredictable geopolitical environments in which they need to be able to operate [1]. The Dutch Ministry of Defense (MOD) surface combatant's main task is to perform combat operations, while in times of peace, they are used for operations such as humanitarian aid after natural disasters or counter-narcotics operations [2]. Another challenge most navies face, including the Royal Netherlands Navy (RNLN), is a limited budget [1]. Modern vessels are required to carry out missions and be ahead of opposing actors. These opposing actors are evolving their gear and weapon systems as well, so the vessel must be able to make adjustments to continue its missions during its lifetime. But making a vessel suitable for all predicted future requirements at the same time will result in, at least some degree of, overdesign, resulting in inevitable compromises on the vessel's performance [3, 4]. A possible way to mitigate these challenges is to make the vessel adaptable to handle a wider variety of tasks. However, designing and building these modern vessels is extremely expensive and time-consuming. In short, the Dutch MoD is seeking less time-consuming ways to design future naval vessels that can handle a variety of tasks [5].

A method that could mitigate these challenges is modularity. For example, when the vessel can install and exchange modules for each purpose, the modules needed for the designated mission can be installed so the vessel retains -or at least decreases the negative impact on- its mission capability without making significant changes to the platform. This form of modularity is called mission modularity, which is often applied by containers on deck [5]. Another possibility is to create space that can be rebuilt very easily. By designing standardized parts of the vessel, time and money could be saved. This type of modularity would probably look more like buying a car. The basics are the same for everyone, but certain options, like the type of radar or weapon system used, can be selected by country. This would be a form of payload modularity [5]. Another way to reduce design time is by applying geometrical modularity, which aims to design generic parts of the vessel, such as hotel facilities, to reduce design time [5, 6]. So, modularity might be able to mitigate the challenges being addressed by modern navies.

It is mentioned that modularity influences design freedom since modules usually require volume margins and have increased weight, and can lead to a decrease in performance and additional required training for the crew to handle the different modules [7]. However, no clear frameworks are found in the literature to evaluate whether modularity would be beneficial for a naval vessel. It is important that a framework is developed to couple the naval vessel's effectiveness to the application of modularity to assess the impact on the vessel's design. Therefore, it is important to shift the focus on modular naval vessels from the detailed design phase to the concept exploration phase and ask ourselves if we should apply modularity in the first place.

2. BACKGROUND INFORMATION

This section will give background information on the problem being addressed and the research gap found after a literature review. Section 2.1 discusses the challenges in naval vessel design for the RNLN, after which section 2.2 discusses the research gap.

2.1 WHAT ARE THE MAIN CHALLENGES IN NAVAL VESSEL DESIGN FOR THE RNLN AND HOW COULD MODULARITY MITIGATE THESE CHALLENGES?

After analyzing literature and consulting experts in naval vessel design from COMMIT, the following challenges for the RNLN have been found [5]:

- *Complexity*: Some details, for instance interfaces, are already needed in early-stage design, while the design knowledge is still low. This makes it time-consuming to select the most suitable systems and possibly results in a less optimal system when evaluating the final design.
- *Maintenance*: The time a vessel is in port/dock for maintenance, it cannot be deployed on operations.
- *Short-term adaptability*: Vessels are required to be adaptable to perform various missions and the mission profile itself can change during the lifetime of a vessel as well. Therefore, the equipment on board may vary during the vessel's lifetime, but it can also be mission-dependent. When, for example, a vessel needs to assist during disaster relief, it needs to be equipped quickly with medical aid and/or search and rescue equipment.
- *High design and production time for small series*: When every part of the vessel needs to be engineered from the beginning, significant time is consumed. For small production series, this is relatively time-consuming since this optimized equipment cannot easily be reused on other designs.

Applying modularity has three main functions: 1) making complexity manageable, 2) enabling parallel work, and 3) accommodating future uncertainty [8, 9]. The last function shows how modularity could be used as a design style solution to the issue of adaptability, which has been an objective to the fore of naval ship acquisition since the end of the Cold War [10]. Modularity could mitigate the challenges mentioned above in the following ways [5]:

- *Complexity*: Modularity can help to make complexity manageable [8, 9]. For example by making standardized interfaces that can handle various systems. This way, the selection of detailed systems is postponed so the proper system can be selected at the end of the design phase and the production can start earlier since the standardized interface can be used.
- *Maintenance*: By using modules, it can become easier to conduct maintenance [11, 12]. When modules can (relatively) easily be replaced, it is possible to change the broken module in port to a new one so the vessel is quickly back in operation again. This will only be useful when the maintenance time is more than the time required to change the module to a new one.
- *Short-term adaptability*: By using exchangeable modules on board, it is possible to change the equipment needed to conduct certain missions quickly [12].
- *High design and production time for small series*: When basic modules are applicable for different vessel types, future designs require less design time since already existing modules can be used in the models, and parallel production becomes possible [8, 11, 12].

2.2 RESEARCH GAP

As mentioned above, modularity can help mitigate several challenges in naval vessel design. A literature review has been performed by Stam [5], and it was found that the literature often assumes using modularity will improve the design without investigating whether the application of modularity is the best solution for the design problem. Therefore, this paper will develop a framework for the concept design phase to investigate the impact of modularity with respect to a non-modular solution. The main research question of the paper is thus:

How can the decision-making for modular systems in naval vessel design be performed in a structural way and evaluated on its effectiveness compared to a non-modular solution?

The framework aims to give a structured way to assess which systems have a high potential to be modular and evaluate the technical impact on the design of modular systems. If a system is a feasible module, the last step is to assess the impact on the vessel's effectiveness to check if the modular system will actually improve the vessel, or if the non-modular solution is preferred.

3. FRAMEWORK

The framework developed in this paper aims to give valuable insights for the naval architect to asses whether modularity could improve their design. The framework will be used to explore new opportunities for modularity and modular systems. It consists of a Modular Function Deployment (MFD), Analytical Hierarchy Process (AHP), Knowledge Based Engineering (KBE) model, and a final evaluation of the effectiveness using AHP again. It focuses on the application of payload modularity.

This choice has been made since it is deemed most suitable to show the impact of modularity on the platform, whereas geometrical modularity would focus on designing standardized generic systems without changing the systems on board, mission modularity adds containers to the platform rather than changing the platform itself, and software modularity focuses more on the IT side of the problem. It is important to understand that to easily replace modular systems, all systems must use the same interface with the vessel. During the MFD model, the vessel's requirements, functions, systems, and modularity drivers are mapped by matrices. These must be formulated well since they are required during the AHP and KBE models. With the AHP model, the reciprocal importance of the vessel's functions is determined using expert opinion. The technical impact of modular systems on the design will be modelled in the KBE model. Different configurations of systems will be used, so the non-modular systems will be compared to their modular variants. This way, the technical feasibility of modular systems is assessed. So, the KBE model translates from operational performance to technical performance. When the modular systems, resulting from the MFD and AHP, are deemed feasible by the KBE, a final evaluation of the vessel's effectiveness is done to assess if the vessel would benefit from a modular variant of the system. A flowchart of the framework can be seen in Figure 1.

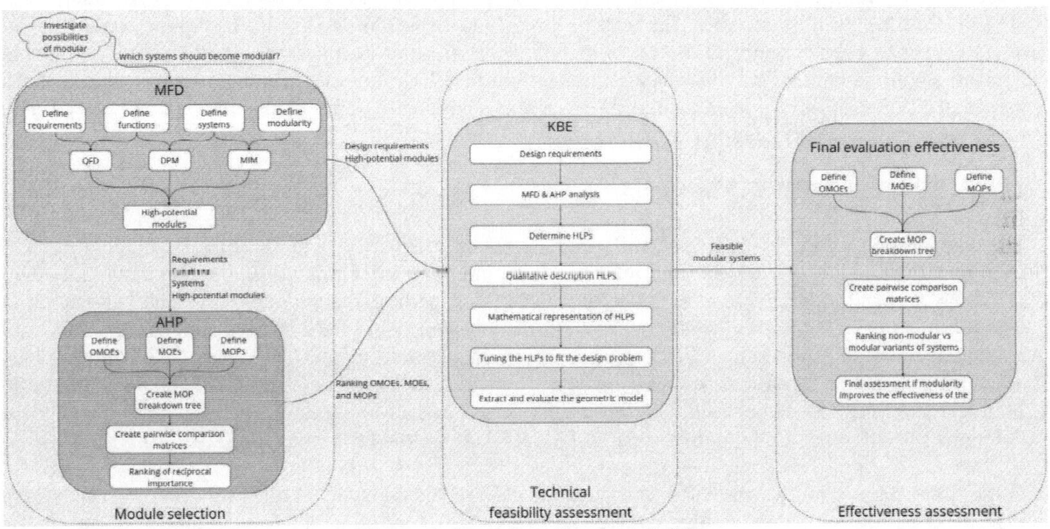

Figure 1: Flowchart framework [5].

4. CASE STUDY

When selecting a case study, it is crucial to choose a vessel that represents a realistic design for current operations but is expected to benefit from modularity. Another important aspect is that the vessel should not have been originally designed for modularity application. This way, the impact of modularity on a non-modular design can be explored. Four projects within COMMIT and one already operational vessel, the LPD-2, a landing platform dock and the first advanced commando platform in Europe, are examined to determine which vessel is deemed most suitable for a case study. These projects include Mican, Future Air Defender (FuAD), Anti-Submarine Warfare Frigate (ASWF), and Amfibisch Transportschip (ATS, an amphibious transportation vessel), which are still in the design or building phase, and the LPD-2, which has already proven itself as a multi-functional vessel [5]. An overview of the projects for the case study selection is shown in Table 1.

Table 1. Overview projects case study selection

Project	Vessel type	Comparable predecessor in the RNLN?	Already in operation?	Designed with intention of using modularity?	Suitable for case study?
MICAN	Small modular platform	No	No	Yes	No
FuAD	Frigate	Yes	No	No	Yes
ASWF	Frigate	No	No	No	Yes
ATS	Amphibious operations	Yes	No	No	Yes
LPD-2	Amphibious operations	Yes	Yes	No	Yes

The case study selection will consist of two parts. The first part, during the MFD and AHP modelling, two types of vessels will be used. This will be a frigate and an amphibious operations vessel. By applying the model on different vessel types, the most suitable vessel for the KBE phase can be selected based on the outcomes of the early analyses of the MFD- and AHP-models. For both vessels, it is important the particular vessel could potentially benefit from at least two types of modularity to show the framework's flexibility.

For the case of the frigate, the FuAD is deemed most suitable. This choice is made because this frigate type will replace the LCF, whereas the ASWF is a new vessel type replacing the M-frigates. As a result, more public information can be found on the LCF, which can be used to make assumptions for the FuAD because it will be a larger, but mission-wise comparable, vessel than the LCF. The FuAD will be a larger vessel than the LCF. Therefore, the vessel will be compared to the LCF, and its dimensions are compared to the similar German project the F127, having a displacement of 10,000 tonnes, a length of 160 meters, and a beam of 21 meters [13]. The cruise and maximum speed will be taken over from the LCF since the tasks, such as escorting allied vessels with speeds up to 30 knots, will not change. For the case of the amphibious operations vessel, the LPD-2 is deemed most suitable. The vessel's design and operations are known in real-life conditions rather than the expected performance of the yet-to-be-built ATS. Besides, this vessel is suitable for looking at both payload modularity for SEWACO systems and mission modularity for the different missions, such as hosting the staff to command up to 50.000 soldiers on land and conducting amphibious operations [14], which are never performed at the same time [5].

4.1 MODULAR FUNCTION DEPLOYMENT

The goal of applying the MFD is to identify high-potential modular systems for different modularity drivers chosen by the designer. To achieve this, the vessel's requirements, functions, systems, and modularity drivers need to be defined. It is important to notice that during this method, the assumption is made that modularity will be used, and the MFD will indicate how much a system could benefit from modularity. This can be computed into a collection of matrices known as the Product Management Map (PMM) [15]. A PMM consists of a Quality Function Deployment (QFD, interaction between requirements and functions), a Design Property Matrix (DPM, interactions between functions and systems), and a Module Indication Matrix (MIM, impact of functions/systems on modularity drivers). For the full PMM for the case study, see Stam [5].

In the MIM, the systems are rated for their potential modular advantages. The higher the score, the larger the expected benefits from applying modularity. The five highest-ranking modularity drivers are shown in Table 2. By analyzing the ranking, it can be checked if the expected benefits can actually be achieved. It can be seen that for both the FuAD and LPD-2 the same modularity drivers, although in a slightly different order, form the top-5. The three modularity drivers purchase, common unit, and process and/or organisation indicate that the challenges for the 'high design and production time for small series' can be mitigated by the application of modularity. This indicates that using modular systems across the fleet will result in increased scale advantages.

Table 2. Top modularity drivers.

Rank	Modularity driver FuAD	Score	Rank	Modularity driver LPD-2	Score
1	Purchase	266	1	Different specification	268
2	Common unit	262	2	Purchase	261
3	Different specification	236	3	Common unit	226
4	Service and maintenance	220	4	Service and maintenance	214
5	Process and/or organisation	208	5	Process and/or organisation	212
	Total points FuAD	**1833**		**Total points LPD-2**	**1812**

The five highest-ranking systems are shown in Table 3. From these results, it can be seen which systems are the most important for completing the vessel's main missions. It could be seen that the most potential benefits of modularity come from using systems across the fleet, which results in increased scale advantages. This is in line with the results from the top-5 modularity drivers. From the average points per system, it can be seen that the FuAD most likely benefits more from modularity than the LPD-2.

Table 3. Top systems.

Rank	System FuAD	Score	Rank	System LPD-2	Score
1	Horizon surveillance radar	73	1	Horizon surveillance radar	73
1	Air surveillance radar	73	1	Air surveillance radar	73
3	Surface-to-surface missile	71	3	Tracking radar	65
4	Tracking radar	65	4	Heavy machineguns	63
5	Heavy machineguns	63	4	Electronic Support Measures	63
5	Electronic Support Measures	63	4	Electronic Counter Measures	63
5	Electronic Counter Measures	63			
	Total points FuAD	**1833**		**Total points LPD-2**	**1812**
	Average points per system FuAD	**36.66**		**Average points per system LPD-2**	**33.56**

When applying the MFD, it is important to keep in mind how modular systems already are. An example is the Vertical Launching System, which is designed to handle any missile type in any cell, making it an extremely flexible system [16]. Since this is already a modular system, there are very few benefits to making it modular which results in a low score on the MFD. As mentioned before, this framework is used to explore possibilities for new modular systems, so the low scoring of modular systems is no issue.

4.2 ANALYTICAL HIERARCHY PROCESS

The next step is to assess the reciprocal importance of the vessel's functions. When developing an approach to assess a naval vessel's effectiveness, the following terminology is used [17]:

- **Overall Measure of Effectiveness (OMOE)** — Index (scale 0-1.0) describing the vessel's effectiveness for all assigned (and expected future) mission types.
- **Measures of Effectiveness (MOE)** — Index for specific missions or performance categories (such as offensive capabilities, mobility, survivability).

- **Measures of Performance (MOP)** — Specific vessel or system performance metric independent of the mission (such as speed, range, seakeeping, vulnerability, reliability).

The AHP is a method to asses numerical values for these parameters by directly using expert opinion [18]. When conducting the AHP on a vessel, the results indicate which MOPs are the most important for a specific vessel type. Comparing the AHP results of varying vessel types indicates which important functions share commonality and have the potential to increase scale advantages. Another reason the accompanying systems of the most important MOPs could be designed as a module is to reduce maintenance time by making the module easily replaceable. If the system is destroyed during naval operations, the vessel cannot continue its operations as required. When the system is designed as a module that a new one could easily replace, the vessel can continue its operations faster.

When applying the AHP, it is important to make a well-structured MOP breakdown tree as shown in Figure 2. The next step is to make pairwise comparison matrices and assess the reciprocal importance of the vessel's OMOEs, MOEs, and MOPs. It is important to use real experts for grading the reciprocal importance of different systems. The definition of an expert used in this paper is defined as 'someone whose expertise in a particular area makes his assertions reliable—more likely to be true than false' [19]. At COMMIT, naval architects who design the vessels for the RNLN are deemed experts in naval vessel design. Therefore, the case study will use the expert opinion of multiple naval architects at COMMIT to compare their opinions. An example of such a pairwise comparison matrix is shown in Table 4. The right column shows the relative weight (Wt) of the components of each matrix. This relative weight is calculated by normalizing the principal eigenvector of the matrix [20]. The final ranking of the AHP will give insights into the most important functions of the vessel. These functions can be coupled to their accompanying systems, which gives an indication of the most important systems.

Table 4. Pairwise comparison matrix AAW neutralization FuAD.

MOE5.2 - AAW neutralization FuAD	MOP8: Track enemy threat	MOP9: Control fired projectile	MOP13: Evade incoming enemy threat by EW	MOP14: Long-range air defense	MOP15: Short-range air defense	Wt
MOP8	1.00	1.00	0.32	0.71	0.41	0.31
MOP9	1.00	1.00	0.71	1.00	0.71	0.22
MOP13	3.17	1.40	1.00	1.00	1.00	0.15
MOP14	1.40	1.00	1.00	1.00	3.00	0.22
MOP15	2.44	1.40	1.40	1.00	1.00	0.11

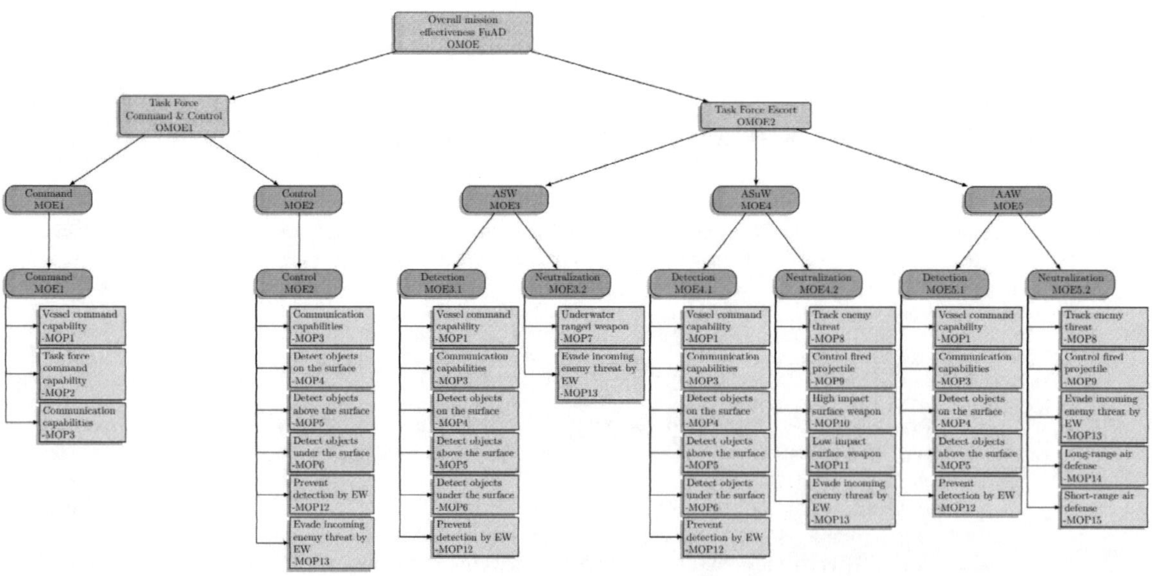

Figure 2: MOP breakdown tree FuAD.

The top-5 results, shown in Table 5, show that 'communication capabilities' is the most important MOP for both FuAD and LPD-2. In this case, the MOP cannot be coupled to specific accompanying systems. The second highest ranking MOP for FuAD is MOP5 'Detect objects above the surface', which is coupled to the system 'Air surveillance radar'. This shows the importance of a working air surveillance radar. The MFD results for the FuAD show that the air surveillance radar is already a high-potential modular system based on modularity drivers.

Table 5. Top-5 results AHP.

MOP FuAD	%	MOP LPD-2	%
MOP3: Communication capabilities	19.85	MOP3: Communication capabilities	25.34
MOP5: Detect objects above the surface	14.51	MOP1: Vessel command capability	11.01
MOP1: Vessel command capability	12.67	MOP18: Personnel LC transferring	9.35
MOP4: Detect objects on the surface	9.63	MOP17: Vehicle LC transferring	7.36
MOP12: Prevent detection by EW	7.45	MOP16: Vehicle storage	7.15

When both the MFD and AHP show possible advantages of modular designing, the system is particularly interesting for investigating modular applications. By developing this ranking for the naval architect, the AHP can improve insights into the importance of the vessel's functions and accompanying systems, which can help select systems as high-potential modules. It is important to note that the ranking focuses on the main operations of the vessels, while the lower-ranked MOPs are still indispensable for the ability to complete their missions. The AHP can give insights into the most important MOPs of a vessel, but the lower-scoring MOPs cannot be neglected.

4.3 KNOWLEDGE BASED ENGINEERING

The next step in the framework is the KBE model. By reusing available information, multiple design possibilities can be generated and evaluated, after which the best-performing design can be chosen. This way, a more suitable starting point (first design) can be generated, which would reduce the required iterations for the final design, thus reducing design time. As mentioned before, the design time reduction is an important factor for modern navies. KBE is chosen as a method to reduce design time because it is an already proven advantageous methodology in other sectors, such as aerospace, automotive, and architecture [21]. KBE is a method developed in the artificial intelligence (AI) sector designed to solve multidisciplinary problems by a rule-based reasoning process that gives an output in computer-aided design (CAD) software [21]. When using the parametric design procedure, (sub)systems will be automatically coupled to an algorithm calculating various parameters chosen by the designer [22]. The KBE model consists of parametric building blocks, called 'High Level Primitives' (HLPs) [21]. These HLPs can incorporate and reuse relevant knowledge, such as the calculations for the vessel's stability and center of gravity. This way, KBE can be used to investigate changes in these parameters when changing non-modular systems into modular variants and an assessment can be made if the modular variant is technically feasible.

For building the KBE model, the method as proposed by Charisi [21] is adapted:

- **Identification of the design requirements** The first step is to determine the design requirements, which are dependent on the vessel type and the specific design problem. The design requirements correspond to the input variables, which lead to the tuning of the model to fit different design problems. The design requirements are already defined during the MFD phase.
- **Main drivers analysis** The main drivers are analyzed to identify the way the vessel should be parameterized. These main drivers are already identified by the AHP method.
- **Determine the HLPs** This step consists of the determination of the HLPs of the specific vessel. The developed HLPs form the toolkit from which different parts can be combined to form the different vessel alternatives.
- **Qualitative description of the HLPs** The qualitative description will be the guideline for the mathematical representation. In the context of developing the geometric model of the vessel, space reservation should be ensured.

- **Mathematical representation of the HLPs** The mathematical description of the HLPs is defined based on their qualitative description. Hence, their interrelations are also defined, and thus, the vessel's architecture can be created.
- **Define the HLPs for each 'total ship' architecture** The selected HLPs from the vessel's toolkit are combined to form the 'total ship' architecture according to the design decisions of the naval architect. The total of the different design decision combinations leads to the different solutions in the design space.
- **Tuning the HLPs to fit the design problem** The selected HLPs are being tuned to fit a specific design problem. The design requirements are being used as the guidelines to form a feasible and suitable design solution for the design problem.
- **Extract and evaluate the geometric model** The output of the described framework is the geometric model of the vessel for the naval architect to visualize his ideas and use the model as an input for analysis. This design can be evaluated by using parametric design software.

The KBE model is used to map the technical impact and assess the feasibility of making systems modular. The model consists of a Python script implemented in Rhino. This results in a model that delivers a visual representation, besides the output of the Python model. From this Python output, the vessel's dimensions, weight, GM, estimated draft, and Center of Buoyancy (COB) can be retrieved. By analyzing the changes, the impact of modularity for each system can be assessed, and it can be checked if the design remains within its margins as defined by the naval architect. The first step in this model is to select systems as HLPs. This is done by combining the MFD results, indicating which systems are high-potential modules, and the AHP results, indicating the most important systems of the vessel.

For the case study, performed on the FuAD as reference vessel, the resulting AHPs are an air surveillance radar and Laser-Directed Energy Weapons (LDEWs). As mentioned before, using modular systems will result in additional weight and volume requirements. Therefore, the weight of the modular variants of the systems is increased by a factor two. Since these systems are placed outside of the vessel's hull and superstructure, the additional required volume is not taken into consideration. The weight of the hull and KG of the total weight of the LCF are taken as a starting point for the FuAD model, after which they are scaled to the dimensions of the FuAD. The application of LDEWs results in a significant increase in electric power demand. Therefore, the FuAD reference model is made in two configurations: one for the initial design and one for the design that requires more electric power. As a result, more powerful generators are required, and the vessel needs to be extended by 0.5 meters to make it fit. The visual output of the KBE model is shown in Figure 3.

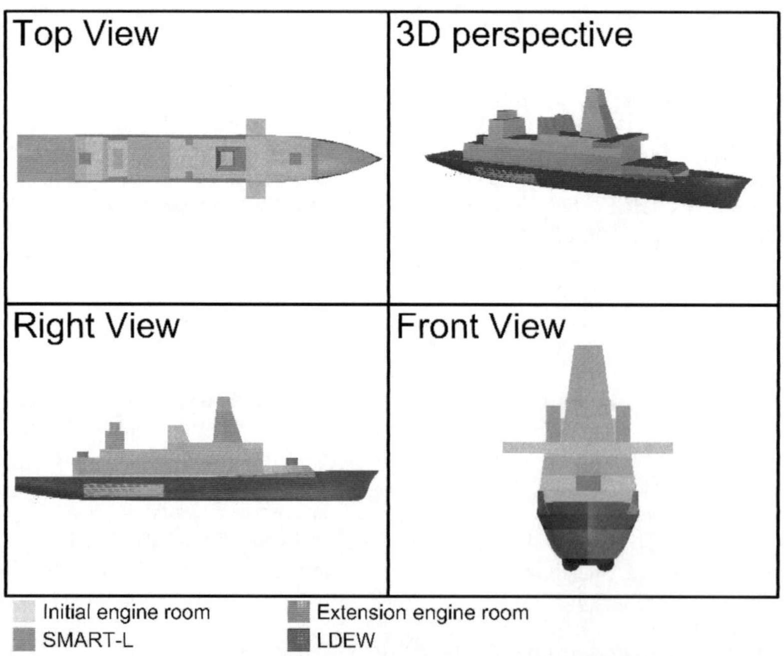

Figure 3: FuAD model including LDEW and SMART-L.

For the first configuration, the SMART-L air surveillance radar was first implemented as a non-modular system and then as a module. For the second configuration, the LDEW was first implemented as a non-modular system and then as a module. This way, an assessment could be made of the impact of making each of these systems modular. It was found that the weight is slightly increased, the draft remains the same, the COB remains roughly the same, and the GM remains in a range of 1.646-1.667 meters, keeping it well within the acceptable range of 1.4-1.8 meters [5]. With these results, the SMART-L and LDEWs are deemed feasible modules. The full results are shown in Table 6.

Table 6. Output KBE model FuAD reference model.

Model	Configuration 1	Configuration 1	Configuration 2	Configuration 2
Modules	Initial design	Modular SMART-L	Initial LDEWs and modular SMART-L	Modular LDEWs and modular SMART-L
Weight [t]	10000	10008	10062	10072
GM [m]	1.664	1.646	1.667	1.666
Draft [m]	6.6	6.6	6.6	6.6
COB (x,y,z) [m]	74.49, 0.00, 3.98	74.49, 0.00, 3.98	74.73, 0.00, 3.98	74.72, 0.00, 3.98

4.4 EVALUATION OF EFFECTIVENESS MODULES

The final evaluation of the effectiveness of the modules is done by application of the AHP. When formulating the OMOEs, MOEs, and MOPs, the possible advantages of modularity should be taken into account. During the final evaluation, the OMOEs are the systems, the MOEs are possible advantages of modularity, in this case study the purchase costs, common use across the fleet, quick changing of systems, maintenance time in dock, and number of systems required, and the MOPs are the non-modular and modular variant of the system. The computed pairwise comparison matrices are filled in by a naval architect with expertise in frigates at COMMIT. When evaluating the results, it is important to focus on the reciprocal importance of the non-modular and modular variant of a system. The results are shown in Table 7.

Table 7. Results module effectiveness.

System	**MOP**	**Absolute importance**	**Reciprocal importance**	**Outcome**
LDEW	MOP1 - Modular LDEW	41.8%	62.7%	Modular
	MOP3 - Non-modular LDEW	24.9%	37.3%	
Air surveillance radar	MOP1 - Non-modular Air surveillance radar	17.4%	52.1%	Non-modular
	MOP2 - Modular Air surveillance radar	16.0%	47.9%	

For the LDEW, the modular variant is preferred, while the non-modular air surveillance radar is slightly preferred. This difference is caused by the high weight of the 'common use across the fleet' and the 'quick changing of systems' for the LDEW, while the air surveillance radar has a low weight for these categories. These results show the importance of looking at the intentions of the end user. The LDEW results show that the focus is on adaptability rather than purely the economic aspect, while the air surveillance radar does not require high adaptability, and therefore, the focus is on the economic aspect. These results show that even when a system scores high on the MFD, its function scores high on the AHP, and it is deemed a technically feasible module in the KBE, it does not mean it will improve the vessel or the organizational process. This shows the importance of assessing whether a feasible modular system improves the design compared to a non-modular solution.

5. CONCLUSIONS

The developed framework aims to give the naval architect guidance on the added value of modularity. When a system scores high on the MFD, it indicates that the system could benefit from modularity. When its accompanying function scores high on the AHP, it indicates the system fulfils an important function. The technical impact of making the system modular can be assessed with the KBE model. If the technical impact is deemed acceptable, the system can be labelled as a 'feasible' module. The final effectiveness assessment will indicate if the feasible module actually improves the vessel's design or the organizational process.

The framework can be applied to a wide variety of vessels, including commercial vessels. The framework uses generally applicable methods, and by combining them in a structural way, the decision-making on whether to use modularity or not can be improved. For naval vessels, the effectiveness depends on the intentions of the end user. By qualifying the expert opinion using AHP, both by identifying the most important functions and the final evaluation, this intention is integrated into the framework. This also personalizes the outcome: based on the intentions of the end-user, completely different outcomes can be retrieved from this framework. After computing the framework for a specific vessel type, the results can partially be applied to other vessel types, within the same organization, due to the end-user intentions mentioned above, as well. If a modular system is preferred over the non-modular variant, this implies the other vessel types will also benefit from this system as a module. Since this system is already defined as an HLP in the KBE model, the system can easily be imported into another design. The hull form can be easily changed as well since this already is an external Rhino file imported into the Python file. These factors make the framework applicable to a wide variety of different vessel types.

Future work is recommended for the integration of modularity in early-stage ship design. The framework investigated the impact of modularity on several possible configurations, while real designs could benefit from larger data sets. Therefore, it is recommended to investigate the possibilities of using parametric designs in order to create large data sets of possible designs to have a more varied selection of system variants to assess the impact of modularity on the design and effectiveness. Besides the creation of large data sets in the design-space exploration, it should be investigated how the added value of modularity could be generated and evaluated in a more numerical way than using expert opinion only.

6. ACKNOWLEDGEMENTS

This work was performed as a part of the academic thesis of the first author (Stam [5]) which was conducted at TU Delft in collaboration with NL COMMIT.

7. REFERENCES

1. ANDREWS, D., 'Adaptability – the key to modern warship design', *International Conference WARSHIP'01, Future Surface Warships*, 2001.

2. MINISTRY OF DEFENCE, 'Fundamentals of maritime operations: Netherlands maritime military doctrine', *https://english.defensie.nl/downloads/publications/2014/02/13/netherlands-maritimemilitary-doctrine*, 2014.

3. FUCHS, C. & GOLENHOFEN, F. J., 'Mastering Disruption and Innovation in Product Management', *Springer*, 2019.

4. NIEUWENHUIS, J.J., 'Evaluating the appropriateness of product platforms for Engineered-To-Order ships', *PhD thesis Delft University of Technology*, 2013.

5. STAM, M.L., 'Modular naval vessels and the impact on the design and effectiveness', *MSc thesis Delft University of Technology*, 2025.

6. UK MINISTRY OF DEFENCE, 'maritime modularity concept', *https://assets.publishing.service.gov.uk/media/63dcece38fa8f57fc8061089/Maritime_Modularity_Concept.pdf,* 2022.

7. SMITH, D.R., 'Multi-role vessels: Design, manning & cost drivers', *Warship 2019: Multi-Role Vessels, 1*, 2019.

8. BALDWIN, C.Y. & CLARK, K.B., 'Modularity in the Design of Complex Engineering Systems', *Springer*, 2006.

9. STEPANDIC, J., WOGNUM, N., & VERHAGEN, W.J., 'Concurrent Engineering in the 21st Century', *Springer*, 2015.

10. ANDREWS, D., 'The multi role combatant - jack of all trades, master of none?', *Warship 2019: Multi-Role Vessels, 1*, 2019.

11. BREKKE, Ø., 'Modular Capabilities on Offshore Support Vessels', *MSc thesis Norwegian University of Science and Technology*. https://ntnuopen.ntnu.no/ntnu-xmlui/handle/11250/238186, 2012.

12. TVEDT, H., 'Modular approach to offshore vessel design and configuration', *MSc thesis Norwegian University of Science and Technology*, 2012.

13. PERUZZI, L., 'Euronaval 2024 – tkMS presents the Meko A-400 AMD', *https://www.edrmagazine.eu/ euronaval-2024-tkms-presents-the-meko-a-400-amd*, 2024.

14. MARINESCHEPEN.NL, 'Johan de Witt Landing Platform Dock', *https://marineschepen.nl/schepen/ johandewitt.html*, 2024

15. SIMPSON, T.W., JIAO, J., SIDDIQUE, Z., & HOLTTA-OTTO, K., 'Advances in Product Family and Product Platform Design', *Springer, 2014.*

16. HILL, J., 'US increases MK41 vertical launching system production', *https://www.naval-technology.com/news/us-increases-mk41-vls-production/?cf-view*, 2023.

17. PARSONS, M.A., ROBINSON, K.M., KARA, M.Y., STINSON, N.T., SNYDER, D. J., WOODWARD, D.C., & BROWN, A.J., 'Application of a distributed system architectural framework to naval ship concept and requirements exploration', *NAVAL ENGINEERS JOURNAL, 132(4):105–124*, 2020.

18. BROWN, A., & SALCEDO, J., 'Multiple-objective optimization in naval ship design', *Naval Engineers Journal, 115*, 2003.

19. WALTON, D., 'Appeal to expert opinion: Arguments from authority'. *University Park, PA: Penn State University Press*, 1997.

20. SAATY, R., 'The analytic hierarchy process—what it is and how it is used', *Mathematical Modelling, 9(3):161–176,* 1987.

21. CHARISI, N., 'Parametric Modelling Method based on Knowledge Based Engineering: The LNG Bunkering Vessel Case', *MSc thesis Delft University of Technology*, 2019.

22. KANELLOPOULOU, A., KYTARIOLOU, A., PAPANIKOLAOU, A., SHIGUNOV, V., & ZARAPHONITIS, G., 'Parametric ship design and optimisation of cargo vessels for efficiency and safe operation in adverse weather conditions', *Journal of Marine Science and Technology, 24,* 2019.

8. AUTHOR BIOGRAPHIES

Micha Stam holds the current position of PhD Candidate at Delft University of Technology. His research interests involve developing new early-stage design methods and tools for complex ships. The Author's previous experience includes his MSc thesis, on which this paper is based.

Austin Kana is an Associate Professor of Early-Stage Complex Ship Design at Delft University of Technology in the Department of Maritime and Transport Technology. His research interests involve developing new early-stage design methods and tools for complex ships, including naval vessels. He obtained his PhD in Naval Architecture and Marine Engineering from the University of Michigan in 2016.

Jurjen Streng holds the current position of Naval architect at COMMIT. He obtained his MSc in Naval Architecture from Delft University of Technology in 2021.

Warship 2025: The Future Fleet: Smart Technology, Sustainability and Autonomy
16th – 17th June 2025
Glasgow, UK

MODEL-BASED SUSTAINMENT: DIGITAL SOLUTIONS FOR IMPROVING ASSET AVAILABILITY

N Danese, S Crook and **G Goulanian**, SSI, Canada

SUMMARY

Modernizing the sustainment of the future and existing naval fleets through digital solutions is critical to ensuring long-term readiness, operational efficiency, and cost-effective lifecycle management. This paper examines the implementation of shipbuilding-specific Product Lifecycle Management (PLM) systems to configuration-manage naval assets across the program lifecycle. Examining strategies deployed in Canadian, Australian, New Zealand, and US Navy programs that leverage model-based sustainment and the digital twin to improve asset availability starting from an on-time entering of service.

By integrating engineering, procurement, production, and in-service support (ISS), a model-based approach establishes an accessible digital thread and maintains digital twins for both new-build and in-service vessels. This enables comprehensive Maintenance, Repair, and Operations (MRO) through intelligent data capture, precise configuration management, and improved information flow across stakeholders. We demonstrate how this strategy reduces total program lifecycle costs.

Key challenges in modernizing legacy fleets and new builds include fragmented data silos, complex maintenance planning, and aging equipment. Our approach emphasizes a central, shipbuilding-specific PLM environment to unify and connect data sources, ensuring a single, accurate source of truth for asset condition, configuration changes, and maintenance planning.

A case study on how SSI is working with a major global naval force on a digital transformation initiative to enhance the efficiency of docking work periods for a fleet of frigates is included. This initiative significantly reduces lead times, improves schedule predictability, enhances collaboration, and supports strategic defence readiness through the application of the model-based sustainment practices as described. This approach supports the current fleets of the world's navies while preparing for emerging missions and technological demands.

1. INTRODUCTION

Naval programs are increasingly complex and expensive, both to build and maintain, driving a need for new approaches to maintenance and support that maximize readiness. Navies worldwide are turning to digital transformation initiatives to enhance asset availability and lifecycle efficiency. For example, the Royal Canadian Navy (RCN) acknowledges that despite a history of leveraging advanced technology, it is now "challenged to keep pace with the accelerating rate of disruptive changes" and requires a more deliberate digital approach to remain effective in the modern era [1]. This shift aligns with broader defence modernization priorities: model-based sustainment has emerged as a key concept, referring to the use of comprehensive digital models to manage and optimize the support of assets throughout their service life.

In essence, model-based sustainment applies digital models to design, engineering, construction, and maintenance activities for the purpose of enhancing vessel sustainment. Such an approach moves maintenance from reactive to proactive and is strongly endorsed in recent military strategies. For instance, the Royal Navy's latest digital plan emphasizes adopting a "system of systems" digital backbone and agile practices to achieve the "required operational, availability and cost benefits" for the fleet [2]. Major allied navies – including those of Canada, Australia, the United Kingdom, and the United States – have all launched digital navy initiatives or policies that prioritize data-driven decision-making, integration of new technologies, and improved sustainment outcomes. These efforts establish the context for model-based sustainment as not just a technical upgrade, but a fundamental component of defence modernization focused on increasing the mission availability of naval assets.

2. EMERGING TECHNOLOGIES IN DIGITAL TWIN AND PLM FOR NAVAL SUSTAINMENT

Modern navies are increasingly adopting digital twin technology and advanced Product Lifecycle Management (PLM) systems to enhance sustainment of ships and submarines.

Navies are linking digital design models created during ship design and shipbuilding with maintenance and operations data to form a continuous digital thread across the asset's lifecycle.

The U.S. Navy's Naval Sea Systems Command (NAVSEA) has made digital transformation a priority, using digital twins for critical systems like the Aegis Combat System and gas turbine engines to test upgrades and diagnose problems virtually. This approach has already saved significant time and cost – in one case a ship's gas turbine fault was remotely analysed via its digital twin, allowing a replacement part to be sent out without flying a technician to the ship, saving weeks of downtime [3].

3. CURRENT APPROACHES AT GLOBAL NAVIES

Naval forces around the world are at different stages of adopting digital sustainment, influenced by their unique policy goals, budgets, and industrial bases. We compare how several navies – including Canada, Australia, New Zealand, the United States, and the United Kingdom – are implementing model-based sustainment and highlight best practices from each.

3.1 UNITED STATES

Funding for digital transformation is substantial – Navy budgets allocate money for "digital shipyard" improvements, cloud migration for maintenance data, and R&D in digital twin tech. One challenge in the U.S. has been the sheer scale of legacy systems, but Navy leadership (NAVSEA, NAVAIR, etc.) has been vocal and open about the need to modernize, helping secure support. The U.S. model illustrates the value of institutional commitment and pilot-driven expansion, as well as the integration of new technology into doctrine (i.e. condition-based maintenance is written into maintenance policy).

A key takeaway is the importance of measuring and publicizing results: the U.S. services have shared tangible metrics (like millions saved and days of availability gained), which builds momentum and justifies continued investment [4].

3.2 UNITED KINGDOM

The Royal Navy has embraced digital sustainment somewhat more gradually, in part due to different budget pressures and fleet size. The RN's flagship effort is Programme NELSON, a digital and data transformation project. Instead of starting with specific maintenance tech, the UK focused first on building a robust data infrastructure – a common data platform to aggregate information from ships and shore [2].

On the new Type 31 frigate program, Babcock built a fully digital production facility and is integrating sensors and digital management from day one to minimize through-life costs. The Type 23 frigate life-extension program also saw ships converted into "digital platforms" during upgrades, showing the UK's intent to retrofit older vessels with modern data systems [5]. Best practices from the UK include the focus on data coherency (getting the digital foundations right) and close Navy-industry partnerships to drive innovation despite budget limits.

3.3 CANADA

The Royal Canadian Navy, although smaller, has been proactive in charting a course for digital sustainment. In 2019, it launched the Digital Navy Initiative, aligning with Canada's defence policy, which calls for innovation [1]. For example, through Canada's Ocean Supercluster (a government-supported industry consortium), the RCN and industry are developing a Digital Ship In-Service Support system that brings together sensor data, AI, and interactive 3D models [6].

Key takeaways from Canada's approach include the importance of having a clear digital roadmap and using innovative procurement to involve agile small companies. Canada's policy-level support (explicitly calling out digital twins, AR, etc., as targets) provides top-down momentum.

3.4 AUSTRALIA

The Royal Australian Navy (RAN) is undergoing a major modernization, and it has identified digital engineering as a cornerstone for both acquiring and sustaining new capabilities. With plans to introduce advanced platforms (like the AUKUS nuclear-powered submarines and Hunter-class frigates), Australia is essentially building digital sustainment into the capability development process. The Australian Department of Defence has advertised for a "digital strategy chief" to lead the use of a digital twin for the new submarines, indicating how critical they view model-based engineering for delivering and sustaining that capability [7].

They are using a digital thread to link design, manufacturing, and maintenance, effectively ensuring that when these ships are in service, there's an accurate, live digital twin to inform upkeep and upgrades [8].

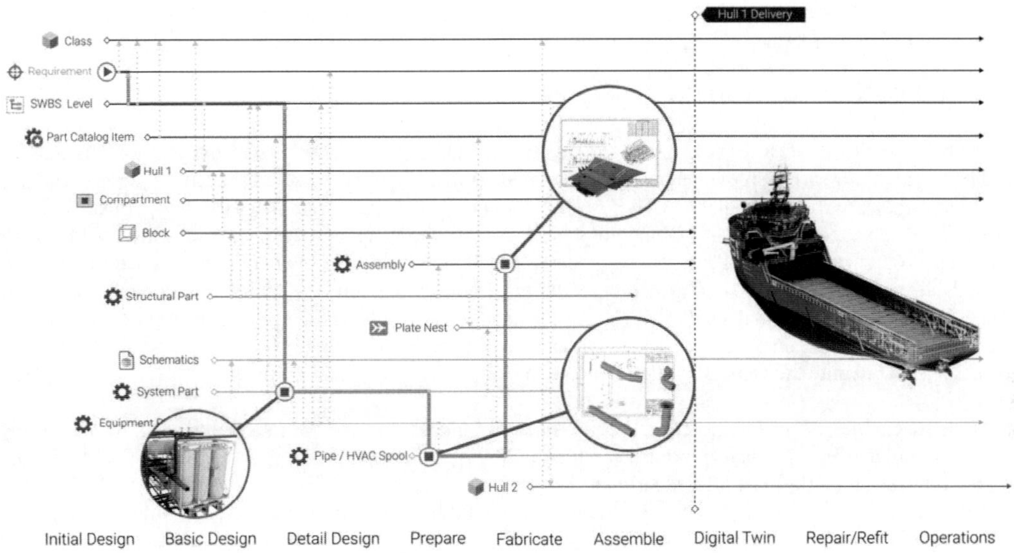

Figure 1: Digital Thread across disciplines

Best practices from Australia include integrating sustainment considerations early in design (so the digital twin concept isn't an afterthought) and dedicating organizational roles for digital innovation. Australia also shows the value of political support and funding when tied to new capability: the high-profile nature of AUKUS and Hunter programs secures funding not just for the platforms but for the digital ecosystem to support them.

4. CHALLENGES AND SOLUTIONS IN IMPLEMENTING MODEL-BASED SUSTAINMENT

4.1 TECHNOLOGICAL INTEGRATION AND LEGACY PLATFORMS

Legacy vessels often lack comprehensive digital models or have data locked in obsolete formats, complicating efforts to establish an accurate "digital twin" for in-service assets. The absence of standardized data interfaces between design, construction, and maintenance systems has historically hindered information flow across a ship's lifecycle [8]. Many naval programs also rely on aging software tools that are now at end-of-life, resulting in fragmented silos of information that cannot easily interoperate. Moreover, supporting IT infrastructure in defence has not kept pace with modern technology; a U.S. Government Accountability Office report highlighted that data collection and IT development were lagging behind the rapid growth of tech-enabled assets, creating a "digital disconnect" in mission-critical sustainment systems [9]. These gaps limit the effectiveness of digital sustainment unless a concerted modernisation effort is made.

To address these issues, navies and shipbuilders are investing in unified digital data strategies and modern toolsets. A key step is adopting common data standards and platforms so that information flows seamlessly from ship design into sustainment. For example, the U.S. Navy's new Model-Based Product Support (MBPS) mandate requires that design and maintenance data be managed in a standardized, shareable environment, ensuring that today's ship programs capture technical information in formats ready to feed tomorrow's sustainment systems [8]. By considering sustainment needs from the design phase – applying model-based systems engineering (MBSE) principles early – programs like the USN's Constellation-class frigate, built at Fincantieri Marinette Marine, are creating a digital thread from navy to shipyard that supports through-life upgrades and maintenance. Shipbuilders have also tackled the legacy data problem by converting old drawings and models into modern 3D formats en masse. In one case, a major U.S. naval shipbuilder saved over US$160 million in engineering labour by translating legacy design data (from decades-old CAD systems) into a new platform rather than recreating it from scratch. Such data migration initiatives not only yield immediate cost avoidance but also future-proof the shipyard against software obsolescence. Additionally, many organizations are deploying shipbuilding-specific product lifecycle management (PLM) systems as a backbone for sustainment data. A tailored PLM can manage the complex configuration of a fleet's digital twins in a controlled, revision-managed repository – something not feasible with ad-hoc spreadsheets or generic tools. By establishing a "common language" PLM environment used by both the Navy and its industry partners, some ship programs have eliminated the need for custom data translators and interfaces, greatly simplifying data exchange. In short, through technology solutions like standardized data models, modern PLM/ERP/MES ecosystems, and bulk data conversion efforts, stakeholders are mitigating integration problems and laying a solid foundation for model-based sustainment.

4.2 ORGANIZATIONAL AND CULTURAL RESISTANCE

Beyond technology, cultural resistance poses a significant challenge to digital sustainment initiatives. Naval organizations and shipyards often have deeply ingrained practices, with personnel accustomed to traditional maintenance methods and documentation. There can be scepticism of new digital tools among those who have successfully relied on paper manuals or 2D drawings for decades. In some cases, organizational silos and hierarchical structures impede the free flow of data and ideas – designers, builders, and maintainers may be used to working in isolation rather than collaboratively sharing a common model. Furthermore, military institutions tend to be risk-averse; the culture prioritizes reliability and safety, which can translate into reluctance to adopt unproven processes. All these factors create inertia against change, even when the long-term benefits of model-based approaches are clear. If the workforce does not buy into the value of a digital thread, the best tools will go underutilized.

Another critical enabler of an integrated PLM approach is stakeholder collaboration and culture. As multiple strategies have noted, digital transformation is as much about people as technology [1] [2]. Navies and industry partners need to foster a culture of information sharing and trust to make a unified environment work. The Royal Navy's plan underscores the need for "a spirit of cooperation underpinned by common standards" and highlights that collective action with allies and industry is "vitally important" [2]. In practice, this might involve co-locating Navy sustainment staff with shipyard teams during new build projects or jointly developing the digital twin updates post-delivery. It also means training personnel to use and rely on digital systems. Decision-makers must be comfortable moving away from paper technical manuals to interactive electronic technical manuals, and sailors need to trust digital feedback (from AI, for instance) in maintenance decisions. Change management programs and demonstrations can help in this regard. Canada's approach of starting with pilot projects is useful: the RCN explicitly uses "short-term pilot projects" to explore digital tech in a controlled way, with the understanding that some will fail but will generate lessons to build upon [1]. This iterative, fail-fast mentality can accelerate learning and buy-in. For example, after a few successes that show time or cost savings, it becomes easier to get broader support for scaling the integrated PLM approach across more vessels or departments.

4.3 CYBERSECURITY CONSTRAINTS

Increased digitization of sustainment activities brings significant cybersecurity concerns. Connecting ships, shore facilities, and contractors via digital threads creates a broader attack surface for adversaries. Naval platforms and their support systems are high-value cyber targets, and any vulnerability in the sustainment IT environment could be exploited to steal sensitive technical data or even disrupt operations. For instance, if a shipyard was compromised, the digital twin (containing detailed design and performance data, as well as the state of the vessel either as it was built, or as in-service) may be accessed by bad-actors. Shipyards, which often interface between secure naval networks and commercial networks, are particularly vulnerable if proper safeguards are not in place.

Addressing cybersecurity is, therefore, an important consideration of implementing model-based sustainment. Navies are developing dedicated cyber strategies to protect their digital initiatives. The RCN's Digital Navy plan, for example, includes a Security focus area aimed explicitly at defending naval digital systems and sensitive data from unauthorized activity [1]. This involves close collaboration with national security agencies and adherence to broader defence cyber standards. On the industry side, shipbuilders and contractors are being required to meet stringent cybersecurity requirements as a condition of program participation. In the U.S., the Department of Defense has introduced the Cybersecurity Maturity Model Certification (CMMC) framework to ensure that contractors handling naval technical data have robust cyber safeguards in place. Practically, solutions include the use of secure, encrypted networks for data exchange between ship and shore, as well as the segmentation of critical systems. For instance, NAVSEA - Naval Sea Systems Command, is exploring how to securely expand the information that can be shared with shipyards ahead of condition-based maintenance activities [10].

Solution vendors must ensure that their solutions, as well as shipyard data, can be hosted in government-authorized cloud infrastructures to balance accessibility with security. For example, SSI solutions natively support the use of industry-standard APIs and serverless technologies, allowing them to be hosted on AWS GovCloud or Microsoft Azure for US Government.

4.4 FINANCIAL AND CONTRACTING CONSTRAINTS

Financial considerations present another significant challenge to adopting model-based sustainment. Building the digital infrastructure for sustainment requires upfront investment in software licenses, IT hardware, data conversion efforts, and training programs. Defence budgets are finite, and funds allocated to new digital tools might compete with other immediate needs like personnel or physical maintenance. The return on investment (ROI) of digital sustainment can be diffuse – improved asset availability or streamlined processes save money in the long run, but those savings might not be immediately evident to budget decision-makers.

To overcome these challenges, naval programs and shipyards are increasingly adopting Total Cost of Ownership (TCO) strategies and innovative contracting mechanisms. Embracing a TCO mindset means evaluating options based on long-term sustainment and upgrade costs, not just the initial price. By accounting for lifecycle expenses from the outset.

A practical step is embedding digital deliverables and data requirements into contracts. For instance, the U.S. Navy plans to mandate that shipbuilders provide a Model-Based Product Support (MBPS) environment as part of new ship programs, ensuring that vessels like the upcoming Constellation-class frigate and DDG(X) destroyer are delivered with complete digital twins and accessible lifecycle data [11].

Naval shipbuilders and operators are leveraging shipbuilding-specific PLM platforms to reduce configuration costs and the time required to see benefit from a solution implementation. Industry-tailored PLM systems are designed to handle the immense complexity of warships (often millions of parts and numerous configurations across a class), providing a single source of truth for design, as-built configuration, and in-service modifications. With all technical data centralized, stakeholders can more accurately forecast maintenance needs and costs, plan mid-life upgrades, and avoid surprises. This digital thread from design through sustainment enables "upgrade-ready" vessels. For example, designing ships with Fitted for But Not With (FFBNW) provisions and maintaining detailed digital twins has proven invaluable during retrofits. This approach was successfully implemented on the RAN Anzac-class frigates. The program was provisioned for the future upgrade to a more complete suite of combat systems. The upgrades were subsequently executed through a mid-lifecycle retrofit for both RAN and Royal New Zealand Navy (RNZN) ships. The approach reduced costs associated with typical challenges to performing an in-depth upgrade post vessel launch [12]. The 2014 modernisation of the RNZN frigates was performed using SSI ShipConstructor [13]. This approach is another strategy that is served by approaching program procurement, design, and construction through a TCO lens.

5. SHIPYARD PERSPECTIVE

Leveraging the digital twin throughout a ship's lifecycle provides many benefits to all project stakeholders - reducing downtime, keeping vessels in service longer, and improving operational efficiency.

5.1 BENEFITS OF DIGITAL TWIN FOR MAINTENANCE, REPAIR, AND OPERATIONS (MRO)

From the shipyard perspective, the availability of a detailed digital twin of a naval asset is a game-changer for maintenance, repair, and overhaul activities. Having a current 3D ship model with all its systems and structures allows engineers to perform feasibility studies for complex repairs or upgrades entirely in the digital realm before touching the vessel. This means potential design clashes or access problems can be identified early on. For example, during a mid-life modernization of the ANZAC-class frigates using SSI solutions, the Australian design team had only legacy 2D drawings at first. They used those drawings combined with laser-scanning of the actual ship to build an accurate 3D model of the vessel. Working from this digital model, they were able to integrate a new combat management system and other upgrades virtually, verifying that everything would fit and function without interfering with existing structures. This model-based approach prevented costly rework by catching issues in simulation. It also improved coordination: all stakeholders (the navy customer, the shipyard executing the refit, and subcontractors) could visualize the planned changes. In the ANZAC refit case, leveraging a common product model for the design meant that no subsystems were overlooked – the digital twin served as the single source of truth for arrangements, which avoided clashes and aided communication between teams. The result was a notable improvement in the quality of refit documentation produced. Navy personnel and shipyard planners received highly detailed, up-to-date drawings and work packages derived from the 3D model, and because the 2D outputs remained associatively linked to the model, any design change was automatically reflected across all documentation. This was a vast improvement over traditional methods where a change might require manually updating dozens of separate drawings.

Leveraging the digital twin of the vessel means shipyards can also conduct MRO activities while vessels remain in service. For example, feasibility studies and impact assessments or work packages. When considering new equipment, the shipyard team can quickly analyse the impact on the vessel – its location, operation, impact on the ship's weight and moment, and how it interfaces with connected systems such as piping, electrical, and HVAC. They can generate work packages before the vessel comes for scheduled maintenance/upgrades. This allows equipment foundations and systems to be prefabricated and delivered dockside ready for installation.

5.2 INFORMATION SHARING WITH NAVY

The digital twin provides navies with a powerful platform that supports not only predictive maintenance but also training and performance optimization across the vessel's lifecycle. By maintaining an accurate, up-to-date representation of the physical ship and its systems, the digital twin enables simulation of realistic operational scenarios. This capability allows naval personnel to engage in immersive, scenario-based training without the need to take actual vessels offline or dedicate live equipment, significantly reducing cost and operational disruption while improving crew readiness.

In terms of performance optimization, the digital twin can be used to analyse the operational behaviour of shipboard systems in various conditions. By studying data from sensors and simulations, naval engineers can assess whether specific components are operating within optimal parameters, identify inefficiencies, and make informed adjustments to enhance performance. This can lead to improvements in fuel economy, system responsiveness, and mission effectiveness.

For predictive or condition-based maintenance, the digital twin offers a dynamic means of simulating system usage to anticipate failures before they occur. By applying usage-based analytics and predictive algorithms, navies can forecast when components are likely to degrade or fail, enabling maintenance teams to schedule interventions just in time—neither too early (wasting component life) nor too late (causing mission downtime). This shift from reactive to proactive maintenance greatly improves asset availability and reduces overall lifecycle costs.

Additionally, the digital twin streamlines the retrieval of technical information, offering a highly navigable interface for exploring the ship's design and maintenance records. Users can interact with the model to locate parts, view associated schematics, consult Vendor Furnished Information (VFI), trace usage across systems, and access service history—all within the same digital environment. This centralized access to information enhances decision-making, shortens troubleshooting time, and ensures consistency in maintenance execution.

Figure 2: Using AR for as-built verification. Image courtesy Incat Crowther

6. INTEGRATED PLM APPROACH

Achieving the full promise of model-based sustainment will require an integrated PLM approach that connects all stakeholders across a naval asset's lifecycle. In practical terms, this means establishing a common digital environment – often centred on a shipbuilding-specific PLM system – where design data, engineering models, maintenance records, and supply information coexist and inform each other in a continuous digital thread. Research and industry experience indicate that implementing fit-for-purpose digital twins in conjunction with a common-language PLM system is foundational to aligning shipbuilders and navies on a single source of truth [8]. By unifying around shared digital models, the traditional gaps between design, construction, and in-service support can be bridged, creating a seamless flow of information from concept to decommissioning.

In an integrated PLM approach, a naval acquisition program would begin with model-based systems engineering and carry those models forward into production and sustainment. For example, the US Navy has demonstrated a workflow wherein the master requirements model is handed off to an integrated PLM that the shipyard also uses, thus "providing a digital thread from requirements to production" with no loss of fidelity [8]. When the shipyard's design and production occur on a platform that "integrates seamlessly with the Navy's PLM system," it "eliminates a significant portion of the implementation work" that would otherwise be needed to map and exchange data between disparate systems.

In essence, the Navy and industry partner are working in a shared digital workspace for the program. This has immediate benefits: as reported, such an architecture means changes are tracked from the beginning and unexpected modifications are virtually eliminated, mitigating the risk of costly rework. Moreover, because the model-based processes and standards are considered from the outset, there is a "clear path from Navy to shipyard" for any through-life upgrades, simplifying future modernizations and tech insertions. These outcomes – no missing information, controlled change, and easier upgrades – directly translate to higher operational availability and lower lifecycle costs for the Navy.

6.1 SHIPBUILDING-SPECIFIC PLM

A shipbuilding-specific PLM system provides the backbone for this integrated environment. As noted earlier, such PLM platforms (e.g., SSI's ShipbuildingPLM built on Aras Innovator) come with domain-specific data models and workflows tailored to ships. When configured as the central hub, the PLM contains the vessel's complete digital twin (structural models, systems schematics, software configurations, maintenance history, etc.) throughout its service life. All participants in the sustainment enterprise can access this hub. For instance, naval engineers can review the as-built 3D model and service bulletins before planning an overhaul; maintainers at a Fleet Maintenance Facility can input the results of inspections or component replacements, instantly updating the configuration record; contractors can retrieve accurate part information for repairs; and logisticians can query support data to forecast spares needs.

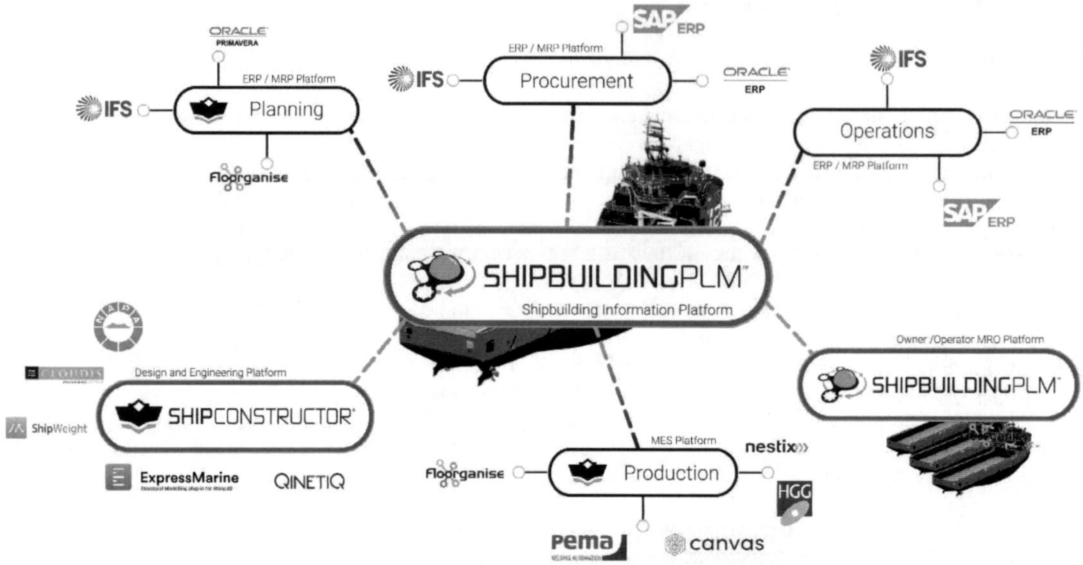

Figure 3: Sample Shipyard technological ecosystem

The PLM thus acts as a central source of truth for all manufacturing and sustainment information and can connect other platforms like ERP/MRP, planning, or owner/operator systems into one coherent ecosystem. In an integrated setup, the Navy's own maintenance management system (or ship's onboard health monitoring system) could be linked to the PLM, ensuring that any condition data or failure reports become part of the asset's digital twin data. This high level of integration means decisions at all levels are informed by up-to-date, accurate data. A Navy commander can trust that the digital twin reflects the ship's current state; a shipyard planner can see the exact baseline configuration before modifying something. The result is fewer errors (since data doesn't need to be re-entered in multiple systems) and faster decision-making (since stakeholders are looking at the same information).

6.2 FUTURE ROADMAP FOR PLM-DRIVEN MODEL-BASED SUSTAINMENT

Implementing an integrated PLM approach across an enterprise is not trivial – it requires not only technology, but also alignment of process and policy. One best practice is to insist on open standards and interoperability. Naval programs should favour PLM and related tools that are open-architecture or vendor-neutral, allowing data exchange with legacy systems and avoiding proprietary lock-in. This ensures that as technology evolves, the digital thread remains intact and accessible. Another important factor is data governance: establishing clear rules for data ownership, updates, and validation among the Navy, shipbuilders, and equipment OEMs. In the integrated model, who is responsible for maintaining the single source of truth? Leading navies have tackled this by formal agreements – for example, the US Navy's product data strategy assigns configuration control responsibilities and requires delivery of data in neutral formats for ingestion into the Navy's systems.

Similarly, Australia's Plan Galileo emphasizes standardizing data and interfaces as part of its enterprise approach [14]. Going forward, we can expect to see contract language that explicitly enforces integrated data environments. The USN's move toward contractual MBPS requirements is a case in point: new ship contracts will likely stipulate that industry must populate the government's chosen PLM or data repository with all the requisite sustainment data (technical manuals, 3D models, sensor schemas, etc.). This essentially makes the integrated digital environment a built-in deliverable of the ship, alongside the ship itself.

7. CASE STUDY

SSI is engaged in ongoing work with a major global naval force on a significant digital transformation initiative to enhance the efficiency and effectiveness of docking work periods for a fleet of naval frigates. The project focuses on the deployment of a shipbuilding-specific Product Lifecycle Management (PLM) solution (ShipbuildingPLM) developed to operate alongside an existing digital shipbuilding design environment (ShipConstructor). This initiative was undertaken through collaboration between a major defense ship designer and the naval force, aiming to modernize maintenance processes and improve operational readiness.

The primary objectives of the implementation were twofold. The first was to accelerate production and enable earlier collaboration among stakeholders. The solution facilitated the development of detailed production packages early in the project lifecycle, enabling pre-fabrication activities to commence well ahead of scheduled docking periods. Key success indicators include the timely distribution of specifications to shipyards, the optimization of review processes, and the expedited procurement of materials. These advancements significantly reduce lead times and enhance schedule predictability.

The second objective centered on achieving continuous and integrated production. By collecting and managing part information, Bills of Materials (BOM), assembly data, and production documentation within a unified digital platform, the implementation ensures a consistent and real-time flow of information across all involved parties. This cohesive digital environment enables improved collaboration among designers, naval personnel, and partner shipyards, resulting in coordinated workflows, greater transparency, and minimized downtime during maintenance periods.

This case represents a notable advancement in the digitalization of naval maintenance programs. By integrating a purpose-built PLM platform with established design tools, the project demonstrates how digital innovation can drive efficiency, reduce drydock durations, and support the strategic goals of national defense operations.

8. CONCLUSION

An integrated PLM approach supported by digital twins, IoT data, and advanced analytics stands out as a transformative solution for naval sustainment. It aligns well with the modernization priorities and examples discussed in earlier sections. The RAN's vision of considering sustainment from day one of design, the Royal Navy's push for data-driven decisions, the RCN's experiments with digital twins, and the USN's drive to embed digital requirements all point toward the same end goal. By learning from each other's initiatives and investing in an integrated, model-based sustainment infrastructure, navies can significantly improve asset availability and lifecycle efficiency.

The opportunity for navies and naval shipbuilders is to reinforce the benefits of a Total Cost of Ownership (TCO) approach. How easily a ship can be maintained and retrofits performed is strongly influenced by the quality of the digital information created as a part of the vessel's design, engineering, and construction. It follows that the success of a program, and its related costs, equally benefit from a strong digital foundation.

The digital shipbuilding platforms that will enable this work are available; the task now is to implement it in a coordinated way. As navies continue this path, we can expect warships that are not only more advanced in their capabilities but also better supported by a responsive, data-rich sustainment system that keeps them ready for the mission required.

9. REFERENCES

1. Royal Canadian Navy, 'Digital Navy: A Strategy to Enable Canada's Naval Team for the Digital Age', *Department of National Defence*, 2020.

2. Ministry of Defence, 'Digital and Data Plan 2022-2025', *Ministry of Defence, London*, 2022.

3. ECKSTEIN, M., 'NAVSEA Seeing Ship Sustainment Successes With Digital Twins, But Wants A More Comprehensive Tool', *USNI News*, 14 May 2021. [Online]. Available: https://news.usni.org/2021/05/14/navsea-seeing-ship-sustainment-successes-with-digital-twins-but-wants-a-more-comprehensive-tool. [Accessed 23 April 2025].

4. SHERBININ, A., and WESTER, T., 'Better, Faster, Cheaper Ship Maintenance—Now!: Augment sailors with predictive maintenance and virtual reality tools', *Proceedings*, p. 427, January 2022.

5. COWPER, I., 'Interviewee, Talking business: Babcock on Type 31, digital twins and more',. [Interview]. 15 April 2021.

6. MORRISON, L., 'Canada's National Shipbuilding Strategy: Revitalizing an Industry and Driving Innovation', 13 December 2024. [Online]. Available: https://www.ssi-corporate.com/blog-lighthouse/canadas-national-shipbuilding-strategy-revitalizing-industry/. [Accessed 23 April 2025].

7. Australian Government, 'National Defence Defence Strategic Review', Commonwealth of Australia, 2023.

8. CROOK, S., ZACHRISSON, C., and GOULANIAN, G., 'Adapting Australian Shipbuilding to the New Defence Reality', *in IMC 2023 International Maritime Conference, Sydney*, 2023.

9. United States Government Accountability Office, 'DOD Software Acquisition: Status of and Challenges Related to Reform Efforts', *Washington*, 2021.

10. NAVSEA - Naval Sea Systems Command, 'Efficient Data Management to Improve Navy Maintenance and Ship Operational Readiness', 14 January 2021. [Online]. Available: https://www.navysbir.com/n21_1/N211-074.htm. [Accessed 23 April 2025].

11. GOULANIAN, G., 'Requirements for Shipbuilders to Support US Navy MRO and Sustainment', 17 December 2024. [Online]. Available: https://www.ssi-corporate.com/blog-lighthouse/requirements-for-shipbuilders-to-support-us-navy-mro-and-sustainment/. [Accessed 23 April 2025].

12. DAVIES, A., and THOMSON, M., 'An Enterprise-Level Naval Shipbuilding Plan', *Australian Strategic Policy Institute*, 2015.

13. SSI, 'Navy Refit Case Study', 2015. [Online]. Available: https://www.ssi-corporate.com/content/navy-refit-case-study. [Accessed 21 September 2023].

14. Australian Government, 'Surface Ship Sustainment in Australia', *Australian Government*, 2025.

15. Australian Government, 'Naval Shipbuilding and Sustainment Plan: Evolving the Enterprise', *Australian Government*, 2024.

10. AUTHOR BIOGRAPHIES

Nick Danese is Senior Advisor for SSI and principal of NDAR, France. He is a Naval Architect / Marine Engineer (MS UofM 1983) with over 40 years of design office, shipyard, and industry experience. Today, Nick focuses on process optimization, the goal being to help our clients maximize ROI while pursuing a digital transformation.

Simon Crook is a Senior Solution Specialist at SSI with over 20 years of experience as a Designer and CAD Manager at leading shipyards in Australia. Simon is a RINA member whose goal is to digitize the marine industry by eliminating the double handling of work, removing information silos, and creating an efficient, effective end-to-end process with a single source of truth for manufacturing information.

Greg Goulanian is Head of Marketing for SSI, where he is responsible for showcasing SSI's innovative software solutions, global success stories, and the future of shipbuilding through papers, presentations, and various forms of media. Greg is committed to sharing how the latest in shipbuilding technology and innovation can help shipbuilders focus on the business of shipbuilding.

SUSTAINABILITY OF NEXT GENERATION POWER PLANTS OF DEFENCE ASSETS

P Karvounis*, **G Theotokatos**, **A Avras**, and **K Jennett**, University of Strathclyde, UK
M Ashley, and **M Reilly**, BAE Systems, UK

*Corresponding Author

EXTENDED ABSTRACT

The growing demand for energy efficiency and sustainability in defence assets necessitates advancements in power plant technologies. Traditional power systems for defence platforms often fail to incorporate waste heat recovery (WHR) and energy storage (ES) technologies, resulting in substantial energy losses of up to 30–40% of fuel energy. This research aims to enhance the energy efficiency of next-generation power plants for defence assets by integrating advanced WHR and ES technologies tailored to their unique operating conditions and profiles. A key objective is to identify and customize the most efficient and feasible WHR and ES technologies capable of achieving efficiency gains exceeding 10%. To address this, a systematic methodology is adopted. The study involves a comprehensive taxonomy and evaluation of WHR and ES technologies, based on determined KPIs, focusing on those with high potential. Individual and hybrid solutions, such as high energy–high power ES systems combined with high-temperature and low-temperature WHR systems, are investigated. The study identified eight WHR and six ES technologies suitable for defence asset integration.

A weighted KPI analysis reveals that maximum efficiency (9.3%) and technology readiness level (8.5%) contribute the most to overall power plant efficiency improvements, followed closely by naval requirements (8.4%) and weight (7.9%), suggesting that mature and efficient technologies aligned with operational constraints are preferred. Relative area and CAPEX, although important, present relatively lower impact factors (5.7% each), highlighting that footprint and cost are constraints but not the primary drivers in defence settings. Moreover, assessment of specific WHR and ES options shows that among ES technologies, the heat pump (0.057 m^2/kW), isobaric expansion pump (0.055 m^2/kW), and absorption chiller (0.051 m^2/kW) exhibit the highest integration density, while from the WHR side, ORC (0.053 m^2/kW) and SRC (0.042 m^2/kW) lead in spatial efficiency, with Kalina cycle (0.036 m^2/kW) offering a compact, albeit lower-impact solution. Extensive mathematical models for WHR and ES technologies are developed, and their integration revealed significant efficiency savings in the magnitude of 5–20%, depending on the specific platform and operational profile. These results provide a pathway for the design of sustainable and efficient power systems, addressing the energy demands of future defence platforms while reducing their environmental impact.

1. ACKNOWLEDGEMENTS

The authors greatly acknowledge the funding from DNV AS and RCCL for the MSRC establishment and operation. The opinions expressed herein are those of the authors and should not be construed to reflect the views of Innovate UK, DNV AS and RCCL.

2. AUTHORS BIOGRAPHY

Panagiotis Karvounis holds a research assistant position at the Maritime Safety Research Centre at the University of Strathclyde. The Author is responsible for drafting the extended abstract and conducting the analysis pertaining to the work presented. The Author's previous experience includes decarbonisation through alternative fuels and thermodynamic modelling.

WHAT RISK?

J P Osmond, RPS Consulting Services Ltd, UK

SUMMARY

A perspective on how we understand 'risk to life' with modern-day challenges.

How do we ensure we understand risks to life posed by new systems where the main focus is removing the people? Autonomous systems will still need to interface with people, not always through obvious or established mechanisms. Military systems still need to follow the chain of command. How will learning systems distinguish between benefits to themselves and to their masters?

This paper investigates how we might identify and manage risk to human life with foreseeable future technologies in the military context.

1. INTRODUCTION

My presentation on this subject draws on my personal knowledge and experiences gathered over a 25-year career spent on defence and maritime projects.

To provide some context to my knowledge and experience, I am a Naval Architect and started my career on the MoD Graduate training programme (DESG). This was followed by several years on early stage RFA procurement projects before moving into the private sector. I have spent the last 17 years working across defence, in a variety of roles, all of which have had a focus on technical, safety or environmental assurance. Mostly as a consultant. The majority of the projects I have worked on have been on the procurement and therefore boundary pushing side of defence, with a few looking at older systems ranging all the way back to HMS Victory. Currently with more of a focus on unmanned systems across land, air and maritime domains, almost exclusively in the defence space, though a couple of deviations into commercial projects.

2. CURRENT APPROACH TO RISK MANAGEMENT IN DEFENCE

Current practice with risk management, and in particular risk to life, follows one of a number of recognised approaches and models to identify, understand, manage and then demonstrate to those responsible that all reasonable effort and cost has been expended to reduce the risks to an acceptable level. This is generally referred to as the ALARP principle. For the MoD, this is achieved through the development of a Safety Argument, to demonstrate a thorough Safety Case, articulated in a Safety Case Report. In the civil context the safety case is often only required in high-risk areas, but the ALARP principal is at the centre of our Health and Safety Executive (HSE) policies.

A key element of the Safety Case approach, which is fundamental when introducing new technologies to trials teams or operators, is the clear articulation of boundaries. Typically, through limitations on how the technology can be used, where it can be used, who can use it and other considerations to provide context to ensure the safe operation of the equipment. In the case of novel systems this is particularly relevant as risks may not align with expectations, or the confidence may need building, or the usual mitigations may be insufficient.

This is then used as part of the decision-making process to support go / no go procurement decisions, or good enough to accept off contract, or safe enough to give to operators to use, all the way through to safe to dispose. The "ALARP" mechanism is well established and based on proven principles that over time can be seen to incrementally improve our systems, solutions, capabilities or outputs. This is just as useful in any industry and is commonly seen through HSE guidance for much less complex systems and through high-risk industries such as aviation and nuclear and more recently trough the requirement for Building Safety Cases.

3. HOW WE GOT HERE

There are many examples of where engineering principles have evolved, alongside safety principles, economic, medical, education, and even software development. At a societal level it is fairly easy to prove that we have fewer fatalities since the introduction of the Health and Safety Executive, however it is not always so easy to prove that the same can be said in the defence. The Aberfan disaster in 1966 led directly to the founding of what we know today as the HSE, the Titanic 1912 led directly to the founding of the IMO, the Nimrod disaster of 2006 led to the Haddon Cave enquiry. The outcome from the Grenfell Tower disaster in 2017 is still to be fully understood.

As with most areas of engineering, the ALARP mechanism borrows from historic precedents and typically takes big steps following significant incidents. In the ship building context the ability to scale proven hull forms and modelling resistance with Froudian principles has steadily allowed for huge energy efficiencies through ever better hull design. Similarly, the introduction of the IMO and their subsequent SOLAS rules following the Titanic continues to improve the safety of life at sea.

We understand and manage the risks that we face in the context of what we have experienced. Humans learn predominantly through pain or reward. As a child, we learn not to touch hot things from burning ourselves, sharp things from cutting ourselves or fast-moving things from bruising ourselves. We learn that we like sweet things from the pleasure we derive and then we get rewarded with sweets for surviving the pain at the doctor. As life progresses, we avoid circumstances that we think might cause us pain and pursue activities that we believe we will enjoy based on our experiences.

Similarly, we delegate the responsibility to our governments and public servants to keep us away from dangers that are outside of our control, or more effectively managed at a higher level. This is applied through rules, regulations, standards and other control mechanisms developed and evolved to reflect experience and lessons learnt. The highway code is regularly updated to reflect lessons learnt from accidents and to accommodate technological changes.

The rules on how we build ships are updated to reflect lessons, all the way back to those first meetings at Lloyds coffee shop, through major disaster like the Herald of Free enterprise, Piper Alpha even principles that pre-date the Mary Rose. Now we use the same basic model for designing and building Warships to Lloyds or DNV rules. Make sure you have transparency of responsibility for safety decisions all the way to the top of the organisation, always ensure there are several well drilled ways to escape, and don't make them top heavy.

4. STRENGTHS

As our society progresses alongside advancements in technology and evolving aspirations, the rules and regulations adapt accordingly. In a time of peace (relatively since 1945) we have grown in more ways than we could have imagined 80 years ago. As our expectations evolve so do our technologies. We expect to live beyond our working life into retirement, so we have state and private pensions, healthcare as we expect to have fewer life changing or ending accidents and incidents. We expect fewer car accidents to be fatal, and our beaches to be cleaner. Our current ALARP and rule-based society supports this, as broadly speaking the systems in place keep up with the technology and our ability to outgrow our systems. The ALARP model allows for incremental step changes, improvements and recognition of errors that can be corrected at the next issue.

5. SHORTCOMINGS OF CURRENT APPROACHES

Current regulations are unable to keep pace with the rapid development of many emerging technologies. You walk out of the hotel door and navigate around commuters using various technologies such as electric bikes, scooters, skateboards, monocycles, or even drone taxis. These technologies are evolving so fast the potential for systematic failure, abuse, corruption or mistakes is significant. Especially when the rules are unpoliceable or still pending.

How should we then identify, understand, and manage risks for technologies we have no real context for? How can we establish historical data for unprecedented scenarios and demonstrate that we have effectively minimized risk to an acceptable level, given the lack of relevant context for comparison?

6. THE CHALLENGES WE FACE

I believe the challenges we face are evolving faster than our readiness, tools, techniques, and budget. Regulation often lags behind innovation to avoid hindering industrial progress. However, this has also led to a lack of clear guidance of what good should look like.

Typically, at least from my own experience, software developers are given an outcome to be achieved by contracting authorities, but not provided with clear, unambiguous, structured guidance on what the regulatory authorities need to see. This is more common in small scale projects, often with a quick timeframe, as opposed to on a major capital procurement program which has deeper projects, longer timescales and close regulatory support and scrutiny. With the smaller projects, such as in the sub 25kg drone, or S2 scale of projects, where the budgets are limited, the timeframes are short, and the scrutiny is often light touch. As a result of abbreviated front-end processes, in many instances, software is not being developed in a clear auditable manner to enable a high rigour of scrutiny to build confidence that it will perform as intended. Instead, a more modular approach has become the norm, earlier packages are borrowed from, or outright copied, blending and integrating the different modules together, stacking uncertainty on top of uncertainty.

When the technology is new, and evolving rapidly, it also means that there are not enough case studies or running hours of systems to give a high degree of confidence in the systems integrity. These under-demonstrated, under-tested, and under-proven systems are then often integrated together into packages that may or may not get the opportunity to be proven before they are then further integrated into the next generation of system.

For software, the Spiral Development model is often quoted, in theory much like the traditional ship design spiral. Where each revolution around the design spiral for a ship takes it closer to the middle, to the focus point, challenging decision making to ensure the risks or problems are identified. Where after each iteration the outcome is refined to a more focused design that has identified and resolved more problems through each cycle.

The Spiral software development in defence programmes, often starts from a singular problem or need statement, that may be accompanied by a few traditional context pieces such as Concepts of Operation or Use documents, maybe a few requirements, maybe a few preferred standards. As the solution is developed it may become apparent that additional benefits are available. That the capability can achieve more. A few problems are identified to be factored into the next step up or specification. The process becomes more of a starting point to see what the "Art of the possible" might be.

In practice it seems to allow systems to be proven and accepted, just in time to become obsolete and overtaken by the next spin around the spiral, gradually broadening in scope and reach. For high pace software intensive projects, this appears to work in the opposite direction to the traditional model.

Each iteration around the spiral generates more options, more potential outcomes, and more complex dilemmas that still have yet to be resolved. It may be that multiple solutions are identified, spin off projects with further, more complex requirements added in to the mix, along with more regulations and updates, maybe with new operators and additional suppliers. The spiral development principal allows for a more responsive solution, however if the guidance and rules on how to keep the boundaries effectively managed throughout the process, to define the acceptance criteria, the horizon will eternally keep slipping out of reach.

It is very difficult to come to the end of a project that does not have effective criteria for acceptance, especially if there has been no guidance or clear regulation to comply to along the way.

In the defence space, there appear to be fewer competent engineers available to support projects make decisions, or to recognise when a decision needs to be made. Far too often we see project suffering from time, resource and cost overruns, sometimes leading to closures of projects and costly commercial and litigious proceedings. The traditional project team difficulties of resourcing on all levels being inversely proportionate to prestige or size of project. Sadly, this often leads to lost opportunities as projects that had great potential but lacked clear directions.

7. IS WHAT WE HAVE ENOUGH?

I would argue no. Do we have risk management processes that enable us to avoid the "Rumsfeldian dilemma" of not knowing what we don't know? Do we have good enough systems in place to identify or recognise many of the risks and problems effectively?

In the UK at least we are not in a position of fighting for our survival, when accepting very high-risk systems to increase the chance of achieving high gain outcomes may be acceptable.

Can we, in the UK, with our strong safety and risk management culture accept this level of uncertainty.

Having recently come to terms with the horrifying reality that the typical new joiner (with degree) could be the same age as a child of mine, has led me to appreciate that maybe some of the miscommunication we often suffer within our own families could be indicative of the different understanding and appreciation of the younger generations who are now developing and using live systems at sea today.

It can be hard to grasp that generational changes occur over a rolling timeframe. At any given point in time, those that make the procurement or engineering decisions are at least one generation removed from those that will be developing, designing, building and using the systems.

People involved may have diverse ages, understandings, experiences, and attitudes towards risk, each bringing unique perspectives with differing contexts and risk appetites on what the "Next generation" technology bring with it. The less we have witnessed events that impacts us directly, the more we believe it is only ever going to happen to someone else.

8. IS CHANGE NEEDED?

Our children and future generations may be less directly exposed to the risks we manage today, or at least many of those risks. So how do we distinguish between the risks that we understand and manage today to acceptable ALARP principles, against the future where unmanned systems remove the operator from direct consequences of their actions? How do we ensure that we don't throw the baby out with the bath water? Afterall there will still be people on vessels even if purely for leisure.

Many maintenance tasks will require human input for many years to come and Cyber Security concerns mean that we will always demand assurance that an incident such as the one recently experienced by Marks & Spencer's, losing their IT for several weeks, cannot be repeated through a deliberate, malicious intervention. Or the potential for a real-life consequence, who is to say that the next 9/11 scale event could not be from an unmanned system being hacked?

The Post Office, Horizon scandal has taken over a generational time frame to be exposed, it will take at least another to be fully resolved. It started as an accounting aid and has now led to an undisclosed number of suicides and an unknown number of lives ruined. How much would this have been improved if the institution was prepared to be more transparent and had listened to the early concerns. I think the MOD has a good culture of transparency, when possible, with a willingness to listen, however it can be extremely difficult to be heard as an outsider when the transparency is very inward facing.

The benefits are largely obvious: reduced exposure to operational risks, less space/weight needed for humans and their hotel services, less fuel to achieve the same output, reduced opportunity for human error, longer deployments, more assets for same price, and reduced support costs.

From a RN perspective the operating model for current capabilities will need to be revised from the ground up. Removing the human from many of the highest risk activities, such as mine clearance, subsea surveillance, traffic monitoring, threat detection, over the horizon observations. However, introducing new risks such as recovery of unmanned systems to host platforms or to shore facilities, gathering and processing vastly larger volumes of data than ever before, re-thinking rules of engagement, reliability of data sources, operating in spaces previously inaccessible. The risk to life becomes more due to the inter-operation of manned and unmanned in the same space. The unmanned system may remove the man from the immediate harm of a hazardous activity, but the unmanned system will still need to interface with manned systems, and maybe many of them concurrently that have been developed to different standards, languages and operating protocols.

9. CONCLUSIONS

I would like to leave you with a few questions, which I believe if addressed would enable our industry to better understand and manage the risks and challenges that are already on the horizon.

Reform is needed to a historically slow and cumbersome procurement process. For smaller scale projects, the process needs to be nimble enough to keep up with the pace of change in the assessment and management of risk. Is it possible to do so with the current models that are, to a great extent, optimised for large capital procurements, which look to precedent to provide benchmarks and acceptance criteria?

In my view investment is needed to create sufficient "Sandbox" type environments to enable experimentation away from accidental third parties. Such that high-risk systems can be trialled without being fettered by over restrictive safety, or risk mitigations and limitations on use. Do sufficient mechanisms and facilities exist to enable novel technologies that may have unanticipated behaviours to be tested, trialled and proven?

In my opinion the Ministry of Defence is not sufficiently informed to develop effective requirements especially for more detailed smaller projects. Is it possible to develop a better tool set for basic components, such as software assurance models and tools, that expedite the acceptance process while prioritising user safety and risk reduction? The MoD has in recent years been very aware of the problems that arise from not listening and has therefore, at least in the past, actively sought to engage a broad spectrum of stakeholders to support making difficult decisions, however in more recent years it has become much harder to get a voice, to reach the people that really need to hear. How is the MoD going to return to a more open and engaging customer to industry and wider stakeholder communities?

Finally, a big all-encompassing question, as we look ahead of us, with the political uncertainty coming from many directions, the technology developing faster that we can keep up, with threats from both the natural world as well as human. Are the current tools and processes encouraging us to look at the right problems and are we adequately equipped to tackle the next challenge?

10. ACKNOWLEDGEMENTS

I would like to thank my colleague Peter Hutchison and my wife Lauren Osmond for their help and support collating this paper and the presentation to the conference.

This paper has not been published before and is not currently being considered for further publication.

11. AUTHORS BIOGRAPHY

Paddy Osmond holds the current position of Business Stream Leader at RPS Consulting Services Limited. The Author is responsible for the technical output of a team of technical assurance consultants specialising in cross domain safety and environmental management services to the MOD and to the wider defence industry. The Author's previous experience includes the MOD Graduate Training Scheme followed by 6 years across a variety of design roles, all including safety management responsibilities for the procurement of RFA ships. This was followed by period as a Senior Safety Consultant in the WS Atkins defence team. Subsequently 4 years as the HP Defence UK Safety Assurance lead, delivering safety assurance across several large IT infrastructure projects including DII. The last 10 years have been with RPS, initially as a Principal Consultant responsible for maritime projects, this role expanded to include Business Development, and now is pan domain. Project managing, delivering or supporting technical assurance to a broad range of Defence programmes with the occasional non-defence project.

ADVANCING SMART, SUSTAINABLE, AND AUTONOMOUS FLEETS WITH DIGITAL INNOVATION TOOLS

R Pérez Fernández, Siemens Digital Industry Software, Spain

SUMMARY

The maritime industry is undergoing a profound transformation driven by the integration of smart technologies, the pursuit of sustainability, and the development of autonomous operations. Advanced digital tools and platforms are enabling shipbuilders, operators, and designers to address these challenges and opportunities through innovative solutions. This paper explores how digital twin technology, advanced simulations, and lifecycle management systems are empowering stakeholders to design energy-efficient, low-emission ships equipped with cutting-edge automation. The discussion also highlights the role of predictive analytics and artificial intelligence in optimizing fleet performance, meeting stringent environmental regulations, and ensuring safety. By leveraging these technologies, the maritime sector can achieve a future of enhanced operational efficiency, reduced environmental impact, and revolutionary advancements in autonomous navigation.

NOMENCLATURE

AI	Artificial Intelligence
AIS	Automatic Identification System
COLREG	International Regulations for Preventing Collisions at Sea
GPS	Global Positioning System
LiDAR	Light Detection And Ranging
ML	Machine Learning
XAI	Explainable AI

1. INTRODUCTION

The maritime industry stands at the cusp of a profound transformation with the advent of autonomous maritime systems, an innovation that is redefining how vessels are designed, operated, maintained, and embedded within the global supply chain. Traditional paradigms of manual navigation and crew-dependent operations are rapidly giving way to intelligent, automated vessels capable of self-directed navigation, advanced situational awareness, and seamless integration into complex logistical networks. This shift is driven by an imperative to significantly enhance operational efficiency, minimize human error, reduce operational costs, and adhere to ever-tightening environmental regulations aimed at mitigating the sector's ecological footprint.

At the heart of this transformation lies a convergence of cutting-edge digital technologies (Benayas-Ayuso, et al., 2024). These include sophisticated sensor arrays, *Artificial Intelligence* (AI), *Machine Learning* (ML), digital twin frameworks (see *Figure 1*), and high-fidelity simulation platforms, all interconnected through robust communication networks. Together, they enable vessels to perceive, analyse, and respond autonomously to dynamic maritime environments, delivering real-time decision-making capabilities and system-wide optimization that were previously unattainable.

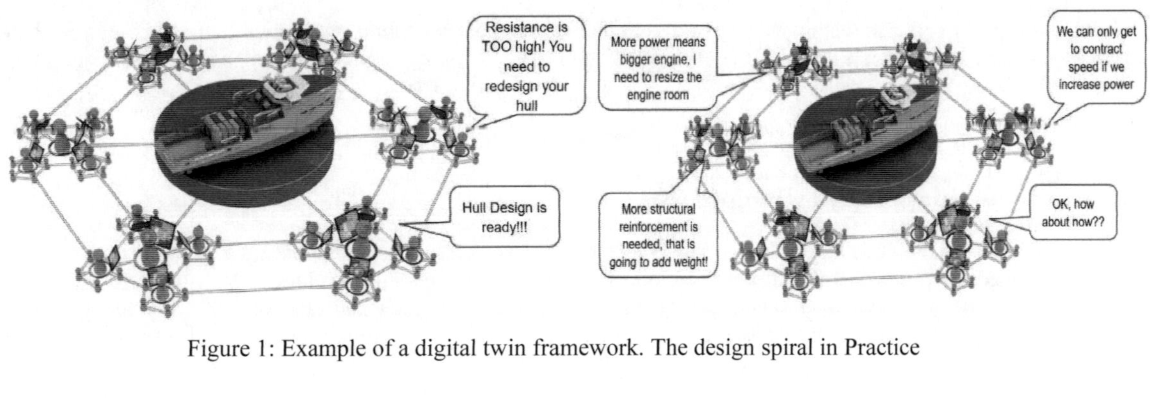

Figure 1: Example of a digital twin framework. The design spiral in Practice

However, the transition to autonomous maritime operations is not simply a matter of integrating new technologies into existing frameworks. It demands a radical rethink of the entire lifecycle, from initial vessel design and construction through to deployment, operation, maintenance, and decommissioning. New design philosophies must account for the integration of AI-driven control systems and digital twins, while engineering workflows must evolve to incorporate iterative simulation and virtual validation. Operational strategies require adaptation to incorporate remote monitoring, predictive maintenance, and adaptive autonomy levels based on environmental and regulatory contexts.

We are seeing an increasing demand for high value-added vessels, multi-role or multi-purpose vessels (especially in the naval, mega yachts and offshore markets), and a growing interest in autonomous shipping (especially in the commercial and naval sectors) as explained in the *Figure 2*. Nowadays navies are often requested to deal with military operations other than war, for example providing humanitarian relief and emergency medical care after natural disasters or establishing a helicopter platform for replenishment or evacuation. The yachting and cruising industries have seen a broadening of horizons: customers do not want to limit themselves to the Mediterranean or the Caribbean. This means that vessels need to be able to operate in a wider range of conditions. Autonomous shipping is of particular interest for short-sea shipping (commercial waterborne transportation utilizing inland or coastal waterways to move freight from major domestic ports to smaller harbour inaccessible to big containerships), ferries & naval operations (Pérez Fernández, 2022).

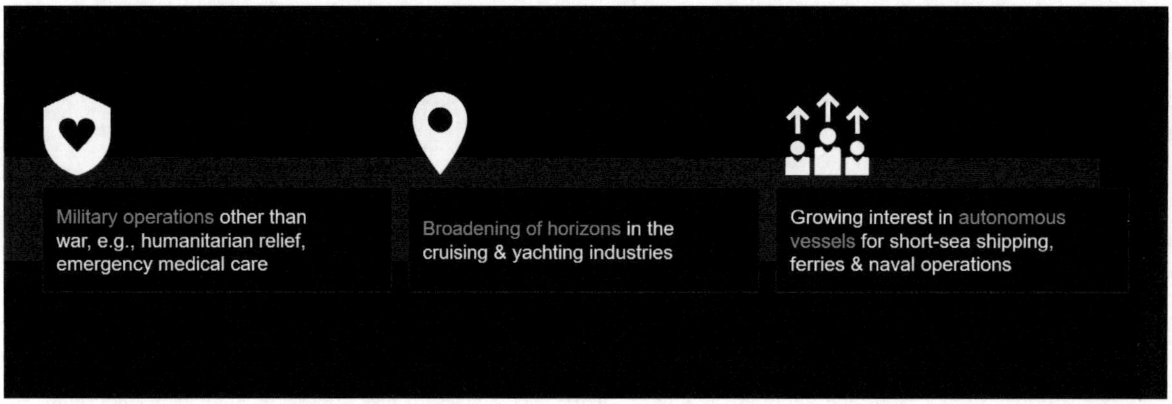

Figure 2: Increased demand for high value-added, multi-purpose, and autonomous vessels

Moreover, the deployment of autonomous vessels introduces complex challenges including ensuring safety and reliability in unpredictable maritime conditions, achieving interoperability among diverse systems, navigating evolving regulatory landscapes, and addressing cybersecurity threats in increasingly connected platforms. Addressing these challenges calls for interdisciplinary collaboration among shipbuilders, technology developers, regulatory bodies, and maritime operators.

Autonomous maritime systems represent not just an incremental innovation but a systemic revolution that promises to redefine maritime logistics, environmental stewardship, and economic competitiveness.

This paper explores the pivotal role of digital innovation tools in enabling this revolution, detailing how their strategic integration is essential to unlocking the full potential of autonomous, smart, and sustainable maritime fleets.

2. ENABLING TECHNOLOGIES FOR MARITIME AUTONOMY

2.1 DIGITAL TWIN TECHNOLOGY

Digital twins serve as high-fidelity virtual representations of vessels, enabling real-time monitoring, predictive maintenance, and operational optimization. These digital models are developed using data from sensors, design specifications, operational parameters, and historical records, creating a continuously updated virtual counterpart of the physical vessel. They are foundational to the digital transformation of maritime systems, especially in the context of autonomy, as they provide the computational environment necessary for modelling, simulation, analysis, and decision-making.

In the development of autonomous vessels, digital twins play a pivotal role in system design and validation. They enable virtual prototyping of autonomous behaviours, allowing engineers to simulate scenarios such as obstacle avoidance, dynamic rerouting, and emergency manoeuvres under diverse environmental conditions. This not only reduces the need for costly physical trials but also improves the fidelity and robustness of control algorithms before deployment. By testing these behaviours virtually, designers can iterate more rapidly and fine-tune performance with reduced risk.

Moreover, digital twins facilitate real-time data integration and analytics, transforming static models into dynamic systems capable of evolving with operational experience. Through continuous feedback loops, they ingest live data from onboard sensors-such as *Global Positioning System* (GPS) or *Light Detection and Ranging* (LiDAR), radar, and engine performance metrics-to update the vessel's virtual state. This enables predictive maintenance by identifying component wear and potential failures before they occur, thereby reducing unplanned downtime and enhancing safety. In addition, digital twins can assist in anomaly detection, alerting operators to irregular patterns that may signify malfunctions or cyber threats.

Another key advantage of digital twins in autonomous vessels is their role in human-machine collaboration. While full autonomy may reduce the need for crewed operations, human oversight remains essential-particularly in supervisory control models. Digital twins provide an intuitive interface for operators to visualize vessel status, simulate decisions, and intervene when necessary. They also support training environments for remote operators by mimicking real-time vessel behaviour in a virtual setting.

In fleet-wide applications, digital twin technology allows for the aggregation of data across multiple vessels, creating a fleet-level operational picture. This enables fleet managers to optimize routing, resource allocation, and maintenance schedules based on predictive insights derived from collective behaviour patterns. Additionally, digital twins aid regulatory compliance by maintaining detailed operational logs and facilitating transparent audits of safety and performance standards.

The integration of digital twins across the vessel lifecycle-from initial design through decommissioning-ensures that every phase is informed by accurate, context-rich data. This lifecycle approach enhances traceability, improves the reliability of autonomy-enabling subsystems, and supports continuous improvement in vessel performance. As the maritime industry moves toward more complex and interconnected autonomous systems, digital twins will remain a cornerstone technology, offering the precision, adaptability, and situational awareness required for safe and sustainable operations.

Regarding sustainability, during the past years have seen the emergence of a global effort to control climate change and intensify the actions and investments needed for a sustainable low carbon future. The historical signature of the *Paris Agreement* by 196 nations and the European Union following COP 21 in December 2015 crystallised this intention and started an era of international cooperation. Although the *Paris Agreement* makes no specific mention of the marine industry, greater environmental awareness and public pressure have targeted the industry, which is responsible for more than 3% of global CO_2 emissions, comparing to major carbon-emitting countries, while the industry continues to grow rapidly.

As a result, the United Nations *International Maritime Organization* (IMO), the main international shipping regulatory body, adopted new ambitious regulations targeting pollutants and *GreenHouse Gas* (GHG) emissions from ships, aiming to phase them out as soon as possible in this century. IMO released its initial GHG strategy in 2018, introducing very strict fuels and emission standards. The strategy has been revised in 2023. By 2050, ships are expected to produce close to 100% less GHG emissions than they did in 2008 to and reach net-zero emissions.

Figure 3 presents an overview of international initiatives aimed at addressing climate change and promoting a sustainable, low-carbon future.

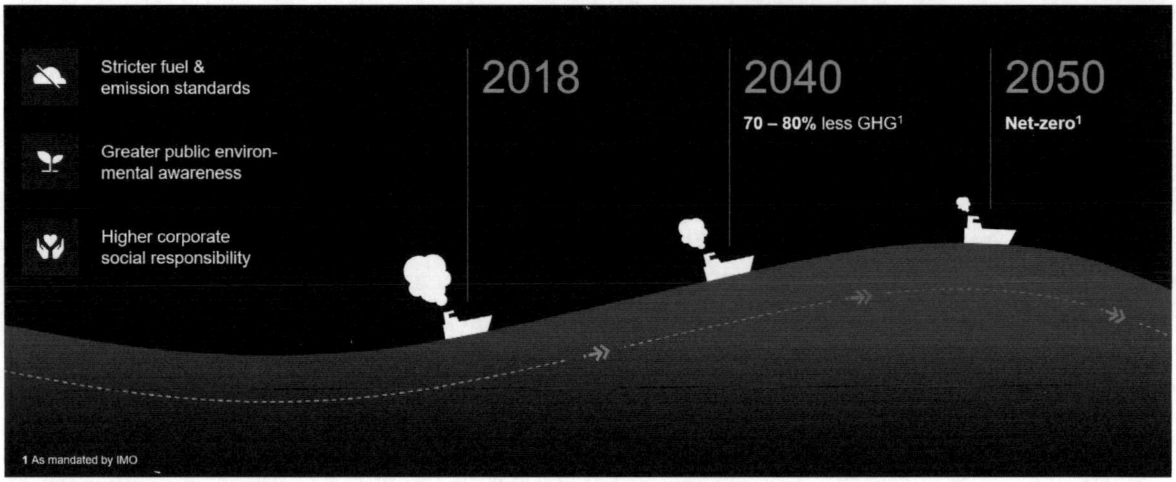

Figure 3: Global effort to control climate change and build a low-carbon future

As a result of greater public environmental awareness and general outcry, the pressure has been mounting for retailers and companies across the globe to become more conscious. More and more, consumers are asking for traceability in the supply chain as they want to ensure that the products they purchase have not only been produced in a sustainable manner but transported without polluting the planet too. As manufacturers make a point to demonstrate their corporate social responsibility to their customers, they want to associate themselves with clean shipping companies only.

2.2 ADVANCED SIMULATION AND MODELING

Simulation platforms are essential in evaluating the performance of autonomous navigation systems, control algorithms, and onboard subsystems. These tools offer a controlled, repeatable, and scalable environment to develop, test, and validate critical technologies that underpin autonomous maritime operations. By enabling exhaustive testing of various operational conditions, simulations reduce uncertainty and increase system resilience before actual deployment.

High-fidelity simulations replicate the dynamic marine environment with intricate detail, including sea states, wind and current interactions, vessel hydrodynamics, sensor fusion effects, and environmental disturbances. These simulations provide developers and engineers with the ability to assess how autonomous control systems will behave in real-time under diverse scenarios. This includes navigating narrow waterways, avoiding collisions in congested routes, docking under variable wind loads, and responding to sudden system failures or external threats. Such granularity ensures that autonomous behaviour is not only feasible but robust under edge-case conditions.

Advanced simulation frameworks integrate physics-based modelling with synthetic sensor data to emulate the inputs received by real onboard systems, such the already mentioned LiDAR, radar or GPS, and *Automatic Identification System* (AIS), and optical cameras. These virtualized inputs enable testing of perception algorithms and decision-making logic in a range of visibility and traffic conditions, including low-light or high-fog scenarios where sensor performance may degrade. As a result, developers can improve situational awareness algorithms, calibrate fail-safes, and verify redundancy protocols before real-world implementation.

In addition to vessel-level simulations, system-of-systems modelling is increasingly used to simulate interactions between multiple autonomous and traditionally crewed ships. This allows testing of cooperative behaviours, such as coordinated manoeuvres, dynamic re-routing, and swarm intelligence in fleet operations. Multi-agent simulations facilitate the development of intelligent decision protocols that align with the *International Regulations for Preventing Collisions at Sea* (COLREGs), as well as emerging maritime autonomy regulations.

Another important application of simulation is digital verification and certification. Regulatory agencies and classification societies are beginning to recognize high-quality simulation as a complementary pathway to physical trials. Through rigorous virtual testing, developers can demonstrate compliance with safety and performance standards, providing traceable records that support formal approval processes.

Simulations also enhance human-machine interaction design by enabling the prototyping of operator interfaces, supervisory control systems, and remote piloting dashboards. By visualizing how human operators will interact with autonomous systems under stress or abnormal conditions, designers can improve usability, reduce cognitive load, and prepare for transitional models of autonomy that require part-time human intervention.

Furthermore, simulation platforms are vital in training and education. Maritime academies, shipbuilders, and operational teams can use them to build proficiency with autonomous technologies in safe, low-risk environments. They allow for iterative learning and skills development without the cost, danger, or environmental footprint of real-world testing.

In summary, advanced simulation and modelling serve as both a proving ground and a developmental accelerator for autonomous maritime systems. By facilitating comprehensive, scalable, and repeatable testing environments, they not only enhance system safety and performance but also significantly reduce development timelines and costs. Their role in enabling autonomy is foundational, especially as the industry moves toward increasingly complex, multi-vessel, and environmentally integrated operations.

2.3 ARTIFICIAL INTELLIGENCE AND MACHINE LEARNING

AI technologies underpin the cognitive capabilities required for autonomous operations, serving as the decision-making core that transforms raw sensor inputs into actionable behaviour. At the heart of AI applications in maritime autonomy is ML, a subset of AI that enables systems to learn from data, identify patterns, and improve over time without being explicitly programmed for every scenario (Gómez-Ruiz, et al., 2023).

In autonomous vessels, AI facilitates real-time perception, decision-making, and control. Perception systems powered by AI ingest data from an array of sensors, such as optical cameras, and inertial navigation systems, to construct a comprehensive situational awareness model. Through sensor fusion techniques, AI algorithms reconcile conflicting or overlapping data sources to provide a coherent understanding of the vessel's environment, including the identification of nearby objects, shorelines, other vessels, weather patterns, and underwater obstacles.

AI-driven anomaly detection is critical for ensuring safety and reliability. By continuously monitoring equipment performance, vessel behaviour, and environmental conditions, ML models can detect deviations from normal patterns and alert operators to potential faults or hazards. These predictive capabilities reduce reliance on manual inspections and allow for proactive maintenance, helping avoid costly downtime and improving overall system health.

In navigation and routing, AI models optimize voyage planning by accounting for variables such as ocean currents, weather forecasts, traffic density, fuel consumption, and port schedules. Reinforcement learning, where systems learn optimal actions through trial and error, can be applied to dynamic routing scenarios to continuously improve voyage efficiency and minimize emissions. Additionally, AI systems can respond adaptively to changing maritime conditions, rerouting vessels autonomously in real-time to avoid congestion or dangerous situations.

Autonomous vessels also leverage AI for behaviour modelling and rule-based decision logic, particularly in complying with COLREGs. ML classifiers and deep learning models are trained on large datasets that include vessel movement patterns, legal constraints, and historical incidents, allowing the autonomous system to choose legally compliant and contextually appropriate manoeuvres during complex interactions with other maritime traffic.

Beyond core navigation, AI supports auxiliary functions such as energy management and environmental monitoring. Smart energy management systems powered by AI can balance power loads, optimize fuel usage, manage battery charging cycles, and adjust propulsion strategies to achieve peak efficiency. Similarly, AI-driven environmental monitoring can detect pollution levels, measure compliance with emission regulations, and adjust operational parameters to reduce ecological impact.

An important feature of AI in autonomous maritime systems is adaptive learning. Using continuous feedback loops, AI algorithms refine their performance over time, integrating new data from real-world operations to improve their accuracy and responsiveness. This evolutionary capability ensures that autonomous systems remain robust even as external conditions change, or unexpected situations arise.

Explainable AI (XAI) is also gaining relevance in maritime autonomy, particularly in systems that require human oversight or regulatory certification. XAI tools make the decision-making process of AI systems more transparent and interpretable, enabling operators, auditors, and regulators to understand the rationale behind autonomous behaviours. This not only builds trust in AI-driven operations but also facilitates the development of accountability frameworks and safety assurance processes.

As maritime AI systems become more sophisticated, edge computing technologies are being deployed to bring computational power directly onboard vessels. This allows for low-latency decision-making even in remote areas with limited connectivity and supports real-time autonomy without dependence on shore-based control centres. Combined with cloud-based learning frameworks, this hybrid approach allows vessels to benefit from both localized autonomy and fleet-wide knowledge sharing.

Finishing this section, AI and ML technologies are indispensable enablers of autonomous maritime operations. Their application extends across perception, decision-making, energy optimization, predictive maintenance, and regulatory compliance. As the field evolves, continuous innovation in algorithm design, data processing, and human-AI collaboration will be essential to realize the full potential of smart, self-governing vessels.

2.4 INTEGRATED LIFECYCLE MANAGEMENT SYSTEMS

Lifecycle management platforms are essential for orchestrating the full continuum of activities associated with maritime assets, from initial concept and design, through construction and commissioning, to operation, maintenance, and eventual decommissioning. In the context of autonomous vessels, these platforms provide a digital thread that ensures continuity, coherence, and coordination across all phases of the vessel's existence.

At the design and engineering stage, integrated lifecycle systems enable the centralization of technical specifications, digital models, and regulatory requirements. By linking *Computer Aided Design* (CAD) 3D models, engineering simulations, and compliance documentation into a unified data environment, these platforms facilitate collaborative design processes and reduce errors due to information silos. Version control, requirements traceability, and real-time design validation ensure that every component and subsystem is configured correctly for autonomous operation and meets international standards.

During ship construction, lifecycle platforms provide a framework for quality assurance and component certification. Digital records of materials, manufacturing tolerances, and subsystem integration steps are maintained, creating a comprehensive construction log. This digital provenance supports post-construction troubleshooting and streamlines the certification process with class societies and maritime regulators, particularly for novel autonomous systems that may lack conventional benchmarks.

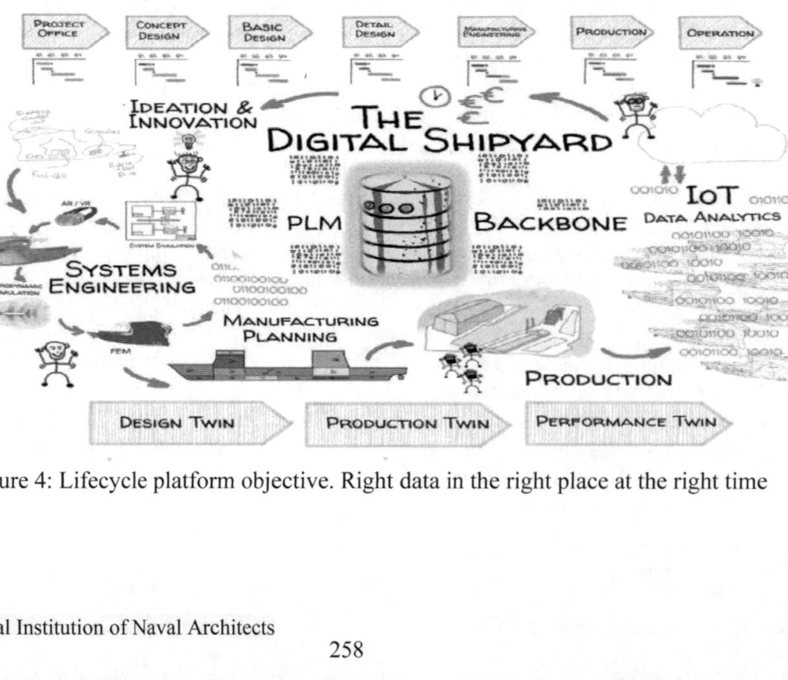

Figure 4: Lifecycle platform objective. Right data in the right place at the right time

Once in operation, lifecycle management platforms serve as the backbone for condition-based monitoring, predictive maintenance, and operational analytics as it can be shown in *Figure 4*. Integrated with digital twin systems and onboard sensors, they collect and organize operational data, such as propulsion efficiency, fuel consumption, software performance, and system diagnostics, into accessible dashboards. This facilitates early detection of performance degradation, automates maintenance scheduling, and reduces downtime through targeted interventions.

For autonomous vessels, this operational transparency is especially critical. Lifecycle platforms help manage the complexity of distributed systems that span propulsion, navigation, communication, and AI control modules. By consolidating telemetry and health status reports from multiple systems into a centralized repository, they enable holistic diagnostics, remote fault management, and secure software updates over the air. In turn, this supports a higher level of automation and reliability without direct human oversight onboard. Another key benefit of lifecycle integration is compliance management. Regulatory landscapes for autonomous vessels are evolving rapidly, and lifecycle systems can track and document compliance with safety protocols, cybersecurity measures, emissions regulations, and functional performance standards. This audit trail is essential for gaining and maintaining approvals from classification societies, flag states, and international bodies.

Lifecycle management also supports sustainability goals by enabling detailed environmental impact assessments and end-of-life planning. By tracking emissions, resource utilization, and system wear over time, stakeholders can make informed decisions about retrofitting, refurbishing, or retiring assets. End-of-life data, such as material recyclability and disposal pathways, can be incorporated early in the design process to minimize environmental harm.

Furthermore, integrated lifecycle platforms facilitate knowledge retention and institutional learning. As autonomous vessels accumulate operational history, lessons learned can be codified into future design iterations, software updates, and crew training protocols. The platform serves as a living knowledge base, reducing duplication of effort and enabling continuous improvement across fleets.

Finally, these systems are foundational for enabling fleet-wide optimization. When multiple autonomous vessels are managed through a common lifecycle management framework, fleet operators can compare performance metrics, benchmark best practices, and coordinate interventions at scale. This not only enhances the economic efficiency of operations but also ensures uniform safety and compliance across geographically distributed assets.

In summary, integrated lifecycle management systems are indispensable for achieving the long-term viability, safety, and environmental sustainability of autonomous maritime fleets. By providing a unifying digital infrastructure that spans the entire asset lifecycle, these platforms enable smarter design, proactive maintenance, robust compliance, and data-driven decision-making in an increasingly complex maritime environment.

3. STRATEGIC BENEFITS OF DIGITAL INTEGRATION

Lifecycle management platforms are essential for orchestrating the full continuum of activities associated with maritime assets, from initial concept and design, through construction and commissioning, to operation, maintenance, and eventual decommissioning. In the context of autonomous vessels, these platforms provide a digital thread that ensures continuity, coherence, and coordination across all phases of the vessel's existence.

At the design and engineering stage, integrated lifecycle systems enable the centralization of technical specifications, digital models, and regulatory requirements. By linking CAD models, engineering simulations, and compliance documentation into a unified data environment, these platforms facilitate collaborative design processes and reduce errors due to information silos. Version control, requirements traceability, and real-time design validation ensure that every component and subsystem is configured correctly for autonomous operation and meets international standards.

During vessel construction, lifecycle platforms provide a framework for quality assurance and component certification. Digital records of materials, manufacturing tolerances, and subsystem integration steps are maintained, creating a comprehensive construction log. This digital provenance supports post-construction troubleshooting and streamlines the certification process with class societies and maritime regulators, particularly for novel autonomous systems that may lack conventional benchmarks.

Once in operation, lifecycle management platforms serve as the backbone for condition-based monitoring, predictive maintenance, and operational analytics. Integrated with digital twin systems and onboard sensors, they collect and organize operational data, such as propulsion efficiency, fuel consumption, software performance, and system diagnostics, into accessible dashboards. This facilitates early detection of performance degradation, automates maintenance scheduling, and reduces downtime through targeted interventions.

For autonomous vessels, this operational transparency is especially critical. Lifecycle platforms help manage the complexity of distributed systems that span propulsion, navigation, communication, and AI control modules. By consolidating telemetry and health status reports from multiple systems into a centralized repository, they enable holistic diagnostics, remote fault management, and secure software updates over the air. In turn, this supports a higher level of automation and reliability without direct human oversight onboard.

Another key benefit of lifecycle integration is compliance management. Regulatory landscapes for autonomous vessels are evolving rapidly, and lifecycle systems can track and document compliance with safety protocols, cybersecurity measures, emissions regulations, and functional performance standards. This audit trail is essential for gaining and maintaining approvals from classification societies, flag states, and international bodies.

Lifecycle management also supports sustainability goals by enabling detailed environmental impact assessments and end-of-life planning. By tracking emissions, resource utilization, and system wear over time, stakeholders can make informed decisions about retrofitting, refurbishing, or retiring assets. End-of-life data, such as material recyclability and disposal pathways, can be incorporated early in the design process to minimize environmental harm.

Additionally, integrated lifecycle platforms facilitate knowledge retention and institutional learning. As autonomous vessels accumulate operational history, lessons learned can be codified into future design iterations, software updates, and crew training protocols. The platform serves as a living knowledge base, reducing duplication of effort and enabling continuous improvement across fleets.

Finally, these systems are foundational for enabling fleet-wide optimization. When multiple autonomous vessels are managed through a common lifecycle management framework, fleet operators can compare performance metrics, benchmark best practices, and coordinate interventions at scale. This not only enhances the economic efficiency of operations but also ensures uniform safety and compliance across geographically distributed assets.

Integrated lifecycle management systems are indispensable for achieving the long-term viability, safety, and environmental sustainability of autonomous maritime fleets. By providing a unifying digital infrastructure that spans the entire asset lifecycle, these platforms enable smarter design, proactive maintenance, robust compliance, and data-driven decision-making in an increasingly complex maritime environment.

4. CHALLENGES AND FUTURE DIRECTIONS

Despite the transformative promise of digital innovation in the maritime domain, several critical challenges must be addressed to realize the full potential of autonomous, smart, and sustainable vessels. These challenges span technical, regulatory, operational, and organizational dimensions, and overcoming them will require coordinated efforts across the global maritime ecosystem.

One of the most pressing issues is data interoperability. Autonomous vessels generate massive volumes of heterogeneous data from onboard sensors, control systems, and external sources such as satellite feeds and maritime traffic services. Ensuring seamless integration and standardized communication between disparate systems remains a significant hurdle. Proprietary data formats, legacy systems, and inconsistent interfaces can hinder effective data sharing, slowing down innovation and reducing the effectiveness of integrated digital platforms. The development and adoption of open standards and maritime-specific data ontologies will be crucial in enabling robust, scalable digital infrastructures.

Cybersecurity poses another substantial risk. As vessels become increasingly connected and reliant on digital control systems, the potential attack surface for cyber threats expands. Autonomous systems must be secured against unauthorized access, data tampering, spoofing of navigation signals, and malicious software intrusions.

Cybersecurity frameworks must evolve to address these maritime-specific threats, incorporating multi-layered defences, secure-by-design engineering principles, and continuous monitoring of system integrity.

Furthermore, the establishment of international cybersecurity certification standards will be essential to ensure global compliance and trust in autonomous maritime operations.

Regulatory complexity is a third major challenge. Existing maritime laws, safety protocols, and liability frameworks were developed for conventionally crewed vessels and are not readily applicable to autonomous systems. Classification societies and flag states are actively working to define new guidelines, but regulatory harmonization across jurisdictions remains a long-term goal. Ambiguities around the legal status of autonomous vessels, human oversight responsibilities, and accountability for machine-made decisions complicate deployment. To navigate this landscape, continuous dialogue among industry stakeholders, regulators, and legal experts is required to develop adaptive, future-proof governance models.

Validation and verification of autonomous functionality present technical and procedural challenges. Unlike conventional systems, AI-driven behaviours cannot always be exhaustively tested using traditional deterministic approaches. The dynamic, probabilistic nature of ML algorithms demands new validation methodologies, such as scenario-based testing, model checking, and digital simulation in complex virtual environments. These methods must be rigorously defined and accepted by regulatory authorities to ensure safety and reliability.

Operationally, the transition from crewed to uncrewed or hybrid vessels introduces logistical and cultural challenges. Maritime personnel must be reskilled to manage remote operations, interpret AI decision-making outputs, and intervene in edge cases. New operational protocols must be established to support remote command centres, autonomous docking, and maintenance regimes for highly digitized platforms. Furthermore, resistance to change among traditional stakeholders may hinder the adoption of innovative solutions, necessitating change management strategies and stakeholder engagement.

From a technology readiness perspective, several enabling technologies are still maturing. Reliable long-range communications, high-precision situational awareness sensors, and resilient edge computing platforms are essential for safe and autonomous navigation in real-world conditions. Edge computing, in particular, will play a critical role in ensuring that decision-making remains reliable even in environments with limited connectivity. Ongoing advancements in sensor fusion, satellite communication, and embedded AI systems will be instrumental in closing these technological gaps.

Looking ahead, the future direction of digital maritime innovation will likely involve increased hybridization of operations (Toman, et al., 2024). Rather than a binary transition to full autonomy, vessels may operate under variable levels of autonomy depending on mission type, location, and regulatory context. Human-in-the-loop and human-on-the-loop models will provide oversight during high-risk scenarios while allowing autonomous systems to manage routine tasks.

Fleet-level intelligence and collaboration will also shape future developments. Vessels will increasingly operate as nodes within larger interconnected networks, sharing data, coordinating routes, and optimizing energy usage across entire logistics chains. Cloud and edge computing infrastructures will support this distributed intelligence, enabling real-time responsiveness and long-term optimization.

Environmental imperatives will further drive innovation. Digital tools will be leveraged to support decarbonization strategies, optimize fuel alternatives, and monitor compliance with increasingly stringent environmental regulations. Integration with renewable energy systems, autonomous waste management protocols, and lifecycle-based environmental impact assessments will become standard features of sustainable autonomous fleets.

So, while the road to autonomous, smart, and sustainable maritime fleets is filled with challenges, it also presents an unparalleled opportunity for systemic transformation. Addressing these challenges demands a multi-disciplinary approach involving technical innovation, regulatory foresight, and cultural adaptation. Through sustained investment, international collaboration, and strategic foresight, the maritime sector can overcome current barriers and usher in a new era of intelligent, efficient, and resilient operations.

5. CONCLUSION

Digital innovation tools form the foundation for the successful realization of autonomous, smart, and sustainable maritime fleets. These tools are not only technical enablers but also strategic instruments that align with the broader imperatives of environmental stewardship, economic competitiveness, and global maritime resilience. The integration of digital twins, AI, advanced simulations, and integrated lifecycle platforms enables a paradigm shift in how vessels are conceived, constructed, operated, and maintained.

Digital twins provide a persistent, high-fidelity digital mirror of the physical vessel, offering real-time insights into system health, predictive maintenance needs, and performance optimization opportunities. In autonomous systems, they are indispensable for validating behaviour in diverse operational scenarios and ensuring the integrity of critical control systems throughout the vessel's lifecycle.

AI and ML inject intelligence into maritime platforms, equipping them with the ability to learn from environmental feedback, optimize decision-making processes, and adapt dynamically to changing operational contexts. These technologies allow autonomous vessels to go beyond automation, transforming them into cognitive agents that respond with agility and precision to real-world conditions.

Advanced simulation and modelling environments offer a virtual proving ground for innovation, allowing engineers and operators to refine control logic, evaluate safety protocols, and stress-test integrated systems across thousands of synthetic scenarios. These simulations significantly reduce risk, compress development timelines, and minimize the cost of iterative refinement.

Integrated lifecycle management systems ensure that the wealth of data generated across the vessel's operational continuum is effectively harnessed to support regulatory compliance, operational transparency, and continuous improvement. These platforms enable the coordination of multi-disciplinary teams, facilitate remote diagnostics and software updates, and lay the groundwork for data-driven fleet management strategies. Taken together, these technologies empower stakeholders to design and deploy vessels that meet the highest standards of safety, efficiency, and environmental responsibility. They also support novel business models, such as autonomy-as-a-service, and prepare the industry for evolving market demands, including green logistics and smart port integration.

Importantly, the benefits of digital integration extend beyond individual vessels to encompass entire fleets and global supply chains. Through shared digital ecosystems, data interoperability, and cloud-based collaboration platforms, maritime enterprises can synchronize operations, streamline logistics, and respond more effectively to global challenges such as climate change, labour shortages, and geopolitical disruptions.

As the maritime sector faces mounting pressure to decarbonize, increase resilience, and embrace automation, the continued integration and advancement of digital technologies will be vital. This requires sustained investment in research and development, forward-looking regulatory frameworks, and an inclusive approach to workforce transformation.

Ultimately, digital innovation is not merely a toolkit, it is the architecture upon which the next generation of maritime operations will be built. Through thoughtful implementation, industry-wide collaboration, and strategic foresight, digital tools will unlock a future in which autonomous maritime fleets operate safely, sustainably, and intelligently across the world's oceans.

6. REFERENCES

1. BENAYAS-AYUSO, A., et al., 'CAD Sensitization, an Easy Way to Integrate Artificial Intelligence in Shipbuilding', *Computers*, 2024. ISSN 2077-1312

2. GÓMEZ-RUIZ, M.A., et al., 'Application of Machine Learning Techniques to the Maritime Industry', *Journal of Marine Science and Engineering*, 2023. ISSN 2077-1312

3. PÉREZ FERNÁNDEZ, R., 'Industry 4.0 in Shipbuilding. Application in Autonomous Ships', *20th International Conference on Ship & Maritime Research*, 2022. ISBN 978-1-64368-296-9.

4. TOMAN, M., et al., 'Backing shipbuilding from the metaverse', *Revista Ingeniería Naval, 2024.* ISSN 0020-1073.

7. AUTHORS BIOGRAPHY

Rodrigo Pérez Fernández holds both PhD and MSc in Naval Architecture & Marine Engineering. He has also completed an MBA and several postgraduate programs in Shipbuilding, Maritime Transport and Executive Management. He is currently Senior Director for Software Engineering at Siemens Digital Industries Software. He previously worked at Siemens as Global Portfolio Development Director.

EVERY FRIGATE A SQUADRON: STUDIES OF LARGE UNCREWED SURFACE VESSELS

R J Pawling, University College London, UK

SUMMARY

Naval Uncrewed Surface Vehicles (USVs) to date have generally been around the size of existing Rigid Inflatable Boats and workboats. The main exceptions to this are the US Navy's Medium and Large USV (M/LUSV), in the form of the trimaran Sea Hunter and monohull Nomad craft. These experimental vessels have displacements of hundreds of tonnes and are being used to develop both single role large USVs (Sea Hunter) and the concept of a general purpose MUSV as a kind of "force adjunct". This paper describes UCL studies into the use of M/LUSV to support a range of naval tasks in mixed forces of crewed and uncrewed vessels, through spreadsheet based initial sizing and a more detailed examination of the adaptation of existing commercial vessels for use as rapidly produced MUSVs.

1. INTRODUCTION

Naval Architecture and Marine Engineering (NAME) and Marine, Autonomy and Robotic Systems (MARS) are subdivisions of the UCL Department of Mechanical Engineering [1]. NAME provides postgraduate and undergraduate education to naval architects and marine engineers including short courses, whilst MARS is a research grouping. Research and teaching interests overlap, with student design projects featuring drawing from MARS areas of interest. Formally known as the Marine Research Group (MRG) the new title MARS indicates the increasing importance of autonomy in all aspects of maritime activities and operations. The author has previously contributed to papers looking at ship design impacts [2], whilst recent research in MARS has seen particular attention to questions of control e.g. [3] [4]. USVs and UUVs are also becoming a part of the education offering at UCL via student projects.

2. UNCREWED SURFACE VEHICLES

Like so many technologies, uncrewed surface vessels are not a new idea, obsolete battleships having been converted to radio control for use as mobile targets in the 1920s [5]. However they are a new implementation. With smaller USVs of a few tonnes displacement having long been in service as minehunting drones [6] the latest developments have included various levels of autonomy. Improved equipment reliability being possibly the key technology enabling current experimental deployments of medium and soon large displacement USVs.

Before proceeding, a note on taxonomy is required. The US Navy (USN) adopted a taxonomy broadly based on size to describe the types of USVs that might be carried and deployed by other vessels [7]:

- X-class: Small, non-standard craft launched from manned RiBs to support special forces.
- Harbour Class: Based on an in-service 7m RiB and thus compatible with most USN vessels.
- Snorkler Class: A semi-submersible craft in the 7m range.
- Fleet Class: A high capability USV with similar overall characteristics to the existing 11m USN RiB.

This taxonomy highlights the constraints on these craft of being compatible with existing boat handling systems as they are generally the payload of some otherwise conventional vessel. The size envisioned for USVs has increased rapidly in recent years, with two new definitions: Medium (M) and Large (L) USV. These are more simply defined based on size [8]:

- Medium, MUSV: less than 200ft (61m) in length and 500 tons (454 tonnes) in displacement
- Large, LUSV: 200-300 ft (61-91m) in length and 1000 tons (910 tonnes) in displacement

As with the development of smaller USVs, whilst bespoke M/LUSVs have been built, e.g. the Sea Hunter trimaran [9], much of the early work has made use of conversions or derivatives of existing commercial vessels, in particular the fast crew supply vessels used to support oil and gas platforms in the Gulf of Mexico [10]. These vessels retain the design features for crewed operation, with their large cargo deck simplifying the fitting of experimental equipment via containers.

Within the LUSV category is NOMARS (No Manning Required, Ship). Whilst the size of the first demonstrator, USX-1 Defiant, is in the MUSV range (55m, 218 tonnes) larger derivatives are already being proposed [11]. Most notably, the NOMARS programme is not a technology demonstrator for payloads or operations, but rather an effort to develop autonomous vessels from scratch [11] with a range of claimed advantages including survival in high seas, hydrodynamic efficiency and lower cost. The latter remains to be proven as, at a claimed $25 million USD excluding mission systems [12], Defiant appears to have a similar cost per tonne as most conventional warships when a typical combat systems proportion of 40% is assumed (87k£/te for Defiant, 105k£/te for Type 26 Batch I, 68k£/te for Type 26 Batch II). However the hydrodynamic efficiency aspects are apparent in the slender hulls, as illustrated in Figure 1.

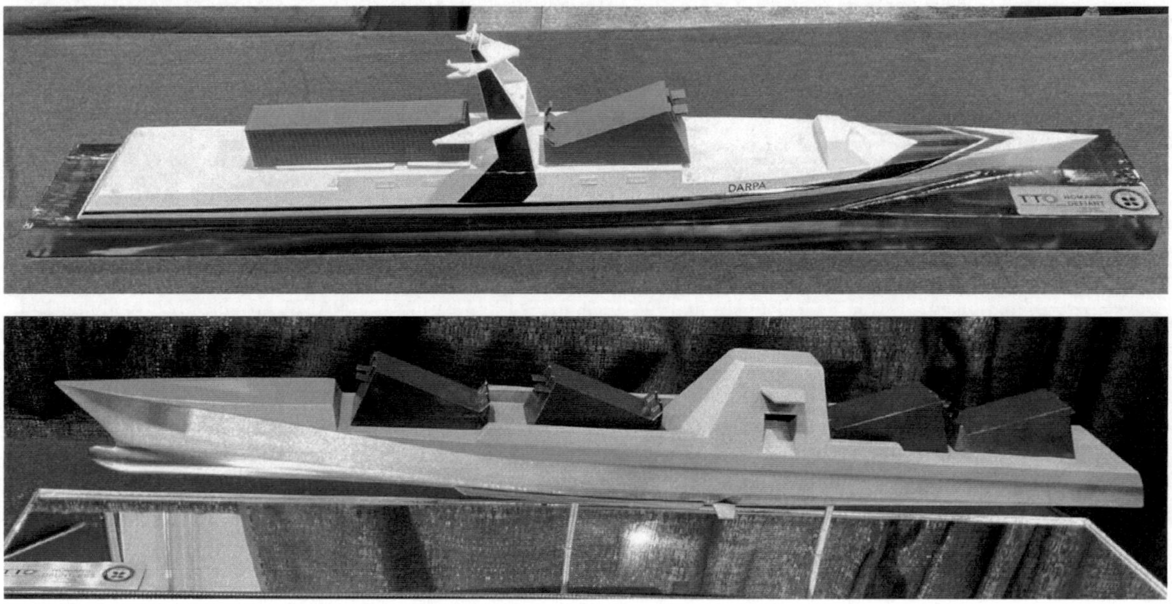

Figure 1: Top; A model of the Defiant USV and Bottom: The larger Dauntless proposal, showing four Adaptable Deck Launchers (ADL) [13]

3. SIZING MODEL STUDIES OF USV FLEETS

The studies described in this paper began with the use of the UCL sizing database to examine bespoke monohull LUSVs, along the lines of NOMARS. Whilst this model is largely based on frigate designs from the 1980s it has been demonstrated to produce acceptably accurate results at the high level [14]. Inspired by the use of high payload fraction platform support vessels as the basis for USN MUSV experiments, the roles selected for examination revolved around carriage of additional materials, in magazines or in stores.

3.1 LUSVS AS MAGAZINES

One possible role for LUSVs is to provide additional magazine capacity for crewed vessels. Of course, this increases both UPC and operating cost when compared against buying a single larger vessel due to the efficiencies of scale. Table 1 compares some characteristics of three approaches to providing an air defence destroyer capability with a total of 144 VLS cells. The first is a combination of a crewed ship with 48 cells and an LUSV with 96 cells, the second is an enlarged destroyer with all 144 cells in one ship and the third is a baseline DDG with two LUSVs each carrying 48 cells. A group of three 48 cell DDGs was also included as a reference point. Unsurprisingly, the multi-vessel options have a higher UPC and fuel burn.

It should be noted that for this study, the LUSV was a single-zone vessel with dual-shaft CODAD propulsion and maintenance access reduced to 50% that of the crewed vessel. Although systems were stripped down, weights and costing were generally based on naval standards, due to the small vessel size limiting design for producibility features such as access for outfitting. The overall cost per tonne displacement compare reasonably well with the much smaller Defiant.

The overall density value is included to illustrate a design issue with monohull LUSVs; without the volume requirements of access and accommodation, they may be very dense, with a reduced reserve of buoyancy and thus survivability against flooding. In this example the weapon systems were very heavy (MK41 VLS) however it should be noted that, despite the heavy exhaust plenum required to withstand a restrained firing, Mk41 is still a similar mass to the Adaptable Deck Launcher (ADL) shown on the NOMARS models in Figure 1 - 4.55 tonnes per Tomahawk for an 8 cell MK 41 vs 5.13 tonnes for a strike length ADL with LRASM [15], and the latter does not require internal volume, further increasing the effective overall density of the vessel.

Table 1. Comparison of numerical models of LUSV options

		48 cell DDG	96 cell LUSV	Combo 1	144 cell DDG	Combo 1/ Large DDG	Combo/3 x 48 cell DDG	48 cell LUSV	Combo 2	Combo 2/ Large DDG
UPC	**M£**	535	230	766	711	**1.08**	**0.48**	147	830	**1.17**
Internal Vol	**m3**	21139	3512		23022			2888		
Deep disp	**te**	5634	1560	7194	6377	**1.13**	**0.43**	1272	8178	**1.28**
Density	**te/m3**	0.267	0.444		0.277			0.440		
Cost Excl Payload	**k£/te**	56.6	62.9		56.0			63.0		
		MWshaft						**MWshaft**		
Speed, knots	**10**	0.74	0.35	1.09	0.80	**1.37**	**0.49**	0.31	1.36	**1.71**
	15	2.65	1.38	4.04	2.83	**1.42**	**0.51**	1.24	5.13	**1.81**
	28	32.48	17.66	50.15	33.80	**1.48**	**0.51**	15.70	63.89	**1.89**

One approach to address this is to add void volume. As with previous UCL studies into alternative fuels [14] this was added as pure void volume – contributing only structure and coatings, but not lighting, access, ventilation etc. Table 2 below shows how the addition of a relatively modest amount of void volume improves freeboard to values of a missile carrying LUSV to those similar to an OPV.

Table 2. Effect of adding void volume on LUSV proportions

		Baseline	+1000 m3	+2000 m3	+3000 m3
Lwl	**m**	80.5	83	84.9	86.7
Bwl	**m**	11.1	11.4	11.7	11.9
D	**m**	5.9	7.2	8.3	9.4
T	**m**	3.4	3.5	3.6	3.7
F	**m**	2.5	3.7	4.7	5.7
Density	**te/m3**	0.444	0.359	0.31	0.277
UPC	**M£**	230	235	239	243
Displacement	**te**	1560	1706	1827	1946

Some reporting on LUSVs has focussed on their low UPC compared to conventional crewed vessels and of course this is deceptive. Figure 2 compares the cost breakdowns including payload for the three vessels in the AAW DDG example. Unsurprisingly, being a very basic vessel the cost of the LUSV is dominated by the payload cost, which reporting on current programmes does not include. This provides us with a potential explanation as to the focus on arguably technically and financially inefficient modularised systems such as the ADL and Mk70 for USN LUSV programmes; they allow the true cost of the LUSV capability – which must include their payload – to be distributed across other procurements, such as adoption of ADL on destroyers etc.

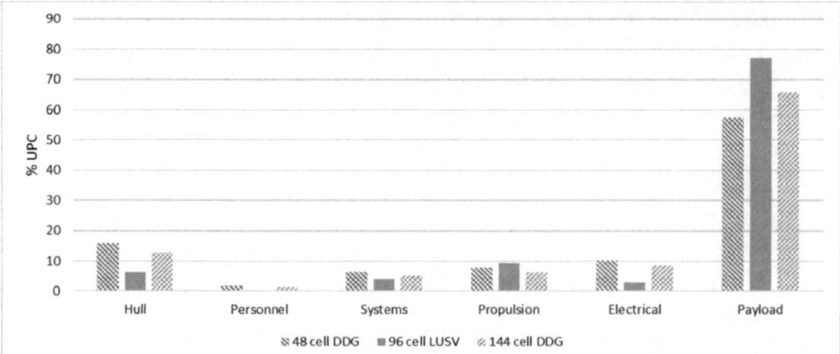

Figure 2: Cost breakdowns for destroyer and LUSV designs in the numerical study

3.2 LUSVS AS RAS SHUTTLE

Another potential role for LUSVs is as an intermediary for Replenishment At Sea (RAS). Whilst the US Navy has demonstrated the refuelling of MUSVs at sea, a more interesting option may be the use of the USV to transport fuel and stores from a replenishment vessel to combatants dispersed and "up threat". UCL has examined the replenishment ship design aspects of such concepts through the medium of the MSc Ship Design Exercise but those made use of generic "shapes" for UXVs based on existing vessels and aircraft.

Table 3 below compares some principal particulars for a range of possible USV concepts for logistics operations. Tankers were considered with two RAS rigs and 1000m3 of tankage for F-76. The baseline tanker included 2000m^3 of additional void space. The dry cargo vessel carries 1000m^3 of palletised cargo, has 2000m^3 of void volume and a stabilised crane based on the BM-T40 [16]. In all cases the endurance on ships fuel was reduced to 15 days.

In the case of fuel tankers, the issue of high density and low freeboard appears again. However, the resistance of a tanker to flooding can potentially be quite high, as the change in density from diesel fuel to seawater is relatively small. In practice the concern would be regarding seakeeping, and in particular the ability for RAS gear to function. One possibility might be to adopt a semi-submersible / low freeboard "whaleback" configuration, with the upperworks concentrated into narrow superstructures that could be proportionally taller as illustrated in Figure 3 This configuration was operated on the Great Lakes and was reasonably successful [17] but ultimately poorly matched to the dry bulk trade in ores that dominated that route.

Table 3. Characteristics of some logistics LUSV concepts

		1000m3 Tanker	**1000m3 Whaleback**	**1000m3 Dry Cargo**	**"Commercial" Whaleback**
Lwl	**m**	87.6	81.0	80.3	81.0
Bwl	**m**	12.1	11.1	11.0	11.1
D	**m**	8.0	4.5	8.6	4.5
T	**m**	3.7	3.4	3.4	3.4
F	**m**	4.3	1.1	5.2	1.1
Density	**te/m3**	0.336	0.518	0.279	0.522
UPC	**M£**	82	60	83	48
Displacement	**te**	2011	1588	1545	1631

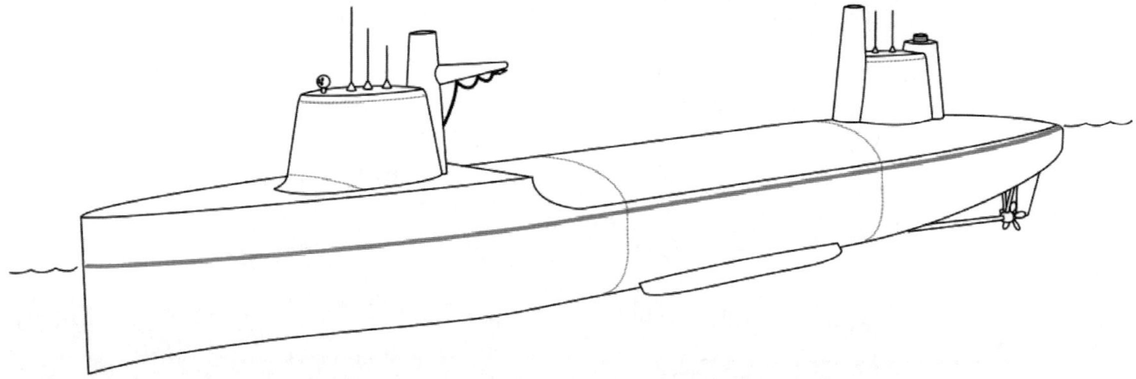

Figure 3: Illustration of a low profile USV carrying 1000m^3 of fuel with RAS rigs forward and aft

Whilst these vessels might be technically viable, the UPCs are very high, being based on naval and small vessel construction costs. The "commercial" costings are based on passenger vessels, the lowest costing rates in the UCL database. These compare poorly with typical costs of newbuild commercial tankers, where the same price would buy a vessel of tens of thousands of tonnes capacity [18]. Those are, however, producibility optimised vessels where lightweight is of little concern due to the very high deadweight capacity and the size allows for easy access for outfitting. They are also built in yards with lower labour rates and much lower overheads for supervision and quality control. It remains as a future exercise to determine just how much cheaper an LUSV logistics vessel acceptable for naval use could be made.

3.3 SO WHY USE BESPOKE LUSVS?

Whilst the payload cost accounting described above may initially seem deceptive (and given the nature of defence procurement it is this authors view that that is probably an "added bonus"), pragmatically, accepting a higher price tag for some discrete items (such as modularised weapons systems) may still result in a lower cost to deliver a capability at sea within a finite timeframe. There are several reasons why an LUSV may be attractive, but these must be considered at not just whole fleet but potentially whole-force level:

3.3 (a) Survivability of Mission Capability

Even though the LUSV (or more specifically, its payload) is not cheap, when it is used as a magazine it may be less mission critical than the sensors and crew on the DDG. This can improve survivability of the critical capabilities by simple dilution of enemy attack over more targets, keeping them further back from the front line etc.

3.3 (b) Sustainment and Reloading

Whilst the DDG+LUSV combination costs more to buy and run, this expenditure is providing extra capability. The weapons on the LUSV can be expended first, with it then returning to base to reload. This leaves the DDGs sensors and decision makers in theatre to either use the ships own weapons or those of other LUSVs.

3.3 (c) Exploiting Modularisation

Modularised weapons or sensors can potentially be used by multiple projects or branches of the armed forces – modularisation will make the launcher heavier and more expensive (due to the cost of the module) but the cost is now shared. However, key to this is making the complete launcher or sensor modularised, not just the weapon or components thereof (e.g. CAMM, a nominally common missile used in two very different air defence systems – Sea Ceptor and Sky Sabre). And this is likely to further decrease the physical or financial efficiency of the implementation at the unit level.

LUSV are likely to be more suitable for this type of nearly "stand-alone" modularisation in that they lend themselves to a "container ship" type approach featuring a minimalist core vessel. The economics of crewed or high-capability vessels will inevitably introduce many additional upperdeck weapons and electronics, all with physical, electromagnetic and safety interactions. As Figure 4 humorously illustrates, this has long been the case, with all warships gaining many additional systems over the course of a conflict.

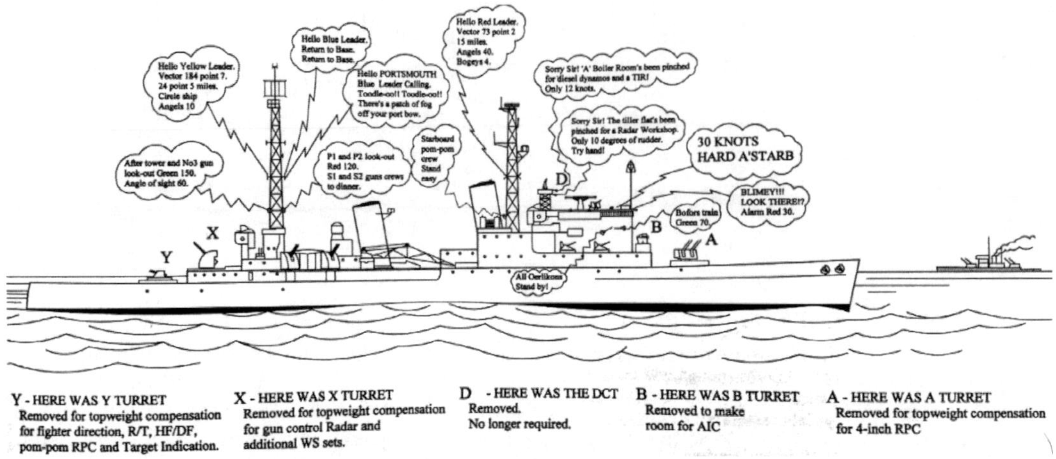

Figure 4: Even in an industrial war of numbers, individual unit capability has always led to warships gaining extra equipment (From [19])

3.3 (d) Production Benefits

Modularisation can bring production benefits by separating detail design processes through the barrier of the module limiting design interaction. USVs however may bring additional production benefits if they allow non-traditional suppliers to feed into military procurement. The author examined this with regards to warship production in a previous paper [20] where full containerisation of an otherwise conventional frigate was proposed as a means to allow alternative suppliers to be used in the event of "grey zone" and unconventional warfare against military prime contractors. This has a historical precent; the Ford Motor Company built large numbers of Eagle class submarine chasers in the First World War, International Harvester (a farm equipment manufacturer) produced trucks and rifles for the US Army in the Second. If autonomous vessels can accept more commercial standards then they may have a substantially wider production base. Some key aspects of this would be avoiding significant changes to existing products, and not requiring commercial builders to adopt methods or equipment they are unfamiliar with.

4. USVS ADAPTING EXISTING DESIGNS

With the changing security situation in Europe it is not inconceivable that the UK might need to rapidly increase the mass of naval forces through increasing numbers. USVs may have a role to play, but if they are highly bespoke, optimised systems, production rates are likely to still be low. A possible solution is to use existing commercial vessel designs with the minimum change, accepting that the design is inefficient compared to a bespoke design, but is available and can be built rapidly. The UK has several small craft manufacturers (< 24m) and one vessel type of interest is the catamaran Crew Transfer Vessel (CTV), used to transport maintenance crew to offshore wind turbines.

4.1 THE BASELINE MODEL

To examine the possibilities for repurposing CTV production as MUSVs for use around the UK and northern Europe, a baseline design was developed in Paramarine, roughly based on the Diverse Marine 24m CTV [21]. Figure 5 shows the baseline design and Table 4 includes some principal particulars. With the general arrangement fixed by the reference, weight was calculated using outfit densities derived from UCL data. Resistance estimates matched the installed power (2014kW vs 2160kW) giving some confidence in calculations of range at varying speed.

Figure 5: The baseline CTV model

Table 4. Baseline characteristics

Characteristic	Value	Calculated / Reference
Tonnes deep	86te	C
Waterline length	21.5m	C
Overall beam	8.63m	R
Payload	10te	C
Speed, max	27 knots	R
Speed, service	23 knots	R
Range at 27 knots	820nm	C
Range at 23 knots	984nm	C
Range at 14 knots	1527nm	C
Electrical generation	2 x 27kW	R
Material	Aluminium	R

The Paramarine model was then used to evaluate the impact on total weight and longitudinal balance of not installing the superstructure and accommodation. Whilst the deck payload of 10 tonnes is useful by itself, what was of particular interest was the weight saved by not fitting the superstructure and its extensive internal fittings for the turbine technicians including seating, galley etc. Table 5 compares the baseline with a stripped version, showing that the weight reduction is significant, with approximately 30 tonnes now available within the displacement limit of 86 tonnes.

Table 5. Weight changes after removal of CTV role specific items

	Weight, tonnes	
	Baseline	**Stripped**
Hull	35.9	22.7
Structure	29.8	18.4
Accommodation	1.0	-
Ship Systems	3.0	2.8
Main Propulsion	13.1	13.4
Electrical	2.7	2.1
Mission Systems	1.6	-
Variables	29.1	14.4
Fuel	14.4	14.4
Deck Cargo	10.0	-
Total	86.5	55.4
LCG, m fwd of mid	-1.19	-2.03
VCG, m abl	2.69	1.57

4.2 THE USV SEAFRAME

To make use of existing manufacturing ability, the minimum changes should be made to the design. And where changes are required, they should be limited in extent (which modularity can help with) and in particular use solutions or methods that are familiar with the manufacturer or do not require learning new skills. There are however some basic alternations and additions that are needed to make the stripped down design suitable as an MUSV "seaframe".

4.2 (a) Endurance

CTVs usually operate at high Froude numbers, with a maximum speed of 28-32 knots and a cruise speed of 18-22 knots, with hullforms ranging from planning catamarans to displacement SWATH and semi-SWATH shapes. They have a relatively high installed power to weight ratio with high bollard pull for towing, but the use of truck-derivative high speed diesel engines leads to higher fuel consumption than other civilian vessels. Some relevant manufactures such as Scania are already active in the fields of autonomy [22] which demands high levels of reliability with minimal maintenance. However, in their COTS form, these engines usually need some form of manual intervention after around two weeks of running [23].

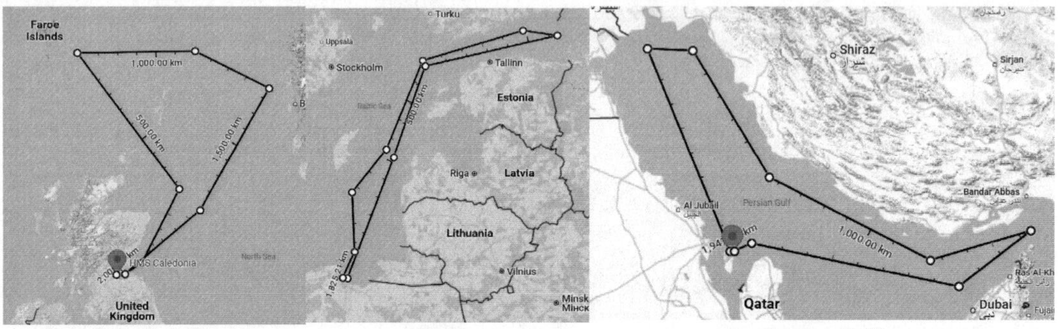

North Sea Baltic Persian Gulf

Figure 6: Examples of missions with 2000km / 1000nm range; maps from https://www.google.com/maps

As Figure 6 shows, whilst the baseline range is sufficient for some useful missions, an increase in range or more specifically endurance could be desirable compared to the baseline. This could be implemented by self-contained fuel barrel-shaped tanks added in the hull. Being self-contained COTS items these should be cheaper and easier to fit compared to enlarging the integral tanks. Placing them high in the hull allows for gravity drainage, potentially removing the need for a pump. To avoid interference with the vertical access amidships for the Seaframe these were located in the volume previously used for crew cabins, increasing endurance by 60%.

4.2 (b) Deck Loading

Another aspect is deck loading. Whilst CTVs will have some structural reinforcement in areas such as seatings for the bow fenders, small craft structural rules generally specify a minimum design deck loading of only 1te/m^2 [24] Some vessels are designed for higher loadings of up to 2te/m^2, but values of 1.5te/m^2 are more common. This is acceptable for the relatively low density cargoes carried by a CTV, where the empty weight of a ten-foot container may be much greater than the weight of equipment within it, however many naval weapons and sensor systems are much denser. Table 7 compares some modules and their possible deck loadings. These calculations assume non-containerised systems require an ISO-container compatible base frame, and the loadings calculated based on overall area and also on an assumption of 0.25*0.25m pads placed on a 10 by 8ft grid.

Table 7. Some possible module size and weights. Green cells are lower than 1.5te/m²

			Mass		**Deck Load**	
	Length	**Width**	**Equip**	**Base**	**Distributed**	**Pads**
Item	m	m	te	te	te/m2	te/m2
500kW DG	6.05	2.44	9.3	9.3	0.6	4.6
800kVA DG	6.06	2.44	16.0	16.0	1.1	8.0
1250kVA DG	6.06	2.44	19.6	19.6	1.3	9.8
1400kVA DG	12.19	2.44	25.4	25.4	0.9	8.5
CAPTAS NANO winch	3.50	3.20	5.0	6.3	0.6	3.1
CAPTAS handling system	6.40	4.40	15.0	17.5	0.6	7.0
Multistatic VDS HPU	1.60	2.60	2.2	3.5	0.8	2.3
Multistatic VDS Tow body winch	7.00	2.30	9.5	12.0	0.7	4.8
Multistatic VDS passive winch	2.70	2.30	5.0	6.3	1.0	4.2
Quad Mk141 with Harpoon lightweight	1.91	1.42	3.9	5.1	1.9	3.4
Quad Mk141 with Harpoon normal	1.91	1.42	6.3	7.6	2.8	5.0
Adaptable Deck Launcher (tactical + 16 ESSM)	7.21	3.86	18.8	21.3	0.8	8.5
Adaptable Deck Launcher (tactical + 4 VLA)	7.21	3.86	14.2	16.7	0.6	6.7
Adaptable Deck Launcher (strike + 4 LRASM)	8.08	3.86	20.5	23.0	0.7	9.2
NSM quad	1.91	1.42	3.9	5.2	1.9	3.4
MK41 8 cell strike with 8 tomahawk	2.62	3.43	36.4	37.6	4.2	25.1
MK41 8 cell tactical with 8 VLA	2.62	3.43	26.3	27.5	3.1	18.3
Mk70 with 4 Tomahawk	12.19	2.44	22.2	22.2	0.7	7.4

Although CTV decks are usually fitted with reinforced padeyes, presumably with a local strength effectively exceeding 1.5te/m², twistlock fittings are rarer and Table 7 indicates that point supports may not be sufficient to support weapons and sensors without the use of wide spreaders. Relying on uniform contact between a module and deck is dubious unless they are welded, but a possible solution might be to use steel-backed rubber sheets, readily available as wear liners, to transfer the load between containerised equipment and the deck. The weight of these was added to the USV seaframe. A downside to this approach is that equipment retention may have to be by lashing bars rather than simple twistlocks at each corner.

4.2 (c) Hotel Loads and Supply

An important consideration in the design of UXVs in general is the hotel services demands of the mission systems. For UUVs this is particularly important with regards to the total electrical demand and energy storage. For UUVs and USVs with long endurance, increasing hotel service demand leads to more complex hotel systems with their own reliability questions. Hotel load demands of combat systems can be complicated. Unlike a domestic refrigerator, for example, which using simple on-off control might only have two states (compressor running and compressor stopped), a simple fixed upperdeck missile launcher may have multiple loads, including:

- Missile self-test, programming and initialization electrical loads
- Launcher controller electronics including network and self-test hardware
- Tube door opening and closing motors, or ignition pulses for pyrotechnic actuators
- Tube door de-icing

- Tube barrel and interior heating
- Tube interior cooling via chilled water, cooling air supply or electrical supply to a local chiller
- Lighting for maintenance tasks

For example, the four-cell Adaptable Deck Launcher (Formerly the Cocoon launcher) requires [15]:

- 115 Vac 60 Hz 1 phase (Lighting) 2 kW
- 115 Vac 60 Hz 1 phase (Backup power for 440 Vac) 4 kW
- 115 Vac 60 Hz 1 phase (Launch control unit) 5 kW
- 440 Vac 400 Hz 3 phase 40 kW
- Cooling 10,000 Btu/hour
- Heating 10,000 Btu/hour
- Recirculation 560 ft3/min

Difficulties in accurately characterizing electrical loads are well known [25] [26] but the situation is not as bad as the numbers above would appear. Many electrical loads are likely worse-case transients that would be better handled by batteries than oversized generators. However those related to weather (heating and cooling) cannot be ignored and will need to be provided continuously for operation in anything other than a very limited range of latitudes.

4.2 (d) Seakeeping and Stability

CTVs are generally limited to significant wave heights of 1.5m for the actual personnel transfer operation, but for aluminium catamarans slamming on the underside of the cross deck is an additional limitation on transit. Some CTVs with SWATH hullforms or increased wet deck clearance offer slam free operation at up to 20 knots in 2m waves [27]. It was assumed that designers of CTVs will generally aim to prevent slamming at cruise speeds in at least 1.5m waves, to align with the transfer limitations. This could be represented as a speed reduction with wave height shown in Figure 7. Wave height data from [28] is summarised for five operational regions in Table 8, showing that the North Atlantic would remain challenging, with even larger vessels possible forced into a survival mode around 15% of the time. It should be noted that catamarans are generally very resistant to slamming in anything other than head seas, with much lower speed reductions experienced [29]. Another aspect of seakeeping is deck wetness. It was assumed that the CTV derived USV will retain the existing bulwarks, with an additional breakwater added on the foredeck. This was assumed to be aluminium and potentially secured via twistlocks.

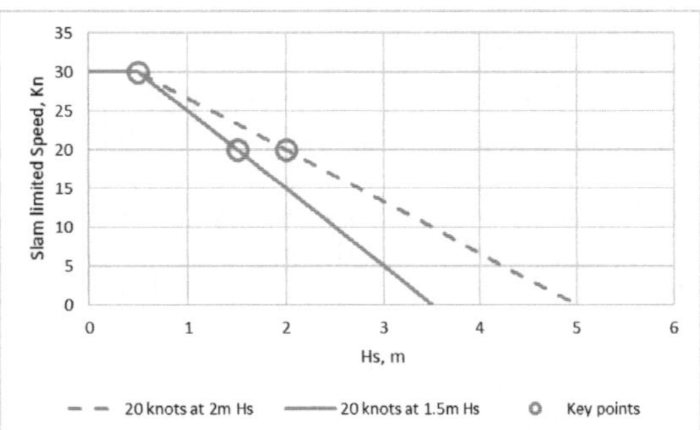

Figure 7: Representative CTV speed reduction with sea state

Table 8. Wave height occurrence data, all seasons / all directions

	North Sea	North Atlantic East	North Atlantic West	Irish Sea	Med
<=1.5m Hs	69	43	44	66	80
<=2m Hs	80	57	58	77	89
<=3.5m Hs	95	85	84	94	98

With a full-waterplane catamaran hullform, CTVs have a very high GM, between 5-10m. Some designs feature ballast water tanks in the box structure to reduce this, but in practice a large mass or free surface will be required to significantly reduce such a high GM. This will limit application of the CTV to certain roles: Recovery of UAVs may be difficult due to high roll rates; mechanical pointing devices will need a higher response speed; radars will need electronic stabilisation.

4.2 (e) Damage Control, Survivability and Structures

One possible advantage of the catamaran configuration is the inherent duplication and separation of propulsion and hotel systems. Each hull contains a propulsion drivetrain, diesel generator and half of the tankage. The baseline CTV should be able to make around 15 knots on one engine, although in practice heavy but survivable flooding of one hull leads to large angles of heel in catamarans, which would be more limiting on mobility.

A review of product datasheets for various CTVs showed that fixed firefighting systems are usually only fitted in the machinery spaces, and these would be retained. Whilst concepts such as using inert gas to create an anaerobic atmosphere in the hull are interesting, the added complication to the build, requiring airtight intake ducting to the engines and a generally higher quality of build, plus the addition of a system unfamiliar to CTV builders. For this study, the fixed firefighting system mass was scaled to protect the entire hull and was assumed to only consist of systems familiar to the builders (usually CO2 smothering or water spray systems). Glockling and Savage [30] have discussed several aspects of firefighting systems for USVs, with particular detail on sensors, but in the context of the adaptation of a commercial vessel, the solutions must fall within the experience of the builders. The addition of a large number of sensors will also lead to an increase in cost, as running cables requires a lot of labour hours.

Of course, commercial CTVs are almost all constructed of aluminium, which is generally unacceptable for warships due to loss of strength and even collapse in fires. However, this must be considered in the context of a very small vessel and broader survivability. With fixed firefighting systems likely to prevent the spread of accidental fires, and the small size of the MUSV leading to its loss after all but the smallest of missile impacts, what is the appropriate threat to design to survive, and will the structural material actually be important? A more significant concern with aluminium is possibly the maintenance of structural strength over time, particularly if operations are regularly extended into rough seas.

4.3 EXAMPLE APPLICATION

As per the title of this paper, the example application of the CTV conversion examined replacing the capability set of a patrol frigate operating in waters around the UK, extending west into the North Atlantic and east into the Baltic. The capability set was similar to that offered by both in service RN vessels (Type 23) and also various proposed designs (Type 24, OPV3, Castle class OPV upgrades) and is summarised in Table 9 below.

Table 9. Capability areas considered for the CTV-derived USV (Y = Yes, N = No, P = Partial)

Capability Area	Function	Viable on CTV USV	Notes
ASW	Detection	Y	Towed array or USV deployed UUVs.
	Localisation	P	Needs to be relatively fast response. Requires an air vehicle or another USV.
	Prosecution	Y	Air launched or rocket thrown torpedo, wire guided HWT.
	Processing	P	Fractional TFLOPS possible on board using a Beowulf cluster. Balance of onboard processing vs. raw data transmission TBC.
ASvW	Light strike	Y	Replaces helicopter launched ASM. Horizon-range weapons possibly derived from land systems e.g. FPV. Onboard targeting.
	Medium strike	Y	MOTS SSM.
	Light strike via guns	N	Not practical.
	OTH targeting	P	Surface duct ESM. Third party assistance required.
AAW	Self-defence (includes onboard ECM)	P	Difficult with a single USV due to deck space limitations. Requires operation in groups.
	Protection of escorted vessels	Y	Point defence missile limited in capability against crossing targets.
	Early warning to homeland	Y	Radar, ESM and acoustic sensors for raid detection and relaying to land based air defence.
SIGINT	Comms ESM	Y	Mast height and vessel length for triangulation need to be considered.
	Radar ESM	Y	
	Processing	P	Fractional TFLOPS possible on board using a Beowulf cluster. Main purpose is to feed some other decision making centre.
Marking	Marking of naval vessels	P	Lighter construction of CTV-USV and greater involuntary speed reduction limit application.
	Marking of grey zone vessels	P	

VBSS	Persuasion to halt	Y	Loudhailers, small calibre guns.
	Deployment of boarding teams	N	Not possible
Endurance	~30 Days at sea without RAS	N/P	Difficult with COTS equipment. Engines generally designed for manual maintenance activities after ~ 2 weeks
	~90 Days at sea with RAS	N	Not possible with COTS equipment.

Three functions have been examined in more detail; Anti Surface Vessel Warfare (ASVW), Anti-Submarine Warfare (ASW) and Anti Air Warfare (AAW). The additional modular systems required for each function are summarised in Table 10 and Figure 8 compares the three configurations proposed. The notional layout has all the modules on the centreline to keep existing vertical access and equipment hatches in the hulls amidships and aft clear, and also allow for lashings, fairings to reduce RCS etc. Longitudinal stacking also allows the widest arcs for sensors etc. This concentrates all the mass on the centreline however, which may exceed the design loadings for the cross deck structure. If this is the case then some containers would be staggered port and starboard, which is illustrated with the AAW option.

Table 10. Mission system fits for the options

Function	ASvW	ASW	AAW
Control and comms	1 x control quadcon 1 x SATCOM quadcon 1 x radio quadcon		
Sensors	1 x nav radar 1 x ESM quadcon	1 x nav radar 1 x lightweight TAS 1 x processing quadcon	1 x nav radar 1 x TRS-3D radar TEU
Weapons	2 x 4 MM40 SSM	2 x SeaHake HWT	8 – 12 CAMM, MICA or IRIS-T SL in 1 x quadcon
Infrastructure	Additional fuel tanks	Additional fuel tanks 27kW generator TEU Booster pump for fuel	Additional fuel tanks 27kW generator TEU Booster pump for fuel

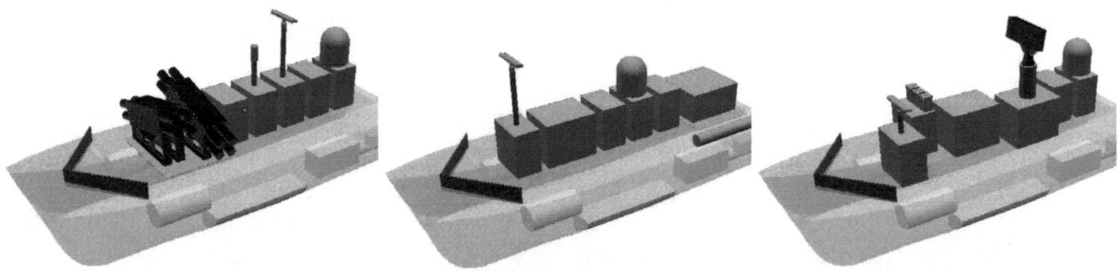

ASvW option ASW option AAW option

Figure 8: The three main options represented in Paramarine

The cramped nature of the CTV upperdeck, specifically the limitations on arrangement to limit changes to the basis design will have implications for mutual interference between systems. For example the generator module may cause vibration or exhaust ingestion issues for nearby modules (although the latter could be reduced at the cost of additional weight for mountings). In the AAW option, a soft-launched SAM such as CAMM could allow closer proximity to electronics, but the location of the navigation radar would no doubt see it damaged by repeated launches of hot launched weapons such as VL-Mica.

Table 11 summarises some high level parameters of the options compared to the baseline and stripped versions. This relatively simple study indicates that some of the main components of the capability set for a patrol frigate could be met by COTS CTV hulls with modular mission systems. Weight, stability and electrical supply are broadly within the limitations of the basis vessel. However the capability match is not one-for one.

Table 11. Principal characteristics of the CTV USV options

Variant	Baseline	Stripped	ASvW	ASW	AAW
Light, te	57.4	41.0	62.6	65.2	60.9
Deep, te	86.5	55.4	84.4	88.1	85.6
Mission systems, te			28.6	32.3	29.8
Containers only, te			5.2	6.46	8.76
Fuel, te	14.4	14.4	22.3	22.3	22.3
Generating capacity, kW	54	54	54	81	81
VCG, m abl	2.69	1.57	2.35	2.30	2.38
LCG, m fwd of midships	-1.19	-2.03	-0.67	-1.47	-0.77

Whilst the surface strike option is closely aligned, AAW requires the use of a small-warship 3D surveillance radar, has limited magazine capacity and a low mast height giving a short radar horizon (~60% that of a frigate in this example). The ASW option has limited magazine depth with only two weapons, and in all cases the utility of the MUSVs is highly dependent on their communication to some controlling centre, be it ashore, afloat or airborne.

5. DISCUSSION AND CONCLUSIONS

This paper has described two types of studies into medium and large USVs recently conducted at UCL as part of the ongoing operator-industry-academic "conversation" regarding these craft. These provide some interesting numerical comparisons of overall parameters, but caution must be exercised in using simple metrics to compare LUSVs vs crewed warships, or combinations. Remembering the series of "Comparative Naval Architecture" papers [31] that showed how design rationale and high level goals influenced the solution and performance in many ways, we note that simple comparisons can miss holistic or stylistic aspects that can have significant impacts on overall performance in ways not immediately obvious. Where style here is used as per Pawling et al [32]; a type of design decision (such as using aluminium structure) that can have cross-cutting impacts over a range of design and functional aspects.

Some of those stylistic aspects have appeared in the studies described here; for example the need for LUSVs to contain proportionally significant amounts of void volume to achieve adequate freeboard, or the CTV derived catamarans having limitations on layout (to reduce changes) that may propagate through to modular arrangements with mutual interference, difficulty in lashing etc.

The spreadsheet based comparison of some LUSV options indicates that, despite the efficiencies of scale that work against solutions using multiple craft, the dominance of combat systems in overall cost means the percentage impact on UPC may be acceptable (10-20%). Table 1 includes a comparison of a DDG plus LUSV to illustrate a key point; one has to consider what is actually needed in theatre. If the main need was for VLS cells rather than radars (for a strike mission for example) than the LUSV is a much cheaper option than additional fully equipped destroyers.

Regarding unit costs, it was observed that the bespoke LUSVs currently under development in the US appear quite expensive per tonne, as would be expected from a compact naval vessel, and much more than commercial vessels. This raises the question of how much cost reduction can really be expected of LUSVs intended to carry sophisticated combat systems. Of course Defiant's cost may be inflated by development costs, novel systems and being a smaller and "tighter" ship, but the comparison illustrates that claims of cost savings provided by USVs must be examined carefully. Notably the CTV study has not included cost, primarily because the UCL design database handles the cost of very small ships poorly. However at around £4.5 Million per unit [33] a CTV has a cost of around 40k£/te, meaning we need to consider what design features - or lack thereof - make these vessels cheap, and how this would impact their use as naval USVs. This may include aspects such as spartan furnishing but also minimal duplication of control and service lines, significantly increasing vulnerability (and potentially endurance in the face of finite reliability).

For the CTV studies, the mission systems chosen were those known to have a relatively low level of hotel systems dependency, reducing the "seaframe" systems required, but it was noted that the stylistic aspects of (combat) survivability and endurance (in the broadest sense) may be impacted. This is a trade-off that needs to include the operational benefits of having additional units, and consideration of whether it would actually be worth risking additional assets to recover a USV with more than light damage.

The need to link technical options with operations appears at several levels within the CTV studies. That concept arose from the specific consideration of how the UK could rapidly increase the number of sensor and weapon systems at sea using minimum change designs from non-traditional manufacturers – itself an operational concept (albeit a strategic one). What needs to be understood is how a notional gestalt "frigate of squadrons" would fare in operations. One obvious approach is wargaming, feeding into operational analysis.

7. REFERENCES

1. https://www.ucl.ac.uk/mechanical-engineering/research/marine-autonomy-and-robotic-systems-mars

2. PAWLING, R J., and ANDREWS, D J., 'Large Uninhabited Vehicles and the Minor War Vessel', *International Conference Warship 2013: Minor Warships*, RINA, Bath, June 2013.

3. GRANDE, D et al, 'Passive Fault-Tolerant Augmented Neural Lyapunov Control: A method to synthesise control functions for marine vehicles affected by actuators faults', *Control Engineering Practice*, Vol, 148, 2024.

4. ZHUANG, J et al, 'Navigating high-speed unmanned surface vehicles: System approach and validations', *Journal of Field Robotics*, Vol. 38, Issue 4, 2021.

5. Pamphlet No. M.129, On the Distant Control of a Target Ship (H.M.S. "Agamemnon"), Admiralty Pamphlet, 1920

6. BERTRAM, V, Unmanned Surface Vehicles – A Survey, 2007

7. USVMP Team, The Navy Unmannned Surface Vehicle (USV) Master Plan, July 23rd 2007

8. O'ROURKE, R., 'Navy Large Unmanned Surface and Undersea Vehicles: Background and Issues for Congress', CRS report R45757, December 2024, 2024.

9. https://www.leidos.com/capabilities/integrated-systems/autonomy-and-autonomous-solutions accessed on 2025-05-07

10. https://www.bairdmaritime.com/security/naval/unmanned-naval-systems/vessel-review-mariner-ranger-us-navy-launches-new-fleet-of-unmanned-test-vessels accessed on 2025-05-07

11. https://www.darpa.mil/research/programs/no-manning-required-ship accessed on 2025-05-07

12. https://nationalinterest.org/blog/buzz/the-navys-usx-1-defiant-drone-is-the-future-of-sea-warfare accessed on 2025-05-07

13. https://www.twz.com/sea/darpas-defiant-fully-uncrewed-demonstrator-ship-will-hit-the-seas-later-this-year accessed on 2025-05-07

14. PAWLING, R J., 'Examining the Impact of Future Alternative Fuels on Naval Vessels', *Int. Conf. Warship 2022: Designing Warships for the Future*, RINA, Bristol, June 2022

15. Product brochure, available from https://www.baesystems.com/en/product/adaptable-deck-launching-system accessed on 2025-05-07

16. https://www.barge-master.com/products/bm-t40/ accessed on 2025-05-07

17. LENGIEZA, J T., 'Vernacular in Curves: The Mythologizing of the Great Lakes Whaleback', East Carolina University, April 2016

18. https://www.rivieramm.com/news-content-hub/tanker-newbuilding-orders-continue-to-roll-in-83037 accessed on 2025-05-07

19. BROWN, D K., 'From Nelson to Vanguard: Warship Design and Development 1923-1945', Chatham Publishing, 2000.

20. PAWLING, R J., and BRADBEER, N., 'Pushing the Limit: The Role-Less Multirole Warship', *Int. Conf. Warship 2019: Multi-Role Vessels*, RINA, Bristol, 2019

21. 'Chartwell 24 Crew Transfer Vessel', product brochure, available from https://www.diversemarine.co.uk/crew-transfer-vessels accessed on 2025-05-07

22. https://www.scania.com/group/en/home/innovation/autonomous-solutions.html accessed on 2025-05-07

23. https://www.sbmarinemechanic.com/marine-diesel-engine-maintenance-schedule/scania-marine-engine-maintenance-schedule/ accessed on 2025-05-07

24. DNV-GL Rules for Classification; High Speed and Light Craft, Part 2 Chapter 1

25. HATZILAU, I K., TSEKOURAS, G J., PROUSALIDIS, J., and GYPARIS, I K., 'On electric load characterization and categorization in ship electric installations', *Conference Proceedings of the Institute of Marine Engineering, Science and Technology - INEC 2008: Embracing the Future,* 2008.

26. WORK, F W., 'Opportunities for Improved Warship Energy Efficiency: A Canadian Patrol Frigate's Operational Energy Use Patterns', Master's thesis, Harvard Extension School. 2016

27. Diverse Marine Storm Class CTV, product brochure, available from https://www.diversemarine.co.uk/crew-transfer-vessels accessed on 2025-05-07

28. HOGBE, N., and LUMB, F E., 'Ocean Wave Statistics', *London: HMSO*, 1967

29. CTV Performance Plot (P-Plot) Development: Research Project Summary, Carbon Trust, June 2017

30. GLOCKLING, J., SAVAGE, I., 'Fire Fighting Systems for Autonomous Vessels', *Conference Proceedings of EAAW 2023,* 2023.

31. KEHOE, J W., GRAHAM, C., BROWER, K S., and MEIER, H A., 'Comparative Naval Architecture Analysis of NATO and Soviet Frigates – Part I', *Naval Engineers Journal*, October 1980.

32. PAWLING, R J., ANDREWS, D J., PIKS, R., SINGER, D., DUCHATEAU, E., and HOPMAN, H., 'An Integrated Approach to Style Definition in Early Stage Design', *12th COMPIT*, Cortona, Italy, 15-17 April 2013.

33. "Mainstay Marine Unveils UK-built Crew Transfer Vessel With Financing Offering", https://workboat365.com/mainstay-marine-unveils-uk-built-crew-trans accessed 2025-05-09

8. AUTHORS BIOGRAPHY

Dr. Rachel Pawling is Programme Director for the MSc Marine Engineering course and a Lecturer in Ship Design at UCL, teaching the subject of ship design to undergraduate and postgraduate students. She obtained her PhD in computer aided ship design in 2007 and has subsequently continued her research in projects funded by the EC, UK and US governments.

ON THE BENEFIT OF ADJUSTABLE STERN ENERGY SAVING DEVICES FOR FAST WARSHIPS

S Brizzolara, Virginia Tech, USA, and **U Parajuli, R Wiltshire, L Morris, B Ferrier,** Fincantieri Marinette Marine, USA

SUMMARY

This paper presents the main results of a theoretical investigations based on and virtual/physical data aimed at quantifying the advantages of stern energy devices on fast warships, such as corvettes, frigates or cruisers. The goal of the study is well defined and motivated by the following fundamental questions:

- What is the sense of energy saving devices on warships?
- Why focusing the attention on the ship stern?
- What advantage do adjustable devices have on fixed ones?

From the analysis of a systematic set of virtual towing tank tests, obtained by high fidelity computational fluid dynamics simulations and from their comparison with experimental results available in the open literature, the above questions are critically addressed in the paper. Results obtained in the case of a notional fast frigate provide evidence on which device is able to provide the biggest energy saving across the whole ship operational profile, while ensuring the best performance at full speed.

NOMENCLATURE

υ	Kinematic viscosity (N s m^{-2})	$Fn_\nabla = V/\sqrt{g\nabla^{1/3}}$	Volume based Froude number
ρ	Density of water (kg m^{-3})		
p	Pressure (N m^{-2})	$C_F = R_F/(0.5\rho S_W V^2)$	Friction Resistance Coeff.
τ_w	(local) Wall shear stress	$C_{F_{ITTC}} = 0.075/\sqrt{\log(Rn) - 2}$	ITTC'57 C_F correlation line
g	Acceleration of gravity (m s^{-2})		
V	Ship speed (m s^{-1})	$C_\tau = t_w/(0.5\rho V^2)$	(local) Skin Friction coefficient
R_F	Friction resistance (N)		
S_W	Wetted Surface of the Hull	$C_P = P/(0.5\rho V^2)$	Pressure Coefficient
$Fn_L = V/\sqrt{gL}$	Length based Froude number		

1. INTRODUCTION

Transom stern hull forms, universally preferred for the design of fast naval vessels, offering advantages in wider deck areas and improved resistance characteristics over cruise stern hull forms, at certain length Froude numbers (typically in the range $0.25 < Fn_L < 0.35$), right beyond complete ventilation. Since the times of NPL series [1], transom stern hull forms with relatively high transom depth appeared to be inefficient in the high-speed displacement regime, close to the hump of the resistance coefficient ($0.45 < Fn_L < 0.55$), where the stern tends to sink into the water for dynamic pressure effects. NPL series investigated also the effect of a 1.5·L_{PP} stern wedge (represented in Figure 1-a), with a limited number of inclination angles. NPL model scale results, extrapolated to hull lengths of 30.5m, highlighted the benefit of mitigating the large dynamic trim by stern observed for the bare hulls obtained with the fixed wedge transom appendages, in the high-speed displacement and semi-displacement regime (corresponding to Froude number range $1.5 < Fn_\nabla < 2.0$).

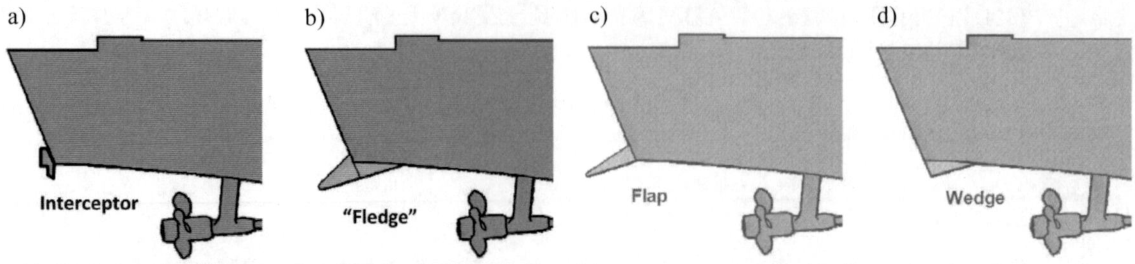

Figure 1: Proposed interceptor (a) as an alternative adjustable stern energy saving device to fixed wedge, flaps or integrated flap/wedge ("fledge") currently used in US Navy combatant hull forms. Figure adapted from [4].

Stern wedges were appeared on large combatant ships knowingly [2] few years after the NPL series was published, first in Soviet cruisers and then on Italian [2] destroyers of the Maestrale class. Across the Ocean, inspired by this emerging trend, the US Navy started a systematic study [4] that considered not only wedges [5] but fixed flaps [6] (essentially wedges originating at and protruding off the transom as pictured in Figure 1-c) and eventually 'evolved' in a hybrid combination of the two devices, which is nowadays referred to as "fledge" (Figure 1-b). Eventually all US Navy frigates and destroyers new build have been equipped with wedges and even the existing units have been retrofitted with fixed flaps.

The study presented in this paper aims at quantifying the potentials of adjustable transom-stern devices, such as interceptors (Figure 1-a), to reduce the hull drag at high speeds, as alternative energy saving devices to traditional fixed wedge or fledge solutions, and it was motivated by the large effort made in the US around the new frigates and destroyers programs.

Interceptors are well known in the boating/yacht world, as they are commonly used as adjustable stern devices to control the dynamic trim angle in the hump region (pre-planning regime) of planing hulls, as a simpler to install alternative to well known trim tabs or flaps. Interceptors were first installed on small/medium size yachts and then extended to larger ships such as fast ferries. Their hydrodynamic principles were first theorized in [11-12] and then verified by results of full scale trials on fast ferries [11] and by virtual towing tank tests on smaller sized

Hence, taking inspiration from a previous study [9] that for first analysed the hydrodynamics of interceptor placed on the transom of a type FFG-7 frigate towing tank model [10] and their resistance reduction properties versus the bare hull, the current study intends to expand on the advantages of interceptor with adjustable height (or stroke) over fixed fledges designed to optimize the fuel consumption throughout the whole ship operational profile.

For the design and assessment of the efficiency and efficacy of the energy saving devices for naval vessels, the following considerations have to be considered:

1. US naval vessels spend a considerable amount of time crossing the oceans to address the ever-increasing needs of peace keeping and war deterrence operations around the world. The fuel burnt at sea represents a quite significant portion of the running cost. Reduction of fuel cost and carbon emission are gaining importance among the various design optimization objectives, also for high-speed naval combatant crafts.
2. Almost all class of warships hull are of transom stern type. The flow leaving the transom contains a useful amount of vertical momentum which can be used to correct dynamic stern sinkage at high displacement speeds (corresponding to length Froude numbers range [0.35:0.55]) by appropriately design stern energy saving devices, such as wedges, flaps, and interceptors (intruders).
3. The amount of time that a fast warship spends at full speed amounts only to few percent points of the whole operational time. Navigation at sea happens for the most part at low or medium speed. A fixed (non-adjustable) energy saving device would reduce fuel consumption only in a narrow speed range around the design speed, either penalizing the maximum ship speed (which is not an option for a warship).

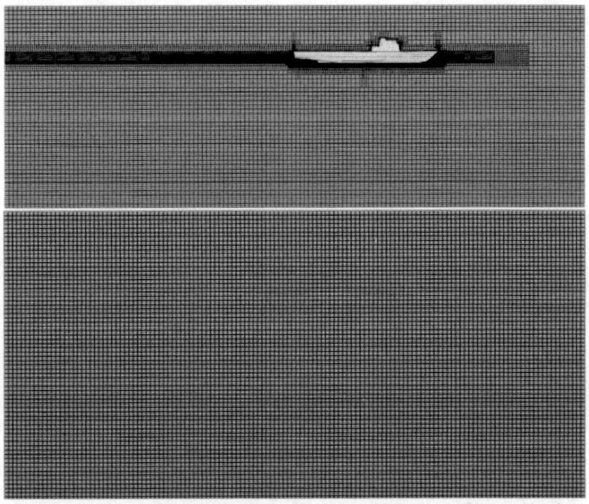

Figure 2: Domain extension and medium mesh resolution

In the following sections, in addition to a full analysis of the effectiveness and efficiency of different stern energy saving devices when applied to a modern notional frigate (of interest to the US Navy), the paper provides quantitative resistance and power reduction results derived from physics based numerical simulations, backed up by experimental results. Conclusions on which device is able to provide the biggest energy saving across the whole ship operational profile, while ensuring the best performance at full speed, will be drawn in the conclusion section.

2. APPLICATION CASE AND VIRTUAL TOWING TANK (CFD) MODEL

This study consider a notional frigate hull form, of type and size that are of interest to the US Navy with principal characteristics given here below: hull length~142m; moulded breadth~19m; draft~5.9m; block coefficient 0.52; twin shaft with V-brackets, twin canted rudders, single skeg and bilge keels. Following good practice, we provide the main details of the CFD model used to carry out the sets of virtual experiments and the results of the mesh sensitivity analysis which provides the level of confidence on the presented results.

All CFD simulations considered the fully appended hull with its full scale dimensions, and were setup and run using Rhinoceros/Orca-CFD and Simerics pre/post processing tools and URANS solver. The standard k-ω turbulence model with wall functions and volume of fluid model with sharp interface were used. All computations ran on laptop computers with 4 to 16 physical cores and at least 16GB RAM. Depending on the mesh resolution, ship speed and laptop configuration, each simulation took from 16 to 36 hours to reach the steady state values. The hull was left free to heave and trim to achieve the dynamic equilibrium at each speed. The discretized fluid domain extends one ship length upstream of bow, two ship lengths downstream and laterally as presented in Figure 2. The same figure shows the unstructured, prevalently cartesian mesh automatically generated by Simerics using standard settings valid for high-speed displacement hull forms and selecting only the resolution level of the mesh. All other settings were kept at their standard values. The physical simulation time required to ensure the convergence of the global quantities of interest (resistance, trim and heave) onto stabilized values, largely depends on the ship speed, even if the time step was adapted to satisfy the CFL condition at each speed. As characteristics of every volume of fluid free surface solver for steady state ship hydrodynamics in calm water, the hydrodynamic force components and dynamic trim and sinkage, show periodic oscillations after the hull is left free to move which dampen with time. At some ship speeds, a longer time was necessary to achieve convergence. In any case any quantity of interest, such as resistance, trim and sinkage, was averaged on an integer number of oscillation periods to be critically evaluated and reported in this paper.

2.1 MESH CONVERGENCE STUDY AND RESULTS VALIDATION

Following good practice, we provide the main details of the CFD model used to carry out the sets of virtual experiments and the results of the mesh sensitivity analysis which provides the level of confidence on the presented results.
The results of the mesh convergence study is presented in Table 1, which records the percentual errors of the predicted total resistance for the three difference mesh resolutions with respect to the full scale resistance extrapolated from model tests as a function of speed. As desirable, the error is generally reducing with an increase of mesh resolution and for the finest mesh, the it remains bounded by a maximum of about 3% with the exception of the case at higher speed (Fn_L) where the it reaches 4.4%. The level of uncertainty reached with the fine mesh was deemed acceptable and the predicted values well validated with the experimental results. Some of the ship speeds were not analysed with the normal mesh in the advantage of time. Absolute differences in the predicted dynamic trim angle and sinkage were generally very small,

Table 1. CFD prediction error on total ship resistance at full scale

Fn	Fine	Medium	Normal
(-)	(-)	(-)	(-)
0.166	0.9%	1.8%	5.8%
0.221	3.2%	6.4%	9.4%
0.277	2.3%	5.5%	7.7%
0.332	0.5%	2.2%	
0.360	2.8%	4.3%	
0.387	4.4%	8.0%	

practically insignificant, in the order of one hundred of a degree and few millimetres, respectively. So the dynamic equilibrium seemed insensitive to mesh resolution.

3. HYDRODYNAMICS OF TRANSOM STERN ENERGY SAVING DEVICES

3.1 FIXED WEDGE, FLAPS AND FLEDGES

The design principles of fixed transom wedges or flaps when installed on high-speed displacement vessels is covered in a series of papers that describes the experience of the US Navy [4-8]. Here, we will briefly try to summarize the main findings and conclusions from those studies and integrate their conclusions with some more insights to justify their beneficial effect on the drag experience by transom stern hull forms at, from a pure hydrodynamic point of view. In fact, while the effects of these transom devices are well known, no clear fundamental understanding has been provided about the drag reduction mechanism that they are able to realize when positioned at the stern. Often the drag reduction achieved with these devices is simply explained by the reduction in dynamic trim at high speed they cause, based on the clear correlation that is obtained between these two parameters.

The hydrodynamic mechanisms of these drag reduction, though, are not completely explained and fully understood. A common understanding is that the wave-making resistance changes caused by the ch ange in dynamic sink and trim which in turn changes the shape of the submerged volume of the hull [14]. But wave-making resistance alone cannot explain the whole resistance reduction especially at lower speeds. Another part of the savings comes from a change of friction drag which is not simply due to a reduction of wetted surface from the change in trim but also from a reduction of the friction itself. In a previous study [9], it was found that the friction drag (or global coefficient) increases, due to a sudden acceleration of the flow on the bottom upstream of the transom, as shown in the graph of Figure 3 that shows results of CFD simulations made on the FFG-7 hull model. The acceleration and consequent friction drag increase is correlated with the change of the transom flow regime, from a wetted to ventilated, as shown in the four snapshots on the left side of Figure 3. At lower speeds when the transom is still wet, the dead flow in the wake behind it forces the fluid particles to slow down. As speed increases (Fn_L=0.32 to 0.35 in the example of Figure 3) and the transom gets fully ventilated (Fn_L=0.37), the fluid particles flowing outside the boundary layer on the bottom have to accelerate to reach atmospheric pressure at the transom edge (for Bernoulli law). This causes the wall shear stress (or skin friction coefficient) to increase, as shown in the comparison of skin friction coefficients plot in Figure 4.

Figure 3: Left figures: Change of predicted wetted area of the FFG-7 model transom at different Froude numbers (blue fully aeriated, red fully wet, green 50% wet). Right Plot: Friction drag coefficient C_F predicted by CFD (red line) compared with ITTC'57 correlation line (blue line) for the same hull at different Froude numbers.

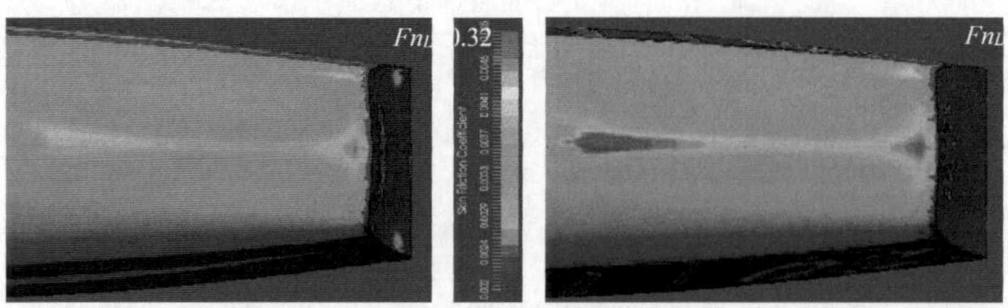

Figure 4: Skin friction coefficient predicted by CFD below (left) and above (right) ventilation speed

When a wedge is installed, its effect is to create a high-pressure region around the leading edge of the wedge, i.e. the line at which the inclined surface of the wedge joins the bottom surface of the hull. The pressure peak generating along the wedge leading edge can be clearly distinguished by comparing the two plots of Figure 5 (baseline hull without on the left and with the wedge on the right): from its maximum value, the high-pressure contours extend upstream with a gradually decreasing value until they blend into those of the stagnation pressure region at the trailing edge of the rudders. This high dynamic pressure, when integrated along the pertinent area, create the lift force which is responsible for the change in trim and sinkage of the hull.

Also visible in Figure 5 is the negative pressure region at the edge of the transom that corresponds to the acceleration of the flow needed to reach atmospheric pressure, since the transom is fully ventilated and the flow separates from the bottom leaving the characteristic free surface hollow. As it appears, the flow accelerates only in close proximity of the lower edge of the transom or of the wedge as the negative pressure area extends only a short distance upstream. At lower speeds when the transom is wet and shedding a highly turbulent separated flow downstream (in the case of the notional frigate for Fn_L<0.20), the sudden acceleration and negative dynamic pressure at the transom edge vanishes.

Figure 6 compares the wave through (or hollow) created aft of the transom in case of the baseline hull (left) and the version with the wedge, simulated at a speed corresponding to Fn_L=0.22. This speed is slightly above the threshold of incipient transom ventilation of the original hull. In fact, in case of the wedge (Fig. 5 right plot), the free surface rises into a very steep front to form the first crest. This almost vertical wave front is at the limit of causing wave breaking, i.e. the speed is very close to the critical ventilation speed. A slightly lower speed would cause the front to break and the flow to reattach to the transom which would lose its ventilated properties.

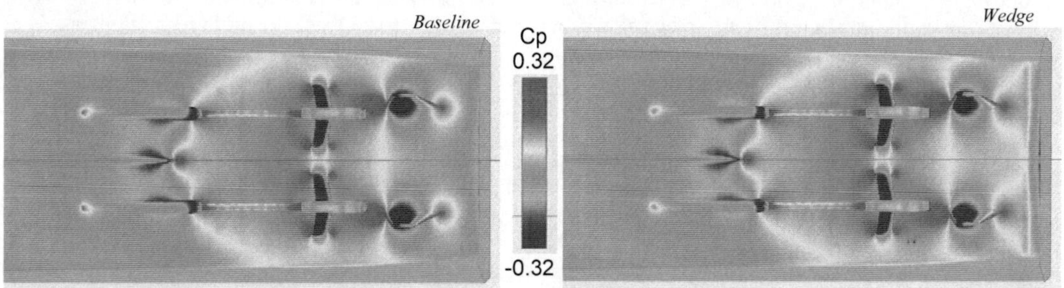

Figure 5: Dynamic pressure coefficient CP on the hull bottom upstream of the transom for the notional frigate, baseline hull (left) and hull with stern wedge (right). Hull forward speed corresponds to Fn_L=0.22.

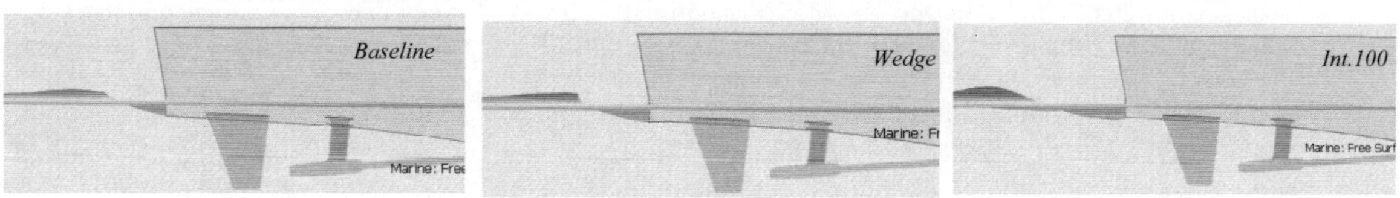

Figure 6: Free surface wave pattern behind the transom (section at symmetry plane view) of the notional frigate, baseline hull (left), hull with stern wedge (mid) and interceptor at 100mm. Forward speed corresponds to Fn_L=0.22.

From a pure hydrodynamic perspective, the main difference between the flap or fledge versus the wedge, is that the former increases the effective length of the hull, hence reducing the effective Froude number at the advantage of a lower wave resistance.

3.2 INTERCEPTORS

Interceptors, schematically presented in Fig. 1, develop a lifting force at the stern using another fluid dynamic mechanism. The hydrodynamic principles of interceptors were first theorized in [11] and [12], with relevant laws of similitude to study their effects and an approximate method to estimate the force induced in the stern of a fast ships [11] or planing boats [13].

Wedges and flaps generate lift by deflecting downward a large mass of the water flowing around the transom (both in and outside the boundary layer). Interceptors, on the other hand, generate lift by stagnating the incident flow, which can be visualized by taking a control volume as high as the interceptor blade which is extended upstream of it [12]. Both interceptors and wedges/flaps are bottom appendages and hence immersed in the boundary layer of the hull, but the difference in operating principles makes interceptors more dependent on the boundary layer characteristics of the hull (momentum thickness, primarily). Both types of stern appendages suffer from imperfect Reynolds scaling, when tested in model scale, but interceptors are inherently more affected [12].

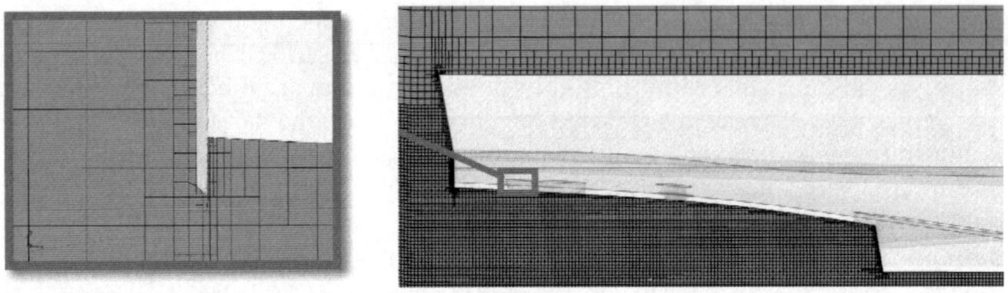

Figure 7: Longitudinal section on symmetry plane of the mesh around the Interceptor deployed of 100mm

In this study, an interceptor spanning the whole bottom edge of the transom up to the rounded bilge (excluded) has been designed and simulated at different speeds (including low speeds) and at two different blade (or stroke) heights: 50mm and 100mm (full scale dimension). The mesh and relative size of the interceptor blade out 100m from the transom edge is represented in Figure 7. For questions of maximum mesh size and memory available on the computers used for the study, it was not possible to increase the mesh resolution upstream of the interceptor the same way it was made in [9]. The wall functions used by the RANS solver may be sufficient to guarantee the correct momentum of the flow insisting on the interceptor height. Additional tests are being performed to verify the quality of the flow resolution in the boundary layer and around the interceptor blade.

The dynamic pressure generated on the hull bottom by the interceptor blade set at the two investigated strokes is compared in Figure 8 with the corresponding distributions predicted for the baseline hull and the baseline+wedge version. The comparison is made for high-speed, Fn_L=0.36, where the dynamic lift of stern appendages is mostly required to correct the tendency of the hull to squat by stern. The shape and magnitude of the pressure contours are as expected: the max pressure on the bottom is recorded just upstream of the interceptor blade out at 100mm (Cp~0. 55). From its maximum value, at stagnation, the pressure decreases hyperbolically (as found in [12]) upstream of the interceptor for a length in excess of 10 blade heights. Three-dimensional effects can also be noted by looking the sharp pressure drop toward the side, just before the bilge radius, where the interceptor blade ends. In fact, no end plates were fitted at the side edge of the interceptor in this case, differently from the application presented in [11]. It can be noted that the high dynamic pressure generated by the 100mm stroke interceptor is the highest, while the designed wedge seems to have generated the weakest pressure increase, with the interceptor at 50mm stroke being in between these two. This ranking is confirmed by the amount of trim by stern correction that these three, as it will be shown in the next section (Figure 11).

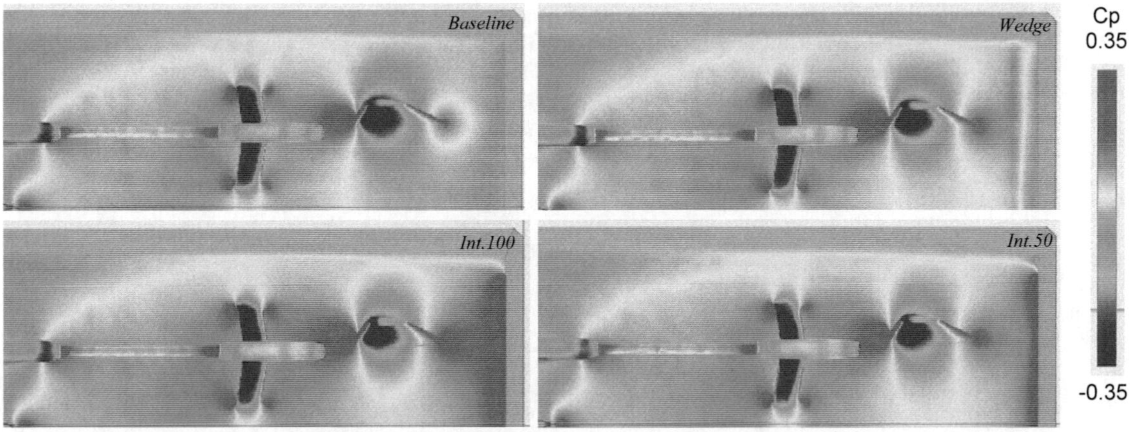

Figure 8: Dynamic pressure coefficient C_p on the hull bottom (only half symmetric hull is presented) at the stern of the notional frigate equipped with various stern energy devices. Simulated speed corresponds to Fn_L=0.36.

As already noted at lower speed, at high speed too the two alternative stern appendages are able to modify the nearfield wave patters mostly behind the stern, as presented in Figure 9. The deepest hollow is created by the interceptor at its maximum investigated stroke. It can be noted that, while the flow tends to leave the transom with a velocity almost tangent to the slope of the surface upstream of the transom edge, in case of the interceptor the flow separates at the tip of the blade with a downward velocity component which depends on the magnitude of the stagnation pressure created upstream. In case of the int.100, the wave hollow just aft of the stern is much more pronounced of the one created by the wedge, causing a larger increase of virtual ship length that further reduces the wave-making resistance at high speeds.

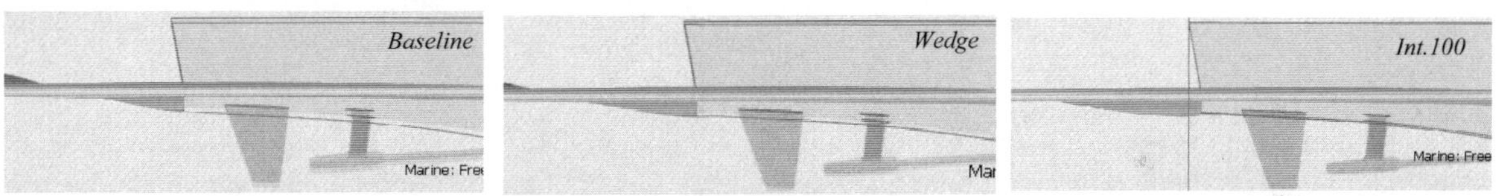

Figure 9: Free surface wave pattern behind the transom (section at symmetry plane view) of the notional frigate, baseline hull (left), hull with stern wedge (mid) and interceptor at 100mm (right). Forward speed corresponds to Fn_L=0.36.

4. GLOBAL RESULTS

This section is dedicated to the discussion and comparison of global results, i.e. the integral parameters that are of practical interest to naval architects, as opposed to hydrodynamicists. The ultimate goal of the stern devices discussed in the previous sections is to achieve a meaningful resistance reduction, possibly not only at high speed, but rather across the whole ship operational speed profile.

In fact, while naval vessels have stringent requirements of max speed reached at combat displacement with the max available propulsion power, they spend most of the time at cruise or even lower than cruise speed (depending on Navies and type of ship). Hence, the optimization of the design of a stern energy saving device with respect to maximum speed only, would turn in to a barely effective device. Indeed, the stern wedge was designed to provide resistance reduction benefits

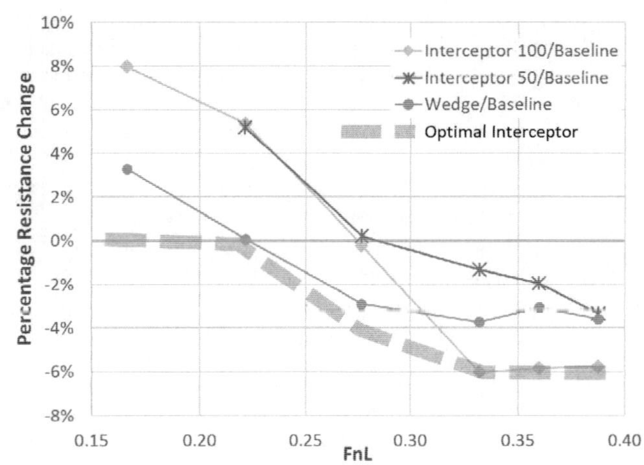

Figure 10: Resistance change (negative = reduction) predicted for different energy saving devices versus speed. The cyan dotted line represents the best performance of an adjustable interceptor

over the largest possible speed range. In fact, as it can be seen in Figure 1o, the percentage resistance change achieved by the wedge extend over a wide speed (or Froude number) range. This at some sacrifice of efficiency of the wedge in reducing the resistance at top speed. In fact, the resistance reduction of the wedge at top speed, Fn_L=0.39, is just shy of 4% and it maintains around 3-4% moving down to Fn_L=0.28, to hit the break-even point at Fn_L=0.22. For ship speeds lower than Fn_L=0.22, the wedge generates added drag with respect to the baseline hull form, hitting more than 3% at Fn_L=0.17. In this respect, the wedge would not be ideal stern energy devices for a ship spending most of her time loitering or cruising at low speeds.

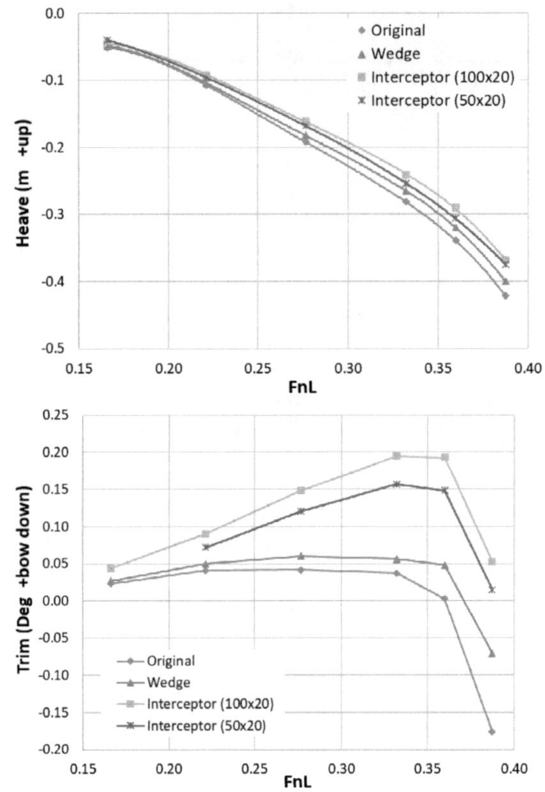

Figure 11: Dynamic Sinkage (heave) and Trim

The interceptor out at 100mm (yellow line in Figure 10), on the other hand, is able to reduce the resistance of 6% at high speeds, 0.33<Fn_L<0.39, hitting the break-even point at Fn_L=0.28. Consistently with expectations, the same interceptor out at 50mm is able to reach lower resistance reductions in the high speed range and reaches the same breakeven point. The higher interceptor resistance reduction effect predicted at higher speed is reflected into the greater ability of the interceptor to change dynamic sinkage and trim, as evident from the comparison presented in Figure 11. The interceptor out at 50mm and 100mm is the only device able to correct the trim by stern of the ship at top speed (Fn_L=0.39). Both interceptors are able to lift the hull 3.5cm more outside the water than the wedge. This also results in a resistance reduction, thanks to a lower wetted surface. The real advantage of interceptors, though, is to be adjustable: the blade, in fact, can be easily moved up and down along with electric or hydraulic actuators fit on the transom stern. Hence it is desirable and very easy to implement in practice, for the interceptor to reach even higher strokes than the maximum considered in this study, say 200mm, for instance. Under this assumption it is plausible for the interceptor to be able to achieve the cyan dashed line shown in Figure 10, by attaining the following stroke schedule: high speed range (0.33<Fn_L<0.39), 100mm stroke; medium speed range (0.22<Fn_L<0.33), 150-200mm stroke; low speed range, fully retracted. In this respect, it can be seen that an adjustable interceptor can ensure the maximum possible energy saving across the whole ship operational speed range, on the contrary to a well optimized stern wedge which suffers higher drag at lower speeds and suboptimal drag reduction at top speed.

The overall impact of the interceptor (out at 100mm) and wedge in terms of free wave pattern modification is given in Figure 12, as predicted by the CFD solver at high speed (Fn_L=0.36). As anticipated the main difference is in the wave trains generated behind the stern: first the hollow created aft of the transom and then the roostertail like crest which originates a system of divergent waves propagating in the wave. A slight difference for a good eye can be noted also in the main divergent and transverse wave trains created by the hull. This is due to the change of trim and sinkage that these appendages are able to generate in this speed range.

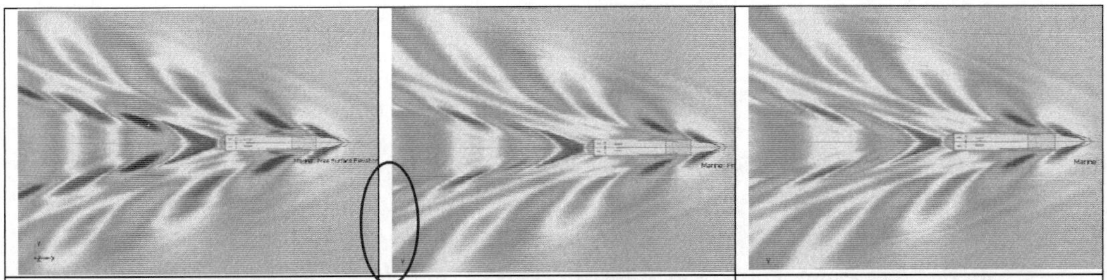

Baseline, Fn_L=0.36	Wedge, Fn_L=0.36	Int.100, Fn_L=0.36

Figure 12: Free wave patterns calculated for the baseline hull and the two appended variants at high speed.

5. CONCLUSIONS

The paper presents a comparative study on the effect of adjustable interceptors and fixed wedges in the stern of a notional frigate. The hydrodynamic principles that lead to a resistance reduction are discussed for both types of devices in terms of field (fluid dynamic) variables, such as pressure distribution, free surface disturbance due to the free waves generated by the appended hull, as well as global variables, such as resistance, dynamic trim and sinkage.

A large speed range has been considered for the analysis and discussion of the CFD predictions, in order to cover most of the operational profile of a naval vessels, hence being able to draw conclusions about the overall efficacy of the different devices. With reference to the main research questions formulated upfront, as reported in the abstract, the following conclusions can be drawn from the presented study:

- What is the sense of energy saving devices on warships?
  Energy saving or resistance reduction devices can be of practical interest for application on naval vessels, if they can reduce the resistance (and save propulsion energy) not only at top speed but in a large range of speed, possibly down to loitering speed.
- Why focusing the attention on the ship stern?
  It has been shown the stern appendages are ideal to correct the natural attitude of transom stern hulls to squat by stern at high speed. A correction of the dynamic trim by stern at high speed results directly into a resistance reduction due to lower submerged volume (lower wave resistance) and lower wetted surface (lower friction resistance). The fluid dynamic mechanism to achieve the dynamic correction is to create high dynamic pressure on the bottom upstream of the transom, where the flow separates from the hull. Interceptors are much more efficient in creating high pressure in the high-speed range than stern wedges.
- What advantage do adjustable devices have on fixed ones?
  Interceptors can be more easily actuated (lower power needed with respect to movable flaps) and hence ensure the best resistance reduction over the whole operational speed profile of a naval vessel, reducing to a bare hull in the lower speed range.

6. ACKNOWLEDGEMENTS

This study is the fruit of a spontaneous collaboration between Fincantieri Marinette Marine and Dr. Stefano Brizzolara during his independent sabbatical research period, dedicated to various advanced research studies with the engineers of the FMM DC office. The authors wish that the promising results obtained in this pilot study may have a continuation in the future up to the application of the best energy saving devices found in this study to a real naval vessels built by Fincantieri. Lastly, the authors would like to thank Rhino-Orca-CFD and Simerics expert support.

7. REFERENCES

1. BAILEY, D., 'The NPL Round Bilge Displacement Hull Series', *Maritime Technology Monograph no. 4, Royal Institution of Naval Architects*, 1976.

2. KEHOE, J W., JR., et al. 'U.S. and Foreign Hull Form', *Machinery, and Structure Design Practices. Naval Engineers Journal*, Nov. 1983.

3. BAU, F C., 'Comment and Discussion of Technical Session Papers', *Naval Engineers Journal*, May 1987, p. 91.

4. KARAFIATH, G., CUSANELLI, D S., and LIN, G M., 'Stern Wedges and Stern Flaps for Improved Powering - U.S. Navy Experience'. *SNAME Transactions*, Vol. 107, 1999, pp. 67-99

5. ZSELECZKY, J., And HAYS, B., 'FFG-7 Model Powering Test with and without a Stern Wedge', *U.S. Naval Academy Report EW-28-84*, December 1984.

6. CAVE, W L III., and CUSANELLI, D S., 'Effect of Stern Flaps on Powering Performance of the FFG-7'. *Marine Technology, Vol. 30*, No. 1, Jan. 1993

7. CUSANELLI, D S., 'Stern Flaps - A Chronicle of Success at Sea (1989-2002)', *SNAME Innovations in Marine Transportation, Pacific Grove CA*, May 2002.

8. CUSANELLI, D S., KARAFIATH, G., 'Hydrodynamic Energy Saving Enhancements for DDG 51 Class Ships'. *ASNE Day 2012, Crystal City, Arlington*, VA February 9-10, 2012.

9. SALIAN, R., and BRIZZOLARA, S., 'Adjustable Energy Saving Device for Transom Stern Hulls'. *SNAME Transactions vol. 126*, pp.13, 2018. ISSN 0081-1661.

10. WHITE J., BRIZZOLARA, S., and BEAVER W., 'Effect of Inverted Bow on the Hydrodynamic Performance of Navy Combatant Hull Forms'. *SNAME Transactions, vol. 124*, 2016.

11. BRIZZOLARA, S., 'Hydrodynamic Analysis of Interceptors with CFD Methods'. *7th Int. Conference on Fast Sea Transportation, FAST 2003*, vol. 3, p. E.49-E.56, 2003. ISBN/ISSN: ISBN 88-901174-0-0.

12. MOLINI, A., and BRIZZOLARA, S., 'Hydrodynamics of Interceptors: A Fundamental Study'. *International Conference* ISBN/ISSN: 88-901174-5-1

13. VILLA, D., BRIZZOLARA, S., 'A Systematic Analysis of Flaps/Interceptors Hydrodynamic Performance', *10th International Conference on Fast Sea Transportation, FAST2009. Athens (GR). Univ. of Athens, vol. 1*, 2009. ISBN/ISSN: 9789602546864

14. BRIZZOLARA, S., and BRUZZONE, D., 'Numerical wave resistance and dynamic trim for high-speed crafts' In *Proc. NAV 2000 Int. Conference, Venice, vol. 1*, pp. 4-2. 2000.

15. UITHOF, K., VAN OOSSANEN, P., MOERKE, N., VAN OOSSANEN, P G., and ZAAIJER, K S., 'An Update on the Development of the Hull Vane', *International Conference on High Performance Marine Vehicles*, 2014.

16. DAY, A H., and COOPER, C., 'An experimental study of interceptors for drag reduction on high-performance sailing yachts', *Ocean Engineering*, 38:983–994, 2011.

17. TSAI, J., and HUANG, J K., 'Study on The Effect of Interceptor on High-Speed Craft', *Journal of The Society of Naval Architects and Marine Engineering, Vol.22*, No. 2, pp. 95- 101, 2003.

8. AUTHOR BIOGRAPHIES

Stefano Brizzolara, PhD, MRINA, MSNAME, is full professor of ship hydromechanics and Crofton faculty fellow in Department of Aerospace and Ocean Engineering at Virginia Tech. Founder and director of the Innovative Ship Design Lab (VT-iShip), co-director of the Center for Marine Autonomy and Robotics. After his early career as ship hydrodynamic designer in Fincantieri Naval Shipbuilding Division and junior professor at the Univ. of Genova, in 2011 he debuted in the US academic world as Peabody Visiting Associate Professor at MIT, where he stayed as assistant director for research of the MIT Seagrant college program. He moved to Virginia Tech in 2016. His managed research portfolio is funded by the ONR, DARPA, DoE and the shipbuilding industry and gravitates around numerical hydrodynamics for the design of advanced marine vehicles and high performance propulsors. Author of more than 250 peer reviewed papers, designated inventor of eleven patents (two pending) for innovative marine crafts and ship systems.

Bernard Ferrier, PhD, FRINA, FRAeS is a Systems Engineering and Technology Assistant (SETA) in NAVAIR PMA-266 with the primary task of integrating and deployment of Fire Scout MQ-8C and its technologies and defining the operational limitations of the deck environment. Ferrier holds an honorary Professor position at the University of Exeter's College of Engineering. He is the Engineering Head of Syntek Technology's Aircraft-Ship Dynamic Interface Lab in Fairfax, VA. Ferrier led the CL227 UAV dynamic interface program at Bombardier in Montréal. Ferrier was a rotor dynamist and project leader of the dynamic interface program of the marinised AH-64 (Apache) and LHX programs at the McDonnell Douglas Helicopter Company. He received his last doctorate in helicopter/ship interface engineering at the École Polytechnique de Montréal (Québec) Canada.

Ross Wiltshire, BSc in Naval Architecture and Marine Engineering at University of Michigan (with a minor in Nuclear Engineering), also holds a Master's degree in Industrial Operations Engineering from the Technical University of Berlin, and a MBA from George Washington University. Mr. Wiltshire has been involved all first-of-class USN ship designs since 2004, and is currently part of the Concept Development and Innovation management team at Fincantieri Marine Group in Washington, D.C. His professional interests include International Technology and Knowledge Transfer for maritime applications, AI in Ship Design and the development of modular nuclear propulsion systems for commercial shipping.

Ujwol Parajuli BSc in mechanical engineering at Kathmandu University in 2014. He worked as a mechanical engineer for multiple hydroelectric projects from 2014 to 2022. Currently, he is working in Fincantieri Marinette Marine's Washington DC wings as a mechanical engineering supporting various in-house design efforts. Mr. Parajuli is a co-author of Artificial Intelligence Utilization for Ship Design Optimization, which was published on SNAME Maritime Conference,2024

Lance Morris holds the current position of Technical Engineer at FMM. Mr. Morris is responsible for CAD modelling and FEA. Mr. Morris is a graduate of the Clark School of Engineering's Mechanical Engineering program at University of Maryland, with prior experience in structural FEA for mechanical design.

THE GREENHOUSE GAS EMISSIONS OF WARSHIPS

S Downie, Babcock International Group Plc., and The University of Edinburgh, UK

SUMMARY

In the commercial world there is a strong justification for the existence of shipping, with greater than 80% of world trade being delivered at a cost of less than 3% of Global Greenhouse Gas (GHG) emissions. There are three statutory requirements from the IMO, which are designed to reduce the GHG emissions of shipping: EEDI, EEXI, and CII. However, a general exemption is made for warships, justified by fact that they make up only 3% of the world fleet.

When comparing the two types of ship: commercial ships tend to travel a set route at a relatively constant speed and spend as much as 95% of the year at sea, whereas warships spend on average only a third of their time at sea, with significant variation in their speed and route dependant on the mission and circumstance. These among other variables associated with their GHG emissions make warship behaviour different and not suitable for commercial GHG models.

The combination of very efficient build-to-print shipbuilding and high repeatability/optimisation of process, along with the relative simplicity of a commercial ship configuration, means that 80% of their emissions are expelled during the in-service fuel use phase of its lifecycle. However, for warships it is necessary to take account of the protracted design and build phase of a bespoke product, a supply chain sourcing the best components, not the: cheapest, greenest, or the closest. With that and more considered, the in-service fuel emissions, as targeted for the commercial world, is no longer the obvious area for reduction or comparison. This paper discusses the differences between warships and commercial ships, the life cycle assessment that this research is undertaking, and how the warship design and build community may reduce GHG emissions for the better.

1. INTRODUCTION

In 2023 Babcock International Group Plc. commissioned a piece of research conducted to model the lifecycle emission of their largest and most complex product the Arrowhead 140 frigate. The research formed part of an ongoing PhD programme with the university of Edinburgh. The project aims were to model and assess the total carbon cost of the product from cradle-to-grave; to model the energy consuming systems and propulsion onboard and to analyse potential to reduce carbon emissions. This work is to assist the company in achieving their PlanZero40 pledge to have net zero emission by 2040.

The Arrowhead 140 design, to date, has been sold to three states: the UK, Poland, and Indonesia, with 5, 3, and 2 hulls for each state respectively.

The uniqueness of this research comes from the warship subject and the level of detail considered. Many studies in the past analysed commercial ships, such as container ships cruise ships or even yachts, and have avoided warships as specialist and non-open-source subject matters, with several behavioural differences. This paper discusses the work conducted.

2. BACKGROUND

2.1 ANTHROPOGENIC CLIMATE CHANGE

The Anthropocene began with the industrial revolution; whereas the Earth's climate has been everchanging, the impact of climate change attributable to humans is termed Anthropogenic Climate Change. This form of climate change is directly attributable to the presence of GHGs in the earth's atmosphere.

GHGs are those gaseous constituents of the atmosphere, both natural and anthropogenic, that absorb and emit radiation at specific wavelengths within the spectrum of thermal infrared radiation emitted by the Earth's surface, the atmosphere itself, and by clouds. This property causes the greenhouse effect [1].

GHGs include Carbon Dioxide (CO_2), Methane (CH_4), Nitrous oxide (N_2O), Hydrofluorocarbons (HFCs), Perfluorocarbons (PFCs), and Sulphur hexafluoride (SF_6) [1] (Figure 1).

The global warming potential (GWP) varies dependant on the gas ranging from a baseline of 1 for CO2 to 22,800 for SF_6 [2] this ratio of potentials is often referred to as CO_2e. CO_2e is used as the reference unit throughout this research.

Often when a GWP is quoted without suffixes, it is referring to GWP_{100} meaning the potential over 100 years. However, there are other indicators, such as $GWP_{20,}$ along with Global Temperature Potentials GTP_{50}, GTP_{100} [3], which are used within scientific analyses.

The key points, heavily reinforced by numerous sources, are that CO_2 is both the most common GHG produced by humans and the GHG with the longest legacy.

An effective corporate climate change strategy requires a detailed understanding of a company's GHG emissions. Until recently, companies have focused on emissions from their own operations under scope 1 and scope 2 of the GHG Protocol [4]. Scope 3 emissions are much harder to quantify and often estimated, this scope includes items such upstream components and travel and commuting. The IMO's strategy [5] is the latest adopted revision of its strategy on the reduction of GHG emissions from ships.

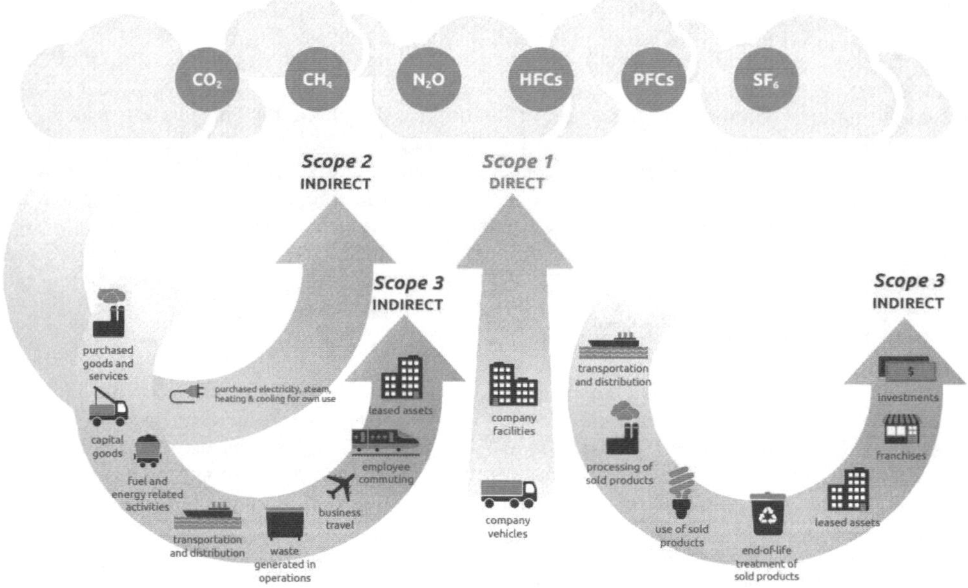

Figure 1: Emissions Scopes [4]

2.2 SHIPS AND WARSHIPS

In 2016 a study was undertaken involving the calculation of world ships versus total warships and concluded that 4% of the world ships were warships [6]. This figure appears stable and applicable to the current period of study given there are 103,000 ships [7] and 3,757 surface combatants from Janes [8]. The UNCTAD [7] is the United Nations Conference on Trade and Development and publishes the worldwide ship numbers while Janes report on military ship numbers. Both Military and commercial ships have an IMO (International Maritime Organisation) registration number and therefore the Janes dataset is a subset of the UNCTAD set. However, an older 2002 source [9] states that there are only 1300 warships from global fleet of 90,363, equating to 1.4%. The doubling of the world's navies over the last 20 years is not supported in any literature. It is concluded that while the exact number of warships could be in dispute the ratio is suitably small, typically less than 4%, compared to commercial ships. There is also a much smaller source of research studies involving warships versus commercial ships, likely to the understandable restricted publication of information and data for such ships and limited access for studies.

2.3 WHOLE LIFE CYCLE

From Fan et al., paper [11] it was outlined well the phases of the life cycle: Ship Construction (SC), Well to Tank (WTT), Tank to Wake (TTW), and Ship Disassembly (SD) phases. A well-known military ship design model of CADMID 'concept, assessment, detailed design, manufacture, in-service, and decommissioning' is also commonly used in UK warships [10]. While not making specific reference to it, the diagrammatic inference is that the in-service section is the largest (in time). The GHG protocol [4] cites another division of phases cradle-to-gate and cradle to grave. A common thread in these definitions is a three-stage lifecycle involving: production, use, and disposal, which can be a simplified version of all the above. The breakdown is shown graphically below (figure 2) alongside the standard breakdown (figure 3). Well To Wake (WTW) is the collective term for the middle section (also considered B- use stage) of the cycle concerned with energy use in service. The terminology comes from burning of fossil fuels (in shipping diesel), where the production of that diesel (now any fuel) is termed WTT, and the consumption of that energy on board as TTW.

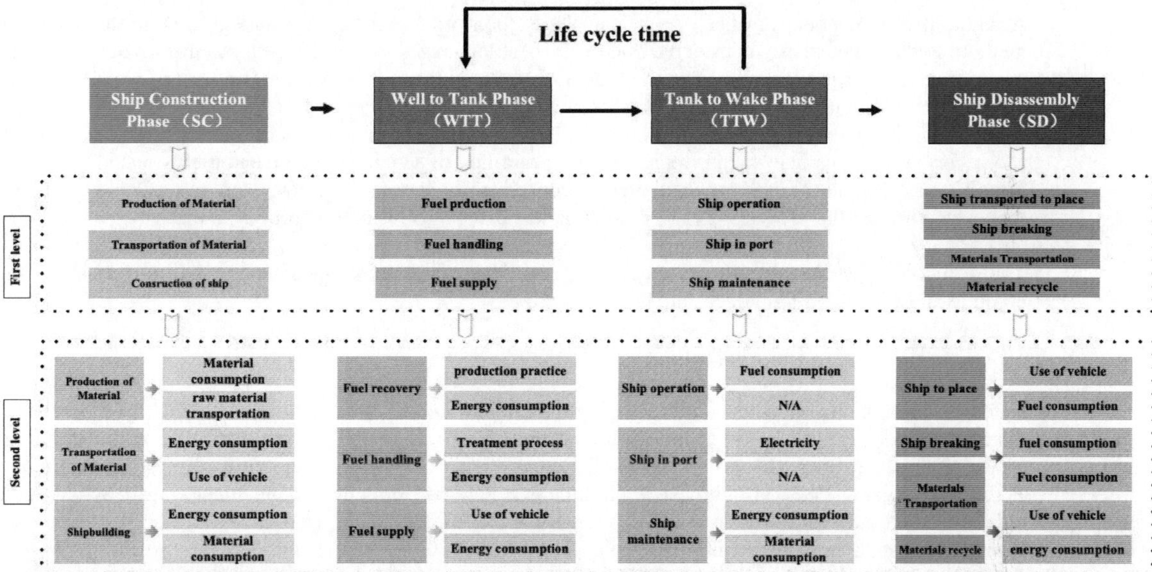

Figure 2: Ship's Lifecycle [11]

Product stage			Con-struction stage		Use stage							End of Life stage				Benefits and loads beyond the system boundary
Raw material supply	Transport	Manufacturing	Transport to the site	Assembly	Use	Maintenance	Repair	Replacement	Refurbishment	Operational energy use	Operational water use	Deconstruction demolition	Transport	Waste processing	Disposal	Reuse-, Recovery-, Recycling-potential
A1	A2	A3	A4	A5	B1	B2	B3	B4	B5	B6	B7	C1	C2	C3	C4	D

Figure 3: Product Lifecycle [12]

2.4 REVIEW

This research has identified the need to improve knowledge in the following areas, and therefore these gaps constitute aims of the research:

- In general, the lifecycle GHG emissions for warships, unlike their commercial counterparts, has a much smaller body of research. For life cycle assessments for bespoke warships the difficulty lies with data and security. There is also a tendency for warships to have a significantly different component makeup with more 'exotic' materials.

- In the build phase the makeup and operation of a naval shipyard is very different from that of the commercial sector and therefore the energy intensity and emissions efficiency.
- In service, the warship's diverse mission sets make tangible emissions calculations difficult, and the entire in-service phase for emissions would have to be a statistical and broad result with many varying factors. Open-source research also lacks the nuances of military operational behaviours.
- For the disposal phase less research and documentation exist, there is a tendency for commercial and military deconstruction to be done by different sites in different ways.
- All phases of the lifecycle have not been brought together into one integrated model to understand the whole process in widespread literature.

3. MODEL METHODOLOGY

A whole lifecycle model has been created in the Python programming language and using the Montecarlo method. Python is currently considered one of the best languages to create a data science model such as this, where there is a large amount of data to draw statistical conclusions from. Initially for the research SimaPro was used to calculate the Life Cycle Assessment as an industry standard of the shelf software. While the database which underpins SimaPro (EcoInvent) is still used to define some data, the requirement to perform Montecarlo with statistical spreads representing uncertainty of actual collated data quickly tested the limits of SimaPro's computation. Python was therefore consequently chosen. The arrays are significantly large such that even with parallel processing in Python complete calculation is in the order of hours.

The decision to use Montecarlo is often an option when the computing power, which would allow more complex machine learning models, are not available.

3.1 THROUGH LIFE MODEL

The through life model is constructed by summation of nodes in categories making up a whole ship distribution. Figure 4 shows the breakdown of the high-level model using a standard notation found in conventional Life Cycle Assessment (LCA). Readers will recognise this notation from an Environmental Product Declaration (EPD) (Figure 3). Each node is generated from either a statistical spread, actual data, or a database and the nodes are summed using the Montecarlo method. An example would be ship availability, which takes actual ship availability (that may only include 20 data points and forms a distribution based on that actual data with a mean of the fleet availability. This data input has extremes at very good availability and very bad availability. All such data, including supply chain data and material makeup of the ship, is based on actual data.

3.1 (a) Steel Example

Steel is sourced from a number of suppliers around the world and with the production of e.g. any steel plate an EPD is sourced from the mill in question, which falls into category A1-Raw Material Supply. After leaving the mill it will be transported to the shipyard with associated emissions under A2-Transport. Upon arriving at the yard that sheet of steel is nested and cut with a certain amount of waste, those cut parts are formed and welded into structure. This is just one example of the processes modelled.

In bounding the LCA for the ship, all the scopes of emissions have been included. This also as aligns with how companies such as Babcock record and report their emissions. For instance, part of scope 3 emissions include the employee commuting contribution. To define the input distribution for the carbon cost of commuting, employee surveys were used to understand habits and extrapolate this data across the whole workforce. Along with this, IT infrastructure and office infrastructure is a known, so therefore it can be possible to estimate the emissions cost of the engineering design hours, which for a warship is considered significantly greater than that of a commercial vessel.

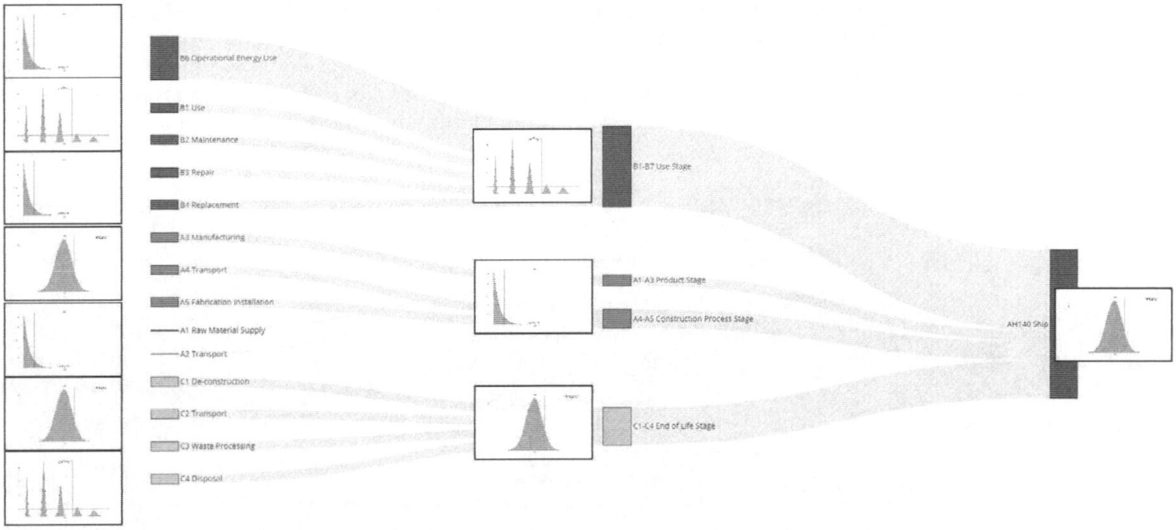

Figure 4: Whole Life Representative Sankey Diagram with Demonstrative Spreads.

3.2 IN SERVICE OPERATIONAL MODEL

While a cradle-to-grave model includes in service support, the in-service operational model is separated away from that main model. The in-service model is constructed using the same Montecarlo method with actual data representing equation inputs and giving a statistical result.

3.2 (a) Inputs

For the purpose of this paper and demonstration of the calculation, the speed profile of RRS Sir David Attenborough (SDA) has been used and is represented in Figure 5. This is for security reasons and is to illustrate an example of actual data without displaying the actual speed data of a warship. Not only does part of this research also include the Natural Environmental Research Council's (NERC's) ships, but research ships have more in common with warships than their commercial counterparts too.

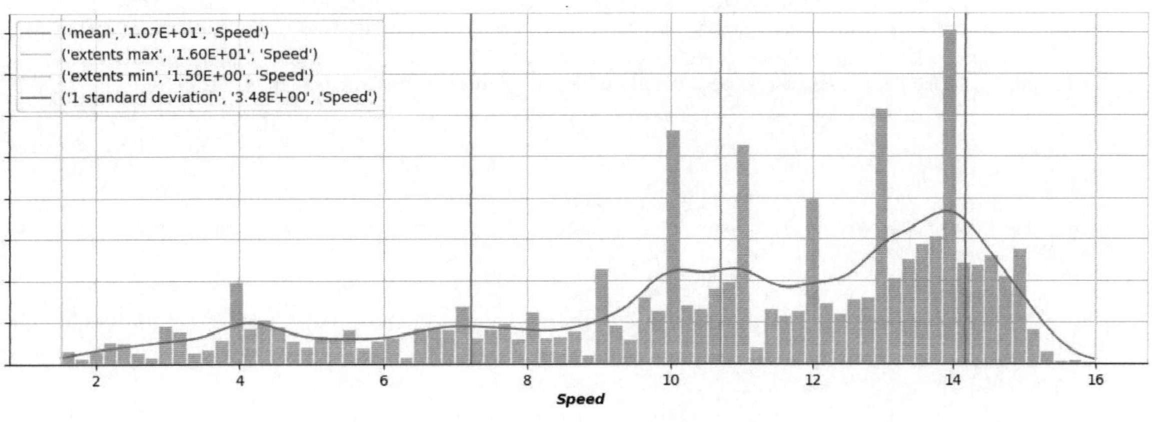

Figure 5: RRS Sir David Attenborough Speed Profile [13]

3.2 (b) Energy modelling

As part of the in-service model the behaviours of all the components are modelled, with interactions and flows between them. The diagram below (figure 6) shows the part of the model focused on the basic components of the propulsion system. In the same way as figure 4, the flow through the model in the Sankey diagrams is the mean. The thicknesses of the lines represent the energy flow between components (the full extent of the spread of results cannot be displayed in this way) the nodes are colour coded (red, yellow, and green) to show the usefulness of the energy to the purpose of the system or ship. The purple represents the energy source, i.e. the diesel tank supplying the four diesel engines through the propulsion chain to two propellers. The yellow components are the turbos as their energy consumption is recirculated in visible closed loops within the system. The red components are non-useful waste energy: exhaust gas, and cooling water etc.

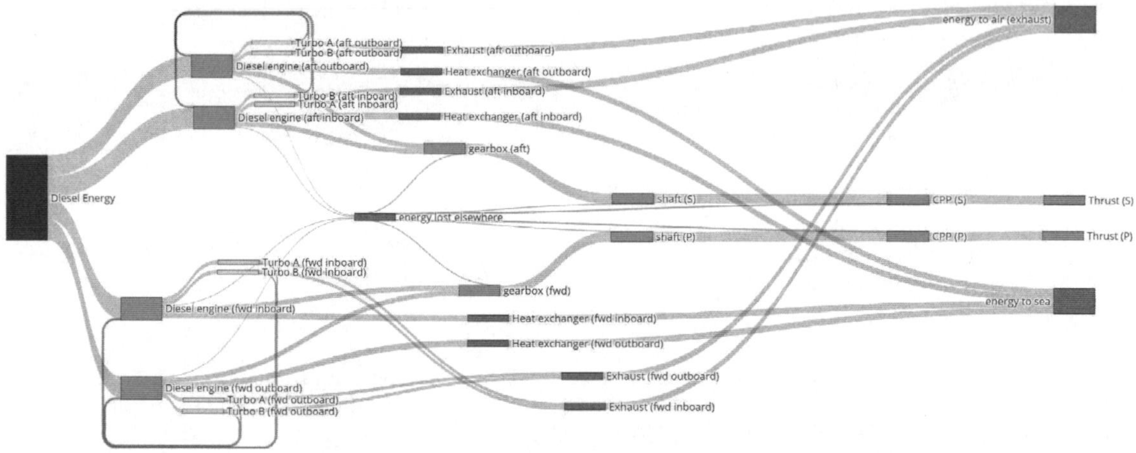

Figure 6: Simplified Propulsion Model

4. PRELIMINARY FINDINGS

The heatmap chart below (figure 8) shows one of the example outputs of the as-built propulsion model. The chart uses the 'hotter' colours to show areas with more data points than those illustrated with 'cooler' colours. Due to the statistical nature of the model the results are a distribution with 1e7 data points. With the number of Montecarlo inputs the model is converging at this value, as more variables are added it may be necessary to go higher than this number of points.

As the complexity and number of variables increases in Montecarlo modelling the sample size must also increase to capture the nuances of each variable. Convergence is being used to assess the variance of the result with increasing sample size (figure 7) shows convergence at >1e6 therefore currently the model is set to 1e7. As more is added the model will be periodically tested for convergence and adjusted accordingly.

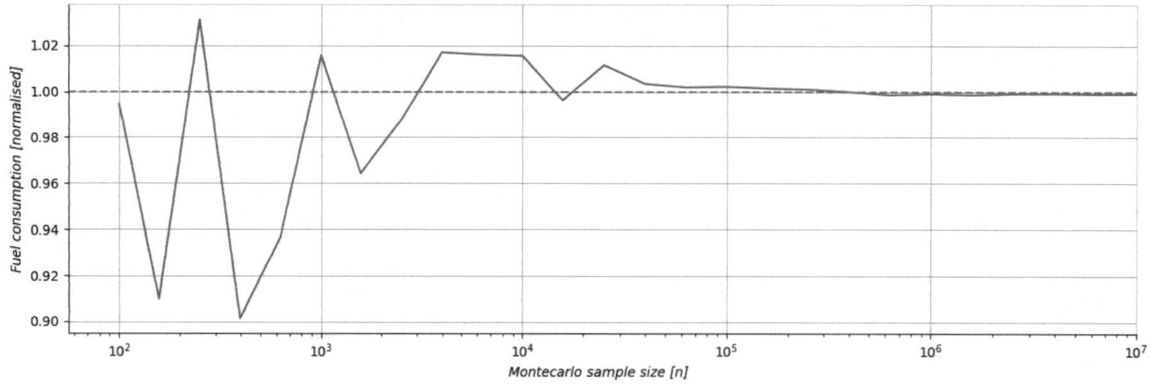

Figure 7: Result of testing for convergence in the model.

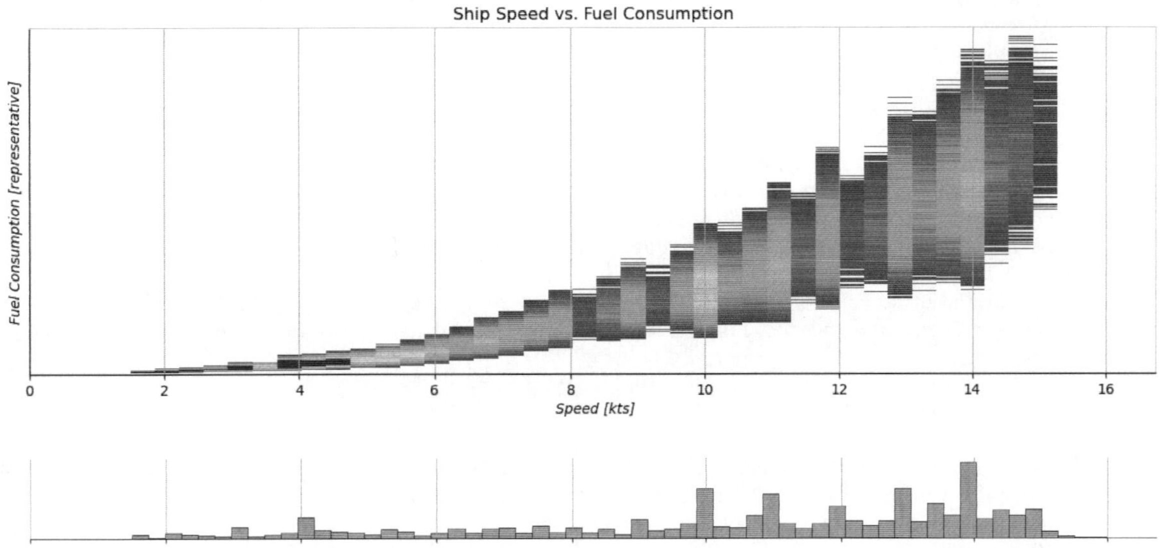

Figure 8: Example heatmap results for propulsion model.

5. DISCUSSION & LIMITATIONS

Figure 8 shows one set of results from the propulsion model speed vs. fuel consumption in the form of a heatmap. The heatmap represents the occurrence of 1e7 Montecarlo points. The pattern of the speed profile can be seen with 'hotter' colours in areas at circa 4, 10, and 14 kts corresponding with the lobes of the speed profile (Lower histogram).

While the appearance is that the values diverge with increase in speed this is just the appearance and there are a proportional number of data points to the amount of time spent at that speed the standard deviation from the mean is still constant. From the chart the relationship of speed vs. power can be seen in the approximate shape of the curved dataset, as fuel consumption is approximately proportional to the power, the trend (as expected) follows that. The large spreads are due to the number and range of the statistical inputs. The results can now be used to determine a WTT value for embedded carbon linked to fuel consumption, but also the information can be used to understand risk in the ship's range calculations, key tool in concept design.

Looking beyond the nuances of the naval datasets, as previously discussed, the mean availability of warships is comparatively low compared to a commercial ship. The commercial expectation is that almost 100% is achievable to maximise return on investment on a trade route. Warships possess the potential for high sprint speeds and basic naval architecture shows that the top end speed of a ship has a very large impact on the design and the emissions profile, again comparing to the commercial world where the concept of 'slow steaming' has been saving shipping money for some years already. Available and speed have the greatest effect on results so far. While results are still preliminary the in-service phase of operation emits less carbon than their commercial counterparts as a percentage. The short disposal phase is relatively stable with comparative values to a commercial ship tonne for tonne displacement, the differences lie in increased complexity of components, however a large percentage of warship specific components are reused before they are recycled. The design and build phase does not leverage the economies of scale like the commercial world so the carbon emitted is higher couple this with a much lower availability of warships, these two phase are becoming more similar in their emissions.

The key to effective modelling is good data and therefore the limitation to any model. As previously discussed, warships make a small subset of ships in the world, so historical data for warships can be limited. While the author has access to all the data for one specific warship (Arrowhead 140) some datasets are interpolated.

The Montecarlo method has its own limitations as the number of input variables is increased the sample size needs to increase; the ideal sample size is achieved effectively through testing for convergence. The

Montecarlo method was chosen in preference to learning models due to the classification and size of the computers used. However, Montecarlo method comes with its own computational burden, the full model with a sample size of 1e7 can take a day to complete the full analysis with parallel processing on a 16-core machine.

6. FUTURE WORK

All the results in this paper are either preliminary or for demonstration due to the incompleteness of this data set. The primary objective of future work to complete this analysis is to collect all data points either, with a credible reference or an informed assumption. All results will be a spread with mean, max, min, standard deviation. This will represent the uncertainty, and hence risk, of the results.

It is desired as a conclusion to create carbon metrics such as: carbon per metre of weld, or carbon per hour of engineer time, carbon per tonne of ship. This way estimating carbon becomes as streamlined as estimating cost.

7. CONCLUSIONS

The analysis currently performed is too early to draw specific conclusions however, the trend is the makeup of the lifecycle of warship production draws focus away from the in-service phase (unlike commercial ships) due to the uniqueness of warship vs. commercial ships.

Early sensitivity analysis shows that some inputs effect the results of the model much more than others, identification of these are partly the point of sensitivity analysis. Combining this with the spreads of those inputs makes a large difference. When using open source past performance data for navies, the results from the low number of ships in the fleet the large spread of availability of those ships is almost less credible than an assumed value.

The research to date has been working well to answer the gaps cited in the review (section 2.4). The research is capturing all materials which go together to make a warship, mapping all components and processes in build. This will help to understand naval shipyard efficiency vs. a well-researched commercial body of evidence.

The diversity of warship operations and behaviours will be fully modelled so that and the builder, and, or the owner, can understand the effect of changes at any stage in the ship's lifecycle and make the most effective changes.

While security aspects will still exist, the aims of the research will be to disseminate all the information required to make global greenhouse emissions less without compromising national security of the states involved.

8. ACKNOWLEDGEMENTS

8.1 DATA

Data used is courtesy of the referenced author and available to the public, or the use is approved by Babcock International Group Plc. or the Natural Environment Research Council.

8.2 FUNDING

This paper is a product of an individual's [Stewart Downie] PhD programme from the University of Edinburgh and is fully funded by Babcock International Group Plc. and the 'City Region Deal' for Edinburgh and southeast Scotland.

8.3 ARTIFICIAL INTELLIGENCE (AI)

AI has been used in the following ways: to assist with streamlining coding, structuring existing code in a way which makes the computational burden less onerous, and it has been used to check errors in code written by the author, when the code did not perform as expected.

9. REFERENCES

1. IPCC. (2007). Climate Change 2007 Synthesis Report. https://www.ipcc.ch/site/assets/uploads/2018/02/ar4_syr_full_report.pdf

2. BRANDER, M., and DAVIS, G., 'Greenhouse Gases, CO2, CO2e, and Carbon: What Do All These Terms Mean?' https://ecometrica.com/assets/GHGs-CO2-CO2e-and-Carbon-What-Do-These-Mean-v2.1.pdf, 2021.

3. ZOU, J., and YANG, B., 'Evaluation of Alternative marine fuels from dual perspectives considering multiple vessel sizes', *Transportation Research Part D*, 115. https://doi.org/10.1016/j.trd.2022.103583, 2022.

4. GHG-Protocol. (2013). Technical Guidance for Calculating Scope 3 Emissions (version 1.0). In.

5. IMO. (2023). 2023 IMO Strategy on Reduction of GHG Emissions From Ships In RESOLUTION MEPC.377(80).

6. CARRASCO, P., BENDAÑA, R., PAREDES, A., MICHINEL, H., PEDRO FERNÁNDEZ DE CÓRDOBA, ARCE, M E., and ZARAGOZA, S., 'Analysis of key variables for energy efficiency in warships', *Journal of Engineering for the Maritime Environment,* 234, 26-36. https://journals.sagepub.com/doi/pdf/10.1177/1475090219864816, 2020.

7. UNCTAD, United Nations Conference on Trade and Development Handbook of Statistics https://unctad.org/system/files/official-document/tdstat47_en.pdf, 2022.

8. JANES. (2023). Janes Fighting Ships 2023-2024.

9. EYRING, V., ISAKSEN, I S A., BERNTSEN, T., COLLINS, W J., CORBETT, J J., ENDRESEN, O., GRAINGER, R G., MOLDANOVA, J., SCHLAGER, H., and STEVENSON, D S., 'Transport impacts on atmosphere and climate: Shipping', *Atmospheric Environment*, 44, 4735-4771. https://www.sciencedirect.com/science/article/pii/S1352231009003379, 2010.

10. FORD, G., MCMAHON, C., and ROWLEY, C., 'Naval Surface Ship In-Service Information Exploitation 2nd International Through-life Engineering Services Conference https://www.sciencedirect.com/science/article/pii/S2212827113005337/pdf?md5=8d5e5d2376fa9ab261e0199b1d835c9a&pid=1-s2.0-S2212827113005337-main.pdf&_valck=, 2013.

11. FAN, A., XIONG, Y., YANG, L., ZHANG, H., and HE, Y., 'Carbon footprint model and low–carbon pathway of inland shipping based on micro–macro analysis', Energy, 263. https://www.sciencedirect.com/science/article/abs/pii/S0360544222030365, 2023.

12. ACHENBACH, H., DIEDERICHS, S K., WENKER, J L., and RÜTER, S., 'Environmental product declarations in accordance with EN 15804 and EN 16485 — How to account for primary energy of secondary resources?', *Environmental impact assessment review*, 60, 134-138. https://doi.org/10.1016/j.eiar.2016.04.004, 2016.

13. Marine Traffic., 'AIS data for RRS Sir David Attenborough', *https://www.marinetraffic.com/*, 2024.

10. AUTHOR'S BIOGRAPHY

Stewart Downie, MSc, BEng, CEng, MRINA is a Principal Naval Architect in Babcock International's surface ship concept design team and lead for Babcock's Clean Maritime Group. He has experience of every single Royal Navy in-service class of ship and submarine, and some sadly now decommissioned. Notably he has been a naval architect on the Queen Elizabeth Class carrier, Type31/Arrowhead 140 frigate, and SSNR (now AUKUS) design phases. A graduate of the Universities of Plymouth and Newcastle, he is now researching a PhD with the University of Edinburgh in greenhouse gas emissions of Babcock's Arrowhead 140 frigate as part of the company's commitment to becoming net zero by 2040.

Warship 2025: The Future Fleet: Smart Technology, Sustainability and Autonomy
16th – 17th June 2025
Glasgow, UK

RECOVERABILITY OF ALTERNATIVE FUELS

T Beard, R Stanley and **I Savage**, BMT, UK

SUMMARY

As navies prepare for the future, they will inevitably encounter alternative fuels, either directly onboard naval vessels or when interacting with commercial vessels. This paper delves into the recoverability implications of three anticipated long-term alternative fuels: ammonia, methanol, and hydrogen. It examines the challenges associated with these fuels, including fire hazards, toxicity, and other risks arising from accidental or hostile damage. The paper also discusses suitable mitigation strategies, detailing how these challenges can be effectively managed and addressed in the event of damage.

Key words: Alternative Fuels; Recoverability

1. INTRODUCTION

The drive for Net Zero within the maritime domain is only viable with the use of alternative fuels, i.e. not from fossil fuels [1]. There is much debate about the route that defence, in particular navies, are going to undertake to maintain operations in the future [2]. However, it is guaranteed that no matter the route forwards, crews of naval vessels as well as Security/Police and SAR will encounter alternative fuels.

Law enforcement operations would result in placing crews into potentially harmful operations when searching vessels [3]. If the current geopolitical landscape continues it would result in these vessels attending incidents to support commercial craft following hostile action.

The IMO IGF Code for Low Flashpoint and Gaseous Fuels [4] should form the basis for all vessel designs, whilst originally for liquified natural gas many of the principles are applied for other alternative fuels including those discussed in this paper. The IGF code utilises a goal-based approach and as such alternative options can be explored on the basis that acceptable evidence is produced. However, this does not account for hostile action which could pose further risks and challenges out with those in the code, for example the loss of gas tightness caused by fragmentation.

The likely scenarios mentioned above, as well as the unknown fuel for naval assets, requires an understanding of the risks and hazards for the plethora of fuel options. These can loosely be drawn into two categories flammability and toxicity, Figure 1 shows high-level outcomes following damage to fuel containment. There is also the storage medium of the fuel that can pose risks and challenges for recoverability.

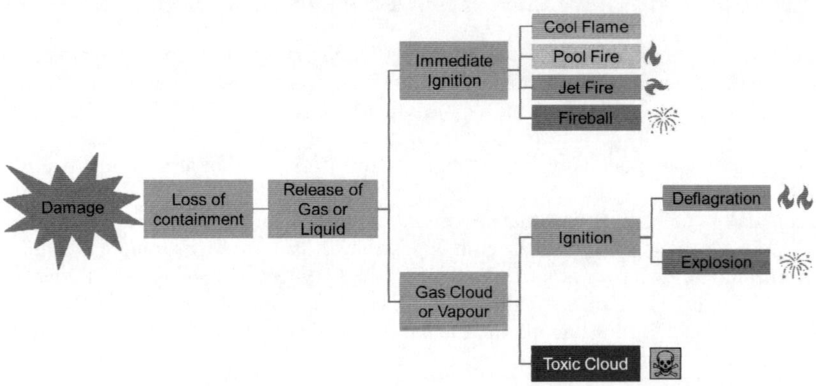

Figure 1: High level damage outcomes for gaseous & liquid fuels [5]

The paper explores the properties of the fuels and the hazards that can arise from them and the storage mediums in a Naval Environment, particularly after hostile action. Finally covering challenges for recoverability and proposing solutions to mitigate some of these challenges.

2. FUEL HAZARDOUS PROPERTIES

The major fuels that are likely to become more common are ammonia, hydrogen, methanol and methane; this paper will examine the first three options. Methane is not covered as currently it is viewed as a transition fuel in the Commercial sector, as such there is taxation involved for the use of methane (LNG) in the future [6]. The transition fuel aspect of methane would make this non-viable to be used in the Naval sector due to the timescales for the transition. Batteries will also become more popular for energy storage, although most likely on smaller vessels for complete power provision.

The following table showcases the various challenges associated with the fuels when compared against marine diesel.

Table 1. Significant properties of the main alternative fuels [7]

	Hydrogen, H_2	Ammonia, NH_3	Methanol, CH_3OH	REFERENCE (MGO / F76)
Flash Point	-253°C	-33°C	+12°C	+61.5
Flammability Limits in air (vol%)	4.0 – 75.0	15.0 – 28.0	7.3 – 36.0	0.7 – 5.0
Minimum Ignition Energy in air (mJ)	0.02	8.0	0.14	20.0
Explosion Risk	Large flammability range with low ignition energy	Medium flammability range with high ignition energy	Medium flammability range with reasonable ignition energy	Small flammability range with high ignition energy
Toxicity	None	Highly toxic to humans and aquatic life	Toxic to humans, but very low-toxicity to aquatic life	

It should be noted that ignition energy is actually a curve with the minimum ignition energy relating to the stoichiometric mixture of a fuel. Either side of this concentration the ignition energy increases until it reaches the flammability limits. At which point it will not ignite as it is either fuel rich or lean.

Table 1 highlights that ammonia and methanol are toxic to humans, this is explored in greater detail below.

Methanol poisoning occurs due to the metabolites that are produced, mainly formic acid. There is a cure for methanol poisoning, in the form of Fomepizole, although it is possible to also use high strength alcohol as a substitute [8]. The UK has set the following workplace exposure limits; long-term is 266 mg/m^3 (200ppm) and short-term is 333 mg/m^3 (250ppm) [9]. Acute inhalation at concentrations of 400 mg/m^3 [10] has been associated with toxicity. Whilst for ingestion the minimal lethal dose is 300 to 1,000 mg/kg [10], this vast range is due to varying metabolisms within people.

There is no cure for ammonia poisoning compared to methanol, whilst called poisoning it is actually a chemical reaction. This is due to the hydrophilic nature of ammonia and how it reacts with the mucous lining within aspects of the human body including lungs. Ammonia reacting with these linings creates ammonium hydroxide, a strong base (alkali). The effects of ammonia on a person are related to concentration and exposure time, this is shown in Table 2.

Table 2. Concentration & exposure limits for ammonia toxicity symptoms [11][12]

Concentration		Exposure Time (mins)	Signs & Symptoms
ppm	Volume %		
50	0.005	120	Irritation to eyes, nose & throat
100	0.01		Rapid eye and respiratory tract irritation
250	0.025	30 – 60	Tolerable by most people
700	0.07	Immediate	Irritation to eyes and throat
>1,500	0.15		Pulmonary oedema, coughing, laryngospasm
2,500 – 4,500	0.25 – 0.45	30	Fatal
5,000 – 10,000	0.5 – 1	Rapid	Fatal due to airway obstruction

3. FUEL HAZARDS

The high-level outcomes from hostile damage are shown in Figure 1. One of the potential failure modes not included in Figure 1 is Boiling Liquid Expanding Vapour Explosions (BLEVE), which is possible for liquids stored in tanks and is covered within the paper. Each fuel is reviewed separately to identify each unique challenge.

3.1 HYDROGEN

Hydrogen is stored as either a compressed gas or cryogenically. Both forms have pressure relief devices on the storage tanks. Due to the extremely low ignition energy, hydrogen is more susceptible to provide jet flames. Although a delayed ignition creates the potential for both deflagration and detonations. Cryogenic hydrogen is also highly susceptible to a BLEVE event although this requires a prolonged external fire.

Hydrogen is colourless & odourless and burns with a near invisible flame in daylight, see Figure 2, & low thermal radiation. This means it is extremely difficult to identify a hydrogen accumulation or fire. A fire signature will be evident but would require training to understand this, i.e. thermal signature, but is detectable by suitable sensors.

3.2 AMMONIA

The standard storage of ammonia is in liquid form via cooling to circa -30°C. Due to the lower flammability limit ammonia is less susceptible to jet flames although these may occur from pressure relief valves. Thus, increasing the chances of deflagration and/or detonations. However, because of the small flammable range and high ignition energy it is less of a fire hazard in normal conditions. Hostile action would likely increase the fire risk associated with ammonia. Due to the storage mechanism for ammonia, it is also susceptible to a BLEVE event similar to hydrogen. Ammonia burns with an orange/yellow flame, which is visible, see Figure 2.

Ammonia is colourless but has a significantly pungent odour at extremely low concentrations. It is also reactive with many chemicals and as discussed in the previous section highly toxic.

3.3 METHANOL

Methanol is stored as a liquid at ambient conditions although with such a low flashpoint, compared to diesel, there is still a risk of a BLEVE if heat was applied to a tank top. Although this is only likely from above depending on the layout of the vessel. Due to the presence of pressure relief valves/vents a jet flame may be possible similar to ammonia and hydrogen. There is a greater risk of deflagration and detonation, compared to ammonia, due to the larger flammable range and lower ignition energy.

Methanol burns with a light blue flame that is almost invisible to the human eye in daylight and with no smoke, see Figure 2. Although similar to hydrogen surrounding material may show evidence of a methanol flame and would require training to understand a methanol fire signature.

Methanol is also toxic to humans as discussed in the previous section, which would need to be considered for both normal compartment SOPs and under damage conditions.

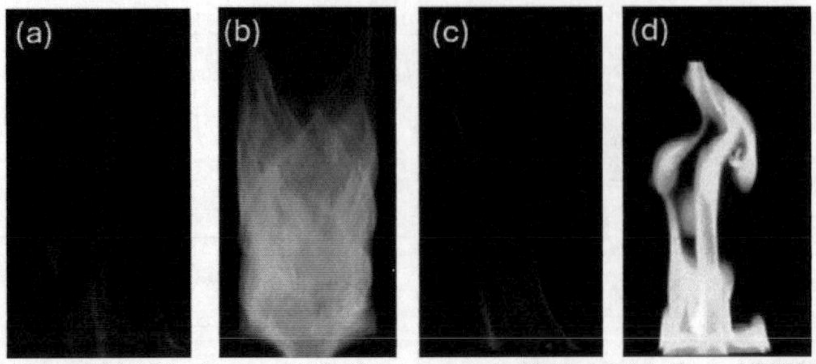

Figure 2: Flame colours of (a) Hydrogen [13], (b) Ammonia [14], (c) Methanol [15] & (d) Diesel [15]

4. RECOVERABILITY

4.1 OVERVIEW

Recoverability is a key component within the broader concept of Survivability, which is defined as 'The capability of a weapon system to continue to carry out its designated mission(s) in a combat threat environment' (NATO, ANEP-43, 2003). It is comprised of the following components:

- Susceptibility – avoidance of damage;
- Vulnerability – limitation of damage;
- Recoverability – containment, control and repair of damage.

Recoverability is typically split into 7 'pillars' [16] as shown in Figure 3 and described below.

Figure 1: The 7 'Pillars' of Recoverability [16]

Situation Awareness - The collection, display and assimilation of information that provides the Command team with a clear understanding of the external threats, mission priorities, material state, ongoing incidents and existing platform defects linked to estimated times to repair.

Management - The effective co-ordination and prioritisation of recoverability activities to best meet the command aim utilising the available data, experience, knowledge and resources.

Containment - The ability to prevent the further loss of capability due to the spread of damage to unaffected areas and sections of the ship whilst also isolating potential sources of incident escalation. Note that inherent redundancy and separation of vital systems with minimised damage footprints fall to vulnerability reduction measures.

Prosecution - The ability to limit the further loss of capability within the damage zone through aggressive action and resolution of a damage incident.

Recovery - The ability to provide resilience in system operation and means of restoring partial or full capability within specified timescales using predominantly onboard resources unless otherwise stated.

External Assistance - The ability to capitalise on external assistance from nearby units or shore-based support facilities to aid recoverability action while still in an area of high risk/threat. Conversely, the ability to render assistance to another unit or vessel that requires assistance.

Escape & Evacuation - The safe recovery of personnel from untenable spaces to a place of relative safety while maintaining the chain of command and mission priorities.

The recoverability mitigations for each of the three fuels assessed (Hydrogen, Ammonia and Methanol) are considered against the pillars of recoverability in the proceeding sections.

4.2 SITUATIONAL AWARENESS

Fundamental to situational awareness is the information that can be gathered to inform decisions on the management of the incident. All three fuels will require suitable leakage / gas sensors to be employed in all spaces that any of the fuels could be present with a greater increase of sensors in locations that the fuel will be present in standard operating conditions. The pipework should also be monitored by the Power/Propulsion Management System or Integrated Platform Management System (IPMS) to identify leaks based on pressure differentials, similar to some existing systems. Sensors should also be located in relevant ventilation trunking such as from zonal distribution as well as a compartment. Whilst air intakes near vents should be immediately closed upon a vent sensor activating it is recommended that sensors are located in other trunking to ensure that the closure has occurred. This would also provide the ability to manage any risk if higher concentrations were prevalent in compartment vent trunking rather than within a compartment due to ventilation conditions. Whilst this goes against the principle of the IGF code, this should be considered as the IGF code does not account for battle damage, for example breaching containment.

Suitable flame detection sensors are also necessary, this is because the flames are either near invisible (hydrogen & methanol) or there is a considerable probability of a toxic atmosphere (ammonia). Standard RN practice dictates that on activation of a fire detector, personnel arrive at the scene of the incident to verify the detection within 2 minutes, they would normally be unprotected However, in the case of ammonia this is not safe. Whilst for both hydrogen and methanol it is unlikely that a human would identify a flame in compartment lighting. Therefore it is necessary that CCTV cameras, ideally with enhanced detection algorithms, and suitable infrared detection systems are employed instead.

A person entering a compartment that has identified a hydrogen leak should wear anti-static clothing, due to the low ignition energy of hydrogen. Whilst for methanol standard PPE would also be eye protection and gloves, which would be required as part of the SPE for the compartment to ensure a continuous rolling attack is achievable without risk to personnel.

4.3 MANAGEMENT

Communication systems would be expected to mirror those of a modern warship, utilizing VOIP for point-to-point and group communication, with redundancy for power loss; any communications equipment within spaces at risk of fuel leakage must have appropriate enclosure ratings to allow continued safe operation. IPMS, fire system, and alarm data should be efficiently conveyed to operators for situational awareness, enhanced by data concerning fuel concentration and system status to overlay the additional hazard status on an intelligent Damage Control (DC) Incident board.

SOPs and EOPs for alternative fuels must cover all leakage scenarios, forming the core of response protocols while integrating with fire/flood procedures. Relevant data should be communicated to operators and the DC incident board, ensuring command teams have the necessary insights for decision-making. In development of operational safety mitigations, emergency shut down of plant on leakage should be considered (as required for UMS vessels by the IGF code). The design needs to consider continued military mission delivery and may require additional levels of redundancy, or greater segregation/enclosure protection, to allow complete machinery spaces to be shut down to avoid ignition of leaking fuel.

Safe and effective incident management would require a thorough understanding of the state of the fuel system onboard. Prioritisation would be to minimise leakage from pipework and shut down ignition sources (non-safe equipment); whilst if the storage tank is damaged then the priority would be removing the flammable or toxic cloud from the ship whilst initiating automated responses. The incident commander must be cognisant of any ventilation outlet locations and wind conditions to enable effective communication to, and therefore safety of, external parties that may be rendering support.

For ammonia and methanol the command aim must factor in hazards alongside fire and flood, as the toxicity poses a risk of mass casualties or mission failure. The strategy should balance military objectives with life safety, considering how this new hazard impacts operational priorities.

For unignited hydrogen the command aim must factor the explosion risk from accumulation of hydrogen within the compartment especially the concentrations of hydrogen stratification and whether a homogeneous mixture is forming.

4.4 CONTAINMENT

A key requirement across all fuels is the ability to prevent escalation by immediately isolating fuel supply systems in the event of either fire or leakage. Generally, this would be from design integration such as blast resistant structure and / or A60 fire protection, however this cannot be guaranteed to be completely effective after hostile action due to possible fragmentation or weapon blast pressures. The emergency shut down of that system and non-safe equipment is something that can be controlled to mitigate escalation, as discussed in the previous section. This will therefore require redundancy onboard a vessel to ensure power and propulsion are maintained, although this is prevalent on many warships already.

Each compartment will have ventilation requirements in normal operations for a specified number of air changes per hour, depending on the fuel and compartment function. This may be deployed for some leaks depending on the fuel and the concentrations although where fuel will always be present then mechanical ventilation at calculated rates should be used. All ships external ventilation openings should have gas monitoring alarms and be located to prevent re-circulation into air intakes. In the event of fire near a storage tank then a water spray system should be operated to keep the tanks cool and minimise the risk of a BLEVE. Regarding containment, each fuel also has unique requirements which are explored below.

Hydrogen requires significant ventilation due to the large flammability range, this would mean the use of mechanical ventilation to aid the removal of hydrogen from compartments, via blast proof trunking, due to the buoyant and diffusive nature of hydrogen; this would need to be explosion proof equipment, alongside all other essential equipment.. A zonal system for hydrogen containment would be the tank and plant compartment(s).

Ammonia should be provided with a tiered containment system due to the levels of toxicity. Primary containment would be the fuel tank, secondary containment would be the plant compartment(s) with tertiary containment being zonal. The tertiary containment should align with a smoke containment zone or ventilation zone, to minimise the additional requirements for ventilation. The zone would need to be as small as practicable to minimise the hazard to personnel; accommodation spaces would not be located within this zone for the same justification. This zonal configuration would be designed to limit the spread of the toxic cloud rather than fire; with traditional fire zones and boundary protection used for the latter. This zone may require closure until the vessel could return to port depending on the cause and quantity of ammonia realise. Therefore, it should be positioned to ensure access to areas outside the control zone. The use of armour should be considered to minimise battle damage, with the priority being the tank which would also be applicable to hydrogen. For minor leaks ventilation could be employed to remove vapours, however in the event of a large leak (tank rupture) then ventilation should be closed and alternative means used which are covered in section 4.5. A separate bilge system should be employed for ammonia containing compartments to reduce potential cross-contamination of bilge water in other parts of the ship.

Methanol should employ zoning as per hydrogen, i.e. the tank and compartment to contain methanol vapours. A separate bilge system is required as per ammonia, although more for condensate than vapour removal.

4.5 PROSECUTION

Water spray systems are required to protect the spaces related to all of the fuel types for different reasons, in addition to compartment fire extinguishing. Hydrogen would use this to maintain cooling on the tanks.

Ammonia and methanol would utilise water sprays for cooling tanks and also removing/controlling the toxic cloud, which is why a separate bilge system is required.

The water spray system should be automatically triggered upon either fire or gas leaks after an agreed concentration has been reached, depending on the fuel, via a dedicated detection system akin to a local application system used on high-risk equipment in Category A machinery spaces. This ensures a fast and proportionate response to the risk and minimising potential for escalation especially after hostile action. This would require any equipment that is deemed to be essential to have a high enough IP rating to withstand the activation of the water spray system, in addition to being explosion rated if it presents an ignition hazard.

In the event of combustion of the hydrogen there is very little that can be done to prevent the hydrogen flame. A premixed fire would result in something akin to a flashover, after which it would be a jet flame until the fuel supply can be isolated. It is therefore important to minimise combustible material within compartments containing hydrogen systems, thereby limiting the potential for escalation through ignition of surrounding materials. However, if a fire does occur then current firefighting procedures should be used to tackle these secondary fires. In addition, given the normal extinguishing strategy for hydrogen is to isolate the fuel source, the design must consider the ability to isolate leakage, and the sub-division of high-risk leakage areas as part of the containment strategy. Dilution based fire fighting systems could be employed however this must be carefully considered to ensure it would sufficiently reduce the region of flammability, for example carbon dioxide above 60% concentration in air would be required to prevent ignition [17].

Ammonia would require Specialist Protected Equipment for anyone entering the compartment/zone after an event, e,g. this would be necessary for casualty recovery.

Water spray systems protecting methanol installations should be provided with Alcohol-Resistant Film-Forming FluoroProtein foam (AR-FFFP), alternatively carbon dioxide can be used to extinguish the fire [18].

4.6 RECOVERY

Recovery is very much a sliding scale depending on the extent of the damage. If the damage is to a fuel tank, then this would need to be replaced and there would also be no fuel remaining, which emphasises the requirement for tankage and other essential equipment to be subjected to vulnerability analysis. However, if the damage was to pipework then the vessel can be recovered, depending on the severity of related collateral damage due to fire. The post damage assessment must focus on identifying where the pipe is damaged. The PMS may provide indication of where this occurred, depending on the cause of the shutdown.

For recovery of pipework following a leak detected in the outer pipe it would be advisable to trace the pipework that has been shut down, starting from the tank, and 'manually' overriding the valve, from the safety of the control room, opening and closing individual isolation valves for each segment of pipework. Thus, minimising the quantity of fuel in the pipework but also allowing the identification of the section with a leak whilst maintaining containment. Portable detection for the fuel would be required and suitable PPE for the specific fuel should be used. If the secondary pipe has ruptured then the process should be conducted with either HP air or with an inert gas such as nitrogen.

If the damage is to the fuel tank(s) then this would not be recoverable and the tanks would need to be replaced. However, operationally if alternative tanks are available, the damaged element of the system should be isolated before switching over to an alternative tank.

Hydrogen sensors within compartments should be utilised to determine whether the concentration is below an agreed level for re-entry into the compartment, this would normally be half of the lower flammability limit (2%); these sensors will therefore need to be fire rated. Additionally, any person entering the compartment should wear anti-static clothing to prevent static discharge causing an ignition.

Ammonia fuelled vessels should consider the necessity to recover a compartment that has undergone a significant leak event until return to port because of the potential presence of toxic fumes. This is because the water spray system that is employed to remove the toxic cloud cannot run indefinitely because the water must be stored in separate bilges as per the IGF code [4].

Methanol does not require a compartment shut down in the event of leakage or fire because of the reduced toxicity (compared to ammonia) and adequate protection afforded by standard PPE. Whilst methanol vapours are likely to accumulate in the bilge due to its dew point of ~64°C, this also means that water spray systems could be used to support the removal of methanol from surfaces.

Due to the potential for significant bilge water to accumulate it could be explored to determine if neutralising the bilge water prior to discharge would be acceptable with the classification society (or IMO more generally).

4.7 EXTERNAL ASSISTANCE

Any external assistance must be made aware of the potential risk in the environment, both air and waterborne, because of the fuel upon arrival and be in communication with the damaged vessel.

Hydrogen and ammonia would require assistance to approach from upwind to minimise potential entrainment of the flammable/toxic cloud into the support vessel air intakes. This would also apply to any helicopter operations near the damaged vessel. If this is unavoidable then air intakes should be closed on the vessel (if fully CBRN capable) and aircrew use suitable breathing equipment. If this is not possible then the attending vessel should remain at a safe distance with assistance provided by boat and air. The attending vessel should also ensure its monitoring regime is maintained to prevent contamination approaching hazardous levels.

Ammonia and methanol would need to offload the bilge system to a dedicated facility once alongside, unless an agreed overboard discharge solution can be used and justified against Best Available Techniques/Best Environmental Practices.

Methanol, due to its low dew point, is likely to remain in solution and therefore present no additional hazards to an attending vessel. However, if conducting prosecution or recovery operations then additional local precautions (e.g. PPE) would be expected to be taken.

4.8 ESCAPE AND EVACUATION

Escape and evacuation requirements vary for the fuels due to the differing hazards associated with them. Although loss of containment from escape routes should be minimised as there is a need for air locks for entrance and exit in normal operation to compartments as per the IGF code.

Hydrogen does not present significant challenges for escape and evacuation, arrangements would be expected to be similar to current design practice because it dissipates quickly and premixed combustion is similar to a flash over. However, considering the loss of compartment containment through the escape routes, together with ensuring low level escape access will be important; additionally, the arrangement of life saving appliances must be deconflicted from hydrogen and hydrogen plant space ventilation outlets and tank locations which could present additional challenges to the layout of a modern warship that tend to have low areas of exposed upper deck.

Ammonia would require escape routes that avoid the ammonia control zone to ensure safety to personnel, with additional PPE required in case of breach of the control zone with all exits being via an airlock. Again, consideration must be given to the potential contamination path that the escape route itself may create. Evacuation of the vessel would be akin to doing so whilst under a CBRN threat with the use of appropriate PPE and handheld monitoring equipment. As with Hydrogen the position of compartment and tank pressure venting arrangements are critical to ensuring that the ship can be safely abandoned in all circumstances.

Personnel entering the compartment, e.g. for search and rescue, will be required to enter the in appropriate PPE including a protected breathing supply. Upon leaving the compartment a full wash down at a safety station would be required, immediately outside the compartment airlock exit prior to traversing any other area of the ship.

Methanol escape procedures would be similar to current operations as standard PPE should be sufficient mitigation, whilst evacuation arrangements will need to have similar considerations as given for Hydrogen, above.

5. CONCLUSION

The use of alternative fuels in the maritime domain will provide several challenges for Navies in the future, regardless of whether used onboard a vessel. The risks vary for the fuels but can be summarised as either highly flammable (hydrogen), highly toxic (ammonia) or a mixture of both (methanol).

Hydrogen and methanol have similar aspects across some of the pillars of recoverability. Whilst the greater toxic nature of ammonia causes considerable challenges to recover a vessel. A vessel using ammonia would need considerable thought to the layout to enable the zonal control required.

Table 3 provides a RAG comparison for the recoverability of the three alternative fuels. The scoring is 1 to 3, with 1 being low recoverability performance. Diesel is the baseline for comparison and as such is scored as a 3 across all criteria.

Table 3. RAG comparison of recoverability

	Hydrogen, H_2	Ammonia, NH_3	Methanol, CH_3OH	REFERENCE (MGO / F76)
Situational Awareness	2	1	2	3
Management	2	1	2	3
Containment	2	1	2	3
Prosecution	2	1	2	3
Recovery	2	1	2	3
External Assistance	2	1	3	3
Escape & Evacuation	2	1	2.5	3

Hydrogen has reduced scores due to the reliance on sensors and stand-alone tanks. This is coupled with no formal means of extinguishing a hydrogen fire, although ventilation is far easier.

Methanol scores comparable to diesel for recovery, external assistance and escape & evacuation. This is primarily due to the integrated tanks and reduced buoyancy of vapour clouds. However, there are challenges for situational awareness primarily due to the difficulty in using a person to confirm incidents. Containment will be more challenging due to the additional segregated ventilation, and prosecution requires suitable firefighting systems.

It is evident that of the three fuels, methanol would be best placed to be a naval replacement fuel based on recoverability.

6. REFERENCES

1. IMO, '2023 IMO Strategy on Reduction of GHG Emissions from Ships', https://www.imo.org/en/OurWork/Environment/Pages/2023-IMO-Strategy-on-Reduction-of-GHG-Emissions-from-Ships.aspx, 2023.

2. UK Ministry of Defence, 'Maritime Operating Concept', https://www.gov.uk/government/publications/maritime-operating-concept-maropc, 2023.

3. UK Defence Journal, 'British naval ship assists in $14m Caribbean drug seizure', https://ukdefencejournal.org.uk/british-naval-ship-assists-in-14m-caribbean-drug-seizure/, 2024.
4. IMO, 'International Code of Safety for Ship Using Gases or Other Low-flashpoint Fuels (IGF Code)', 2023.

5. BEARD, T., 'Alternative Fuel Implications for Navies', *Lloyd's Register Naval Ships Technical Committee Meeting*, 2022.

6. EUROPEAN UNION, 'Regulation (EU) 2023/1805 of the European Parliament and of the Council of 13 September 2023 on the use of renewable and low-carbon fuels in maritime transport, and amending Directive 2009/16/EC', 2023.

7. NEWMAN, S., and BEARD, T., 'The Viability of Low-Carbon Fuels & Green Technologies for the Front-line Naval Vessel', *RINA Warship*, 2022.

8. UK HEATLH SECURITY AGENCY, 'Methanol: toxicological overview', https://www.gov.uk/government/publications/methanol-properties-incident-management-and-toxicology/methanol-toxicological-overview, 2024.

9. HEALTH & SAFETY EXECUTIVE, 'EH40/2005 Workplace exposure limits', 2020.

10. UNITED NATIONS, 'ENVIRONMENTAL HEALTH CRITERIA 196; Methanol', 1997.

11. UK HEATLH SECURITY AGENCY, 'Ammonia: toxicological overview', https://www.gov.uk/government/publications/ammonia-properties-incident-management-and-toxicology/ammonia-toxicological-overview#main-points, 2024.

12. U.S. DEPARTMENT OF HEALTH AND HUMAN SERVICES, 'Toxicological Profile for Ammonia', 2004.

13. INTERNATIONAL FIRE PROTECTION MAGAZINE, 'Hydrogen and fire safety: Detecting the most flammable element on earth', https://ifpmag.com/hydrogen-and-fire-safety-detecting-the-most-flammable-element-on-earth/, accessed May 2025.

14. SHOHDY, N N., ALICHERIF, M., and LACOSTE, D A., 'Transfer Functions of Ammonia and Partly Cracked Ammonia Swirl Flames', *Energies 16*(3), 2023.

15. MISHRA, S., et al, 'Experimental investigation of small-scale CS_2 (carbon disulphide) pool fires', *Journal of Process Safety and Environmental Protection volume* 145, 2021.

16. BARTLET, S., and SAVAGE, I., 'Naval Autonomous Surface Vehicle Recoverability', *Engine as a Weapon*, 2023.

17. LEWIS, B., and VON ELBE, G., 'Combustion, Flames and Explosions of Gases, third edition', 1987.

18. BEARD, T., and GRIFFITHS, D., 'Dual Fuel Technology: A route to reduce emissions', *International Naval Engineering Conference*, 2024.

7. AUTHOR BIOGRAPHIES

Dr. Thomas Beard CEng, MIMarEST, MIMechE is the Clean Shipping Service Lead at BMT UK. Technically responsible for Clean Maritime, including new build & retrofit of vessels as well as infrastructure requirements. He has a doctorate in Hydrogen Safety from Loughborough University and over a decade of experience in alternative fuels. He is the co-chair of the National Maritime Hydrogen & Future Fuels group as well as being a member of several other industry groups and steering/advisory boards for academic research.

Rebecca Stanley CEng, CMarEng, MIMarEST is the Fire Protection Capability Service Lead at BMT UK. She is a Chartered Engineer with a First Class Masters in Mechanical Engineering from the University of Manchester with over a decade of experience in Marine Engineering. She was elected to the IMarEST Council and sits on the Board of Trustees as a council trustee.

Ian Savage CEng, CMarEng, MIMarEST is the Chief Engineer for Transversals and Technical Assurance in BMT. Specialist in Naval recoverability he has undertaken engineering design and leadership particularly focusing on Naval Fire Protection and Damage Control on a wide range of Naval and Naval Auxiliary vessels.

FUTURE BATTERY PROTECTION STRATEGIES FOR NAVAL VESSELS

T Berry MSc MIET MIMarEST and **T Beard** PhD CEng MIMechE MIMarEST , BMT UK

SUMMARY

Thermal runaway is always a concern when considering Energy Storage Systems (ESS), battery technology and the use of Lithium-ion battery modules. Within the maritime industry, existing cooling methods and battery monitoring systems (BMS) monitor the risk of thermal runaway and where possible provide mitigation, however, there is an additional cooling method to be investigated to ensure that the risk becomes as low as reasonably practicable (ALARP). Within the automotive industry and within data centres, immersion cooling is being explored as a cooling medium. The evidence gained so far has demonstrated the benefits of immersion cooling, reducing the impact of a single module going into thermal runaway by eliminating a chain reaction allowing the incident to be managed. This paper explores the research undertaken so far, how this would work for the cell chemistry available, integration considerations, how this can support a de-centralised ESS and how this can be applied to a Naval application.

NOMENCLATURE

ALARP	As Low as Reasonably Practicable
ASW	Anti-submarine Warfare
BMS	Battery Management System
ESS	Energy Storage System
EV	Electric Vehicles
HVAC	Heating, ventilation, and air conditioning
IACS	International Association of Classification Societies
IEC	International Electrotechnical Commission
LfE	Learning from Experience
MCA	Maritime Coastguard Agency
MGN	Marine Guidance Note
MoD	Ministry of Defence
SME	Subject Matter Expert
SoC	State of Charge
SoH	State of Health

1. INTRODUCTION

The application of battery technology within the maritime industry is not a new concept, the first fully electric short-sea ferry powered by Lithium- Ion batteries sailed in 2015, the Norwegian Ferry MV Ampere [1]. More recently, in 2023 it was announced that the first battery tanker 'X', PowerX, Inc., which will be scheduled for domestic and international field testing in 2025 [2].

Whilst there are a variety of different chemistries available for batteries the more favourable is Lithium-Ion. This is due to the increased volumetric and gravimetric energy densities, as shown in Figure 1.

The use of ESS for Power and Propulsion supports the Ministry of Defence's Climate Change and Sustainability Strategic Approach 2050 [3]. Whilst the use of ESS can support Net Zero, it also has the potential to provide operational advantage. It can support engines to operate more efficiently, thus potentially increasing the range of the vessel and reducing the carbon footprint of operations. It also could be used to support missions where noise is of concern, such as anti-submarine warfare (ASW). Finally, ESS could support survivability of a vessel, by having a disaggregated solution for energy storage that could support vessel recoverability.

Currently batteries are used for ESS on Astute class submarines [4], although they use Nickel-Zinc cell chemistry [5] that are not susceptible to thermal runaway. However, if the Ministry of Defence (MoD) were to utilise and lever the benefits of Lithium-Ion on Naval Platforms the safety of adopting such technology on ships will be questioned. The risk associated with storage and charging of such batteries must be mitigated and reduced to ALARP.

Thermal runaway remains a substantial risk [6] relating to the use of Lithium-Ion batteries. If a battery cell is damaged or intense heat applied, then a chain reaction, known as thermal runaway, is likely to occur unless mitigated. This is likely to have a detrimental effect on the vessel and the crew due to the fierce reaction and off-gassing from the effected cells. Noting the operational requirement of Naval vessels and the hostile environments in which the ships may operate, this must also be factored into the design intent of an ESS solution.

Traditionally, battery modules are air cooled through heat dissipation and supported by a sufficient Heating, Ventilation, and Air Conditioning (HVAC) system to control the ambient and module temperature. However, this arrangement is supportive of a centralised ESS system to allow integration and weight management. Due to this, the risks surrounding thermal runaway and the additional risks faced during operation and combat, the suitability of a centralised ESS for a Naval application is not favoured. For the purposes of this paper, an alternative solution to air cooling is sought to support innovative solutions when moving towards a zonal resilient system and its integration.

The challenge with any ESS solution is to minimise the impacts of thermal runaway. Currently this is normally done by continuously removing the excess heat from the damaged cells via water for a prolonged period [7]. This strategy to minimise the risk to the vessel is not viable, due to the significant amount of water used. This has potential to influence ship stability and removal of any surface water would need to be planned to ensure continued compliance with MARPOL Regulations. Therefore, this paper investigates an alternative way to support cell cooling that could minimise the potential for thermal runaway between cells of a battery module.

2. THERMAL RUNAWAY

Lithium-ion batteries have become the most popular cell battery used within the maritime industry due to the high energy density in comparison to its weight as shown in Figure 1 below. However, the negatives include the requirement for a sophisticated BMS and the potential risk of thermal runaway.

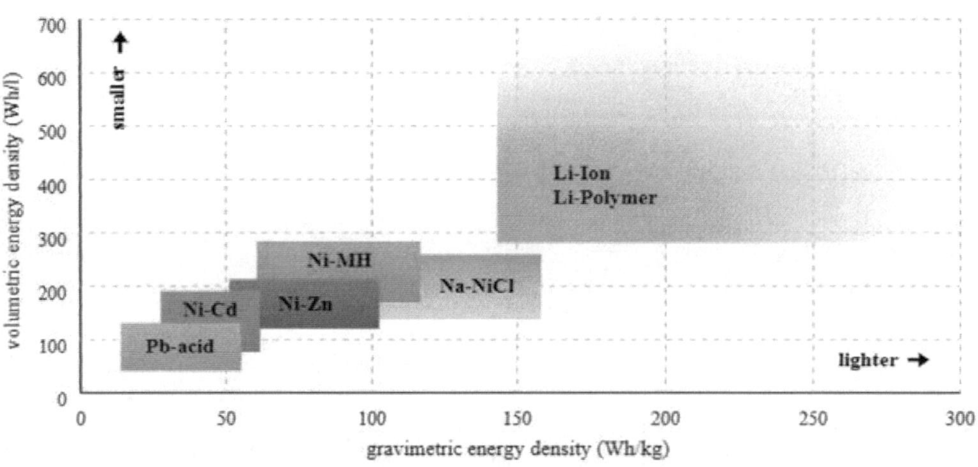

Figure 1: Battery Volumetric Energy Density [8]

To provide context, according to Dai [9], 'thermal runaway is a self-accelerating reaction that can occur when the battery is overcharged, short-circuited, or physically damaged', this is due to certain failure conditions as shown in Figure 2 below.

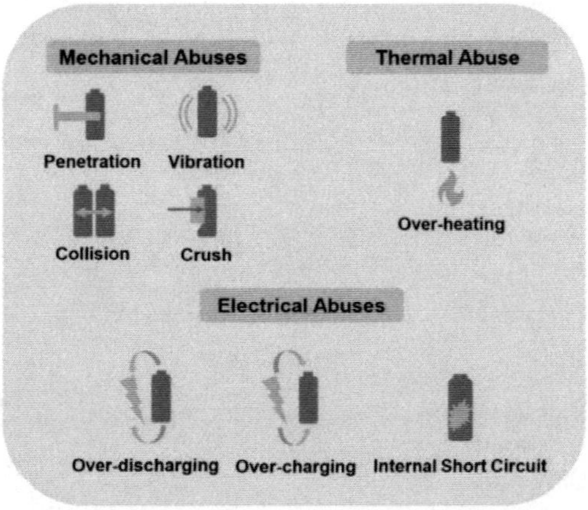

Figure 2: Thermal Runaway Failure Conditions [9]

A naval vessel is at higher risk of mechanical and thermal abuses due to hostile action. Whilst electrical abuses could be possible, this should be mitigated against via the BMS and quality control.

Thermal runaway is a self-sustaining, exothermic reaction within a battery that leads to a rapid increase in temperature and pressure. This can lead to a chain reaction across the battery cells which may result in a battery emitting toxic gasses, catching fire, or even exploding.

To mitigate thermal runaway battery management and heat dissipation is key. This paper concentrates on heat dissipation and how this can be mitigated via immersive cooling.

3. IMMERSIVE COOLING

Immersive cooling is when battery cells (in this instance) are submerged in a dielectric fluid which is non-conductive. This fluid absorbs the heat generated by the battery cells through direct contact between the heat source and the fluid. This technology has predominantly been used for data centres, automotive electric vehicles (EV), deep sea research by protecting electronics from corrosion and act as pressure-compensating fluids in remotely operated underwater vehicles (RoV), medical devices and High Voltage applications to act as a cooling medium for components and cables.

Immersive cooling falls into two categories, single-phase immersion cooling and two-phase immersion cooling, see Figure 3 and Figure 4. The single-phase cooling method is where the fluid remains in a liquid form and circulated in a closed loop system utilising a heat exchanger. The two-phase method is when the fluid evaporates when the heat is absorbed and then is condensed. The gas is cooled by a heat exchanger, such as a condenser coil.

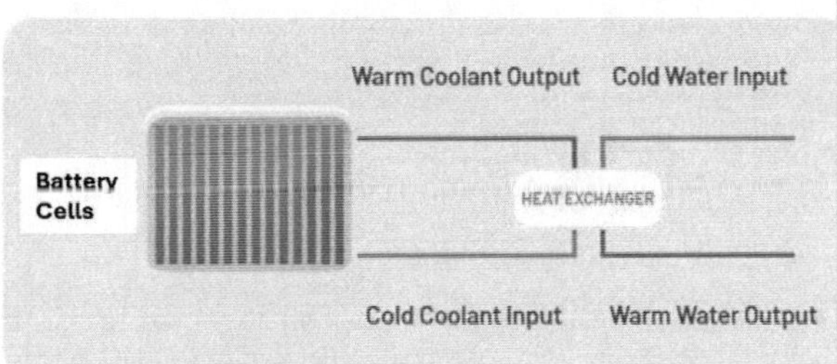

Figure 3: Single-Phase Immersion Cooling [10]

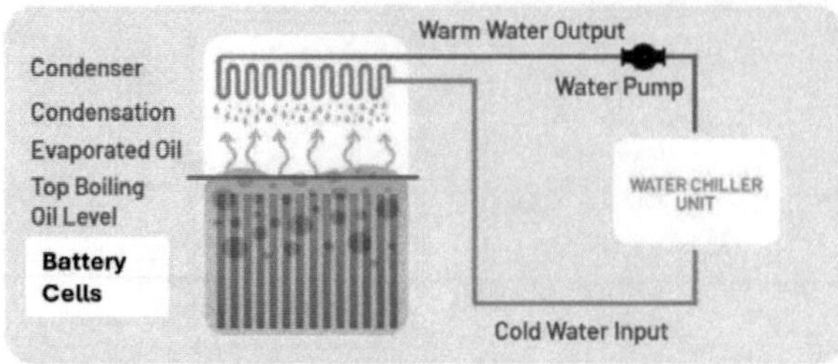

Figure 4: Two-Phase Immersion Cooling [10]

When considering the application of immersion cooling for the Naval sector, the single-phase method would be recommended, this is due to the following factors:

- Less maintenance required
- No fluid losses as it is a closed loop system
- Non-toxic bio-degradable fluids available
- Lower cost than that of the more complicated two-phase system
- Potential to act as a fire suppressant

Immersion cooling is known to dissipate heat more efficiently than traditional air-cooled systems, therefore removing the need for an air-cooled integrated system within the battery bank.

Historically, Li-ion batteries have a limited lifespan (discharge cycles) due to the fast charge/ discharge rates which heats the batteries. By adopting immersion cooling, there is the availability to optimise the temperature of the batteries during operation, hence it is expected to extend the battery life and therefore reducing the carbon footprint over the life of the battery as shown in Figure 5 below.

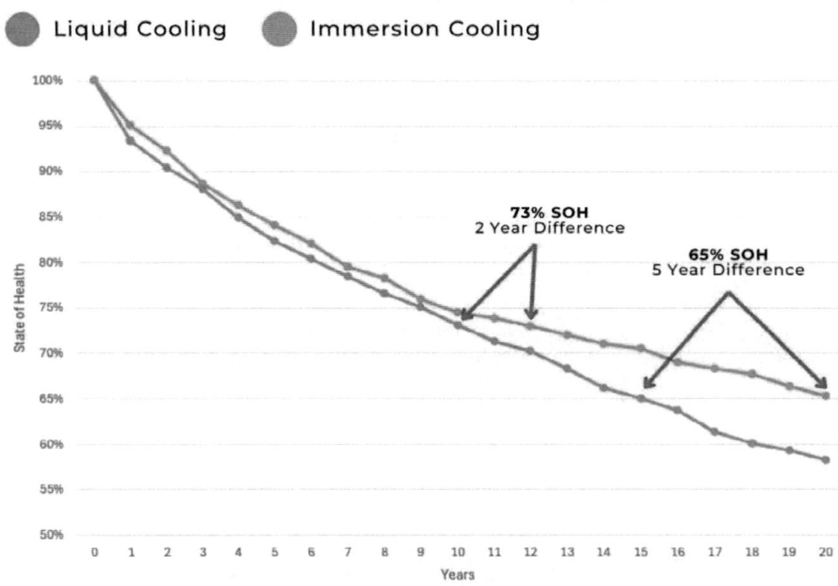

Figure 5: Battery Longevity Through Immersion Cooling [11]

Although some of the benefits are variable depending on the fluid used, there is a renewed focus to investigate/develop this for the automotive sector, therefore providing data and any learning from experience (LfE) for the naval sector.

Figure 6 shows a simplified diagram of the arrangement required for a single-phased immersive cooled battery system. Seawater could be used as the cooling medium for the heat exchanger, but there is also an opportunity to use chilled water to ensure that the temperature can be controlled whilst operating in hotter climates. The dielectric fluid which is contained within the module and circulated through the pump and heat exchanger is within a closed loop system which may also require a dielectric fluid reservoir for the system. The battery bank itself would be sized based on the platform requirements, furthermore, the heat exchanger could be sized such that it provides some form of power back to the system.

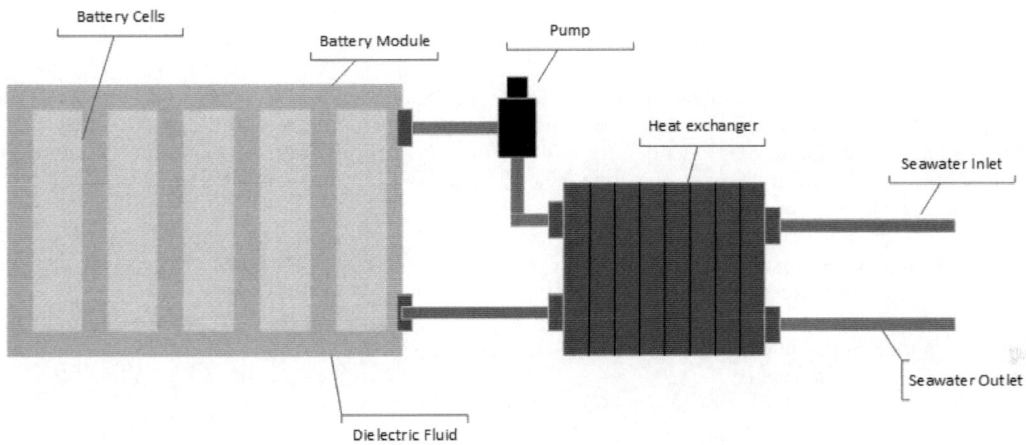

Figure 6: Immersion Cooled Battery Module

4. INTEGRATION CONSIDERATIONS

As mentioned in the previous section, there is no longer a need for an integrated air-cooling system within the battery bank, thus reducing power consumption and weight. However, considerations would be required for the immersion cooling system space and weight impact against other requirements such as survivability. This would vary dependent on the size of the ESS based on the power requirements and the amount of immersion coolant required. This would need to be assessed further.

As the coolant fluid could act as a fire suppressant, it reduces the risk of thermal runaway propagation within the battery module. It is anticipated that immersion cooling will supress a thermal runaway for such a duration in time to allow the BMS to sufficiently react and isolate the affected battery module to reduce the risk. The BMS shall provide real time monitoring of the battery cell temperatures and act accordingly. Therefore, by removing the excess heat from the damaged cell it prevents propagation to the neighbouring cells. The duration would be dependent on the type and quantity of fluid used. The fluid should be contained as much as practical, thus supporting a zonal configuration for the ESS. It may be possible to create the component level to be isolated reducing the risk of fluid. This would mean that in the event of hostile action and loss of containment surrounding a cell that minimal damage occurs. This segregated approach should be near the centre line of the ship to support firefighting water impacts on stability whilst affording as much protection from the vessel as possible. This would therefore support the need for further analysis to be conducted to determine the safety measures required for the ESS and a safety case developed.

As indicated above, a key integration consideration is the BMS, as per the traditional BMS, it is required to monitor battery parameters including voltage, current, state of charge (SoC), state of health (SoH) and temperature of each battery cell. For systems that utilise immersion cooling, this can be further exploited to support the health of the system via gauges on the coolant system. The BMS can also react to any thermal events, if an increase of temperature is detected then the flow of the coolant is increased, if in the event of minimal power discharge resulting in lower temperatures then the flow of the coolant can be reduced.

The electrical architecture of the vessel may need modification to support the easier integration of such a system.

5. HOW CAN THIS BE APPLIED TO A NAVAL APPLICATION

By adopting immersion cooling and mitigating thermal runaway, this will allow the opportunity to utilise ESS. This can support naval operations due to providing an efficient means of storing excess power and thus allowing engines to be operated more efficiently, in the form of a hybrid propulsion architecture. This has been conducted on vessels of varying size and is fundamentally driven by the operating profile of the vessels. There are hybrid vessels operating in the commercial sector from sub 24m to over 100m in length. It is very much driven by the energy demand and the role the system is to play. This can also provide greater resilience to the vessel via another form of energy storage, similar to nuclear submarines. Further, it could be used to support certain operations that have a requirement for a reduced noise signature.

Whilst an ESS could be leveraged to support normal operating conditions, modification to the BMS is possible such that the batteries could replace the emergency generator. Thus, allowing critical systems to be maintained. This could be extended to account for varying operational requirements via the BMS. Therefore, having different operating parameters, such as SoC, on the batteries depending on the operating scenario of the vessel.

For ships such as mine sweepers, ASW and survey ships, where loitering is an operational requirement, using an ESS would be beneficial, the benefits include but are not limited to the following:

- improved operational efficiency
- reduced maintenance
- a reduction of running hours on prime movers
- reduction or even elimination of CO_2
- reduction in noise and vibration

There is a further requirement for survivability of a naval vessel. The deployment of ESS could reduce the susceptibility, whilst the zonal configuration may reduce the vulnerability. The use of immersive cooling should have benefits for the recoverability of the vessel. Although there is a need to ensure that the fluid is continuously contained.

6. DLOD+ REVIEW

The Defence Lines of Development plus (DLOD+) [12] is an enhancement of the traditional DLOD that is used by the UK MoD to assess and manage military capability, programme or strategy normally within acquisitions. Impacts against the DLOD+ are discussed below.

- Training – Greater training will be required for the batteries and the immersion cooling. Whilst the technology will be different, training already exists for various other industries which could be leveraged.
- Equipment – This will be new to defence but as previously discussed is already commercially available.
- Personnel – Skills upgrades would be required to provide understanding of the BMS and the system.
- Information – Development will be required initially as per integration of any new technology. This could provide greater data to commanders in the future due to the use of the BMS.
- Doctrine & Concepts – The use of zonal power networks is not standard onboard vessels, although redundancy is standard.
- Organisation – Unlikely to change organisation structures significantly.
- Infrastructure – Minimal impact on infrastructure for such a change. This may allow a short period of usage of batteries whilst alongside. This may also enable onshore power supplies.
- Logistics – Minimal impact on logistics under normal operations.
- Integration – Potentially challenging to change to a zonal network and battery technology.
- Safety – Batteries are always a safety concern, especially Li-ion, although immersion cooling could improve safety. Choice of fluid is critical to the safety debate.
- Climate Change & Sustainability – Potential to significantly improve sustainability whilst operating. Although would need to factor the whole lifecycle and other environmental implications.
- Regulations & Certification – Commercial regulations already exist for batteries. Naval specific regulations would need to be developed.
- Test & Evaluation – Whilst already a significant TRL, this is for the commercial domain. Further testing would be required for uptake in the naval domain.

- Commercial – Anticipated to be similar to current procurement.
- Finance – There will be an increased CAPEX cost, although potential to decrease OPEX due to fuel savings.

7. STANDARDS

Lithium-ion battery rules and regulations are available within the maritime industry including but not limited to the following:

- Marine Guidance Note (MGN) 550 (M+F)
- Classification Society Rules and Regulations
- International Electrotechnical Commission (IEC) standards including but not limited to:
    - IEC 62619:2022 which specifies requirements and tests for the safe operation of secondary lithium cells and batteries used in industrial applications, including stationary applications
    - IEC 62133-2:2017+AMD1:2021 Secondary cells and batteries containing alkaline or other non-acid electrolytes - Safety requirements for portable sealed secondary cells, and for batteries made from them, for use in portable applications - Part 2: Lithium systems
    - BS EN 60092 Series, Electrical Installations in Ships.

However, none consider the application of immersion cooling as a cooling medium. Therefore, it would be recommended to adopt a risk-based approach whilst not only considering the existing standards but to also recognise and integrate the following guidance and requirements:

- Open Compute Project Design Guidelines for Immersion-Cooled IT Equipment
- Open Compute Project Immersion Requirements revision 2.0

To adopt a risk-based assessment of such a system will require strong stakeholder relationships. It is key that this is adopted from the concept stage to ensure success.

8. FURTHER INVESTIGATION

As mentioned earlier in the paper, immersion cooling is used for several applications within the commercial sector, therefore the technology can be argued to be TRL6 as defined by ISO16290 [13] as the technology has been demonstrated in a relevant environment. To progress to a higher TRL further testing would be required to replicate an integrated military environment.

Compliance of commercial grade fluids for immersion cooling should be explored in greater detail to determine viability for use within the naval domain. Whilst not in scope for this paper, there would be strong positioning that all maritime immersion cooling fluids should be based on mitigating effects after hostile action.

It is likely to be beneficial if further investigations and experiments are conducted in collaboration with a potential client to define the conceptual requirements, a battery manufacturer, and classification societies/ Certification Authorities. This would provide the required subject matter expertise (SME) to influence the design, mitigate identified risks and support in seeking an approval in principle once the design is matured.

This conceptual idea could develop the footprint for immersion cooling onboard all Naval vessels that is scalable depending on the quantity of the batteries.

A further consideration required of ESS for Naval applications would be to adopt a zonal network infrastructure to support resilience, not explored further in this paper but recommended to be investigated further. As the immersion cooling system is a closed loop system, this would allow an ESS zonal configuration to be adopted to align with the damage control zones as the historical limitations forcing a centred ESS have been mitigated, this is visually explained further in Figure 7.

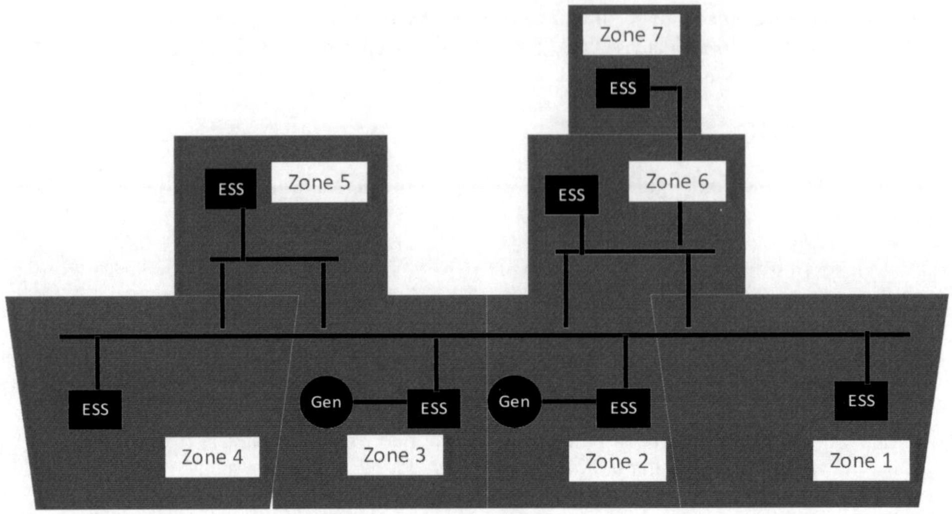

Figure 7: Zonel ESS Network

As noted in Section 7, there are currently no standards that refer to immersion cooling for a marine application, therefore once the design matures it is recommended to share the findings with IACS, MCA and the IEC, plus others where applicable to allow rules and regulations to be developed to support future integration safely into platforms.

A heat transfer analysis would be required to be conducted to an ESS to understand the heat transferred within an immersion cooled system during fast charging and discharging scenarios based on the concept of operation of the system including imitation of hostile environments in which a battery cell or module may be damaged.

9. CONCLUSIONS

This paper has been written to conduct the initial investigations into immersive cooling of ESS, more specifically Lithium Ion, and how it could be integrated into a Naval Platform. Forming part of an energy system based on zonal configuration that could aid survivability of a vessel. This would be a move from current doctrine and has the potential to provide significant operational advantage.

The adoption of immersion cooling will promote design space exploration focussing on the concept of operations of the vessel and the mission capability. Historically ESS has had limitations in optimal design due to the risk of thermal runaway, further investigations into a suitable fluid for this technology in Naval applications is required which may allow this to be a thing of the past.

There are many benefits which have been identified. If the system is sufficiently integrated, mitigated from any identified mechanical and electrical failures including those initiated from a damage scenario, then the immersion cooling solution could provide greater confidence for use within the Naval domain. Also supporting Ministry of Defence's Climate Change and Sustainability Strategic Approach 2050. A concept that although in its infancy within the maritime industry, has proven successful for other technologies and therefore is recommended to be exploited.

10. REFERENCES

1. Techhistorian, 'Which is the First Electric Ferry', *techhistorian.com*, 2015

2. PR Newswire, 'Introducing the World's First Battery Tanker 'X': The Inaugural Ship of Power Ark 100. On Track for Completion by 2025, with Field Testing Set to Commence in 2026', *prnewswire.com*, 2023

3. Ministry of Defence, 'Ministry of Defence Climate Change and Sustainability Strategic Approach', Gov.uk, 2021

4. Haynes, 'Astute Class Submarine Manual', p60, 2018

5. UK Defence Journal, 'Australian firm awarded Royal Navy submarine battery contract', UKDJ, 2021

6. BUGRYNIEC, P J., et.al, 'Assessment of the Risks Posed by Thermal Runaway within Marine Li-ion Battery Energy Storage Systems - Considering Past Incidents, Current Guidelines and Future Mitigation Measures', *SSRN*, 2021

7. FPA, 'How to Control a Lithium-Ion Battery Fire', Fire Protection Association Advice and Guidance Article, 2024

8. PICHLER, B., 'Increasing Electric Vehicle Range with a Recommendation App providing Context Specific Trip Rankings', *Research Gate*, 2016

9. DAI, Y., 'Thermal runaway process in lithium-ion batteries: A review', *Science Direct*, 2025

10. HENSHAW, K., et.al, 'Single-Phase vs. Two-Phase Immersion Cooling, What's the difference and which one is right for you?' *Submer*, 2021

11. EticaAG, 'Immersion Cooling and Fire Suppression for Battery Energy Storage Systems', EticaAG, 2025

12. UK Ministry of Defence, 'Surface Platform Power & Energy Industry Event', Portsmouth, HM Government, 2023

13. ISO, 'Definition of the Technology Readiness Levels (TRLs) and their criteria of assessment', ISO16290, 2013

11. AUTHOR BIOGRAPHIES

Tania Berry MIET MIMarEST holds the position of Electrical Capability Lead at BMT, she has over 25 years of experience working within the maritime industry, 14 of which were serving in the Royal Navy including T45. She also has experience within Class (Lloyd's Register) Design Assurance and the Technical Policy Group. She was also the High Voltage Manager on QEC Class and more recently was within Stage 1 Design as the Principal Electrical Engineer for the Australian Navy's Hunter Class programme within BAE Systems. Tania has an MSc in Systems Engineering Management from the University of Strathclyde.

Thomas Beard PhD CEng MIMechE MIMarEST is a clean maritime expert with over a decade of experience in alternative fuels. He is involved in numerous industry bodies, including as the co-chair for the National Maritime Hydrogen and Alternative Fuels Working Group and several academic research program advisory boards. Thomas has a doctorate in Hydrogen Safety from Loughborough University and a BSc in Chemistry and Mathematics from Manchester Metropolitan University.

INTEGRATING SHIELD AND SOAM FOR SYSTEMIC HUMAN FACTORS ANALYSIS IN MARITIME COLLISIONS: THE HELGE INGSTAD CASE

Y B A Farag, R E Kurt, and **O Turan,** Department of Naval Architecture, Ocean and Marine Engineering (NAOME), University of Strathclyde, Glasgow, UK

SUMMARY

Effective human factors (HF) analysis is essential for understanding complex failures in high-risk maritime operations. While HF classification taxonomies exist, few offer integrated, scalable frameworks that support both structured factor identification and causal mapping across system layers. This study proposes a combined methodology that integrates the SHIELD Human Factors Classification System (SHIELD-HFACS), influence and certainty weighting, and the Systemic Occurrence Analysis Methodology (SOAM), anchored around functional safety barriers.

The framework enables comprehensive classification of human and organisational factors across four levels: Acts, Preconditions, Operational Leadership, and Organisational Influences. It also assesses their impact and evidential strength. These factors are then linked to core collision safety barriers to trace how systemic failures propagate through socio-technical systems. SOAM mapping supports causal visualisation, highlighting interactions between latent conditions and frontline errors.

The methodology is demonstrated through application to the 2018 KNM Helge Ingstad collision, where 32 contributory factors were identified and linked across barrier breakdowns. The analysis revealed how misperception, degraded supervision, interface gaps between VTS and navy protocols, and organisational oversights collectively undermined all active defences. These findings align with broader trends observed in a SHIELD-classified dataset of 106 maritime collisions, which commonly reveal systemic breakdowns in supervision, communication, and organisational oversight.

The proposed SHIELD-SOAM framework contributes a replicable and scalable method for HF analysis in maritime accident investigations. By structuring classification, weighting influence, and mapping causal pathways, it enhances system-wide learning and supports evidence-based safety improvement. Future work will apply the full framework to the broader collision dataset to uncover recurring patterns and inform risk-informed design and intervention strategies.

1. INTRODUCTION

Maritime operations represent complex socio-technical systems where safety outcomes depend on intricate interactions between human operators, organisational structures, and technological components. Despite advances in automation and regulatory frameworks, human-related errors continue to dominate maritime accident statistics, accounting for over 90% of ship collisions [1], [2]. In naval operations, this complexity intensifies due to additional constraints, such as mission-specific imperatives, hierarchical command structures, restricted Automatic Identification System (AIS) use, and deviations from standard maritime practices, leading to heightened vulnerability and increased collision risks [3], [4].

The collision and subsequent sinking of the Royal Norwegian Navy frigate HNoMS Helge Ingstad in November 2018 starkly exemplifies these challenges. Despite rigorous training and clearly defined operational procedures, experienced personnel failed to correctly interpret critical navigational cues, culminating in significant human, technical, and financial losses [5]. The event emphasised persistent weaknesses in team coordination, supervision, visual perception, and communication within complex naval-civilian maritime interactions, which conventional human factors methodologies have not fully captured or resolved.

Traditional human factors frameworks, such as the widely used Human Factors Analysis and Classification System (HFACS), have provided structured insights into accident causation across various sectors. However, their aviation-centric origins and rigid hierarchical structure present limitations when applied to dynamic, high-tempo maritime operations, particularly in naval contexts. Critical elements like team coordination under cognitive load, spatial constraints, and complex command structures are inadequately addressed, necessitating an approach better attuned to maritime operational realities [6], [7].

To address these gaps, the SAFEMODE project's SHIELD Human Factors Taxonomy was developed, explicitly targeting maritime-specific requirements. SHIELD retains a structured multi-layered analytical logic akin to HFACS but significantly enhances domain specificity by incorporating system resilience concepts and detailed maritime-relevant categories. It aligns closely with maritime investigation practices, facilitating clearer identification of contributory factors and deeper systemic learning [8], [9].

This paper applies the SHIELD taxonomy to the KNM Helge Ingstad incident to systematically evaluate the multi-layered human and organisational factors that contributed to the vessel's collision and sinking. The study demonstrates the robustness and applicability of SHIELD in providing structured, actionable insights into complex maritime accidents. It further explores the value of integrating a structured human factors taxonomy within maritime organisations and regulatory frameworks to strengthen risk management and support incident learning. The remainder of the paper is organised as follows: Section 2 reviews existing human factors classification approaches. Section 3 introduces the proposed methodological framework, combining the SHIELD taxonomy, influence and certainty weighting, SOAM causal mapping, and safety barrier alignment. Section 4 presents an overview of the incident, followed by the human factors classification results. Section 5 discusses the implications of these findings and highlights how structured HF analysis can inform safety management practices. Section 6 concludes with recommendations for future application and system-level safety improvement.

This paper introduces a novel methodological integration of SHIELD-HFACS, influence-weighted classification, and SOAM causal mapping, uniquely anchored around functional safety barriers to support systemic analysis and systemic safety learning.

2. REVIEW OF HUMAN FACTORS CLASSIFICATION METHODS

Classifying human and organisational factors in accident investigation and safety reports has been central to safety science, driving improvements from individual-focused explanations to systemic analyses of complex socio-technical systems. Numerous HF frameworks have emerged, each with varying degrees of suitability for maritime operations, particularly naval contexts characterized by dynamic interactions, high operational tempo, and critical team coordination requirements [10]. These tools aim not only to trace the origins of unsafe acts but also to reveal deeper systemic contributors and support safety learning. In the maritime and naval sectors, where operational demands are highly dynamic and team coordination under uncertainty is critical, selecting or adapting the right HF framework is essential for meaningful accident analysis and risk management.

2.1 HFACS (HUMAN FACTORS ANALYSIS AND CLASSIFICATION SYSTEM)

Initially developed for aviation, structures human factors into four hierarchical levels, unsafe acts, preconditions, supervisory factors, and organisational influences, enabling traceability from frontline behaviours to systemic conditions [11]. While widely adopted across aviation, military, and maritime sectors [12], [13], HFACS exhibits limitations in maritime contexts due to its rigid aviation-centric structure, limited vocabulary for maritime-specific operations (e.g., pilot-master interactions), and inadequate accommodation of dynamic cognitive loads common to naval navigation scenarios [6], [7], [14].

2.2 AcciMap

A system-theoretic method developed by Rasmussen, effectively maps causal relationships across multiple hierarchical levels, distinguishing proximal operational factors from latent higher-tier conditions. It has successfully analysed diverse maritime and naval incidents, such as the Sewol ferry disaster and COVID-19 management aboard naval vessels [15], [16], [17]. However, AcciMap's lack of formal classification logic and structured vocabulary limits its application for coding incidents in investigation databases [18].

2.3 FRAM (FUNCTIONAL RESONANCE ANALYSIS METHOD)

The method offers a complementary, function-based approach focusing on variability in functional interactions within complex systems, such as Arctic ship navigation and naval operations [19], [20]. While FRAM provides detailed system insights, its methodological complexity and limited structured guidance challenge its integration with large-scale safety databases and structured learning processes [21], [22].

2.4 TRACEr

The method originally designed for offshore oil and gas, emphasises cognitive errors at task levels, providing structured categories for analysing crew decision-making [23]. However, TRACEr lacks the multi-layered depth needed for analysing broader supervisory or organisational factors prevalent in maritime operations.

2.5 STAMP (SYSTEM-THEORETIC ACCIDENT MODEL AND PROCESSES) AND CAST (CAUSAL ANALYSIS BASED ON STAMP)

These methods focus explicitly on control structures and safety constraints, facilitating structured examinations of systemic control failures. CAST particularly excels at identifying regulatory and organisational breakdowns but lacks standardized taxonomy, complicating its integration into high-volume HF classification and incident databases [24], [25].

Hybrid approaches, integrating systemic methods (CAST) with structured frameworks (HFACS), have been proposed to leverage combined advantages. However, these hybrid frameworks increase analytical complexity, potentially diminishing practical feasibility in routine maritime safety reporting systems as noted by [26].

2.6 SHIELD: A MULTI-DOMAIN HUMAN FACTORS FRAMEWORK

Recognising the limitations of existing frameworks and the need for cross-domain compatibility, the EU-funded SAFEMODE project developed the SHIELD Human Factors Taxonomy. SHIELD integrates systemic insight with structured classification logic and was explicitly designed for application across multiple safety-critical domains, most notably aviation and maritime operations. While building upon HFACS's layered analytical structure, SHIELD enhances granularity, domain relevance, and systemic traceability, enabling its adaptation to diverse operational environments.

During SAFEMODE, SHIELD was specifically refined and validated using extensive datasets from both aviation and maritime investigations. This validation process ensured its effectiveness in capturing human factors in complex operational scenarios, addressing limitations related to sector-specific terminology, cognitive load management, and procedural nuances.

Figure 1 illustrates the positioning of human factors taxonomies based on their methodological orientation (model-based vs. epidemiological) and domain applicability (domain-specific vs. domain-transversal). This taxonomy mapping was a core outcome of SAFEMODE's cross-sectoral review and played a critical role in shaping the design of SHIELD by identifying elements suitable for integrated learning across aviation and maritime contexts [27].

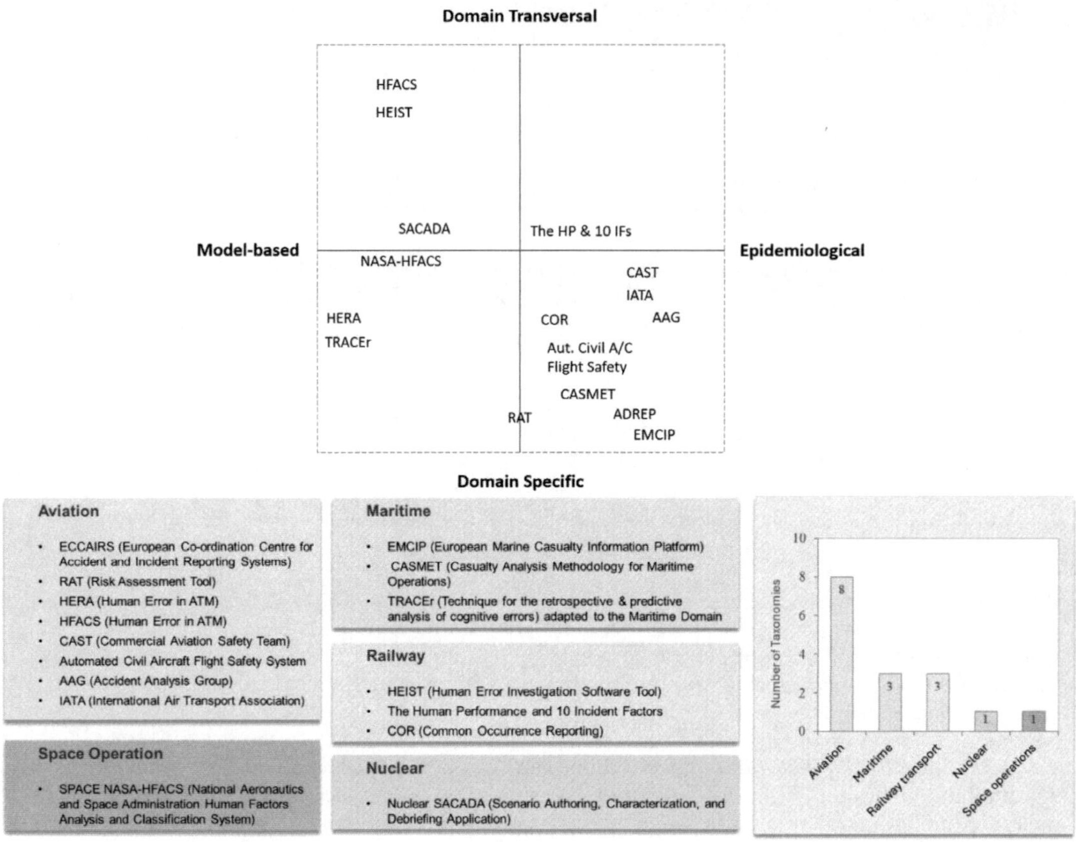

Figure 1: Review of existing human factors taxonomies according to their methodological orientation (model-based vs. epidemiological) and scope (domain-specific vs. domain-transversal, [28].

SHIELD was developed as an enhanced, domain-adapted evolution of HFACS. It draws on a structured review of 16 taxonomies, including HFACS, HERA, NASA-HFACS, TRACEr, and HEIST, to overcome classification limitations in maritime and aviation safety data. Retaining HFACS's four-layer logic, SHIELD expands its scope with more detailed categories tailored to operational realities. Its structure, covering human acts, preconditions, operational leadership, and organisational influences, enables structured database integration and facilitates cross-case learning in both retrospective and proactive safety contexts [28], [29], [30].

3. METHODOLOGICAL FRAMEWORK

This study applies a structured human factors classification using the SHIELD taxonomy to identify the contributory factors involved in the collision between HNoMS Helge Ingstad and the tanker Sola TS. The classification is informed by a set of collision safety barriers aligned with relevant COLREGs standards. SHIELD is combined with a weighting scale to evaluate the relative impact of each identified factor on the incident. Additionally, a certainty scale is introduced to reflect the confidence level associated with each classification, as interpreted from the official investigation reports. The results of the human factors classification are then visualised using a SOAM (Systemic Occurrence Analysis Methodology) chart. The following subsections provide a brief description of each component of the proposed framework and Figure 2 illustrate the proposed framework.

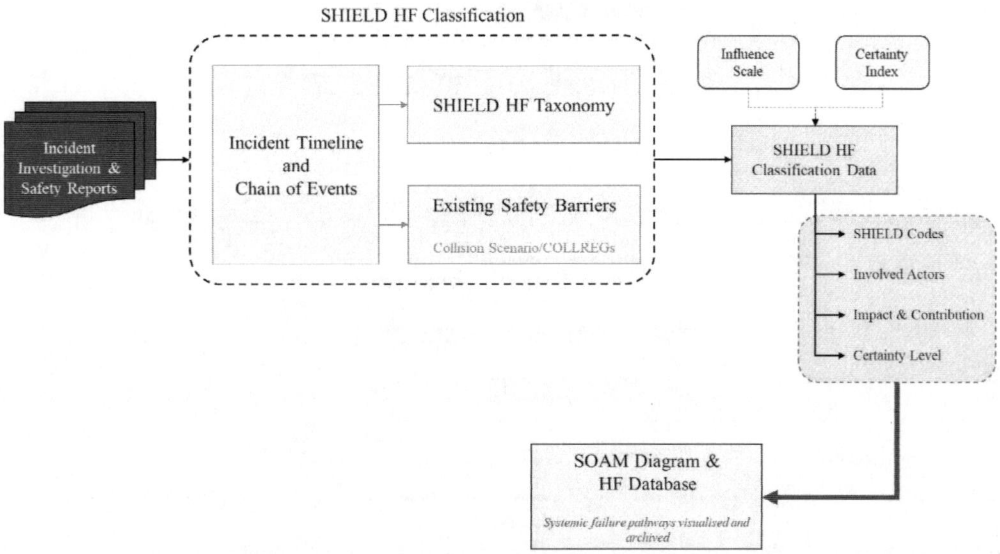

Figure 2: SHIELD-based HF framework for maritime accident analysis. The framework integrates incident timelines, safety barrier mapping, and SHIELD taxonomy to systematically extract, code, and visualise contributory factors using weighted classification and SOAM modelling.

3.1 SHIELD HUMAN FACTORS TAXONOMY

The SHIELD taxonomy consists of four interrelated layers, reflecting increasing levels of systemic influence behind human performance failures. Each layer is broken down into categories, and within each category is a set of SHIELD codes used to classify contributory factors during accident investigation and analysis. The layers are defined as follows:

- **Human Acts**: Immediate frontline actions or inactions, including perception failures, decision errors, response execution issues, and rule violations.
- **Preconditions**: Underlying conditions that affect individual and team performance, such as workload, fatigue, miscommunication, and environmental distractions.
- **Operational Leadership**: Factors related to supervision, task allocation, resource management, and immediate planning and control responsibilities.
- **Organisational Influences**: Higher-level systemic issues including safety culture, training standards, resources, procedural quality, and risk management.

Figure 3 presents a comprehensive overview of the SHIELD taxonomy, illustrating its four-layer structure and associated categories. It demonstrates how human factors are systematically organised, ranging from immediate frontline actions to deeper systemic influences at supervisory and organisational levels.

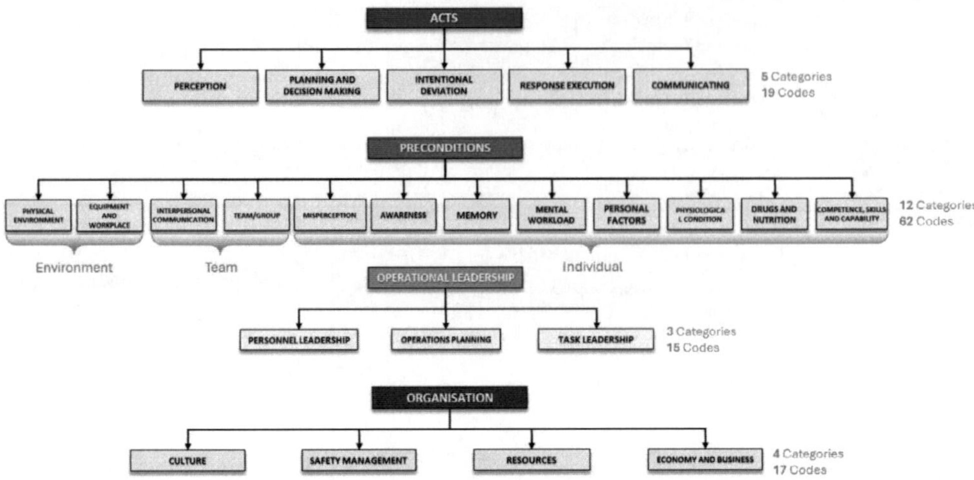

Figure 3: Overview of the SHIELD Human Factors Taxonomy across its four layers, categories, and associated code counts.

To exemplify this structure, Figure 4 focuses on the Acts layer, the top tier of the taxonomy. This layer is subdivided into five key categories: Perception, Planning and Decision Making, Intentional Deviation, Response Execution, and Communicating, which collectively encompass 19 specific failure codes used to classify operator actions and behaviours. Each category captures distinct types of human performance variability observed at the operational interface. A similarly structured categorisation exists for the remaining three layers: Preconditions, Operational Leadership, and Organisation, enabling a consistent and scalable framework for classifying contributory factors across different levels of the socio-technical system.

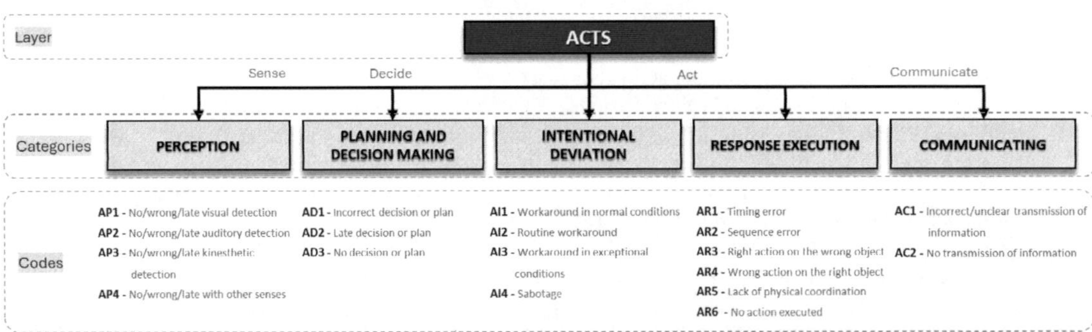

Figure 4: Breakdown of the Act layer to the levels of layer categories and codes

SHIELD was explicitly designed for integration with investigation databases, enabling standardised data collection, structured cross-case comparison, and targeted safety learning. Empirical applications in maritime contexts [9] have demonstrated its ability to identify recurring contributory themes, such as resource management weaknesses, latent design flaws, and procedural deficiencies, offering a more operationally grounded diagnosis compared to traditional methods.

Beyond retrospective use, SHIELD also supports proactive safety learning. Its alignment with real-world maritime operational conditions makes it a strong candidate for integration into risk modelling and resilience assessment frameworks in both civilian and naval contexts [31], [32].

3.2 COLLISION IN RESTRICTED WATER SAFETY BARRIERS

In this study, SHIELD codes were manually assigned through a consensus-based review of official investigation reports. To ensure rigour and comprehensive coverage, the classification was guided by a predefined set of functional collision safety barriers. These barriers were used both retrospectively, to identify failed defences, and prospectively, to structure the classification process across each phase of the incident. By aligning contributory factor identification with barrier functions (e.g., route monitoring, evasive action), the analysis captured human and organisational contributions across the full event timeline, enhancing consistency and traceability.

Collisions in restricted waterways rarely stem from isolated errors. Instead, they typically result from the gradual erosion of multiple, interdependent safety barriers. These barriers are designed to detect, prevent, or contain hazards as operational situations evolve, embodying the dynamic risk management logic of complex socio-technical systems.

To operationalise this logic, the study adopts a functionally layered barrier framework adapted from [32]. A safety barrier is defined as any technical, procedural, human, or organisational element that can interrupt risk propagation. This approach aligns with Reason's Swiss Cheese Model, which conceptualises accidents as the result of concurrent failures across multiple layers of defence.

Table 1. Functional collision safety barriers and their general descriptions

CB Level	**Description**
CB5	Route and transit monitoring by vessel and VTS: Maintains traffic separation and safe lane adherence through bridge team situational awareness and VTS supervision.
CB4	TSS conflict prevention by vessel and VTS: Detects emerging navigation conflicts by monitoring patterns and maintaining shared situational awareness.
CB3	Collision avoidance coordination by VTS: Provides macro-level traffic assessments and advisories to reduce encounter risk.
CB2	Ship-based avoidance (detection, COLREGs implementation and communication): Involves vessel-level traffic detection, risk assessment, rule application, and direct communication.
CB1	Emergency manoeuvring by vessel at final stage: Final evasive actions through helm or propulsion when prior barriers have failed.

The framework also integrates navigational responsibilities outlined in the International Regulations for Preventing Collisions at Sea (COLREGs). While COLREGs prescribe actions such as lookout (Rule 5), risk assessment (Rule 7), and manoeuvring (Rules 8, 15 - 17), the barrier model reframes these behaviours through a system-theoretic lens to support structured observation and classification of failure points.

Although the complete risk model lies beyond this paper's scope, the analysis is organised around five generalised collision barriers (CB1 - CB5), each triggered by precursor events such as incursion or close-quarters situations. Table 1 summarises these core barrier functions, developed with input from maritime officers, VTS personnel, and system safety experts.

Figure 5 illustrates the conceptual alignment of these barriers with escalating collision precursors and highlights their role in structuring both classification and causal analysis in this study.

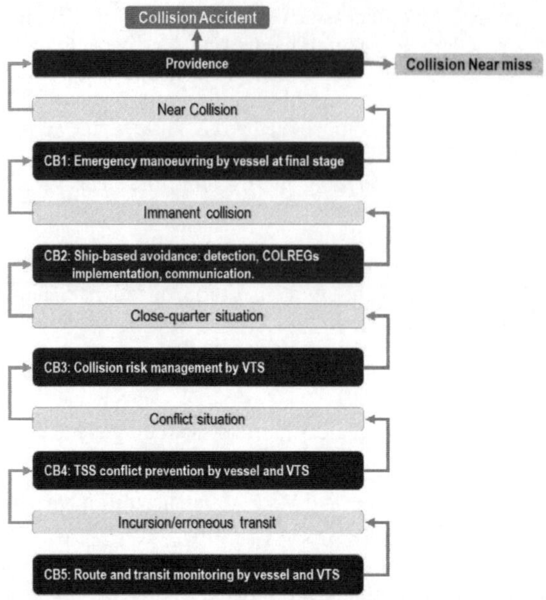

Figure 5: Conceptual structure of functional safety barriers and precursors

This barrier model supports integration with SHIELD classification by mapping human and organisational factors to specific defensive functions and failure pathways. It enables analysts to connect individual acts and decisions to the degradation of system-level protections, offering a systemically grounded view of causation.

3.3 INFLUENCE SCALE

To support the prioritisation and impact of contributory factors, each SHIELD code was evaluated using an influence scale that reflects its relative impact on the accident scenario. This scale, adapted from the NASA HFACS guidelines [33], enables the classification of factors not only by their type but also by their significance in the causal chain. It facilitates more nuanced insights into which failures had the greatest influence on the outcome, and which may indicate latent systemic vulnerabilities. Table 2 outlines the scale, ranging from causal factors (which, if eliminated, would have prevented the occurrence) to minor deviations with no immediate influence but potential future implications.

Table 2. SHIELD Influence Scale

Weight	Description
4	Causal impact. If eliminated from occurrence chain of events, prevents its occurrence
3	Major impact on mishap scenario. Its elimination could have prevented the occurrence
2	Moderate impact on occurrence scenario. It is a departure from process/procedure, however its elimination by itself would not have prevented the occurrence
1	Minimal impact on occurrence scenario. However, it is a noted departure from process/procedure
0	Present but no influence on the occurrence in subject, however, it has a potential for future occurrences or other failures

3.4 CERTAINTY INDEX

During practical applications of SHIELD, feedback from end users indicated challenges in assigning codes with high confidence, particularly when investigation reports lacked clarity or detail. To address this issue and enhance transparency and analytical credibility in the classification process, a certainty index was incorporated to reflect the analyst's confidence in each assigned code. This index enables differentiation between well-supported classifications and those based on limited or inferred evidence.

Table 3 presents the three-tiered certainty scale; High, Medium, and Low, based on the strength of evidence available in the official reports.

Table 3: SHIELD Certainty Index

Certainty Level	Description
High	The factor is strongly supported by explicit evidence in the report, such as formal findings, direct quotations, clearly described actions or omissions, or witness statements. Typical language: "caused by," "due to," "as a result of."
Medium	The factor is reasonably inferred from the sequence of events, actor responsibilities, or technical evidence, even if not explicitly stated. Typical language: "likely," "probably," "appears to have," "may have contributed."
Low	The factor is suggested only vaguely, indirectly, or circumstantially, or lacks supporting evidence due to data gaps or investigative limitations. Typical language: "might have," "cannot be excluded," "possible," or references to unavailable data or untested assumptions.

This index should enhance the transparency and credibility of the HF classification by clearly signalling the strength of the evidence base behind each coding decision.

3.5 SYSTEMIC OCCURRENCE ANALYSIS METHODOLOGY (SOAM)

The Systemic Occurrence Analysis Methodology (SOAM), developed by EUROCONTROL, is a structured framework for incident analysis in high-risk domains. Based on [34] organisational accident model, SOAM focuses on the interplay of latent conditions and active failures across system layers [35]. Its system-theoretic foundation makes it well-suited for analysing complex failures in socio-technical operations such as naval navigation.

SOAM follows a six-step process: (1) data gathering, (2) mapping the occurrence net, (3) identifying root causes, (4) categorising contributory factors, (5) formulating preventive recommendations, and (6) integrating findings into safety management systems [36]. During data collection, SOAM encourages triangulation from sources such as reports, interviews, and technical logs. The SHEL model (Software, Hardware, Environment, Liveware) often supports this phase, guiding the identification of human and system interactions.

In this study, SOAM was applied in parallel with the SHIELD-HFACS framework to explore the systemic contributors to the Helge Ingstad and Sola TS collision. While SHIELD facilitated structured coding of human and organisational factors, SOAM provided a complementary narrative mapping of causal interactions across layers, linking micro-level actions to macro-level system deficiencies. Notably, the SOAM analysis was synchronised with the barrier breakdowns described in Section 3.2, allowing human factor codes to be explicitly mapped to specific safety barrier degradations. This dual-pronged approach, quantitative SHIELD classification combined with qualitative SOAM causal tracing, enabled a more holistic and interpretable account of the accident's underlying causes. It also facilitated the visualisation of how actions, decisions, and latent organisational factors converged to erode multiple lines of defence, ultimately leading to the collision.

4. HF ANALYSIS OF HELGE INGSTAD AND SOLA TS COLLISION

4.1 OCCURRENCE FACTS

At 04:01:15 local time on 8 November 2018, the Royal Norwegian Navy frigate KNM Helge Ingstad (KNM HI) collided with the outbound crude oil tanker Sola TS in Hjeltefjord, near the Sture terminal, Norway. At the time, KNM HI was transiting southbound at approximately 17 knots as part of a routine navigation exercise, following its participation in NATO Exercise Trident Juncture.

KNM HI's Automatic Identification System (AIS) was set to receive-only mode, meaning the vessel was detectable on radar but did not broadcast its identity. This contributed to situational ambiguity for both surrounding vessels and Fedje Vessel Traffic Service (VTS).

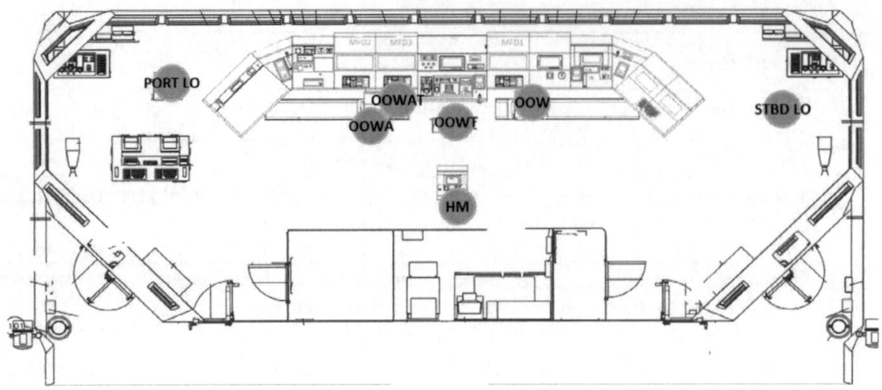

OOW – Officer of the watch
OOWT – Officer of the watch trainee
OOWAT – Officer of the watch assistant trainee
OOWA – Officer of the watch assistant

Bridge watch team:
HM – Helmsman
STBD LO – Starboard lookout
PORT LO – Port lookout

Figure 6: Positions on the bridge of HNoMS Helge Ingstad. Illustration: The Royal Norwegian Navy

The bridge team aboard KNM HI, comprising the officer of the watch (OOW), trainees, assistants, and a helmsman, visually misidentified the fully illuminated Sola TS, escorted by tugs, as a stationary structure, possibly part of the Sture terminal. This critical misperception persisted during the OOW handover, leading to a failure to recognise the developing collision risk. Figure 6 illustrates the personnel distribution and roles on the KNM HI bridge at the time of the incident.

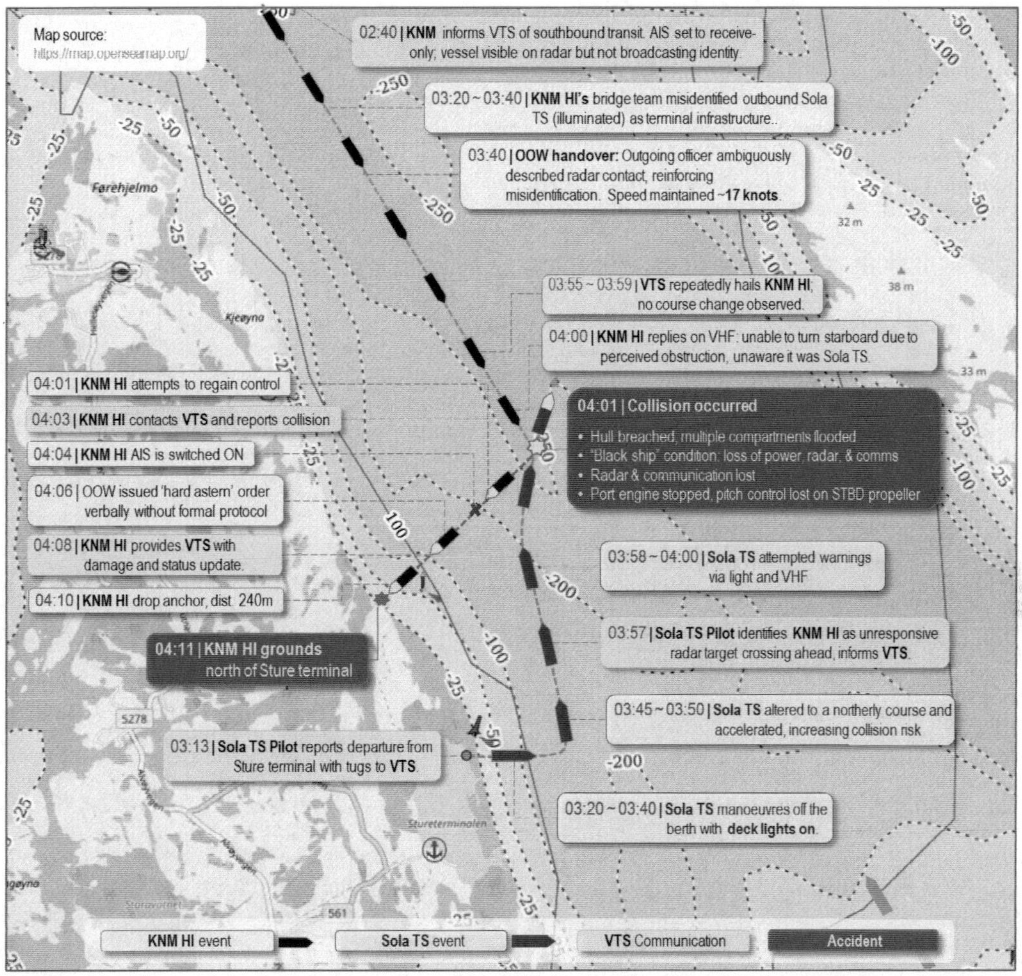

Figure 7. Timeline and spatial reconstruction of the collision between KNM Helge Ingstad and Sola TS, by authors.

Between 03:57 and 04:00, both Sola TS and Fedje VTS issued repeated warnings over VHF, identifying the approaching vessel as KNM HI and urging course alteration. The frigate's response, hindered by a perceived obstruction, failed to acknowledge that the obstruction was in fact the Sola TS.

At 04:01, the collision occurred. The starboard anchor and bow of Sola TS tore through approximately 46 metres of KNM HI's starboard hull, breaching five watertight compartments. The impact triggered a "black ship" condition, complete loss of power, radar, and communications. Critical systems, including pitch control on the starboard propeller, steering, and the port main engine, were disabled. Attempts to issue a 'hard astern' order and regain control failed.

At 04:11, KNM HI ran aground north of the Sture terminal and was later declared a total loss. Figure 7 presents a detailed timeline and spatial reconstruction of key actions and decisions taken by KNM HI, Sola TS, and VTS, based on official investigation records.

All 137 crew members were evacuated. Seven sustained minor injuries; no fatalities occurred. The incident resulted in an estimated $80 million in salvage and emergency response costs. The projected replacement cost of the vessel was approximately $1.4 billion.

Investigations conducted by the Accident Investigation Board Norway (AIBN, 2018), AIBN and the Defence Accident Investigation Board Norway (DAIBN, 2019), and the Norwegian Safety Investigation Authority (NSIA, 2021) identified several contributing factors:

- Visual misperception and failure to reassess situational cues
- Procedural breakdown during the OOW handover
- Inadequate supervisory oversight on the bridge
- Poor communication between VTS, Sola TS, and KNM HI

These factors collectively indicate a breakdown of critical safety barriers across the operational, supervisory, and organisational levels, forming the basis for the SHIELD and SOAM analyses that follow.

4.2 SHIELD HF CLASSIFICATION RESULTS

The SHIELD human factors (HF) classification framework was applied to the Helge Ingstad–Sola TS collision through a structured process that linked each contributory factor to both its taxonomy layer and the relevant collision safety barrier. A total of 32 SHIELD codes were identified across the four layers, as summarised in Table 4 and detailed in Table 5 (APPENDIX). Codes were assigned through a consensus-based review of the official investigation reports and evaluated using the influence and certainty scales described in Sections 3.3 and 3.4.

Table 4. Summary of HF Classification Results by Taxonomy Layer, Influence Weight, Certainty Level, and Affected Safety Barriers

Layer	Total	Weight				Certainty Level			Safety Barriers					
		4	3	2	1	High	Med	Low	CB5	CB4	CB3	CB2	CB1	PC
Acts	8	2	2	3	1	5	2	1	2	3	1	2	1	3
Preconditions	10	0	3	5	2	3	6	1	1	4	3	7	5	2
Op. Leadership	4	1	2	1	0	1	2	1	1	2	0	2	2	1
Organisation	10	0	3	7	0	1	9	0	3	6	1	2	0	2
Total	**32**	**3**	**10**	**16**	**3**	**10**	**19**	**3**	**7**	**15**	**5**	**13**	**8**	**8**

The classification revealed a predominance of latent conditions, with the highest number of codes recorded in the Preconditions and Organisational layers (10 each), followed by Human Acts (8) and Operational Leadership (4). In terms of influence, 3 codes (9%) were rated as causal (Weight 4), 10 (31%) as major contributors (Weight 3), and 16 (50%) as moderate (Weight 2). Only 3 codes were considered minimal (Weight 1). Certainty levels indicated reasonable confidence in the analysis: 10 codes (31%) were assigned high certainty, 19 (59%) medium, and 3 (9%) low.

When mapped against the five functional collision safety barriers and post-collision (PC) conditions, CB4 (Traffic Conflict Prevention) and CB2 (Ship-Based Avoidance) emerged as the most affected, with 15 and 13 associated codes respectively. These were followed by CB1 (Initial Detection, 8), CB5 (Recovery, 7), CB3 (Coordination, 5), and PC (8), revealing a cascading breakdown of defences from early detection through to post-incident response.

A layer-by-layer breakdown of key issues is as follows:

- Human Acts (8 codes): Included perceptual misjudgements and failure to respond to VHF warnings.
- Preconditions (10 codes): Captured deficiencies in lookout practices, team communication, and mental model alignment.
- Operational Leadership (4 codes): Reflected limited supervisory oversight and poor bridge coordination during a high-risk transit.
- Organisational Influences (10 codes): Highlighted misaligned AIS policies and inadequate preparation for propulsion loss.

Barrier-based reasoning helped clarify causal pathways. For example, communication failures linked to CB3 were traced to both Precondition and Leadership factors, while the misidentification of Sola TS (CB2) involved perceptual and planning errors within the Human Acts layer.

The level of analytical granularity achieved was made possible by the thoroughness of the official investigation reports. However, this level of detail is not consistently available across maritime cases, highlighting the need for transparent and standardised reporting to enable robust HF classification and safety learning.

4.3 HELGE INGSTAD SOAM ANALYSIS - SHIELD HF CAUSAL MAP

The HF classification data was used to construct a SOAM chart, as outlined in Section 3.6. Figure 8 visualises the dynamic interactions between human and organisational factors, tracing how each contributed to the degradation of safety barriers (CB1 - CB5), ultimately leading to the collision. The causal structure mirrors SHIELD's four hierarchical levels, Organisation, Operational Leadership, Preconditions, and Acts, each representing a layer of defence progressively eroded over the course of the incident.

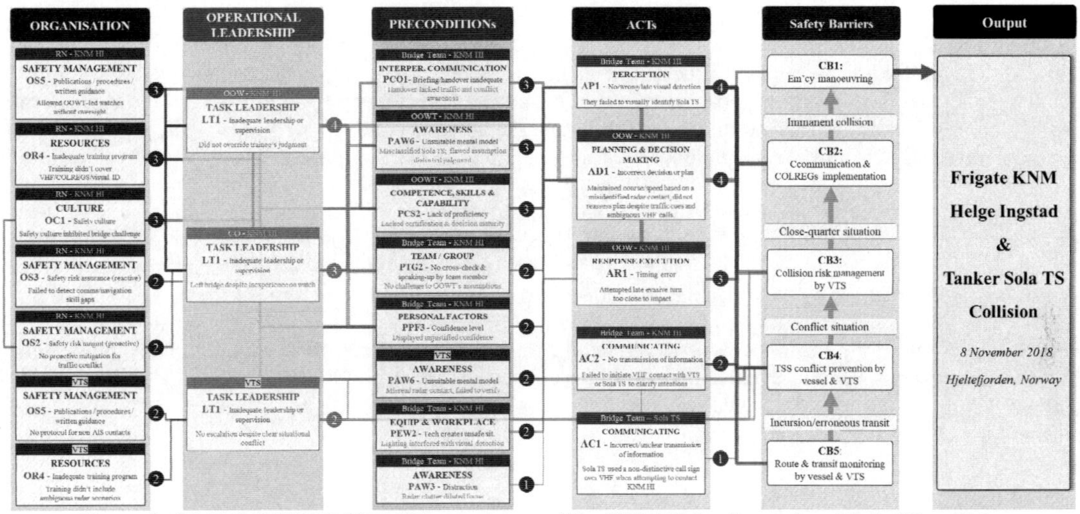

Figure 8: SOAM–SHIELD HFACS causal map of the KNM Helge Ingstad and Sola TS collision

At the Organisational level, systemic flaws weakened safety well before the incident. Inadequate training frameworks (OR4) failed to prepare crews for managing complex traffic situations (OS5), while a culture that discouraged assertive questioning of senior decisions (OC1) allowed latent risks to persist. The absence of escalation mechanisms at both the vessel and VTS levels further limited coordinated responses once risk signals emerged.

Within Operational Leadership, the lack of proactive task leadership (LT1) proved critical. Onboard, the commanding officer left the bridge during a high-traffic phase, placing responsibility on a junior officer with limited support. At VTS, the operator failed to escalate despite multiple signs of convergence, neglecting supervisory duties. These leadership gaps allowed both teams to operate without effective intervention.

At the Preconditions level, interrelated breakdowns degraded situational awareness and decision readiness. The bridge team maintained an inaccurate mental model of the traffic situation (PAW6), reinforced by ambiguous communication during the OOW handover (PCO1). Visual perception was impaired by poor visibility (PEW2) and overload (PAW3), while team members lacked assertiveness (PTG2) and familiarity (PCS2) to challenge assumptions or recover from the escalating threat.

At the Acts level, the failure began with misperception (AP1): the bridge team visually misidentified the fully illuminated Sola TS as static infrastructure. Despite radar and AIS (in receive-only mode), the error was not corrected. This led to flawed navigational decisions (AD1), ineffective internal communication (AC1), and delayed external exchanges (AC2). When evasive action was finally attempted, it was lately executed (AR1), breaching the final barrier (CB1).

Figure 8 illustrates the causal pathways from latent organisational conditions to frontline actions. Arrows represent weighted causal links, as defined in Table 2, providing traceability across levels. This integrated SOAM-SHIELD model supports a structured understanding of the Helge Ingstad incident and offers a foundation for systemic analysis for similar cases.

5. DISCUSSION

To contextualise the Helge Ingstad incident, its findings were benchmarked against a broader SHIELD-classified dataset comprising 106 maritime collision cases. This comparison offers insight into recurring patterns of contributory factors and systemic breakdowns in barrier performance, particularly within constrained navigational environments.

5.1 HUMAN-SYSTEM FAILURES IN CONTEXT

The Helge Ingstad case exemplifies the convergence of human and organisational factors across all SHIELD layers. At the frontline, misperception by the bridge team, specifically, the misidentification of Sola TS, combined with delayed and ineffective communication, triggered the collision sequence. These errors occurred in a context already compromised by concurrent training activities and the absence of supervisory oversight, placing critical decision-making responsibilities on a junior officer.

Onshore, Fedje VTS did not maintain adequate situational awareness or initiate conflict resolution despite clear signs of convergence and prior knowledge of KNM HI's AIS receive-only mode. Contributing further to the misidentification, Sola TS's full deck lighting, while compliant with internal safety policies, obscured her navigational lights, reinforcing the illusion that she was part of terminal infrastructure.

These dynamics reflect broader patterns observed in other restricted water collisions: fragmented mental models, limited team coordination, and disjointed understanding between stakeholders. Rather than resulting from a single point of failure, such incidents typically emerge from a cascade of uncorrected misjudgements and latent vulnerabilities across a multiteam system.

5.2 PATTERNS FROM THE SHIELD COLLISION DATASET

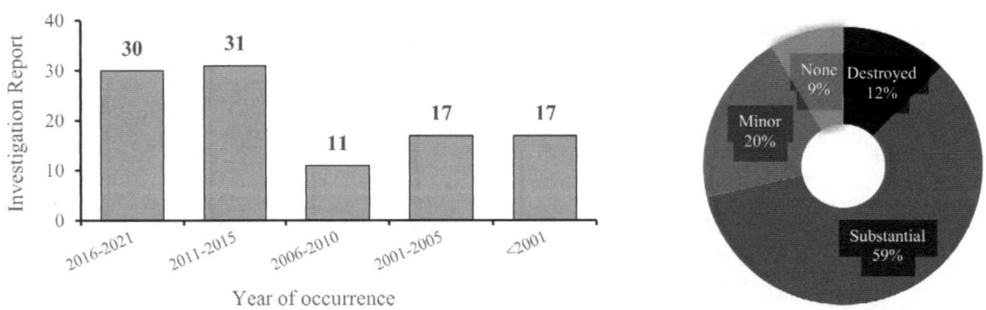

Figure 9: Distribution of data sample by their occurrence year and consequences

The SHIELD collision database (IR=106) reinforces these systemic findings. As shown in Figure 9, most incidents occurred between 2011 and 2021, with over half resulting in substantial damage and 12% in vessel loss Figure 10 ranks SHIELD codes by prevalence, with planning and decision-making (87%), perceptual oversights (55%), and communication breakdowns (51%) dominating the Acts layer, mirroring key contributors in the Helge Ingstad case.

Figure 10: SHIELD HF's categories ranked by coverage percentage in analysed collision accidents (IR=106)

At the supervisory and organisational levels, Figure 11 highlights persistent weaknesses in task leadership (55%), operations planning (42%), safety management (47%), and resource allocation (38%). Similar to the Helge Ingstad case, these deficits impair effective oversight, degrade situational awareness, and limit teams' ability to recognise and respond to emergent risk.

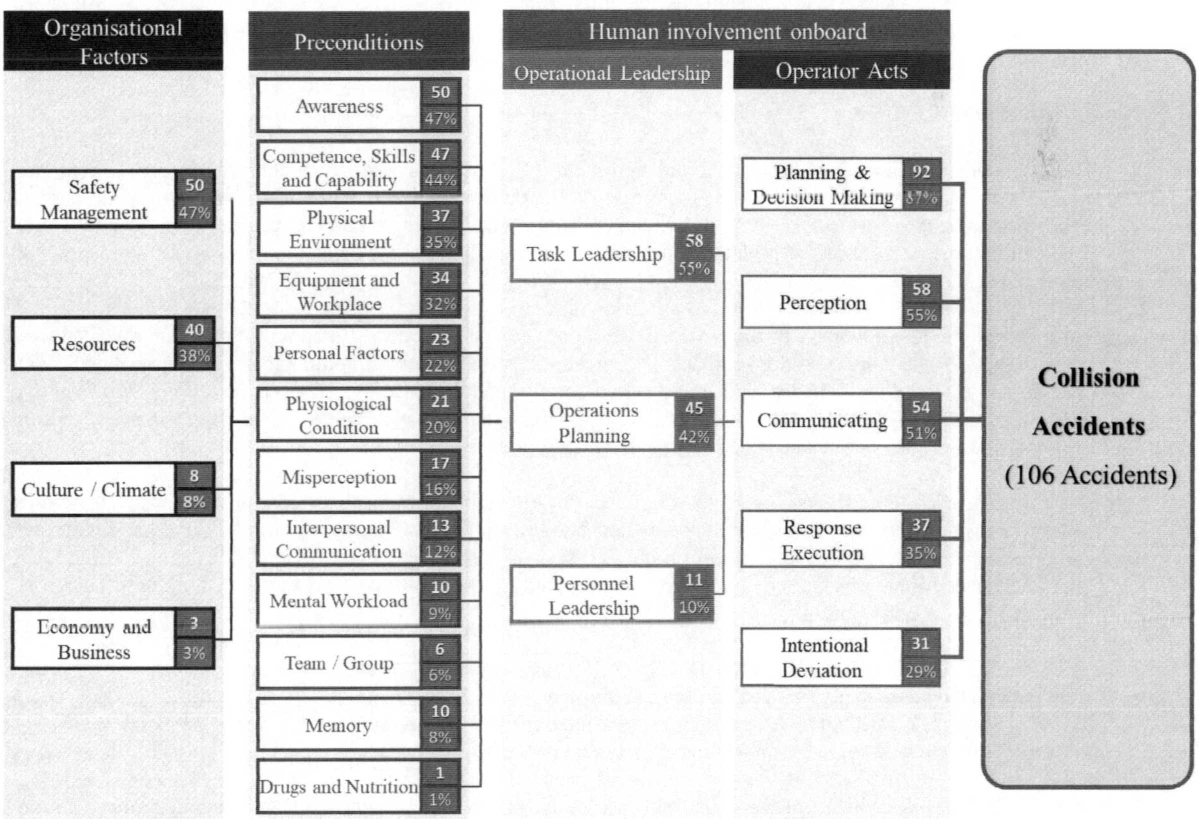

Figure 11: SOAM demonstration of obtained results using SHIELD Collision database (IR=106)

These patterns confirm that the Helge Ingstad incident is not anomalous but representative of systemic features in maritime collisions, particularly in high-density or mixed operational environments.

5.3 CROSS-LEVEL FAILURE PATHWAYS

The collision illustrates that the mere presence of procedures and technology does not guarantee safety. The ability to enact procedures, escalate concerns, or adapt in real time is constrained by human performance variability, shaped by training, leadership, workload, and organisational culture.

Onboard Helge Ingstad, the bridge team operated under cognitive and procedural constraints: tunnelled attention, rotating trainees, and a lack of assertive challenge. At VTS, operators relied heavily on AIS, lacked formal escalation triggers, and did not intervene despite repeated indicators of risk. These failures, while occurring at different levels, were structurally linked.

The SOAM-SHIELD causal map (Figure 8) visualises this interaction, tracing how latent organisational oversights (e.g., insufficient training structures, lack of VTS - navy interface protocols) combined with degraded team cognition and ineffective leadership to erode layered defences. Notably, the OOW trainee's misjudgements (AP1, AD1) occurred within a permissive system that normalised role ambiguity and deferred supervision. VTS inaction, though technically policy-compliant, revealed a misalignment in operational expectations between civilian and military actors.

Rather than attributing failure to isolated acts, this layered analysis reframes the event as a systemic normalisation of vulnerability, where known risks (e.g., passive AIS, inadequate watch handover, degraded VHF clarity) remained unmitigated by design or culture.

The implications are clear. Maritime safety must be reoriented around the realities of human performance. This includes strengthening team-based supervision, ensuring clear escalation protocols across agencies, integrating operational training with contextual complexity, and aligning interface standards in mixed-traffic environments. SHIELD and SOAM together offer the tools to capture these vulnerabilities and structure learning, not just from individual incidents, but across the safety system.

6. CONCLUSION

The collision between KNM Helge Ingstad and Sola TS illustrates a systemic failure pattern common in constrained maritime environments: the progressive erosion of defence layers through interacting human, organisational, and supervisory factors. Rather than stemming from a single point of failure, the incident emerged from routine misperceptions, fragmented situational awareness, and absent escalation across multiple teams, on board and ashore.

Through structured application of the SHIELD taxonomy and SOAM causal mapping, this study revealed how latent organisational vulnerabilities (e.g., insufficient supervision, leadership, policy misalignment between VTS and naval protocols), combined with degraded team cognition and inadequate supervision, culminated in active operational failures. The SHIELD-SOAM integration enabled a multi-layered understanding of the event; tracing how contributory factors breached each safety barrier in sequence.

These findings reinforce that safety is not guaranteed by procedures, training, or equipment alone, but by how well systems support human performance under conditions of uncertainty, high workload, and complex coordination. Enhancing resilience requires robust supervisory presence, clearly defined rules and procedures for inter-agency coordination, particularly between naval and civilian entities, and the human-centred design of navigational tasks and training.

While the broader SHIELD collision dataset (n = 106) has already been classified using the taxonomy, future work will extend this analysis by incorporating influence weighting, certainty scoring, and functional safety barrier mapping. This will allow for a deeper exploration of failure modes and patterns, dominant causal pathways, and cross-layer vulnerabilities, enhancing the analytical power of the dataset.

Several limitations should be noted. The current case study is based on one, albeit well-documented, investigation. The depth and quality of HF classification are dependent on the transparency of official reports, which can vary significantly across cases, particularly in describing latent organisational contributors and the depth of HF analysis within these reports. Additionally, while the SHIELD classification and SOAM mapping followed a systematic process, the assignment of influence and certainty inevitably involved expert judgement.

Nevertheless, the SHIELD-SOAM and similar approaches provide a structured and scalable methodology for analysing systemic failure, supporting more rigorous investigations, evidence-based safety learning, and institutionalised improvements in both risk management and system resilience.

7. REFERENCES

1. KIM, H., NA, S., and HA, W., 'A case study of marine accident investigation and analysis with focus on human error', *Journal of the Ergonomics Society of Korea, 30(1), 137–150.*, 2011.

2. FENG-WU, W., LIANQI, G., and WANG, L., 'Cause analysis of ship collision accident based on Bayesian network', *https://doi.org/10.1117/12.2652600*, 2022.

3. HOLLAWAY, D., and Johnson, J., 'HFACS: Investigatory tool for human factors in offshore operational safety', *Offshore Technology Conference. https://doi.org/10.4043/25130-MS*, 2014.

4. ERGAI *et al.*, 'Assessment of HFACS: Intra-rater and inter-rater reliability', *Safety Science, 82, 393–398.*, 2016.

5. SINGH, A., DALAKLIS, D., and BÄUMLER, R., 'Revisiting the HNoMS Helge Ingstad and Sola TS collision: Discussing the contribution of human factors', *Maritime Technology and Research, 5(3), 262199.*, 2023.

6. BERRY, K., STRINGFELLOW, P., and SHAPPELL, S., 'Examining error pathways: An analysis of contributing factors using HFACS in non-aviation industries', *Proceedings of the Human Factors and Ergonomics Society Annual Meeting, 54(21), 1900–1904.*, 2010.

7. MAYA, B., KHALID, H., and KURT, R., 'Application of card-sorting approach to classify human factors of past maritime accidents', *Maritime Policy & Management, 48(1), 75–90.*, 2020.

8. DE MAYA, B N., *et al.*, 'Human Factors' Contribution Into Maritime Accidents By Applying The Shield Hf Taxonomy', in *the SNAME 14th International Marine Design Conference*, Vancouver, Jun. 2022. Accessed: Feb. 11, 2023. [Online]. Available: https://doi.org/10.5957/IMDC-2022-336

9. STROEVE, S., *et al.*, 'SHIELD Human Factors Taxonomy and Database for Learning from Aviation and Maritime Safety Occurrences', *Safety*, vol. 9, no. 1, p. 14, Mar. 2023, doi: 10.3390/safety9010014.

10. STANTON, N A., SALMON, P M., WALKER, G H., and STANTON, M., 'Models and methods for collision analysis: A comparison study based on the Uber collision with a pedestrian', Dec. 01, 2019, *Elsevier B.V.* doi: 10.1016/j.ssci.2019.06.008.

11. WIEGMANN, D., and SHAPPELL, S., 'A human error analysis of commercial aviation accidents using HFACS', *https://doi.org/10.1037/e420582004-001*, 2001.

12. MIRANDA, A., 'Understanding human error in naval aviation mishaps', *Human Factors, 60(6), 763–777.*, 2018.

13. WANG *et al.*, 'Hybrid causal logic methodology in ship collision risk assessment', *Journal of Marine Science and Engineering, 8(7), 485.*, 2020.

14. WENG *et al.*, 'Evaluation of two-ship collision severity using ordered probit approaches', *Journal of Navigation, 71(4), 822–836.*, 2018.

15. CHASSERY *et al.*, 'A COVID-19 outbreak on board ship: Analysis of the sociotechnical system of epidemiological management in the French Navy', *Safety Science, 140, 105296.*, 2021.

16. TABIBZADEH, M., and MOKHTARI, M., 'An integrated system-oriented methodology for the risk analysis of offshore oil and gas: AcciMap and bow-tie united', *https://doi.org/10.2118/190059-MS*, 2018.

17. KEE, D., et al., 'AcciMap analysis of the Sewol ferry disaster', *Safety Science, 92, 121–131.*, 2017.

18. YOUSEFI, A., RODRÍGUEZ, M., and PEÑA, V., 'Systemic accident analysis models: A comparison study between AcciMap, FRAM, and STAMP', *Process Safety Progress*, vol. 38, no. 2, p. e12002, 2018, doi: 10.1002/prs.12002.

19. SMITH *et al.*, 'Using FRAM to understand Arctic ship navigation', *TransNav, 12(3), 447–457.*, 2018.

20. BOUASLA, S., ZENNIR, Y., and EL-ARKAM, M., 'The use of functional resonance analysis method (FRAM) to model an emergency response plan', *Algerian Journal of Signals and Systems, 7(2), 61–70.*, 2022.

21. SALEHI, V., VEITCH, B., and SMITH, D., 'Modeling complex socio-technical systems using FRAM: A literature review', *Human Factors and Ergonomics in Manufacturing & Service Industries, 31(1), 118–142.*, 2020.

22. ZHAO *et al.*, 'Mapping FRAM to Bayesian networks through AcciMap for system risk assessment', *Eksploatacja i Niezawodnosc – Maintenance and Reliability.*, 2025.

23. GRAZIANO *et al.*, 'Analysis of maritime accidents using TRACEr', *Reliability Engineering & System Safety, 145, 75–83.*, 2016.

24. UNDERWOOD, P., and WATERSON, P., 'Systems thinking, STAMP, and accident analysis: A review of literature', *Safety Science, 68, 55–71.*, 2014.

25. GONCALVES FILHO, A P., JUN, G T., and WATERSON, P., 'Four studies, two methods, one accident – An examination of the reliability and validity of Accimap and STAMP for accident analysis', *Saf Sci*, vol. 113, pp. 310–317, Mar. 2019, doi: 10.1016/j.ssci.2018.12.002.

26. SALMON, P M., CORNELISSEN, M., and TROTTER, M J., 'Systems-based accident analysis methods: A comparison of AcciMap, HFACS, and STAMP', *Safety Science, 50(4), 1158–1170.*, 2012.

27. SAFEMODE Project, 'Strengthening synergies between aviation and maritime sectors through the adaptation of the human risk-informed design approach (SAFEMODE)', 2022.

28. SAVE, L., *et al.*, 'Analysis of reporting systems in safety critical domains to develop the SHIELD HF taxonomy. Results from EU funded SAFEMODE Project', 2021. [Online]. Available: https://www.safemodeproject.eu/

29. MAYA *et al.*, 'Human factors' contribution into maritime accidents by applying the SHIELD HF taxonomy', *https://doi.org/10.5957/imdc-2022-336*, 2022.

30. STROEVE *et al.*, 'SHIELD human factors taxonomy and database for learning from aviation and maritime safety occurrences', *Safety, 9(1), 14.*, 2023.

31. RAMÍREZ-MARENGO *et al.*, 'SAFEMODE's approach for incorporating human factors into risk-informed design', *https://doi.org/10.3850/978-981-18-2016-8_335-cd*, 2021.

32. MACKINNON, S N., *et al.*, 'Using Occurrence Data to Map the Elements of a Risk Model', Research Publishing Services, Aug. 2023, pp. 3151–3156. doi: 10.3850/978-981-18-5183-4_s31-03-408-cd.

33. DILLINGER, T., FASHFA, P D., and KIRIOKOS, N., 'NASA Office of Safety and Mission Assurance Human Factors Handbook Procedural guidance and tools', 2019. [Online]. Available: http://www.sti.nasa.gov

34. REASON, J T., *Managing the Risks of Organizational Accidents*. Aldershot, England: Ashgate Publishing, 1997.

35. LICU, T., CIORAN, F., HAYWARD, B., and LOWE, A., 'EUROCONTROL-Systemic Occurrence Analysis Methodology (SOAM)-A "Reason"-based organisational methodology for analysing incidents and accidents', *Reliab Eng Syst Saf*, vol. 92, no. 9, pp. 1162–1169, Sep. 2007, doi: 10.1016/j.ress.2006.08.010.

36. NAVAS DE MAYA, B., RAMIREZ MARENGO, C., GIAGLOGLOU, E., KURT, R., TURAN, O., and KIRWAN, B., 'Application of a Soam-Based Systemic Method to Conduct a Comprehensive Analysis of a Maritime Accident', *SSRN Electronic Journal*, Apr. 2022, doi: 10.2139/ssrn.4075426.

37. Accident Investigation Board Norway (AIBN), 'Preliminary report collision between frigate KNM Helge Ingstad and the oil tanker Sola TS 8 November 2018', Nov. 2018. Accessed: May 28, 2025. [Online]. Available: https://www.aibn.no/Marine/Published-reports/2019-08-eng?WAF_IsPreview=true

38. Norwegian Safety Investigation Authority (NSIA), 'PART TWO REPORT ON THE COLLISION BETWEEN THE FRIGATE HNOMS "HELGE INGSTAD" AND THE OIL TANKER SOLA TS OUTSIDE THE STURE TERMINAL IN THE HJELTEFJORD IN HORDALAND COUNTY ON 8 NOVEMBER 2018', Lillestrøm, Apr. 2021. Accessed: May 28, 2025. [Online]. Available: https://nsia.no/Marine/Published-reports/2021-05-eng

39. Accident Investigation Board Norway (AIBN) and Defence Accident Investigation Board Norway (DAIBN), 'PART ONE REPORT ON THE COLLISION ON 8 NOVEMBER 2018 BETWEEN THE FRIGATE HNOMS HELGE INGSTAD AND THE OIL TANKER SOLA TS OUTSIDE THE STURE TERMINAL IN THE HJELTEFJORD IN HORDALAND COUNTY', Nov. 2019. [Online]. Available: www.aibn.no

8. AUTHOR BIOGRAPHIES

Yaser B. A. Farag has been a Research Associate at the University of Strathclyde since 2021, with over 20 years of experience in the maritime sector spanning marine engineering, lecturing, and applied research. He holds a Chief Engineer Certificate of Competency and previously served as a Lecturer at AASTMT. His work focuses on human factors, energy efficiency, and maritime education and training. He contributed to the EU-funded SAFEMODE project through the development and validation of the SHIELD human factors taxonomy and led systemic HF analysis in complex maritime operations. He was also involved in the EU-funded GATERS project, where he conducted life cycle assessments of the Gate Rudder System.

Rafet Emek Kurt is a Reader and Director of the Maritime Human Factors Centre (MHFC) at the University of Strathclyde. He served as Principal Investigator and Maritime Coordinator for the EU-funded SAFEMODE project and is committed to advancing the practical application of human factors research in the maritime industry. He has contributed to numerous research initiatives, including SILENV, SEAHORSE, and EXCROSS, and his expertise has supported the work of the IMO, EMSA, and OCIMF.

Osman Turan is a Professor and Director of the Oldendorff Research Centre for Sustainable Shipping (ORCSS) at the University of Strathclyde, and a leading expert in shipping safety. He has been actively involved in shaping safety-related regulations through the IMO and other international bodies. His research focuses on accident investigation and the mitigation of safety-critical conditions. He has contributed to over 20 EU-funded projects, including coordinating the FP7 SEAHORSE project, recipient of the LR-RINA Maritime Safety Award, and serving as Scientific Coordinator for SAFEMODE. In 2018, he was awarded the TRAVISON Senior Researcher Award by the European Commission and is a member of the UK MCA Human Element Advisory Group (HEAG).

APPENDIX

Table 5: Helge Ingstad and Sola TS Accident HF Classification

Impacted CB(s)	SHIELD Layer	Layer's Category	SHIELD Code & Descriptor	SHIELD Code's Definition	Reasoning for Code Selection	Actor	Weight	Certainty Level
CB5, CB4, CB3, CB2	ACTS	PERCEPTION	**AP1** - No/wrong/late visual detection	The operator does not detect (or detects too late or inaccurately) a visual signal necessary to formulate a proper action plan or make a correct decision.	Bridge team persistently **failed to visually identify** Sola TS as a vessel under way with deck lights, despite proximity.	Bridge - KNM HI	4	High
CB5, CB4	ACTS	COMMUNICATING	**AC2** - No transmission of information	The operator does not transmit information which is necessary for other actors to operate safely / effectively.	**Failed to initiate** VHF contact with VTS or Sola TS to clarify intentions	Bridge - KNM HI	2	Medium
CB4	ACTS	COMMUNICATING	**AC1** - Incorrect/unclear transmission of information	The operator transmits to other actors' information, which is incorrect or unclear, e.g. use of wrong callsign.	Sola TS used **a non-distinctive call sign** over VHF when attempting to contact KNM HI, which may have contributed to communication ambiguity, though not in violation of protocol.	Bridge - Sola TS	1	Low
CB2	ACTS	PLANNING AND DECISION MAKING	**AD1** - Incorrect decision or plan	The operator fails to elaborate an action plan or make a decision which is insufficiently to manage the situation.	**Maintained course and speed** based on a misidentified radar contact; did not reassess plan despite traffic cues and ambiguous VHF calls.	OOW - KNM HI	4	High
CB1	ACTS	RESPONSE EXECUTION	**AR1** - Timing error	The operator has planned to take an action which is appropriate for the situation but executes it either too early or too late.	Attempted late evasive turn **too close to impact**; inadequate time to take effect	OOW - KNM HI	3	High
Post-collision	ACTS	RESPONSE EXECUTION	**AR6** - No action executed	The operator has planned to take an action which is appropriate for the perceived situation but does not execute it.	**Did not issue** steering shift order post-collision, preventing activation of emergency steering	Bridge - KNM HI	2	High
Post-collision	ACTS	Intentional Deviation	**AI3** - Workaround in	The operator fails to elaborate an action plan or make a decision which	Commanding officer took grounding action in extreme	CO - KNM HI	3	Medium

Impacted CB(s)	SHIELD Layer	Layer's Category	SHIELD Code & Descriptor	SHIELD Code's Definition	Reasoning for Code Selection	Actor	Weight	Certainty Level
			exceptional conditions	is insufficiently to manage the situation.	emergency, **bypassing standard assessment protocols**			
Post-collision	ACTS	RESPONSE EXECUTION	**AR1** - Timing error	The operator has planned to take an action which is appropriate for the situation but executes it either too early or too late.	Evacuation order delayed despite escalating flooding and inaccessible compartments	CO - KNM HI	2	High
CB5, CB4, CB2	PRECONDITIONS	INTERPERSONAL COMMUNICATION	**PCO1** - Briefing or handover inadequate	Individual or team has not well completed the briefing or handover of an activity.	Incoming watch did not receive full situational update; traffic awareness and TSS conflict not clarified	Bridge Team - KNM HI	3	High
CB4	PRECONDITIONS	AWARENESS	**PAW6** - Unsuitable mental model	Operator uses an unsuitable mental model to integrate information and arrives at a wrong understanding of the situation (e.g. wrong understanding of automation behaviour).	Assumed unknown radar contact was not critical; failed to escalate or intervene as situation developed	VTS Operator	2	Medium
CB3, CB2, CB1	PRECONDITIONS	AWARENESS	**PAW6** - Unsuitable mental model	Operator uses an unsuitable mental model to integrate information and arrives at a wrong understanding of the situation (e.g. wrong understanding of automation behaviour).	Misclassified Sola TS as a stationary terminal; this flawed assumption shaped all subsequent decisions	OOWT – KNM HI	3	High
CB3, CB2, CB1	PRECONDITIONS	COMPETENCE, SKILLS AND CAPABILITY	**PCS2** - Lack of proficiency	Operator capability to accomplish a task does not meet the performance levels expected from her/his skill level.	The trainee, who was actively navigating, lacked full certification and decision-making maturity; misidentified Sola TS and mishandled close-quarters situation	OOWT – KNM HI	3	High
CB2, CB1	PRECONDITIONS	TEAM / GROUP	**PTG2** - No cross-check and speaking-up by team member	No team member (incl. supervising OOW) questioned the trainee's misperception or navigational plan	No team member (incl. supervising OOW) questioned the trainee's misperception or navigational plan	Bridge Team - KNM HI	2	Medium

Impacted CB(s)	SHIELD Layer	Layer's Category	SHIELD Code & Descriptor	SHIELD Code's Definition	Reasoning for Code Selection	Actor	Weight	Certainty Level
CB2	PRECONDITIONS	PERSONAL FACTORS	**PPF3** - Confidence level	Inadequate estimation by the operator of personal capability, of the capability of others or the capability of equipment degrades performance.	Displayed unwarranted confidence in own interpretation despite unclear VHF and visual signals	OOW Trainee – KNM HI	2	Medium
CB4, CB3, CB2, CB1	PRECONDITIONS	EQUIPMENT AND WORKPLACE	**PEW2** - Technology creates unsafe situation	The design, malfunction, failure, symbology, logic or other aspects of technical/automated systems degrade the operator's performance.	Bright deck lighting and background shore lights degraded visual distinction	KNM HI	2	Medium
CB4, CB2	PRECONDITIONS	AWARENESS	**PAW3** - Distraction	Interruption and/or inappropriate redirection of operator's attention.	Radar clutter and contact overlap impaired vigilance, reducing focus on target trajectory	Bridge Team – KNM HI	1	Medium
Post-collision	PRECONDITIONS	AWARENESS	**PAW2** - Confusion	Operator does not maintain a cohesive awareness of events and required actions, and this leads to confusion.	After collision, crew could not assess flooding rate or structural integrity before grounding	Bridge Team – KNM HI	2	Medium
CB1, post-collision	PRECONDITIONS	INTERPERSONAL COMMUNICATION	**PCO1** - Briefing or handover inadequate	Individual or team has not well completed the briefing or handover of an activity.	Insufficient compartment status sharing degraded coordination in post-collision emergency response	Damage Control - KNM HI	1	Low
CB5, CB4, CB2, CB1	OPERATIONAL LEADERSHIP	TASK LEADERSHIP	**LT1** - Inadequate leadership or supervision	Availability, competency, quality or timeliness of leadership or supervision does not meet task demands.	The Commanding Officer (CO) eft the bridge during complex traffic phase, despite knowing the OOW was still in training	CO - KNM HI	3	High
CB4	OPERATIONAL LEADERSHIP	TASK LEADERSHIP	**LT1** - Inadequate leadership or supervision	Availability, competency, quality or timeliness of leadership or supervision does not meet task demands.	The VTS operator failed to escalate or intervene despite observing a developing conflict; no supervisory control was exercised, reinforcing the communication gap.	Operator - VTS	2	Medium
CB2, CB1	OPERATIONAL LEADERSHIP	TASK LEADERSHIP	**LT1** - Inadequate	Availability, competency, quality or timeliness of leadership or	Failed to override OOW trainee's flawed situational	OOW – KNM HI	4	Low

Impacted CB(s)	SHIELD Layer	Layer's Category	SHIELD Code & Descriptor	SHIELD Code's Definition	Reasoning for Code Selection	Actor	Weight	Certainty Level
			leadership or supervision	supervision does not meet task demands.	understanding and late response			
Post-collision	OPERATIONAL LEADERSHIP	OPERATIONS PLANNING	**LO1 -** Inadequate risk assessment	Grounded vessel without full structural damage or flooding assessment; emergency decisions lacked basis	Grounded vessel without full structural damage or flooding assessment; emergency decisions lacked basis	CO - KNM HI	3	Medium
CB5, CB4	ORGANISATION	CULTURE	**OC1 -** Safety culture	There is a poor safety culture or a low company morale in the organisation, leading to a poor attitude to safety in the organisation.	Organisational culture accepted inexperienced bridge teams without enforcing escalation protocols	Royal Navy	3	Medium
CB5	ORGANISATION	SAFETY MANAGEMENT	**OS2 -** Safety risk management (proactive)	Identification of safety hazards and risk assessments and mitigation are not carried out or are carried out but do not identify and make available relevant safety issues.	Lacked process to proactively identify and mitigate civilian-military traffic risk during exercises	Royal Navy	2	Medium
CB5, CB4	ORGANISATION	RESOURCES	**OR1 -** Personnel	The organisation provides insufficient personnel who are suitably qualified and experienced to perform the tasks safely.	Bridge was staffed without a certified OOW during high-risk nighttime operation	Royal Navy	2	Medium
CB4, CB2	ORGANISATION	SAFETY MANAGEMENT	**OS5 -** Publications / procedures / written guidance	Publications and written documentation are not sufficient to ensure that a task can be carried out safely. Procedures may exist but are poor and cannot be used to perform the task they support. Even where good procedures exist, they do not account for all the operational circumstances within which they are expected to be applied (work as imagined versus work as done).	Procedures allowed OOW trainees to take bridge watch without guaranteed qualified oversight	Royal Navy	3	High
CB4	ORGANISATION	SAFETY MANAGEMENT	**OS3 -** Safety risk	Safety risk assurance is ineffective in identifying and repairing safety shortfalls. Safety risk assurance	Weak assurance mechanisms failed to capture recurring	Royal Navy	2	Medium

Impacted CB(s)	SHIELD Layer	Layer's Category	SHIELD Code & Descriptor	SHIELD Code's Definition	Reasoning for Code Selection	Actor	Weight	Certainty Level
			assurance (reactive)	includes performance monitoring (inspections, review of tasks), testing, and incident reporting.	skill gaps in comms and judgment			
CB3, CB2	ORGANISATION	RESOURCES	**OR4** - Inadequate training program	There are inadequate or unavailable training programs (either one-time, recurrent, upgrade, transition or any other local training program).	Training did not ensure proficiency in VHF use, threat ID, or situational interpretation	Royal Navy	3	Medium
CB4	ORGANISATION	SAFETY MANAGEMENT	**OS5** - Publications / procedures / written guidance	Publications and written documentation are not sufficient to ensure that a task can be carried out safely. Procedures may exist but are poor and cannot be used to perform the task they support. Even where good procedures exist, they do not account for all the operational circumstances within which they are expected to be applied (work as imagined versus work as done).	Procedures lacked escalation protocols for non-AIS contacts or abnormal traffic	VTS Authority	2	Medium
CB4	ORGANISATION	RESOURCES	**OR4** - Inadequate training program	There are inadequate or unavailable training programs (either one-time, recurrent, upgrade, transition or any other local training program).	Training did not cover mixed-traffic scenarios or appropriate response to non-AIS contact	VTS Authority	2	Medium
Post-collision	ORGANISATION	SAFETY MANAGEMENT	**OS5** - Publications / procedures / written guidance	Publications and written documentation are not sufficient to ensure that a task can be carried out safely. Procedures may exist but are poor and cannot be used to perform the task they support. Even where good procedures exist, they do not account for all the operational circumstances within which they are expected to be applied (work as imagined versus work as done).	No clear written procedure guided CO decision-making for grounding or prioritising evacuation after severe flooding onset	Royal Navy	2	Medium

Impacted CB(s)	SHIELD Layer	Layer's Category	SHIELD Code & Descriptor	SHIELD Code's Definition	Reasoning for Code Selection	Actor	Weight	Certainty Level
Post-collision	ORGANISATION	RESOURCES	**OR4 -** Inadequate training program	There are inadequate or unavailable training programs (either one-time, recurrent, upgrade, transition or any other local training program).	Emergency training and drills did not adequately prepare crew for progressive flooding assessment and timely evacuation	Royal Navy	2	Medium

THE ROYAL INSTITUTION OF NAVAL ARCHITECTS

THE ROYAL INSTITUTION OF NAVAL ARCHITECTS

8-9 Northumberland Street, London WC2N 5DA

Tel: +44 (0)207 235 4622

events@rina.org.uk

www.rina.org.uk

Stay connected with our social platforms

The Royal Institution of Naval Architects
8-9 Northumberland Street
London, WC2N 5DA
United Kingdom

ISBN 979-8-3313-2657-9

CURRAN ASSOCIATES INC.
proceedings.com

International Maritime Conference (Pacific 2017)

Sydney, Australia
3-5 October 2017

The Royal Institution of
Naval Architects